Engineering and Ecosystems

Bhavik R. Bakshi
Editor

Engineering and Ecosystems

Seeking Synergies Toward a Nature-Positive World

Springer

Editor
Bhavik R. Bakshi
School for Engineering of Matter, Transport
and Energy, School of Sustainability, School
of Complex Adaptive Systems
Arizona State University
Tempe, AZ, USA

Lowrie Department of Chemical and
Biomolecular Engineering
The Ohio State University
Columbus, OH, USA

ISBN 978-3-031-35691-9 ISBN 978-3-031-35692-6 (eBook)
https://doi.org/10.1007/978-3-031-35692-6

This Springer imprint is published by the registered company Springer Nature Switzerland AG
The registered company address is: Gewerbestrasse 11, 6330 Cham, Switzerland

Paper in this product is recyclable.

उपस्थास्ते अनमीवा अयक्ष्मा
अस्मभ्यं सन्तु पृथिवि प्रसूताः ।
दीर्घं न आयुः प्रतिबुध्यमाना
वयं तुभ्यं बलिहृतः स्याम ॥

O Mother Earth
Let Thy bosom be free
From sickness and decay
May we through long life
Be active and vigilant
And serve Thee with devotion
(Atharva veda, 12.1.62)

To Mamta

Foreword

When you think of it, everything in society, our economics, technologies, and lives, all derive from nature. There is no such thing as a piece of modern technology or percentage of economic growth, that has not been generated through the provision of ecosystem functions (like clean air, water, and productive soils) and services (like biomass, Earth metals, and protein). Or, turn it around. Without nature, no economy, no society, no modern engineering, no humanity... It is therefore quite remarkable that we have been able to advance our modern globalized world gradually, over a few centuries (think of Adam Smith, the eighteenth-century father of modern economics, who placed land as the foundation of economic capital), by persistently decoupling our societies from nature. As if wealth, health and technology, can grow in a vacuum. And, this unfortunate decoupling of people and planet, of society and nature, has had (and has) major consequences.

We have entered the Anthropocene. Our linear growth model, where we exploit natural resources in one end, add value along technological value chains, consume, and then pollute, in unsustainable ways, without considering any finite limitations (while living on the finite "Spaceship Earth" to quote the great economic thinker Kenneth Boulding in 1966), has aggregated to the planetary scale. We have caused an entirely new Geological Epoch. We have pushed Earth out of the safe and stable Holocene Epoch (the inter-glacial state over the past 12,000 years when we have developed modern civilizations on Earth), and are now the dominant force of change impacting on the entire Earth system. The human pressures have been mounting rapidly over the past 70 years, and we are now hitting the ceiling of Earth's capacity to remain stable, and thereby support human development. We are approaching tipping points, like irreversibly melting the Greenland Ice Sheet or pushing the Amazon rainforest across an irreversible point towards a dry Savannah state. All with self-reinforcing feedbacks, that risk causing an unstoppable drift towards a Hothouse Earth state (Steffen et al. 2018). The IPCC, in its 6th assessment 2022, confirms that global warming is now threatening human wellbeing and the health of the planet, and that the window to hold the 1.5 °C line is rapidly closing (IPCC 2022).

Our challenge is much broader than the climate crisis alone though. We have transgressed five of the nine planetary boundaries that scientifically have been defined as fundamental in regulating the state and livability of the Earth system. We are beyond the safe operating space for a stable planet on climate change, biodiversity loss, land use change, overloading of nutrients (nitrogen and phosphorus) (Steffen et al. 2015), and, according to a recent publication, also on overloading the biosphere with chemical pollutants (Persson et al. 2022). This means we face a planetary crisis, with a need for urgent transformations (bending the curves of rising pressures to reach zero or tolerable levels within one generation, i.e., within the next 20–30 years). And remember, it is all interconnected. There is no safe landing on climate (holding the 1.5 °C line) without returning back to safe levels for land, biodiversity, and nutrients, as these determine the carbon stocks in nature, and fluxes of non-CO2 greenhouse gases.

The implication is that we need world development within planetary boundaries, i.e., to adopt strong sustainability principles, where technology, economy, production, and consumption occur within scientifically defined boundaries, not only for climate (1.5 °C) but also for biodiversity, land, water, pollutants, nutrients, ozone, and ocean.

We see an interesting trend in this direction. After the successful signing of the Paris Climate Agreement in 2015, the World Resources Institute (WRI) and Carbon Disclosure Project (CDP) together with the United Nations Global Compact and the World Wide Fund for Nature (WWF) launched the Science Based Targets initiative (SBTi) to put a finite boundary with mitigation pathways on climate. Today, the SBTi follows the Carbon Law (of cutting emissions by half each decade) (Rockström et al. 2017) to reach net-zero by 2050. We are now seeing the development of SBTs also for nature and the other planetary boundaries, e.g., through the Science Based Targets Network (SBTN) of the Global Commons Alliance (https://globalcommonsalliance.org/), and the wide network of NGOs, businesses, and science advancing the "Nature Positive" agenda, which aligns with the planetary boundaries science, by defining the SBT for nature as: (i) zero loss of nature from 2020 onwards, (ii) net positive by 2030 (after regeneration and restoration), and (iii) full recovery by 2050 (Locke et al. 2021). All this shows that we are moving towards a new definition of sustainable development. It is no longer about reducing environmental impacts, it is about "prosperity and equity within planetary boundaries," i.e., innovation, transformation, and social distribution of wealth within strong sustainability bounds.

Will this hamper innovation? Slow down development? Reduce pace of modern technological advancements? I say, on the contrary. Constraints is the mother of innovation. Posing, on top of all other user criteria, the additional requirement of respecting finite planetary boundaries, for engineering solutions, I think will only increase the level of ingenuity and innovation. And we have empirical evidence to support this. When the world in the 1980s listened to science, and signed the Montreal Protocol in 1987, forbidding (very hard sustainability measure indeed!) the use of ozone-depleting refrigerants (CFCs), what happened? The industry rapidly invented much more ozone benign coolants and we have now, 30 years later, returned back to a safe operating space of the protective stratospheric ozone layer (Steffen et al. 2015). The larger the pressure becomes on solving the climate crisis, the more innovation we see in wind, solar, biomass, efficiency, and hydrogen technologies.

Our quest is clear, we must rapidly return within Earth's safe operating space to have a chance of handing over a livable planet to our children and all future generations. Including nature in engineering decisions can result in innovative solutions that are economically, socially, and ecologically superior to solutions from conventional engineering. Hard scientific boundaries can unleash even deeper engineering solutions and systems that enable us to work with nature, not against nature.

Potsdam, Germany Johan Rockström

References

IPCC, 2022. Climate Change 2022, Impact, Adaptation and Vulnerability. Summary for Policymakers. Working Group II contribution to the Sixth Assessment Report of the Intergovernmental Panel on Climate Change. WMO, UNEP, 2022

Locke, H., Rockström, J., Bakker, P., et al., 2021. A Nature-Positive World: The Global Goal for Nature. https://f.hubspotusercontent20.net/hubfs/4783129/Nature%20Positive%20The%20Global%20Goal%20for%20Nature%20paper.pdf

Montreal Protocol, 1987, The Montreal Protocol on Substances that Deplete the Ozone Layer, which entered into force January 1, 1989, appears at 26 I.L.M. 1541, 1550.

Persson, L., et al., 2022. Outside the Safe Operating Space of the Planetary Boundary for Novel Entities. *Environ. Sc. Technol.*, 56(3):1510-1521, https://doi.org/10.1021/acs.est.1c04158

Rockström et al. 2017. A roadmap for rapid decarbonization. Science, 355, 1269-1271, https://doi.org/10.1126/science.aah3443

Steffen et al., 2015. Planetary boundaries: Guiding human development on a changing planet. Science, 347, 6223, https://doi.org/10.1126/science.1259855

Steffen et al., 2018. Trajectories of the Earth system in the Anthropocene. PNAS, 115(33). https://doi.org/10.1073/pnas.1810141115

Preface

The purpose of this book is to introduce engineers to the essential role of ecological systems in supporting industrial and human activities, and to encourage them to explicitly account for this role in the development and use of engineering solutions for meeting human needs. This book puts forth the vision that by seeking synergies with nature, engineering solutions will be able to meet human needs while respecting nature's capacity in an economically feasible and socially desirable manner. Such solutions will encourage the protection and restoration of ecosystems and transform engineering from a discipline that takes nature for granted and attempts to dominate it, to a discipline that seeks synergies with nature and respects its limits. It will unleash innovation beyond what can be provided by conventional techno-centric engineering, enhance human well-being, and transform business decisions toward a nature-positive world.

Realizing this vision requires nothing less than shifting the paradigm of engineering from the current, mostly antagonistic relationship between engineering and ecosystems to a synergistic or win-win relationship. For this, it is essential for engineers to be ecologically educated and collaborate with ecologists and environmental scientists. Currently, most engineering disciplines and programs impart little knowledge about ecosystems. At best, engineers practice benign neglect by implicitly taking nature's role for granted, or at worst, their activities actively dominate nature and degrade ecosystems. Most engineers have a strong faith in the power of technology for meeting human needs and addressing environmental problems. Their ignorance of nature is a symptom of the broken relationship between people and nature. The resulting disconnect is behind challenges such as global climate change, loss of biodiversity, air and water pollution, and the impacts of novel chemical entities. Of course, this broken relationship is not solely due to and in engineering: it pervades across virtually every modern discipline. This realization has motivated the emergence of new disciplines such as Ecological Economics and Ecological Engineering. In addition, there is growing interest in approaches such as Nature-Based Solutions, Circular Economy, and Techno-ecological Synergy.

The hope is that if engineers are knowledgeable about the role of nature and the benefits of including this role in their decisions, it is very likely that they will expand their designs and decisions in this direction. With this goal of growing academic research and practical applications, this book introduces readers to the importance of ecosystems in supporting industrial activities and the concept of ecosystem goods and services. Accounting for ecosystems in engineering decisions requires quantification of their support to specific industrial and human activities. In addition, it is also important to understand nature's capacity to self-organize and quantify the extent to which it can provide specific goods and services. This is because nature has a carrying capacity until which it can provide goods and services. If this capacity is exceeded, the risk of loss of nature's ability to support our activities increases, and could be followed by a loss of goods and services due to ecological collapse. Multiple chapters in this book describe existing data, models, and methods for quantifying the demand of ecosystem services imposed by industrial activities through their use of natural resources like coal, oil, and water, and through emissions into the environment while implicitly expecting that nature will take care of these emissions. The capacity of ecosystems to supply various services can be determined by ecological models and data. Sophisticated models and large quantities of data are available as described in chapters written by environmental scientists and ecologists. Ecosystems are complex and often more challenging to model and understand as compared to most engineering systems. However, this should not deter engineers from using ecological models since many models have undergone years of development and can be quite accurate. Ignoring ecosystems, as done currently, is also equivalent to using a model of nature being an infinite sink. This is much more inaccurate than current ecological models. Large quantities of ecological data are also becoming available due to advances in sensors and remote sensing.

About half of this book describes practical and theoretical examples of the approaches and benefits of seeking synergies between engineering and ecosystems. This includes tools to quantify the role of ecosystem services and combine it with the role of technological systems for meeting industrial and societal needs. Applications convey the potential benefits of techno-ecologically synergistic design to urban areas, agro-ecological landscapes, watershed management, enterprise resilience, and industrial activities such as chemicals manufacturing and power generation. The approach for including ecosystem services in popular methods for assessing environmental sustainability such as life cycle assessment is also described. To provide a broader transdisciplinary view, approaches from environmental economics for quantifying ecosystem services in economic analysis and decisions, and social benefits and pitfalls of accounting for nature are also described.

Putting together an edited volume involves contributions from many, to whom I owe a debt of gratitude. This includes, first and foremost, the authors of each chapter, who are central to putting together such an edited volume. Without their high quality research and contributed chapter, this book would not have been possible. Thanks also to Prof. Johan Rockstrom for his foreword to this book. Several anonymous reviewers have provided feedback on the book and individual chapters. Michael McCabe, Brian Halm, A. Thiyagarajan and Karthiga Barath of Springer-Nature have coordinated the steps in converting the final draft into this published version. My family has supported this project in many ways, including through their love and patience. Finally, I want thank you, the reader, for your interest in this work. I hope this book helps in the transformation of engineering and our world toward a nature-positive future.

Columbus, OH, USA Bhavik R. Bakshi
February 2023

Contents

Contributors

Yazeed M. Aleissa Chemical and Biomolecular Engineering, The Ohio State University, Columbus, OH, USA

Bhavik R. Bakshi School for Engineering of Matter, Transport and Energy, School of Sustainability, School of Complex Adaptive Systems, Arizona State University, Tempe, AZ, USA
Lowrie Department of Chemical and Biomolecular Engineering, The Ohio State University, Columbus, OH, USA

Shelly Bogra Environmental Impacts and Sustainability Department, Norwegian Institute for Air Research (NILU), Kjeller, Norway

Kyle D. Boutin Environmental and Conservation Science, North Dakota State University, Fargo, ND, USA

Michael Charles Department of Biological and Environmental Engineering, College of Agriculture and Life Sciences, Cornell University, Ithaca, NY, USA

Martin Dorber Industrial Ecology Programme, Norwegian University of Science and Technology, Trondheim, Norway

Misty Edgecomb Communications for the Protect Oceans, Land & Water Strategy, The Nature Conservancy, Arlington, VA, USA

Joseph Fiksel The Ohio State University, Columbus, OH, USA
The George Washington University, Washington, DC, USA

Tapajyoti Ghosh Resources and Sustainability Group, Strategic Energy Analysis Center, National Renewable Energy Laboratory, Golden, CO, USA

Varsha Gopalakrishnan The EVERY Company, South San Francisco, CA, USA

France Guertin The Dow Chemical Company, Midland, MI, USA

Todd Guidry Global Technology Center Civil Engineering Department Leader, The Dow Chemical Company, Chennai, India

Richard M. Gunton University of Winchester, Winchester, UK

Andrew J. Guswa Smith College, Northampton, MA, USA

Perrine Hamel Asian School of the Environment, Nanyang Technological University, Singapore, Singapore
Earth Observatory of Singapore, Singapore, Singapore

Rebecca J. Hanes Chemical and Biomolecular Engineering, The Ohio State University, Columbus, OH, USA

Bingbing Jiang Marine and Environmental Science, Zhejiang Ocean University, Zhoushan, Zhejiang, China

Alex Jordan Department of Civil and Environmental Engineering, University of Pittsburgh, Pittsburgh, PA, USA

Sami Khanal Department of Food, Agricultural and Biological Engineering, The Ohio State University, Columbus, OH, USA

Vikas Khanna Department of Civil and Environmental Engineering, University of Pittsburgh, Pittsburgh, PA, USA

Alex M. Lechner Landscape Ecology and Conservation Lab, Monash University Indonesia, Cisauk, Tangerang Regency, Banten, Indonesia

Kyuha Lee Systems Assessment Center, Energy Systems & Infrastructure Analysis Division at Argonne National Laboratory, Lemont, IL, USA

Eric Lonsdorf Environmental Sciences at Emory University, Atlanta, GA, USA

Karen T. Lourdes Landscape Ecology and Conservation Lab, School of Environmental and Geographical Sciences, University of Nottingham Malaysia, Semenyih, Malaysia

Robert I. McDonald Nature-Based Solutions, The Nature Conservancy, Arlington, VA, USA

Lianna McFarlane-Connelly The Nature Conservancy, Arlington, VA, USA

Theresa G. Mercer Biogeography and Planetary Health, Department of Geography, School of Life and Environmental Sciences, University of Lincoln, Lincoln, UK

Samuel K. Miller Innovation and Entrepreneurship Studies and Mendoza College of Business, University of Notre Dame, Notre Dame, IN, USA

William J. Mitsch School of Environment and Natural Resources, The Ohio State University, Columbus, OH, USA
The Water School, Florida Gulf Coast University, Naples, FL, USA

Chris Nootenboom Institute on the Environment, University of Minnesota, St Paul, MN, USA

David J. Nowak USDA Forest Service, Syracuse, NY, USA

Alan Randall Sustainability Institute, The Ohio State University, Columbus, OH, USA

Johan Rockström Potsdam Institute for Climate Impact Research and Earth System Science, University of Potsdam, Potsdam, Germany

Martha Rogers The Nature Conservancy, Arlington, VA, USA

Utkarsh Shah William G. Lowrie Department of Chemical and Biomolecular Engineering, The Ohio State University, Columbus, OH, USA

Shweta Singh Agricultural & Biological Engineering, Environmental & Ecological Engineering, and Davidson School of Chemical Engineering (By Courtesy), Purdue University, West Lafayette, IN, USA

Mason Unger Department of Civil and Environmental Engineering, University of Pittsburgh, Pittsburgh, PA, USA

Francesca Verones Industrial Ecology Programme, Norwegian University of Science and Technology, Trondheim, Norway
Arctic University of Norway, Tromsø, Norway

Zhenyu Wang School of Land Science and Technology, China University of Geosciences, Beijing, China
Peking University Shenzhen Graduate School, Shenzhen, China

Andrew Wilson Florida Department of Environmental Protection's South District Office, Fort Myers, FL, USA

Ying Xue Chemical and Biomolecular Engineering, The Ohio State University, Columbus, OH, USA

Li Zhang The Water School Administration, Florida Gulf Coast University, Naples, FL, USA

Ruonan Zhao Chemical and Biomolecular Engineering, The Ohio State University, Columbus, OH, USA

Part I
Introduction and Motivation

Chapter 1
Why Should Engineering Account for Ecosystems?

Bhavik R. Bakshi

1.1 How Does Engineering Depend on Nature?

In today's world, products of engineering such as the food we eat, the clothes we wear, and the electricity that we use to power various activities have become basic necessities for most of us. In addition, we also rely on transportation systems such as roads and automobiles, communication means such as cellular networks and mobile devices, and health care products such as vaccines and medicines. The question posed by the title of this chapter is why the engineering of such technological products should account for the role of nature. Engineering activities relevant to such products include their design, development, manufacturing, use, and end-of-life. To answer this question, we first need to know how engineering activities and the resulting products depend on nature.

By "nature" we mean ecological systems or ecosystems, which are the smallest self-sustaining units that consist of living (biotic) and nonliving (abiotic) components. Examples of ecosystems include a wetland, forest, lake, and grassland. They involve cycling of materials and energy between biotic components such as bacteria, insects, birds, and animals, and abiotic components such as soil, water, sunlight, and air. In ecosystems, such components have inseparable connections due to which adopting a systems view is mandatory for understanding and modeling such systems. Many books describe the characteristics and basic principles of ecosystems [1, 2]. This includes their hierarchical nature, cycling of materials and energy in

---

B. R. Bakshi (✉)
School for Engineering of Matter, Transport and Energy, School of Sustainability, School of Complex Adaptive Systems, Arizona State University, Tempe, AZ, USA

Lowrie Department of Chemical and Biomolecular Engineering, The Ohio State University, Columbus, OH, USA
e-mail: bhavik.bakshi@asu.edu

B. R. Bakshi (ed.), *Engineering and Ecosystems*,
https://doi.org/10.1007/978-3-031-35692-6_1

networks of biotic and abiotic systems, and dynamics that make ecosystems self-sustaining.

For technological products, such as those listed in the first paragraph, it is easy to determine the contribution from ecosystems: nature is the source of the raw materials that are transformed into these products. Thus, nature is the source of fossil resources such as crude oil, coal, and natural gas that are transformed to power our activities and provide raw materials for producing organic chemicals including plastics. Nature also provides minerals that are the source of steel and concrete for our built infrastructure, and metals such as silicon, lithium, cadmium, and cobalt that are essential for electronics and many renewable energy technologies.

In addition to such dependence on natural resources or ecosystem goods, engineering activities also depend on many other services from nature. To appreciate this dependence, we consider a typical manufacturing process. It has inputs of raw materials such as molecules from diverse sources that are converted into products with the help of utilities such as electricity and water. It produces desirable products that are sold to make a profit, undesirable byproducts that may have some monetary value, and waste that may be discharged into the environment, often for no cost, or disposed after paying a disposal fee. Traditionally, engineering[1] has focused on efficient management of these flows for the design and operation of such processes, with the goal of maximizing profit. However, such a process relies on much more than these flows and activities: it also relies directly and indirectly on a large number of goods and services from nature. In fact, without these goods and services, no engineering or human activities are possible. Some examples of ecosystem goods that sustain engineering activities are depicted in Fig. 1.1 and include the following:

- *Water* is used for heating, cooling, and separation. It is usually drawn from rivers and lakes in the local watershed. Sometimes ocean water is also used for tasks such as cooling of power plants or to produce fresh water by desalination. Water is also a unique and environmentally friendly solvent. Increasingly, it is also being used as a source of hydrogen to replace hydrogen from natural gas since electrolysis of water with renewable energy does not result in any direct greenhouse gas emissions.
- *Air* is commonly used for its oxygen content in processes that involve combustion and oxidation. Other uses of air are for pressurization and for resources such as pure nitrogen, oxygen, argon, etc., that are obtained by separation of air. Of course, our survival and that of most living things on earth depends on respiration of air.
- *Minerals* such as various ores and sand are used to make materials for equipment, catalysts, solar panels, magnets, buildings, bridges, and many other products. These are mostly obtained from the lithosphere, but some elements such as sodium and potassium are also extracted from seawater.

[1] By "engineering" we mean activities and outputs of engineering disciplines. Other related terms are technological systems or technosphere.

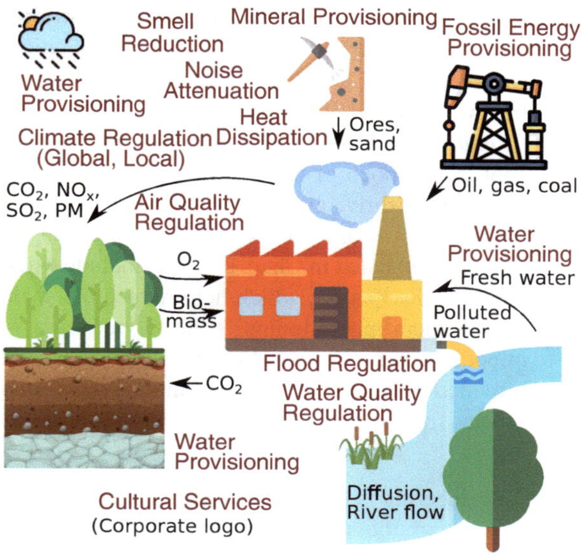

Fig. 1.1 Some ecosystem goods and services that sustain industrial activities

- *Fossil resources* such as crude oil, natural gas, and coal have been the primary drivers of the economy for over two centuries. They are mainly used to generate electricity and transportation fuels, and also as raw materials for the production of various chemicals and other products.
- *Biological resources* include cellulose, various foods, wood and woody biomass, sources of pharmaceutical and medicinal products, genetic resources, etc. These form the input to many manufacturing and industrial activities.

Selected services from ecosystems that sustain engineering activities are also depicted in Fig. 1.1 and include the following:

- *Air quality regulation* is a service provided by various physical, chemical, and biological processes in ecosystems. It includes physical processes such as diffusion, dissipation, and dilution in air that is enhanced by wind. Emissions such as carbon dioxide (CO_2), nitrogen oxides (NO_x), and sulfur dioxide (SO_2) are taken up through the stomata of leaves for use as nutrients. However, excessive concentration of NO_x and SO_2 can hurt the plants. Emissions of particulate matter are mitigated by vegetation due to adsorption on leaves. Also, precipitation scrubs the air to purify it but may result in wet deposition or acid rain.
- *Water quality regulation* is the service that dilutes and detoxifies polluted water. It includes the dilution of pollutants by diffusion in water bodies and by river and ocean currents. In addition, wetlands take up all kinds of water pollutants including those in agricultural runoff, industrial waste, and sewage. Ecological processes for water purification involve microbes, plants, and biogeochemical cycles.

- *Water provisioning* is the service that provides fresh water. It involves ecological processes in the water cycle such as evaporation, precipitation, flow of rivers, and recharge of aquifers. This can be thought of as nature's desalination and freshwater distribution service.
- *Climate regulation* is the service that has maintained the global climate in the geological epoch of the Holocene for the last 12,000 years. This has allowed humanity to thrive and prosper like never before.
- Many other ecosystem services support engineered systems, as shown in Fig. 1.1, and listed in Chap. 2. These include smell regulation, noise attenuation, and heat dissipation, which help reduce the societal impact of industrial activities. Mineral provisioning is the service from geological processes that make minerals and fossil resources available for extraction and use. Cultural services provided by nature are also relevant to industrial systems since corporations use images from nature such as a shell, tiger, bird, crystal, and many others as their logos.

Thus, engineering is heavily dependent on goods and services from nature. Without them, none of the activities related to engineered products and processes are possible. In other words, goods and services from ecosystems are essential for sustainable engineering.

Not just engineering but all human activities rely on the availability of ecosystem goods and services. In addition to ecosystem goods and services, sustainability also requires goods and services from social and economic systems. Social systems include educational, legal, and philanthropic institutions, cultural practices, and consumer behavior, while economic systems include markets and financial institutions. The importance of ecological, economic, and social aspects is often conveyed as the "triple bottom line," implying equal importance of these three aspects. However, in practice, economic activities, including engineering, are nested inside society, which is nested inside ecosystems, as depicted in Fig. 1.2. This triple value model [3] captures the fact that nature provides the foundation for all human activities, and ecological and social systems are essential for the feasibility of economic activities. Given the critical importance of nature for sustaining engineering activities, we will now consider whether engineering accounts for nature.

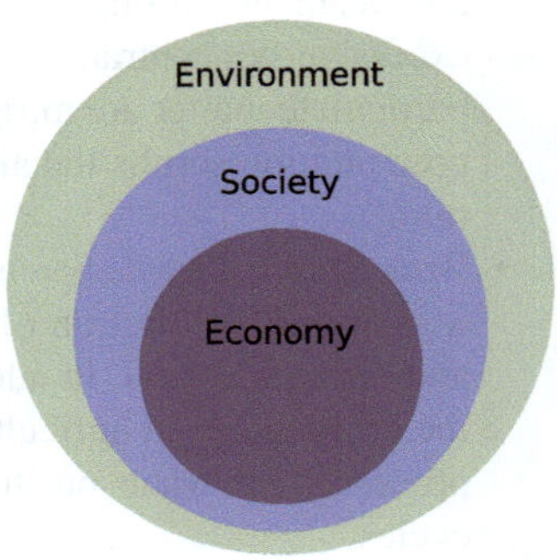

Fig. 1.2 Triple value model conveying that nature provides the foundation for sustaining societal and economic systems

1.2 Does Engineering Account for Nature?

The short answer to this question is mostly, "no." The following quote is from a paper on nature and economics [4], "When asked, economists acknowledge nature's existence, but most would appear to deny that she is worth much." If we replace "economists" by "engineers," the quote would still be valid. This is explained in more detail in the rest of this section.

Engineering decisions including design and operation of various products and processes are primarily based on economic aspects. Therefore, those goods from nature that have monetary value get included in economic analysis. This includes fossil resources, minerals, and biomass. Their monetary value typically reflects their cost of extraction, royalties and taxes, and their demand versus supply in the marketplace. Such costs do not consider the role of ecosystems in producing these resources and making them available for our use. We extract these and other resources from nature as if they are ours for the taking for free and as if nature produced them and will continue to do so for free and forever. Usually, there is nothing equivalent to a monetary transaction with the earth when we take resources from her. Conventional prices also often exclude the impact of resource extraction and use, such as change in land cover due to mining and climate change due to conversion of fossil carbon into carbon dioxide and its emission into the earth's atmosphere. It is often argued that the market accounts for and can adapt to resource scarcities since prices go up when resources are scarce. This is certainly true for resources that are part of the market. However, side effects of resource use such as environmental and societal impacts due to land use change and global climate change are environmental externalities that are routinely kept outside the market. This becomes equivalent to assuming that these impacts have no effect on the economy. Thus, market prices do account partially for the role of some resources, mostly the nonrenewable ones. In contrast, renewable resources like water are greatly underpriced and therefore overconsumed [5].

If we consider ecosystem services, it becomes even clearer that engineering does not account for their role. Services such as carbon sequestration and air quality regulation by terrestrial and aquatic plants; water quality regulation by ecosystems such as wetlands; biogeochemical cycles of carbon, nitrogen, phosphorus, and other materials; and many others are rarely if ever considered in engineering analysis, design, and operation. These services are also not included in conventional economics as conveyed by the quote at the start of this section. Conventional economics keeps ecosystem services outside the market, effectively assuming that they are not worth much. Assuming something is not worth much is equivalent to assuming that it is almost infinite, that is, consuming it will not result in any scarcity. In other words and in terms of economic jargon, ecosystem goods and services that are considered to have no monetary value are treated as "public goods." These are goods and services that are available to everyone and do not decline upon use.

In reality, the "carrying capacity" of ecosystems to provide various goods and services is finite. For example, an underground aquifer can provide water as a renewable resource, provided the withdrawal rate does not exceed the rate of recharge. Similarly, a lake can detoxify pollutants and the biosphere can take up carbon dioxide, but only up to a limit. Exceeding the carrying capacity results in ecological degradation and outcomes such as climate change and harmful algal blooms. Traditionally, engineering decisions have not considered nature's capacity to supply most goods and services. For example, decisions that result in emission of carbon dioxide such as the design and operation of fossil fuel-burning industrial boilers and furnaces in power plants do not consider whether adequate carbon sequestration or climate regulation ecosystem service is available to mitigate these emissions. Similarly, when designing an activity that relies on water such as an urban development or a thermal power plant, the capacity of the watershed to provide water in a sustainable manner without compromising the water needed for ecosystems in the watershed is rarely considered. When products such as plastics and synthetic fertilizers are produced, whether the body of water that will receive the litter and runoff has the capacity to absorb it is also mostly ignored.

Another characteristic of ecosystems that engineering has ignored or attempted to dominate is its dynamic character. At scales larger than that of an organism, ecosystems exhibit homeorhesis [1], which is the tendency to stay in a range of operation, as opposed to a fixed set point. Examples of homeorhesis include the flow of water in rivers due to its cycles of drought and floods, changes in population of species in an ecosystem, and the hourly, daily, and seasonal variation of temperature at a given location. Homeorhesis enables ecological processes to be resilient to perturbations and allows them to maintain their structure and function, including their ability to provide various goods and services. Engineering has routinely attempted to control this homeorhetic or pulsing nature of ecosystems. For example, dams aim to eliminate the effect of flood and drought cycles on the availability of water in a watershed, air conditioning nullifies the effect of variation in outdoor temperature on indoor temperature, and weedicides eliminate the natural diversity of plants in an agricultural monoculture to maximize yield of the desired crop. Imposing homeostasis on naturally homeorhetic systems results in the gradual loss of the system's resilience and adaptive capacity, reduces its ability to supply goods and services, and makes them more vulnerable to being disrupted by perturbations, as described in Sect. 1.4.

The discussion and examples in previous paragraphs convey that modern engineering has developed without taking into account the critical role of ecosystems in sustaining its activities. The traditional boundary of engineering is limited to technological products and processes, and the behavior of engineered systems is desired to be predictable and homeostatic. Not only has engineering ignored the role of nature, but it has contributed, often unknowingly, to ecological destruction and degradation. Historically, the attitude of engineering toward nature has been that of dominance and even antagonism, as conveyed in the quote below from a speech delivered by an award-winning engineer more than a century ago [6].

> "What is Engineering? The control of nature by man. Its motto is the primal one - 'Replenish the earth and subdue it'.... Is there a barren desert - irrigate it; is there a mountain barrier - pierce it; is there a rushing torrent - harness it. Bridge the rivers; sail the seas; apply the

force by which all things fall, so that it shall lift things.... Nay, be 'more than conqueror' as he is more who does not merely slay or capture, but makes loyal allies of those whom he has overcome! Appropriate, annex, absorb, the powers of physical nature into human nature!"

Today, we now know the many undesirable side effects of engineering activities, particularly on the degradation of ecosystems. A natural question that should arise is, why did engineering adopt such an attitude? Why did engineering, along with virtually all other disciplines, take nature for granted? Most importantly, what will it take to abandon the highly anthropocentric paradigm of engineering based on controlling or ignoring nature to an eco-synergistic paradigm that learns from and respects nature, and encourages decisions that protect and restore nature? For this, we will consider the history of the development of modern disciplines in the last three centuries.

1.3 Why Does Engineering Take Nature for Granted?

It is not that engineering is incapable of accounting for the role of ecosystems. However, the scientific principles underlying modern engineering were developed in the seventeenth to nineteenth centuries when nature seemed to be vast and anthropocentrism was popular. As depicted in Fig. 1.3, this includes the work of scientists like Boyle, Newton, Carnot, Gibbs, and Boltzmann. The foundational work in economics of Adam Smith and others was also done at that time and influenced much subsequent development in that discipline. During this period the world was quite "empty": human population was less than two billion and the gross domestic product per person was less than $2000. People and their economic activities occupied a relatively small fraction of the planet. Nature seemed vast and infinite, so making these implicit and explicit assumptions based on ignoring nature seemed reasonable, and it affected fundamental developments in science and in economics. As a consequence, both developed while taking nature for granted. Such an approach also aligns with reductionist thinking which encourages narrowing the focus to systems that are easier to understand, explore, and model. Given the complexity, limited understanding, and difficulty in running scientific experiments on ecosystems, it was easier to keep them outside the system boundary. Thus, both engineering and economics consider nature to be an infinite source and sink that has little economic value [4]. This encourages overuse and degradation of ecosystem goods and services, particularly of those that are not captured well by market prices such as water and soil, and of services such as those which regulate the climate, quality of air and water, pollination, and soil fertility.

Unlike the world of the nineteenth century, today's world is very "full" [7]. As can be seen from Fig. 1.3, human population has quadrupled and so has the GDP. The folly of taking nature for granted and ignoring its limits is better understood and not acceptable anymore. The current engineering paradigm based on ignoring nature and its limits needs to shift to a paradigm that accounts explicitly for the

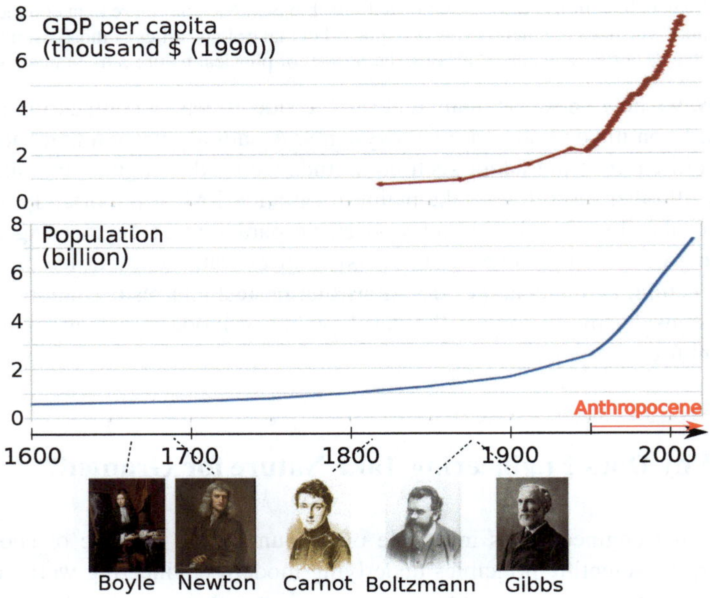

Fig. 1.3 History of scientific development. Most fundamentals were developed when the world was quite "empty." In comparison, today's world is very "full." Photos from the public domain and Wikimedia Commons

role of nature and respects its carrying capacity. Rather than causing ecological degradation and loss of nature, the new paradigm needs to enable decisions that result in a nature-positive world.

1.4 Effects of Taking Nature for Granted

Taking nature for granted has allowed human activities, population, and prosperity to grow in an unfettered manner, but with a large negative side effect of the degradation of ecosystems and their ability to provide goods and services that sustain human well-being. This side effect has been predicted and documented over the last many decades in a large number of studies, journal articles, and reports by governmental and nongovernmental organizations. For example, reports of the Inter-governmental Panel on Climate Change have conveyed for many years, with a high degree of certainty, the anthropogenic nature of global climate change due to greenhouse gas emissions [8]. The Millennium Ecosystem Assessment [9] identified 15 out of 23 ecosystem services as degraded. A visually appealing summary of such findings is provided in the work of Rockström et al. [10] by identifying the "safe operating space" for humanity for various categories of

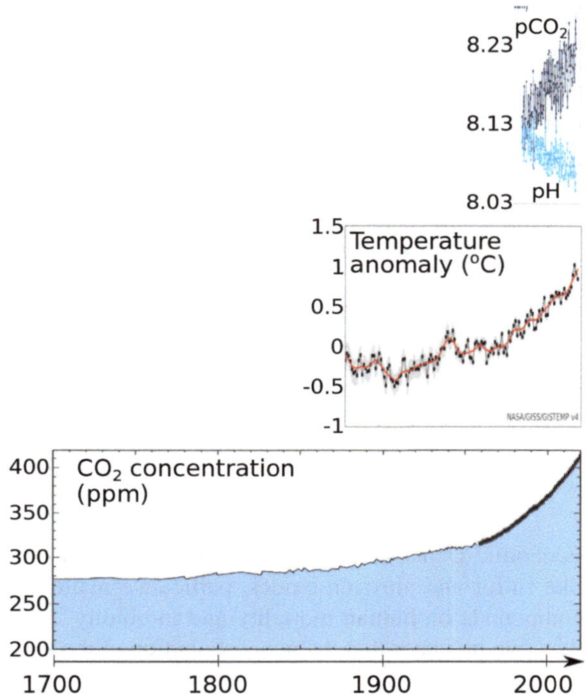

Fig. 1.4 Change in some planetary parameters during the period in which scientific fundamentals underlying engineering were developed (see Fig. 1.3) and their values in recent times

environmental impact. It indicated that we have greatly exceeded the safe operating space for categories such as biodiversity loss and disruption of biogeochemical cycles and moderately exceeded the safe space for climate change. Some of these effects are shown in Fig. 1.4, particularly in the context of the history of scientific advances depicted in Fig. 1.3. As can be seen in these figures, human impact is much more visible since the 1950s, which is considered by many geologists to be the time when we entered the new geological epoch of the Anthropocene or "age of people." The epoch of the Holocene, when the global climate was relatively stable with less variability, which allowed human civilizations to thrive, is behind us. We are entering a period of less predictability and higher variation due to extreme events. This means increasing risk of unexpected and large ecological changes and their impact on the economy, society, and individual well-being.

Another effect of taking nature for granted can be seen by comparing the economic benefits and environmental damages of activities in specific economic sectors. We compare the contribution of economic sectors to the economy versus the damage to society due to their emissions. This can provide insight into which sectors have a net-positive versus a net-negative impact on society. The results of such a study [11] are summarized in Table 1.1. Here, Value Added (VA) is the monetary value addition to the economy from the sector or its contribution to the gross domestic product, which is an indicator of economic prosperity. The Gross

Table 1.1 Ratio of Gross Economic Damage (GED) due to air pollution to Value Added (VA) due to economic activity of selected sectors in the U.S. economy [11]. Ratio larger than one indicates that the environmental damage caused by these sectors exceeds their value addition to the economy, implying their net-negative impact. GED* includes damage due to greenhouse gas emissions from fossil fuel combustion

Industry	GED/VA	GED*/VA
Solid waste combustion and incineration	6.72	
Petroleum-fired electric power generation	5.13	6.93
Sewage treatment facilities	4.69	
Coal-fired electric power generation	2.20	2.83
Dimension stone mining and quarrying	1.89	
Marinas	1.51	
Other petroleum and coal product manufacturing	1.35	
Steam and air conditioning supply	1.02	
Water transportation	1.00	
Natural gas electric power generation	0.34	1.30

Economic Damage (GED) is calculated by monetizing the effect of air emissions like sulfur and nitrogen oxides, particulate matter, ammonia, and volatile organic compounds on human mortality and morbidity. A ratio of GED to VA of greater than one indicates that the net contribution of the corresponding economic sector is negative if the environmental damage was fully internalized through activities such as environmental taxation or emissions trading. Thus, sectors with GED/VA > 1 are found to do more harm than good to society. The second column in Table 1.1 shows the sectors that have such a net-negative impact. Thus, economic activities associated with waste management, mining, and power generation from coal and oil have a net-negative impact on society. If the damage due to carbon dioxide emissions is included, the resulting ratio (GED*/VA) is even larger for sectors that burn fossil fuels, as shown in the last column. Now, power generation with natural gas also becomes net-negative. Such insight does not influence the free market since impacts like those quantified by GED are not fully internalized into the market and do not influence prices. This practice of keeping environmental impact outside the boundary of engineering and economics makes it difficult to determine whether an activity has net-positive or net-negative impact on society. Consequently, engineering decisions are often oblivious of their societal and environmental impacts. This may result in engineering decisions that even encourage net-negative activities while ensuring business profitability. There is little doubt that ecological degradation across the planet is caused by human activities, which confirms the nature-negative character of many human activities.

Another effect on industry and economic activities of ignoring the role of ecosystems is their lack of adequate resilience. As described in Sect. 1.2, engineering's push toward homeostasis reduces the adaptive capacity of the designed systems and makes them more vulnerable to perturbations. At the same time, ecological degradation is increasing the risk of large perturbations, which is already being experienced across the world, as indicated by the increasing frequency of extreme events [12]. For example, extreme weather events have increased from 3656 in 1980–1999 to 6681 in 2000–2019; major floods have increased from 1389 to 3254. This has resulted in a collision course that is already disrupting engineered systems. For example, shifting rainfall patterns are disrupting food production, particularly that is rain fed, and associated industries. More intense rainfall is rendering the existing infrastructure of dams, tunnels, bridges, and subway systems incapable of functioning as intended [13]. Examples include collapse of bridges, flooding of underground roads and trains, and excessive level of water behind dams whose release results in downstream flooding. Extended droughts and extreme temperatures are disrupting power grids and generation systems due to inadequate water for cooling in thermal power plants and for running turbines in hydroelectric facilities. Spikes in energy demand due to heat and cold waves are also taxing the grid beyond its break point in many locations across the world. Zoonotic diseases such as Covid-19 are due to the loss of biodiversity and have caused untold misery due to their health impact, job losses, and supply chain disruptions.

By keeping nature outside its system boundary, engineering solutions fail to benefit from the many goods and services that nature provides. Rather than protecting and restoring ecosystems such as forests and wetlands in the watershed to benefit from nature's ability to provide clean water, the solution from engineering has been to build large dams and water purification systems. Rather than designing buildings and their surroundings to benefit from nature's cooling services provided by trees and wind, engineering solutions rely on expensive and polluting air conditioning systems. Rather than relying on the natural ability of vegetation to detoxify air, industry relies on scrubbers, filters, and reactors to remove pollutants such as particulate matter and oxides of sulfur and nitrogen. This is certainly not to say that technology is not needed. It is very much needed since not all human and industrial needs can always be met by nature-based solutions. That is what motivated the development of many technologies. However, in the pursuit for new and advanced technologies, engineers have lost touch with nature and forgotten about its ability to meet many human needs. What is needed is perhaps a combination or integration of technological and ecological solutions for mutually beneficial or synergistic operation. This would allow society to benefit from nature's services and their smaller environmental impact and larger resilience, while also benefiting from the ability of technological solutions to provide those services that may not be available from ecosystems and in a more controlled and predictable manner. By taking nature for granted, engineering misses out on many opportunities for providing innovative, win-win solutions to meet human needs while protecting and restoring nature. This will be conveyed by several examples and case studies throughout this book.

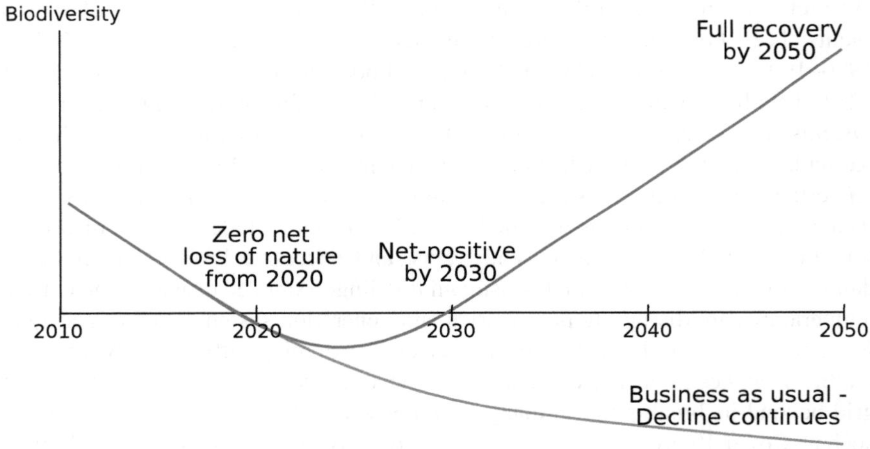

Fig. 1.5 Expected effect of meeting the global goal of nature-positive decisions versus business as usual (Adapted from [14, 15])

1.5 Nature-Positive

Traditionally, many human activities have tended to be nature-negative. That is, our decisions have resulted in a net loss of nature in the form of degraded ecosystems and loss of biodiversity resulting in deterioration of planetary health and increasing risk of destabilizing the life-support system that sustains us. Some examples of such activities are provided in the previous section. The effort toward a nature-positive world is to reverse this ecological decline by decisions that explicitly ensure that we give back more to nature than we take. As depicted in Fig. 1.5, adoption of a Nature-Positive Global Goal for Nature aims to achieve three main outcomes: Zero Net Loss of Nature from 2020, Net Positive by 2030, and Full Recovery by 2050 [14]. This goal has been put forth by a group of scientists and nongovernmental organizations and is meant to complement the Sustainable Development Goals of the United Nations and the net-zero greenhouse gas emissions goal of the Intergovernmental Panel on Climate Change. The three goals of net-zero emissions, nature-positive decisions, and planet-positive decisions are complementary and essential for ensuring well-being of current and future generations.

1.6 Goals of This Book

The World Economic Forum estimates that more than half of the global GDP or $44 trillion is threatened by ecological degradation and the loss of ecosystem goods and services [16]. It also estimates that transitioning to "nature-positive" business activities presents $10 trillion worth of opportunities and 395 million jobs by 2030. Many efforts in this direction are already in progress at forward-looking businesses. The main goal of this book is to encourage and facilitate such efforts by providing a source for engineers and other professionals to learn about the urgent need and many potential benefits of explicitly accounting for and including the role of ecosystems in their activities and products. The resulting solutions should encourage harmony between human and natural systems so that human needs are met in an economically feasible and socially desirable manner, and ecosystems are protected and restored to provide goods and services to current and future generations. Such harmony requires human activities to respect nature's limits and learn to function within ecological constraints. The emphasis of this book on engineering does not mean that this is the only discipline that needs to transform itself toward seeking synergies with nature and striving toward a nature-positive world. Most other disciplines also face this challenge, and the approaches and case studies in this book should be relevant across many disciplines.

This book is organized in five major parts.

- Part I provides the motivation for including ecosystems in engineering decisions. It also introduces the basic concepts of ecosystem goods and services and their status to lay the foundation for the rest of the book.
- Part II describes approaches to account for the demand of ecosystem goods and services for supporting engineering and human activities. Both direct and indirect flows are considered to ensure a life cycle system boundary. Such a large boundary is needed to reduce the chance of unintended harm due to shifting of impacts along the supply or demand network. Chapter 3 introduces the methodological framework that underlies life cycle assessment and footprint analysis. This includes the framework for modeling at the value chain and economy scales with detailed process models and more aggregated input–output models. It forms the basis of the results in many other chapters. Chapter 4 describes methods to account for the demand of the water provisioning ecosystem service. This includes the approach of water footprint analysis, which accounts for multiple categories of water use and calculates the water embodied in various products and through global trade. Chapter 5 focuses on accounting for the role of biogeochemical cycles such as those of carbon and nitrogen. This is related to the ecosystem service such as climate regulation due to the impact of carbon dioxide and water quality regulation due to the impact of nitrogen in agricultural runoff on water quality. It describes models based on monetary and mass flows within the framework of input–output analysis and their application. Chapter 6 considers the ecosystem service of pollination by accounting for the critical role played by insect pollinators in supporting economic activities. It describes this role of pollinators in sustaining economic and industrial activities and the direct and indirect effects of degrading this ecosystem service. Chapter 7 describes the role and importance of biodiversity in the availability and resilience of ecosystem goods and services. It introduces various ways of quantifying biodiversity and including it in decision-making.

- Part III focuses on methods to account for the supply of ecosystem services. The chapters in this section are written mainly by ecologists and environmental scientists, with emphasis on methods, models, data, and tools to quantify the availability of various ecosystem goods and services and nature's capacity. Chapter 8 provides an overview of computational and visualization tools for quantifying the supply of various ecosystem services. It summarizes, illustrates, and critiques existing approaches and identifies challenges for further work. Chapter 9 focuses on hydrological ecosystem services and describes ways of quantifying these services and including them for integrated water resources management. Chapter 10 describes approaches and software for quantifying various ecosystem services provided by vegetation, primarily urban trees. This includes services such as regulation of air quality, aesthetic benefits, regulating water availability, and providing products such as wood. It describes relevant models with emphasis on the software package, i-Tree. Chapter 11 is on services to and from agroecosystems, including soil fertility, pollination, pest regulation, nutrient runoff, and greenhouse gas emissions. It also describes various agro-ecological modeling tools for quantifying these and other relevant services and guides their selection.
- Part IV focuses on methods and case studies on integrating human and natural systems. They demonstrate the challenges and potential benefits of seeking synergies between technological and ecological systems and how it could encourage decisions toward a nature-positive world. Chapter 12 introduces efforts at the intersection of engineering and ecology and how the framework of Techno-Ecological Synergy (TES) can be used to analyze and design integrated networks of technological and ecological systems. It introduces relevant sustainability metrics and an optimization-based framework for synergistic design. It summarizes the findings about the characteristics and challenges of TES designs. Chapter 13 makes the business case for nature-based solutions. It is based on a collaboration between industry, a nongovernmental organization, and academia. It describes the Ecosystem Services Identification and Inventory (ESII) tool and its use for evaluating industrial sites and guiding decisions. Chapter 14 describes how urban planning for ecosystem services could result in people-centric, green, sustainable, resilient, and livable cities. It describes the approach of urban greenprinting to bring information about biodiversity and ecosystem services into spatial planning and to enable more integrated and coordinated approach for urban planning. Chapter 15 describes a novel approach for mitigating harmful algal blooms that occur due to fertilizer runoff. This approach of Wetlaculture (wetland + agriculture) cycles land between farming and treatment of wetland to capture nutrients and then grow crops. Chapter 16 designs agro-ecological landscapes for generating renewable energy while respecting nature's capacity. It uses an optimization-based framework to understand the effect of growing crops, installing renewable energy, or using land for a wetland or forest on the trade-off between economic and ecological objectives. Chapter 17 describes the framework for designing techno-ecologically synergistic supply chains with data and models at multiple

spatial scales. This framework is applied to the design of corn ethanol supply chains for reducing agricultural nutrient runoff by using wetlands, no-till farming, appropriate sources of corn, and location of ethanol manufacturing facilities. Chapter 18 focuses on design of integrated manufacturing processes and vegetation to mitigate the impact of air emissions. It analyzes the demand and supply of the air quality regulation ecosystem service across the United States and identifies counties where ecosystem restoration can be more cost effective than technology for improving air quality. It also describes a framework for spatial design of techno-ecologically synergistic industrial landscapes. Chapter 19 extends the approach of the previous chapter by accounting for the dynamic and intermittent nature of pollution uptake by vegetation. It describes the corporate and social benefits and trade-offs of adapting manufacturing to nature's intermittency, which is the opposite of what is done in traditional engineering. Chapter 20 expands the conventional engineering system boundary by including ecosystem services in the popular technique of life cycle assessment. This results in absolute environmental sustainability metrics that compare the demand and supply of ecosystem services. The benefits of the resulting approach of Techno-Ecological Synergy in Life Cycle Assessment (TES-LCA) are described by application to soybean biodiesel. Chapter 21 describes issues and approaches associated with maintaining resilience and sustainability in corporate enterprises. It includes examples about how accounting for the role of ecosystems could enhance system resilience.
- Part V describes how ecosystem services contribute to and are accounted by other nonengineering disciplines. Chapter 22 describes principles of environmental economics and approaches for accounting for ecosystem goods and services in monetary units. It describes how cost-benefits analysis can assist decisions which include the role of ecosystems and the role of methods such as emissions trading. Chapter 23 describes the view from social science about the opportunities, challenges, and risks in including nature in engineering. Specifically, it focuses on the possibility of unintended harm from socio-ecological interactions and ways of reducing this possibility.

The book ends with an outlook for the future and the hope that more engineers and other professionals will learn from nature and be inspired to develop synergistic designs and strategies that meet human needs while protecting and restoring nature. Such decisions will go a long way in mending the broken relationship between people and nature. It will result in innovative solutions that will simultaneously improve all facets of human well-being and shift the paradigm of ignoring and abusing nature that has been the underlying basis of most disciplines over the last few centuries.

References

1. E. P. Odum and G. W. Barrett. *Fundamentals of Ecology*. Thomson Brooks/Cole, Belmont, CA, 5th edition, 2005.

2. B. R. Bakshi. *Sustainable Engineering: Principles and Practice*. Cambridge University Press, jun 2019.
3. Joseph Fiksel. A systems view of sustainability: The triple value model. *Environmental Development*, 2:138–141, 2012.
4. Partha Dasgupta. Nature in economics. *Environmental and Resource Economics*, 39(1):1–7, dec 2007.
5. Jarmo J. Hukka and Tapio S. Katko. Appropriate pricing policy needed worldwide for improving water services infrastructure. *Journal AWWA*, 107(1):E37–E46, 2015.
6. Rossiter W. Raymond. The new age. *The Journal of Industrial and Engineering Chemistry*, pages 249–251, March 1913.
7. Herman Daly. Economics for a full world. http://www.greattransition.org/publication/economics-for-a-full-world, 2015.
8. V. Masson-Delmotte, P. Zhai, A. Pirani, S.L.Connors, C. Péan, S. Berger, N. Caud, Y. Chen, L. Goldfarb, M. I. Gomis, M. Huang, K. Leitzell, E. Lonnoy, J. B. R.Matthews, T. K. Maycock, T. Waterfield, O. Yelekçi, R. Yu, and B. Zhou, editors. *Climate Change 2021: The Physical Science Basis. Contribution of Working Group I to the Sixth Assessment Report of the Intergovernmental Panel on Climate Change*. Cambridge University Press, 2021.
9. Millennium Ecosystem Assessment. *Ecosystems and Human Well-being: A Framework for Assessment*. Island Press, 2003.
10. J. Rockstrom, W. Steffen, K. Noonel, A. Persson, F. S. Chapin, E. F. Lambin, T. M. Lenton, M. Scheffer, C. Folke, H. J. Schellnhuber, B. Nykvist, C. A. de Wit, T. Hughes, S. van der Leeuw, H. Rodhe, S. Sörlin, P. K. Snyder, R. Costanza, U. Svedin, M. Falkenmark, L. Karlberg, R. W. Corell, V. J. Fabry, J. Hansen, B. Walker, D. Liverman, K. Richardson, P. Crutzen, and J. A. Foley. A safe operating space for humanity. *Nature*, 461(7263):472–475, 2009.
11. Nicholas Z. Muller, Robert Mendelsohn, and William Nordhaus. Environmental accounting for pollution in the United States economy. *American Economic Review*, 101:1649–1675, 2011.
12. United Nations Office for Disaster Risk Reduction. Human cost of disasters: An overview of the last 20 years, 2000–2019, 2020.
13. P. C. D. Milly, Julio Betancourt, Malin Falkenmark, Robert M. Hirsch, Zbigniew W. Kundzewicz, Dennis P. Lettenmaier, and Ronald J. Stouffer. Stationarity is dead: Whither water management? *Science*, 319(5863):573–574, 2008.
14. Harvey Locke, Johan Rockström, Peter Bakker, Manish Bapna, Mark Gough, Jodi Hilty, Marco Lambertini, Jennifer Morris, Paul Polman, Carlos M Rodriguez, et al. A nature-positive world: the global goal for nature, 2021.
15. Natural England, JNCC, Natural Resources Wales, NatureScot & Northern Ireland Environment Agency. Nature positive 2030 – summary report, 2021.
16. World Economic Forum. New nature economy report series, 14 July 2020.

Chapter 2
Ecosystem Goods and Services

Yazeed M. Aleissa, Ying Xue, and Bhavik R. Bakshi

2.1 What Are Ecosystem Goods and Services?

We live on a planet rich with natural resources that are invaluable and irreplaceable. Every species depends on these resources to sustain their lives, and humans are no different. However, how does nature provide food, water, trees, and raw materials? How does it regulate the quality of air and water?

Nature is a complex system consisting of many interlinked environments and ecosystems. Natural ecosystems, such as deserts and forests, are integrated systems that perform chemical, physical, and biological processes to maintain the Earth's natural cycles. Each ecosystem has different characteristics and functions that uniquely contribute to sustaining planet Earth and its interlinked processes.

Ecosystem goods and services (ESs) are the assets generated by the different ecosystems that contribute to human well-being [1]. All people depend on the supply of these benefits since they underpin all human activities and contribute to their well-being. ESs benefit people in numerous ways, directly as material goods like food or indirectly as nonmaterial services such as the pollination service essential for food production.

Y. M. Aleissa · Y. Xue
Chemical and Biomolecular Engineering, The Ohio State University, Columbus, OH, USA

B. R. Bakshi (✉)
School for Engineering of Matter, Transport and Energy, School of Sustainability, School of Complex Adaptive Systems, Arizona State University, Tempe, AZ, USA

Lowrie Department of Chemical and Biomolecular Engineering, The Ohio State University, Columbus, OH, USA
e-mail: bhavik.bakshi@asu.edu

B. R. Bakshi (ed.), *Engineering and Ecosystems*,
https://doi.org/10.1007/978-3-031-35692-6_2

Ecosystems are complex and interconnected; they provide multiple services simultaneously. For example, trees provide goods such as wood and fruits, in addition to regulating air and water quality, provide habitat for maintaining biodiversity, and provide aesthetic value. All economic sectors and industries utilize ES. Agriculture depends on many ESs such as water provisioning, water quality regulation, pollination, soil fertility maintenance, nutrient cycling, air quality regulation, and many more. Likewise, industrial activities depend on the availability of raw materials, climate regulation, water provisioning, water quality, air quality regulation, etc., as described in Chap. 1.

Although it is clear how indispensable ESs are in sustaining human life and well-being, humans tend to exploit natural ecosystems to seek more resources and development. However, nature has a limited capacity to provide these services, and exceeding these thresholds can lead to irreversible damage and degradation of ecosystems. A prime example is the effect of global warming due to the increase in atmospheric carbon dioxide concentration, highlighting that anthropogenic demand is higher than the sequestration capacity of nature.

The consequences of anthropogenic activities usually trigger a ripple effect that impacts different ecosystem processes due to their interconnected nature. The magnitude of the effect can be critical and cause significant damage to ecosystems. For example, the conversion of forest land into agricultural land can increase carbon dioxide in the atmosphere, cause loss of native habitat and biodiversity, disrupt the water cycle, and increase the risk for natural disasters such as wildfires, flooding, and drought. In addition, these changes can impact the availability and flow of ESs, such as the supply of food, clean water, soil fertility, and disease control.

This chapter aims to familiarize the reader with the concept of ESs and highlight the importance of these services for sustaining human activities and well-being. We start with the different definitions of ES, followed by the available methods and tools to quantify ES. Then, a brief discussion of ES classification frameworks followed by a detailed description of the standard categories. The second part of this chapter demonstrates the overall status of ESs and highlights the current status of a specific ES. Finally, a summary concludes this chapter, including the relevant key information and discussion.

2.2 Identification, Quantification, and Valuation

The absence of ESs in decision and policymaking has been an issue due to poor understanding and underestimation of how much human life depends on natural ecosystems. Moreover, modern disciplines of economics, engineering, and others consider ESs not to have a direct role or market value. Therefore, the true worth and importance of ESs have been overlooked in most human decisions. One of the most notable efforts to estimate the monetary value of ESs was made by Costanza et al. [2], paving the way for much subsequent work on evaluating ES. The study evaluated the marginal value of ESs as $33 trillion, which gained a lot of attention and controversy since it was significantly larger than the global gross national product at that time. A more recent estimate by the World Economic Forum valued nature at about $125 trillion [3]. This assessment was based on the gross value added to economies; for example, industries highly dependent on nature generate more than half of the world's gross domestic product (GDP).

Some economists advocate moving from economic indicators to measure human developments such as GDP to more comprehensive ones that include natural and social capital such as the inclusive wealth index [4]. Using monetary evaluation for nature results in metrics that can include the value of ESs along with that of economic goods and services. However, such an approach captures "weak sustainability" since it implicitly assumes substitutability between the aggregated quantities. This implies that ESs are replaceable, which need not be true, particularly if their availability is close to or less than nature's carrying capacity.

A few steps are necessary to preserve natural ecosystems and reduce the chance of unintended harm by human activities. First is the identification of the numerous goods and services generated by these ecosystems. Then, an assessment of the current status and use of models to quantify the capacity of the ecological system to provide these goods and services. Finally, mapping of ESs is essential in understanding and accounting for spatial heterogeneity and the unique attributes that can affect these ecological systems. These steps provide fundamental knowledge that can guide policy and decision-making on local, national, or global scales. They also facilitate all applications that involve ecosystems, such as landscape planning and optimization, environmental resource management and conservation, and risk assessment [5].

The quality of ecological models and data used in accounting and quantifying ESs is essential for better decisions. In addition to the correct use of these methods and the spatial resolution of the application, some ESs are easy to model and measure, such as the provisioning of specific goods such as fuelwoods. However, other services are much harder to quantify due to their interdependence and challenges in modeling, like water quality regulation by fish and coral reefs, or due to the lack of biophysical models for ESs such as cultural services.

There are plenty of efforts in the literature that describe different ecological processes and systems. In addition, tools that utilize these models have gained attention and have been developed by academia and global organizations, such as the Integrated Valuation of Ecosystem Services and Tradeoffs (InVEST) [6], which contain models for carbon sequestration, pollination and crop production, water quality, water supply, and others. Another example is the ARtificial Intelligence for Ecosystem Services (ARIES) [7], which includes models for biodiversity resources, carbon sequestration, flood regulation, water quality, and supply. In addition to more specialized tools like i-Tree, which quantifies the benefits from forest trees such as carbon sequestration, such tools and underlying models are described in Part III of this book.

2.3 Classification of Ecosystem Goods and Services

With the multidisciplinary nature of ESs research, a common vocabulary and structure are necessary for describing, evaluating, visualizing, and communicating about different services. Classification systems aim to define, organize, and group services based on key characteristics and features. Additionally, they identify the relationships within groups and provide a structure that helps measure and account for each group.

There is consensus on defining ESs through the required contribution to human welfare, differentiating them from other natural processes. However, there are different views on how the services are generated and their direct or indirect relation to human beneficiaries [1, 8, 9]. One complicated component in defining ESs is the fine line between ecosystem goods, services, and other ecological processes and functions that support other services. These supporting processes are referred to as intermediate services to differentiate them from final services people utilize. The complexity involving ecological processes extends to the accounting and valuation of ESs as they may contribute to more than one service simultaneously, which might lead to double counting or undervaluation of the service.

In an effort to clarify the different relationships between natural and social systems, the cascade model was developed [10]. This model illustrates how ecosystem functions and processes underpin the final ES, and any interruption in any ecological components will impact the final flow of ES, as shown in Fig. 2.1. The cascade model identifies five main components connecting biophysical processes to the direct value that contributes to human well-being.

These variations have resulted in different classification systems such as the Millennium Ecosystem Assessment (MA) [11], The Economics of Ecosystems and Biodiversity (TEEB) [12], Common International Classification of Ecosystem Services (CICES) [10], and others. These classification systems have some sim-

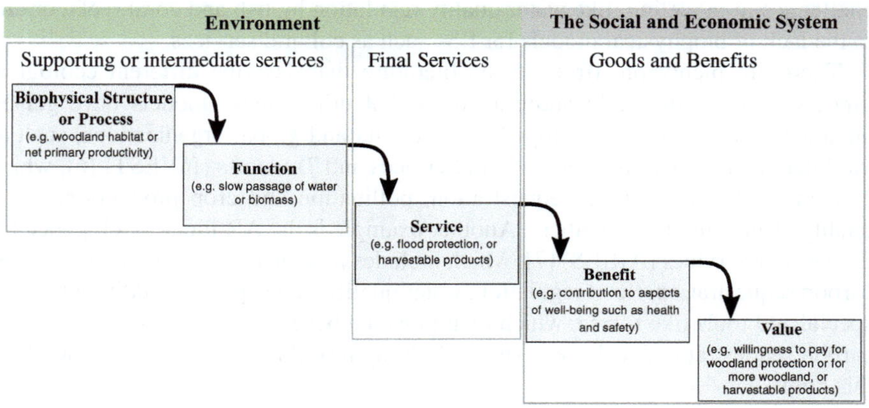

Fig. 2.1 Ecosystem Services Cascade model (Adapted from [10])

ilarities but different levels of detail. The main overlapping thematic categories are provisioning, regulating, and cultural services. Supporting and habitat services are debatable categories that some classification schemes include, while others recognize them as underlying ecological processes. The following subsection and Table 2.1 provide descriptions and examples of some ES based on the CICES scheme.

2.3.1 Categories of Ecosystem Goods and Services

Provisioning Services. For most people, ESs are the direct material benefits humans extract from nature, such as food and raw materials. However, this describes only one type of ES, which is referred to as provisioning services. Ecosystems provide the proper conditions to maintain biodiversity and manage ecological cycles to supply the various goods and services. Examples include the provisioning of water, crops, wood, and plants used in medicine or utilized in the production of clothes. Energy from fossil fuels, fuelwood, and renewable sources are other examples of provisioning services.

Regulating Services are the services that govern natural phenomena and processes that are necessary to maintain a functional and resilient ecosystem. They play an essential role in making life easy for people through direct and indirect benefits. For example, vegetation removes pollutants to provide cleaner air. It can also sequester and store carbon. Trees filter water and reduce risks of soil erosion and landslides, while coastlines are protected from storm damage by coral reefs. Wetlands play a significant role in flood prevention by acting as a natural buffer. They also remove pollutants from water to provide the service of water quality regulation or wastewater treatment. Pollination service is fundamental for growing food, while bacteria and microorganisms decompose waste and reduce water pollution.

Cultural Services represent the nonmaterial benefits humans get from natural ecosystems. These services contribute physically to human well-being by providing areas for recreation and mentally by providing a connection and sense of place. Human connection with nature has helped develop significant knowledge, culture, inspiration, and creativity. In addition, natural places worldwide have a valuable spiritual and religious meaning and are recognized as the driver for ecotourism.

Supporting Services are the underlying processes and functions that enable all other ESs. Although there is no consensus about treating these as a separate class, their fundamental importance is not disputed. The benefits from processes such as water and nutrient cycling, photosynthesis, and soil formation are imperative for sustaining all life forms on planet Earth.

Habitat Services provide the necessary resources such as water, food, shelter to plants, animals, and other species. The different types of ecosystems provide habitat year-round or seasonally for migrating species. In addition, habitat and

Table 2.1 Classification of ESs (Adapted from [10])

Section	Division	Group	Examples of goods and services
Provisioning (Biotic)	Biomass	Cultivated terrestrial plants for nutrition, materials, or energy	Harvested crops, processed timbers, and energy production.
		Cultivated aquatic plants for nutrition, materials, or energy	Vitamin supplement, seaweed for insulation material as a source of energy.
		Reared animals for nutrition, materials, or energy	Meat, eggs, milk production, Hide products, cooking fuel.
		Wild plants (terrestrial and aquatic) for nutrition, materials, or energy	Mushroom, berries for food. Fuel wood.
		Wild animals (terrestrial and aquatic) for nutrition, materials or energy	Food and hide production. Fuel Source.
	Genetic material from all biota	Genetic material from plants, algae, fungi, animals, and other organisms	Species with novel characteristics to increase yields or reduce costs by resisting diseases or pests. Creation of artificial gene products.
Provisioning (Abiotic)	Water	Surface water used for nutrition, materials, or energy	Water used for drinking, as a material, or for the generation of hydroelectric and tidal power.
		Ground water for used for nutrition, materials, or energy	Ground (and subsurface) water used for drinking, as a material or as an energy source.
	Nonaqueous natural abiotic ecosystem outputs	Mineral substances used for nutrition, materials, or energy	Salt for dietary value, pigments for decoration, uranium for energy production.
		Nonmineral substances or ecosystem properties used for nutrition, materials, or energy	Sunlight for vitamin D, renewable energy sources (wind, solar, and geothermal)

Regulation and Maintenance (Biotic)	Transformation of biochemical or physical inputs to ecosystems	Mediation of wastes or toxic substances of anthropogenic origin by living processes	Bioremediation Filtration, sequestration, storage, accumulation by microorganisms, algae, plants, and animals. Smell reduction, noise attenuation, visual screening.
	Regulation of physical, chemical, biological conditions	Regulation of baseline flows and extreme events.	Control of erosion rates, buffering and attenuation of mass movement, hydrological cycle and water flow regulation (including flood control and coastal protection), wind protection, fire protection.
		Lifecycle maintenance, habitat, and gene pool protection	Pollination, seed dispersal, maintaining nursery populations and habitats.
		Pest and disease control	Pest control (including invasive species), disease control.
		Regulation of soil quality	Decomposition of plant residue, N-fixation by legumes
		Water conditions	Regulation of the chemical condition of freshwaters and saltwater by living processes
		Atmospheric composition and conditions	Regulation of temperature and humidity, including ventilation and transpiration. Sequestration of carbon.
Regulation and Maintenance (Abiotic)	Transformation of biochemical or physical inputs to ecosystems	Mediation of waste, toxics, and other nuisances by nonliving processes	Dilution by freshwater, marine ecosystems, and atmosphere. Mediation by other chemical or physical means (e.g., via filtration, sequestration, storage, or accumulation)
		Mediation of nuisances of anthropogenic origin	Mediation of nuisances by abiotic structures or processes
	Regulation of physical, chemical, biological conditions	Regulation of baseline flows and extreme events	Coastal protection, flood protection.
		Maintenance of physical, chemical, abiotic conditions	Maintenance and regulation by inorganic natural chemical and physical processes

(continued)

Table 2.1 (continued)

Section	Division	Group	Examples of goods and services
Cultural (Biotic)	Direct, in situ, and outdoor interactions with living systems that depend on presence in the environmental setting	Physical and experiential interactions with natural environment	Using the environment such as woodlands for recreation, fitness, mental health, and ecotourism. Watching plants and animals in their natural habitat.
		Intellectual and representative interactions with natural environment	Researching and studying nature. Artistic inspiration, tourism, local identity.
	Indirect, remote, often indoor interactions with living systems that do not require presence in the environmental setting	Spiritual, symbolic, and other interactions with natural environment	Cultural and national symbols such as the Bald Eagle, Totemic Species.
Cultural (Abiotic)	Direct, in situ, and outdoor interactions with natural physical systems that depend on presence in the environmental setting	Physical and experiential interactions with natural abiotic components of the environment	Ecotourism for caves.
		Intellectual and representative interactions with abiotic components of the natural environment	Rock faces for climbing.
	Indirect, remote, often indoor interactions with physical systems that do not require presence in the environmental setting	Spiritual, symbolic, and other interactions with the abiotic components of the natural environment	Iconic mountain peaks.

supporting services maintain genetic diversity, representing the variation in genes within species, which is vital for survival, adaptation, and continuity.

2.3.2 Classification Systems

The Millennium Ecosystem Assessment (MA) [11] was the first significant project sponsored by the United Nations to assess the effect of ecosystem change on human well-being and classify ecosystem goods and services. This project included hundreds of scientists and experts from different disciplines who helped identify the current ES conditions, trends, and recommended responses to ensure the sustainability of ecosystems and human well-being. The MA identified four categories of ES: provisioning, regulating, supporting, and cultural services.

The Economics of Ecosystems and Biodiversity (TEEB) [12] is a global initiative that values biodiversity and highlights the potential ecological and economic effects of ecosystem degradation and biodiversity loss. However, unlike MA, the TEEB classification does not include supporting services in its definition of ES. Instead, they identified a new category for "habitat services" along with provisioning, regulating, and cultural services.

The Common International Classification of Ecosystem Services (CICES) [10] is a framework developed by the European Environment Agency aimed to standardize the methods of accounting for ES. CICES follows a systematic and hierarchical approach for the definition and classification of ES. The definition of services focuses explicitly on the final products from ecosystems and clearly distinguishes between ESs and ecological processes. Hence, supporting services are considered an ecological function, not a major category. The classification follows a four-level hierarchical structure with provisioning, regulating, and cultural services at the top. The classification extends in a nested manner based on the similarity and characteristics of the services providing the most detailed scheme available as shown in Table 2.1.

For example, the provisioning of cultivated plants is classified under the biomass group, which is a part of the nutrition division in the provisioning section. In contrast, cultivated crops are classified under the general food provisioning service under MA or TEEB. This level of detail is meant to avoid overlapping classes and provide a comprehensive scheme. In addition, unlike MA and TEEB, CICES classification includes biomass-based energy and biotic and abiotic outputs from ecosystems.

One of the major distinctions between classification systems is the exclusion of abiotic benefits. These are the benefits that originate from nonliving parts of nature, such as sunlight and minerals, which significantly influence living systems. The exclusion of abiotic benefits is due to the fact that the most common definition of ES specifies the services as benefits associated with living processes. There is also disagreement about whether intermediate processes and supporting services should be included as ES class.

CICES includes classes for biotic and abiotic materials in its classification schemes, while MA and TEEB excluded abiotic benefits from service classification. This contradicts the inclusion of water, which is not produced from a living system, in both MA and TEEB. One could argue that other abiotic benefits, such as minerals, should be included for similar reasons [13]. Another debatable topic is incorporating energy services as ES, which include energy generated from both renewable and nonrenewable resources. Only CICES has classes for biotic and abiotic energy sources, including final products such as solar and wind energy. For engineering applications like those considered in this book, abiotic services play a critical role, so including them in the classification system makes sense and is preferred.

Other classification schemes developed by researchers or governmental agencies include the Final Ecosystem Goods and Services (FEGS) [14] and the National Ecosystem Services Classification System (NESCS) [15] by the United States Environmental Protection Agency. Like CICES, these classification systems focus on the flow of final ecosystem services intended to be used by individuals or communities through an online tool.

2.4 Nature's Contributions to People

Sharing similar goals with the concept of ES, which advocates the role of ecosystems and their contribution to humanity, a new approach termed Nature's Contributions to People (NCP) was proposed by the Intergovernmental Science-Policy Platform on Biodiversity and Ecosystem Services (IPBES) [16]. However, NCP has a clear distinction from ESs in determining the links between humans and nature and the associated cultural influence. This accentuates the invaluable knowledge from local and indigenous communities to understand how nature contributes to people [17].

Contributions differ from services in terms of including beneficial and detrimental effects on human well-being. For example, the provisioning of goods, such as water and food, and services, such as air quality regulation are positive contributions. However, disease transmission is an example of negative contributions.

The primary motivation behind the development of NCP is to incorporate the perspective of social sciences and local knowledge in assessing ecosystems and helping the decision-making process. This point of view was lacking in ES approaches which mainly focused on ecological and economic accounting of natural systems. Although all ES classification schemes include cultural services, their accounting and valuation methods are insufficient compared to other services. This results from excluding the point of view of social sciences and humanities in developing the different schemes. Additionally, addressing these services in terms of monetary value can lead to social injustice issues from exploiting nature's services as pure products [18]. More discussion about such issues is included in Chapter 23.

NCP is classified into 18 groups, including the production of food, energy, and genetic resources, which represent material contributions. Nonmaterial contributions include the provisioning of opportunities for physical and psychological experiences, learning and inspiration, and developing a sense of place through landscape and habitat. Ecosystem regulation of air quality, climate change, freshwater, soil, pollination, hazards, and extreme events are examples of the regulating contributions to people.

2.5 Overall Status of Ecosystem Goods and Services

The world faces interconnected crises of biodiversity loss, climate change, and human development inequities [19]. These crises are indications of the great acceleration of human impacts on nature. The development of humanity requires a stable Earth system. However, human pressures on Earth have already caused destabilizing feedback that threatens present and future generations. Living in harmony with nature is the defining task of the twenty-first century which should be the top priority for everyone, everywhere [20]. Relying on biodiversity, ecosystems provide us with oxygen to breathe, drinking water, food, medicine, decomposition of waste, and the resilience of our planet to stay stable when natural disasters happen. More than 50% of global gross domestic product (GDP) moderately or highly relies on ecosystems. This makes biodiversity loss to be one of the top five risks to the global economy [21]. However, research and studies show that there has been a dangerous and worsening decline of biodiversity which would have serious effects on the health of people and the Earth.

Biodiversity loss includes species loss at global level, or at certain regional habitats, it determines the productivity, stability, invasibility, and nutrient dynamics of ecosystem which provide supporting, regulating, provisioning, and cultural services [22]. According to MA, over the last century people have benefited from the transformation of natural-dominated to human-dominated ecosystems and have taken advantage of biodiversity. However, biodiversity loss has resulted in declining well-being and exacerbated poverty in some social communities [11]. Habitat change is one of the most important and direct causes of biodiversity loss and changes in ESs. Anthropogenic activities are indirect drivers for these changes in biodiversity and ESs. Figure 2.2 shows the proportion of species lost due to land-use change, resource extraction, and utilization. Nowadays, ecosystems are facing the most rapid changes. According to the MA report [11], 15 out of 24 (more than 60%) ESs have already been degraded, 5 services are mixed, while only 4 services are enhanced. Degradation means current use exceeds sustainable levels. Enhancement is defined as increasing supply from the service. Mixed status means some components or regions increase while others decrease.

Marine and freshwater ecosystems, temperate grasslands, and tropical dry forests are some of the ecosystems and biomes that have been most significantly altered at global scale by anthropogenic activities. For ESs related to water, carbon, nitrogen, and phosphorus cycling, the atmospheric concentration of carbon dioxide has increased by about 34% since 1750; compared with 1960, the amount of water

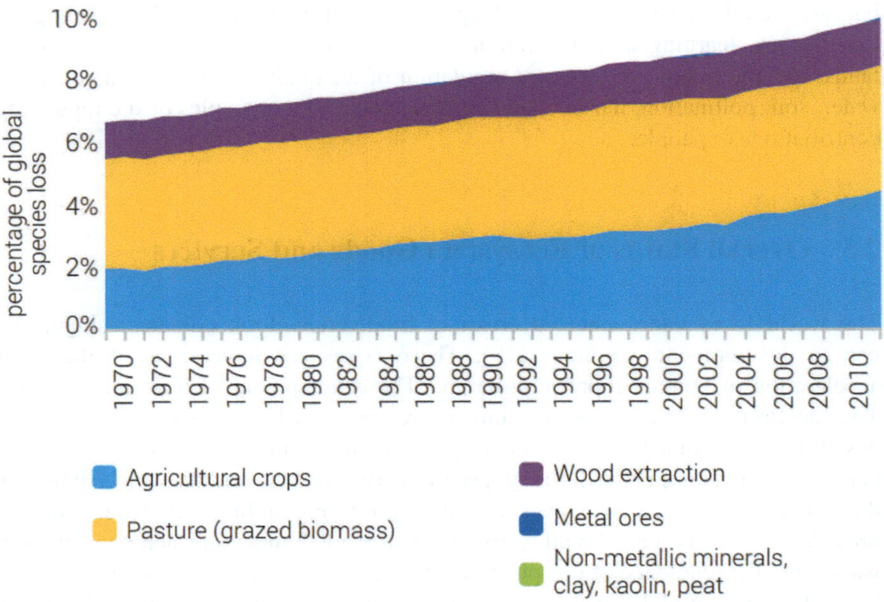

Fig. 2.2 Biodiversity loss (proportion of species lost) at global level [23]

consumption from rivers and lakes in 2000 has been doubled; the total amount of nitrogen created by human activities increased nine times from 1890 to 1990; also, between 1960 and 1990, the use of phosphorus fertilizers and the accumulation rate of phosphorus in agricultural soils tripled [11].

2.5.1 Status of Provisioning Services

Human requirement (demand) for life-sustaining provisioning services like food, water, fuel, material, etc., are growing rapidly and are expected to continue the trend in the near future. Spatial heterogeneity is one significant feature for these provisioning services. For some of these services, demands have overshot supplies at regional and global scales. For current status of ecological supply from provisioning services, between 1960 and 2000, global food yield increased by two-and-a half times; installed hydropower capacity doubled; wood production increased by three times; timber production increased by more than half. These increases in supply are faster than the pace of population growth and slower than economic growth. The trend of demand (resource use) due to human activities is shown in Fig. 2.3. As illustrated in this figure, for most resources, the total use has been increasing while for some resources the per capita use has been decreasing. The supplies of freshwater, phosphate, wild fisheries, wood building materials first increased then

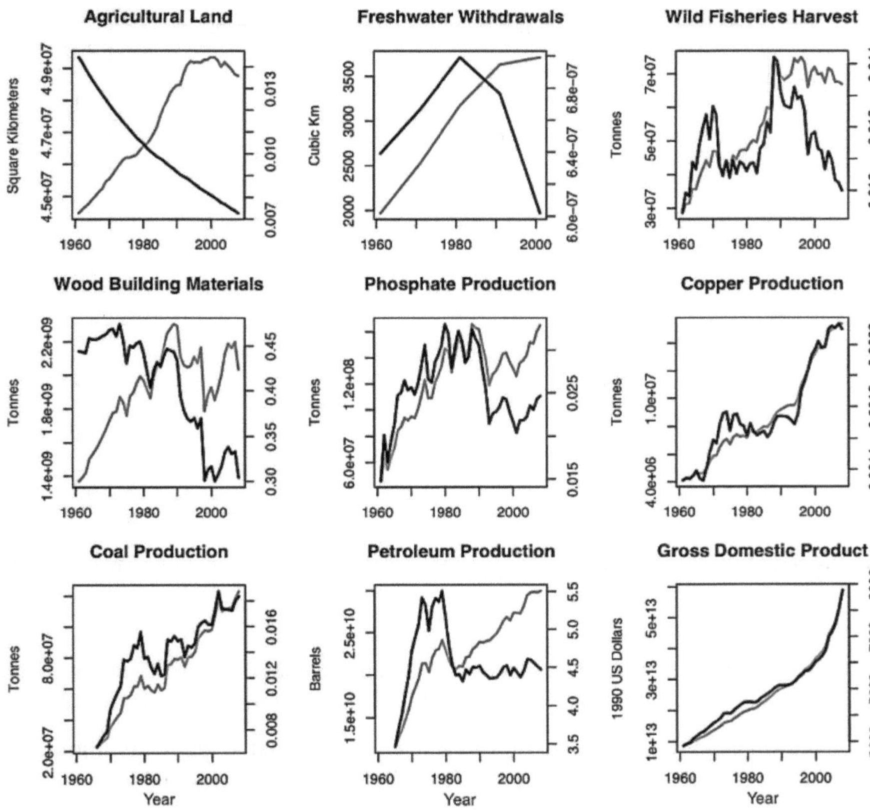

Fig. 2.3 Global resources use/production trends from 1961 to 2008. Grey line represents total use, black line represents per capita use [24]

has decreased. After the peak, these supplies can hardly meet the rapidly increasing demand, and this situation is more obvious for resources without suitable substitutes such as water and phosphate.

Food Food and agricultural productions have increased steadily since the 1950s; there is more food produced today per person than ever recorded. As shown in Fig. 2.4, the world average daily supply of calories per person has been increasing consistently since 1961, but the trends are not the same across the world. Figure 2.4 also shows that although the total production of cereals has been increasing, a decline has been witnessed since 2017. In fact, global cereal production has increased by 280% between 1961 and 2014 while the population increased only 136%, indicating cereal per person is also increasing. Although, the world currently produces more than enough food per capita, there are over 795 million people remaining undernourished. Ensuring sufficient production of food will not solve the inherent imbalances but is still necessary. The Food and Agriculture Organization

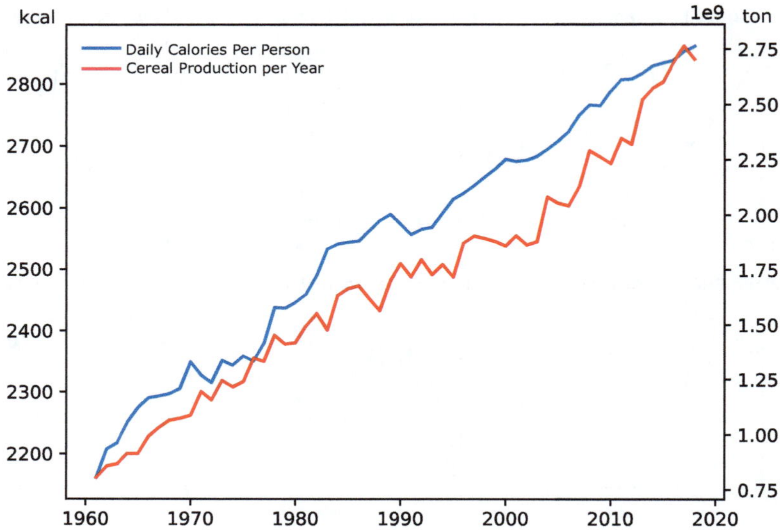

Fig. 2.4 Daily supply of calories per person and global annual cereal production from 1961 to 2018 [25]

of the United Nations (FAO) estimates that due to population growth and the trend of richer diets, we need to roughly double the amount of food we grow by 2050.

Increasing attention to biomass renewable energy has led to the trend of growing nonfood crops for biofuels and biomaterials. Government policies on biofuels influence land-use allocation and other ecosystem resources partition which might result in less resource availability for food production. The food sector contributes the most to environmental and humanitarian impacts. Farmlands occupy around 50% of the plant-habitable surface on Earth and use 68% of the accessible freshwater which is more than twice of industry use (23%). Farmlands also contribute 25%–30% of global GHG emissions.

Among food-related ESs, the ES of capture fisheries has been degraded. Overharvesting or overfishing has resulted in declining production which makes ES of marine and freshwater unsustainable. Many fisheries have already collapsed. According to FAO, the sustainability of global fishery resources continues to decline but at a slower rate in the most recent period. Their sustainability has dropped from 90% in 1974 to 65.8% in 2017. Fish stocks within biologically sustainable levels contributed 78.7% of the global marine fish landings in 2017.

In contrast, there is a substantial increase in production of livestock. This sector has become one of the fastest growing agricultural subsectors as incomes rise, food structure changes, and population increases. According to the FAO, livestock account for 40% of the global agricultural production. This sector provides food and nutrition security and livelihoods to nearly 1.3 billion people.

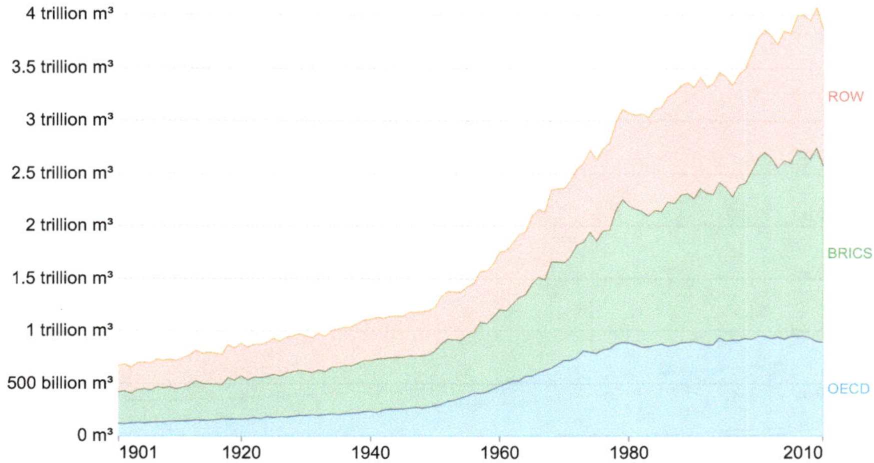

Fig. 2.5 Global freshwater use by regions from 1901 to 2010: OECD nations (38 countries), BRICS countries (Brazil, Russia, India, China, South Africa), ROW which represents rest of the world [26]

Water An increasing number of people are facing the severity of water stress and the risk of water scarcity. Growing population and development have resulted in an increasing demand of water in agriculture, industry, and domestic sectors. Since 1900, freshwater consumption has increased roughly six times as can be seen in Fig. 2.5. Figure 2.6 shows the global renewable freshwater resources from 1962 to 2017 on per capita bases. Per capita renewable resources depend on population size and total amount of renewable water flows. As can be seen from Fig. 2.6, this amount has been declining globally over the past few decades.

The freshwater provisioning service has been degraded and considered as unsustainable in many parts of the world. In detail, around 5%–25% of global freshwater use exceeds long-term available supply. Between 15% and 35% of irrigation withdrawals exceed accessible amount of freshwater. India, which is the world's biggest agricultural water consumer, consumes around 700 billion m^3 annually, this amount keeps growing rapidly. Water supply for drinking and industry has also been overshot. Although the supply of freshwater is decreasing, dams are increasing their ability to use the limited water resource for hydropower.

To stay in a sustainable state of water resources, the rate of water withdrawal cannot exceed the rate of freshwater replenishment. "Renewable internal freshwater flows" represent internal renewable resources (internal river flows and groundwater from rainfall). Resources begin to decline if the renewable internal freshwater flows are below the rate of freshwater withdrawals. In 2018, the global water-use efficiency improved by 9%, from 17.3 USD/m^3 in 2015 to 18.9 USD/m^3 in 2018. For the future, the Sustainable Development Goals (SDGs) on water aim at

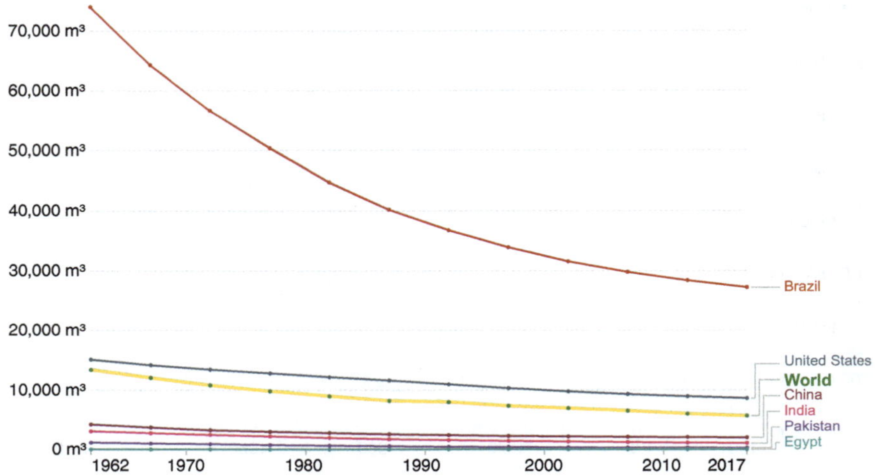

Fig. 2.6 Renewable freshwater resources per capita [26]

substantially increasing water-use efficiency across all sectors by 2030, addressing water scarcity and reducing the number of people suffering from water scarcity.

Genetic Resources The concept of genetic resources is defined as genetic material of actual or potential value to humans. Any animal, plant, microorganism containing genetic functional unit can be considered as biological material. A gene is made of DNA which is the basic physical and functional unit of heredity. Currently, genetic diversity is declining at global scale and the ES of genetic resources provisioning has been degraded. The genetic diversity loss (loss of individual or combinations of genes, loss of varieties of crops) is mainly due to the replacement of traditional landraces by modern, high-yielding cultivars [27].

The increasingly homogeneous distribution of species indicates that the difference between a species in one location and a species in another location is becoming smaller. The homogeneity of species reflects the lack of biodiversity. The loss of unique populations and extermination of species led to the loss of unique genetic diversity.

Over the past few decades, relying on the fast development of advanced biotechnology, humans have taken great advantage of genetic resources which not only changed our understanding of the world but also have driven the development of new products such as vital medicines, and methods that enhance food security which also help improve conservation methods that protect global biodiversity. Deterioration of genetic resources will affect global biodiversity and human well-being.

Fuels A fuel can be any material that releases energy as thermal energy or that can be used for work when reacting with other substances. Fossil fuels are important ecosystem goods produced from ancient biomass that was buried and

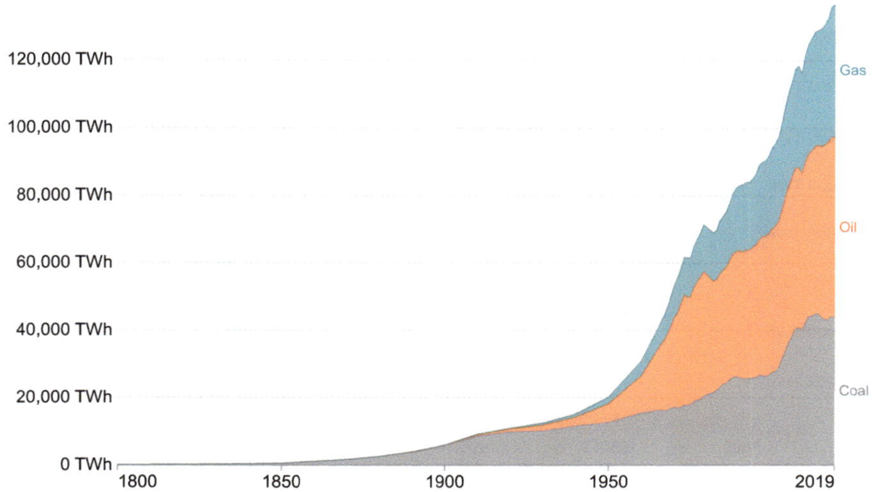

Fig. 2.7 Global fossil fuel consumption by resource [29]

transformed by planetary processes in an oxygen-starved reducing environment. The end products, such as coal, natural gas, and crude oil, are highly concentrated hydrocarbons. Carbon with high fuel value can be easily transformed into other products. Since the industrial revolution, fossil fuels have become the main energy source which have also caused severe environmental impacts. Figure 2.7 shows the change of fossil fuel consumption which has increased significantly over the past decades. Conventional fuels are nonrenewable due to the fact that their extraction rate is much greater than their formation rate. Following this trend, the consumption of nonrenewable resources must inevitably lead to their depletion. The trend of overall energy use (including nonrenewable and renewable) from 1800 to 2019 is shown in Fig. 2.8. Until 1950s, traditional biomass was still the dominant energy source. After that, fossil fuels like coal and oil became the dominant resources. Later, renewable energies take more and more portions. It is foreseeable that the structure of energy resource consumption will gradually change over time.

According to the Statistical Review of World Energy 2021 report [28], energy markets have been affected by the COVID-19 pandemic dramatically. The falling rates of primary energy and carbon emissions are the fastest since World War II. The consumption of primary energy decreased by 4.5% in 2020, which is the largest since 1945. Carbon emissions from energy consumption decreased by 6.3%, and oil consumption fell by 9.3%, both to their lowest level in the last decade. Besides, the consumption of coal and natural gas fell 4.2% and 2.3%. In contrast, renewable energy continued to grow, among which solar power has achieved its largest increase ever.

The COVID-19 catastrophe can be considered as a "black swan" event, and its impact on the world is unpredictable. Energy sector is one of the affected sectors.

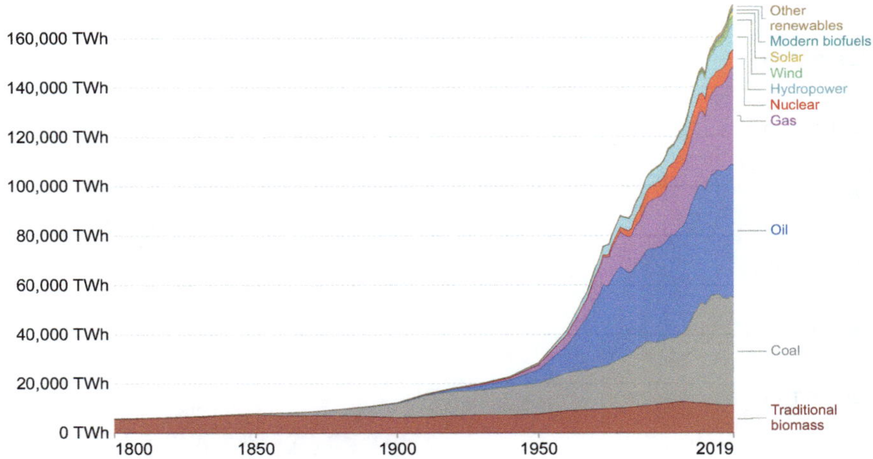

Fig. 2.8 Global energy use by resource [29]

From a historical perspective, the decline in energy demand and carbon emissions is huge and dramatic. But from a forward-looking point of view, the rate of decline for carbon emissions in 2020 is what the Earth needs. To achieve the Paris climate goals, the decline rate of carbon emissions should at least be comparable to that of 2020. If the total carbon emissions decrease at the rate in 2020 for the following 30 years, the overall global carbon emissions would decline by around 85% by 2050 which is roughly midway on the path toward meeting the Net-Zero global goal. Despite the decrease in emissions during the pandemic, the overall atmospheric concentration of CO_2 does not show a decline and continues to break new records.

2.6 Regulating Services

Regulating services include the regulation of climate, decomposition, soil fertility, pests, pollination, soil erosion, etc. This category of ES also includes regulation of the quantity and quality of water, and impact of extreme events on humans and ecosystems. Regulating services have been significantly changed due to human activities. For instance, humans have modified climate and disease regulation to get resources and receive services. Such transformations result in the overshoot of nature's carrying capacity. According to MA, seven regulating services, such as air quality regulation, water purification, and waste treatment, have been degraded.

Air Quality Regulation Ecosystems influence air quality from different aspects. Through a series of complex interactions, wetlands, trees, and soil filter many pollutants in the atmosphere including carbon monoxide (CO), ozone (O_3), particulate matter (PM), and nitrogen oxides (NO_x). Ecosystems are able to regulate air quality

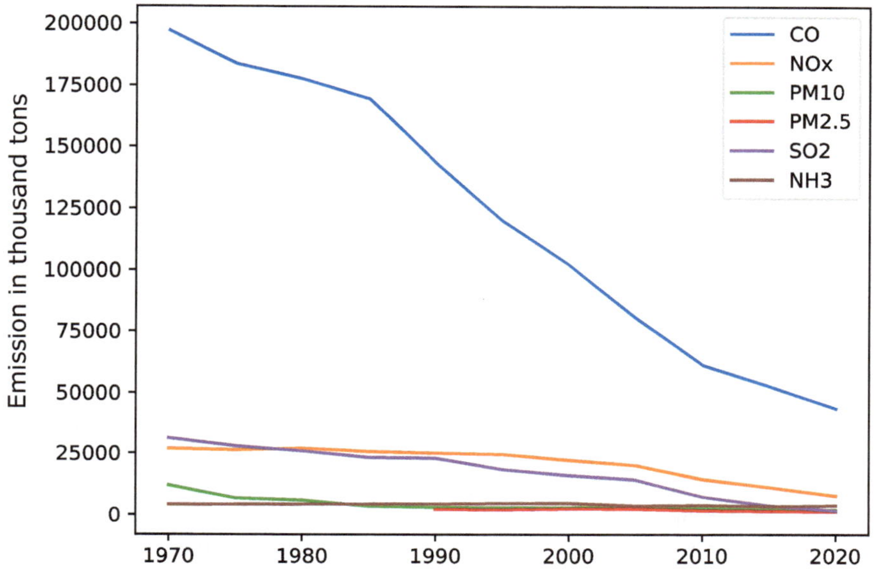

Fig. 2.9 Air pollutant emissions (in thousand tons) in United States from 1970 to 2020 [34]

by transporting, absorbing, and mitigating various emissions; however, the ES of air quality regulation has been degraded. Over the last 10 years, great changes happened in the amount of global and regional sulfur dioxide emission. Continuous increase in the emission of SO_2 since the beginning of the twentieth century mainly in high-income countries [30] resulted in increased public concern about negative effects of air pollution on the environment and human health. This resulted in legislation to reduce emissions [31] due to which emissions in Europe and North America reduced by 70–80% [32]. These large regional reductions resulted in a global decline between 1980 and 2000. A similar sharp rise in emissions followed by a decline due to regulations occurred in China [33]. From 1990 to 2015, global SO_2 emission reduced by 55 TgS (31%). Figure 2.9 depicts the emission of air pollutants in the United States, which shows a significant decrease since 1970; however, the pollution levels in many areas exceed national air quality standards.

Water Quality Regulation The issue of water quality is one of the most important challenges facing society in the twenty-first century. It threatens human health, affects ecosystem functions, reduces food production, and hinders economic growth. Good water quality and sufficient water quantity are essential for achieving SDGs in health, food and water security, and ecosystems. Since the 1990s, water pollution has deteriorated in the majority of rivers in Latin America, Africa, and Asia. Increasing amount of wastewater enters into water bodies, which is the immediate cause of increasing serious water pollution issues. Population growth, rapid development of the economy, expansion of agriculture, and sewerage connections with low levels of treatment are ultimate causes of water quality issues.

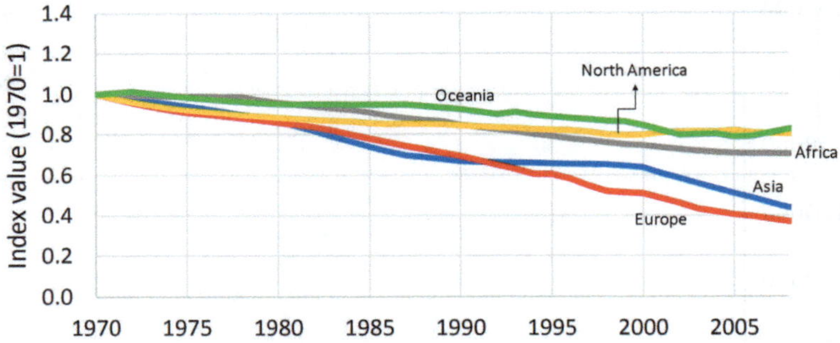

Fig. 2.10 Extent of wetlands from 1970 to 2008 [37]

Ecosystems such as wetlands, rivers, and lakes have a natural ability to regulate water quality [35]. Wetlands have been called nature's kidneys; they provide multiple ESs to humans including water purification, fishing, etc. Wetlands help maintain biodiversity, regulate climate and quality of air, provide recreational opportunities, and have religious and cultural significance [36]. However, despite their essential role, wetlands remain undervalued by policy and decision-makers. Wetlands have been lost all over the world. Around 35% of wetlands on Earth were lost between 1970 and 2015, and the rate of loss is faster year by year since 2000. This is based on the wetland indicator status (WIS) of plant species which can be calculated with species data. This index provides information of relative abundance of species or community types. Figure 2.10 shows the extent of wetlands in different regions, which is declining since 1970.

Climate Regulation Since the preindustrial period, the global average surface temperature has risen about 1 °C, which indicates a huge increase in accumulated heat. Climate change has negative effects on human societies and ecosystems. In recent decades, there is an increase in the frequency and intensity of extreme natural phenomena and weather which negatively affect food security and terrestrial ecosystems. Climate change is also contributing to desertification and land degradation in many regions. As can be seen from Fig. 2.11, the concentrations of various GHGs in the atmosphere are steadily increasing and accelerating in recent decades. The concentration of carbon dioxide (CO_2) has exceeded 420 ppm in 2022, which is roughly 40% more than the highest concentration (280 ppm) since humans started burning fossil fuels at the start of the industrial revolution. Current concentration of methane (CH_4) is much more than double the highest concentration in almost one million years, and the concentration of nitrous oxide (N_2O) has increased by about 14%. These trends indicate that emissions have exceed the carrying capacity of nature. From 1905 to 2016, the composition of GHGs in atmosphere has also changed. Figure 2.12 illustrates atmospheric concentrations of different GHGs in 1950, 1990, and 2016. In this period, 1905–2016, CO_2 occupies a dominant position in GHG.

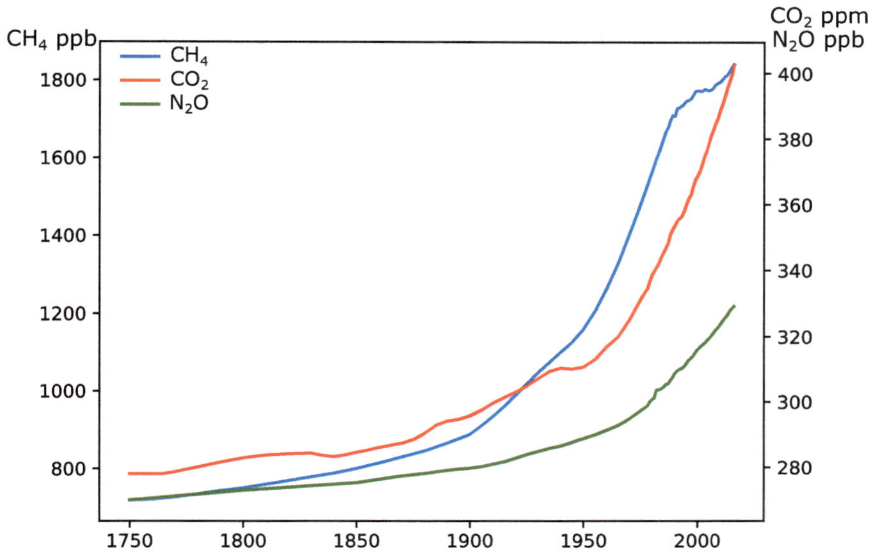

Fig. 2.11 Atmospheric concentrations of CO_2, CH_4, and N_2O from 1750 to 2016 [38, 39]

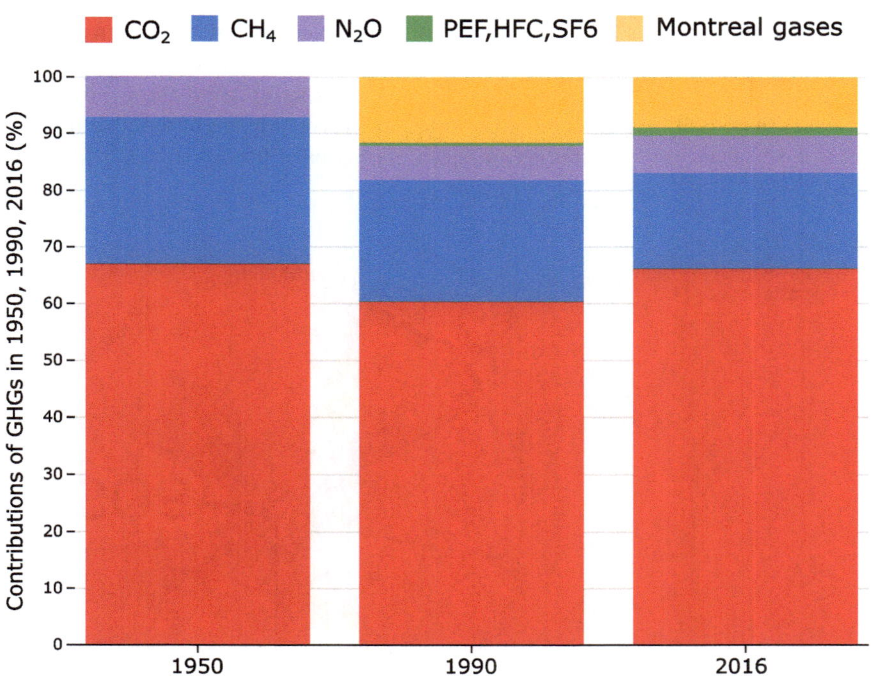

Fig. 2.12 Contributions of GHGs in 1950, 1990, and 2016 (ppm CO_2 eq). The height of the bar illustrates relative increases in greenhouse forcing [38, 39]. The term forcing represents any influence that can shift the climate

The ES of climate regulation regulates processes related to chemical components of the atmosphere such as GHGs, the ozone layer, precipitation, also weather patterns at global and local scales [2]. Forests cover about 30% of the Earth's surface area, and as trees grow, they capture carbon from the atmosphere and store it in wood, plant matter, and under the soil. Without forests and other carbon sequestration processes in soil and oceans, the carbon would remain in the atmosphere in the form of CO_2, which is one of the most important GHGs causing climate change. However, since the end of the last great ice age –10,000 years ago– the world has lost one-third of its forests. Global deforestation has reached its peak in the 1980s, and since then, the deforestation rate has been steadily declining but still remains a concern.

2.7 Planetary Boundary

The conceptual framework of planetary boundaries (PBs) was proposed for assessing absolute sustainability which takes the carrying capacity of nature as a reference value. Planetary boundary is considered as "a bid to reform environmental governance." The PB framework identifies nine processes that regulate the stability and resilience of the Earth system [40, 41]. This framework defines the "safe operating space" (SOS) for human development which is illustrated by the green circle in Fig. 2.13. Within these quantitative ecological thresholds, humans can continue to develop and thrive for generations to come. Crossing over these boundaries increases the possibility of large-scale abrupt or irreversible environmental impacts. Control variables are projections of these ecological thresholds. Each planetary

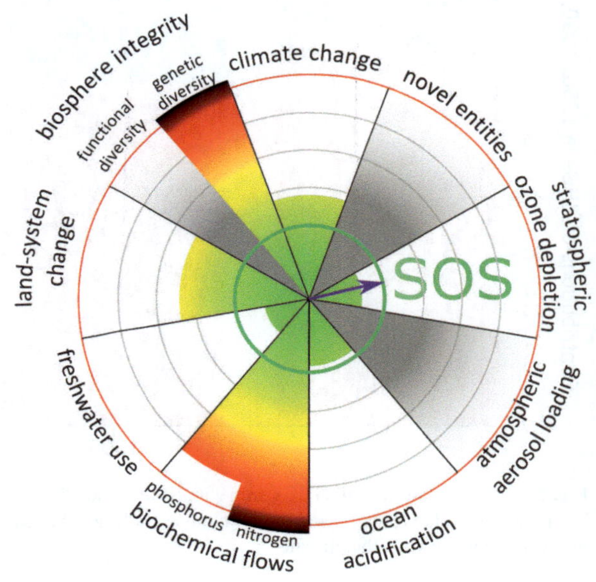

Fig. 2.13 Planetary boundaries and current status, safe operating space [42]

boundary has one or more control variables. Based on the planet's biophysical processes, SOS provides the reference of risks that human perturbations could substantially affect the Earth [42]. The original planetary boundary paper only estimated boundaries at global scale. The Earth system is spatially heterogeneous and some Earth system processes, such as freshwater consumption, happen at regional scales and do not have global boundaries. Later, research has been done in defining sub-global or regional boundaries. As shown in Fig. 2.13, scientists find out that humanity has exceeded five of these boundaries, which are novel entities, biosphere integrity, land system change, biogeochemical flows, and climate change.

Nowadays, a major global challenge that humanity faces is to achieve well-being for all and simultaneously make sure that the biophysical processes and ESs are utilized within their sustainable thresholds. The doughnut economics framework introduces social well-being to the original PB framework through a lower boundary that represents the minimum resources needed to avoid human deprivation [43]. The environmental threshold sets an outer boundary, trespassing the outer boundary will lead to environmental degradation. Ideally, each region lies in between the outer and inner boundaries, which forms a doughnut shape. Operating between the ecological ceiling and social foundation could ensure sustainability by preventing environmental degradation and ensuring human well-being of all [44].

2.8 Conclusions

While nations have succeeded in improving their well-being, it has come at the expense of degrading ecological systems. It is crucial to understand and recognize human dependency on nature. Natural systems provide food, water, materials, clean air, regulate water and nutrient, and much more. It is hard to think of anything that anyone uses that is unrelated to what nature provides. However, the demand for many natural resources and services has exceeded the limit of supply from ecosystems. The trend of how much humans are exploiting ES is alarming, and if the trend continues, ecological systems will be further degraded and humanity will suffer. Reducing current demands is needed to avoid degrading more ecosystems and restoring the impacted ones. In addition to the undeniable need of including ES in decision-making and policy implementation, identifying and quantifying ES is essential to prevent additional harm and provide guidelines for sustainable use. Various chapters in this book describe approaches for quantifying the demand and supply of ES and ways of doing engineering in synergy with nature.

References

1. Gretchen C. Daily. *Nature's services: societal dependence on natural ecosystems*. Island Press, 1997.
2. Robert Costanza, Ralph d'Arge, Rudolf De Groot, Stephen Farber, Monica Grasso, Bruce Hannon, Karin Limburg, Shahid Naeem, Robert V O'neill, Jose Paruelo, et al. The value of the world's ecosystem services and natural capital. *nature*, 387(6630):253–260, 1997.

3. World Economic Forum. Nature risk rising, 1 2020.
4. Partha Dasgupta. *The economics of biodiversity: the Dasgupta review*. Hm Treasury, 2021.
5. Benjamin Burkhard and Joachim Maes. Mapping ecosystem services. *Advanced Books*, 1:e12837–, 3 2017.
6. HT, T Ricketts, AD Guerry, SA Wood, R Sharp, E Nelson, D Ennaanay, S Wolny, N Olwero, K Vigerstol, and others Tallis. Integrated valuation of environmental services and tradeoffs (invest) 3.1. 0 user's guide. *Natural Capital Project*, 2014.
7. KJ, F Villa, GW Johnson, and B Bagstad Voigt. Aries–artificial intelligence for ecosystem services: A guide to models and data, version 1.0. *ARIES report series*, 2011.
8. Rudolf S. De Groot, Matthew A. Wilson, and Roelof M.J. Boumans. A typology for the classification, description and valuation of ecosystem functions, goods and services. *Ecological Economics*, 41:393–408, 6 2002.
9. James Boyd and Spencer Banzhaf. What are ecosystem services? the need for standardized environmental accounting units. *Ecological Economics*, 63:616–626, 8 2007.
10. R. Haines-Young and M.B. Potschin. Common international classification of ecosystem services (CICES) v5.1 and guidance on the application of the revised structure., 2018.
11. Millennium Ecosystem Assessment. *Ecosystems and human well-being*. Island Press, 2005.
12. TEEB. The economics of ecosystems and biodiversity ecological and economic foundations, 2010.
13. E. S. van der Meulen, L. C. Braat, and J. M. Brils. Abiotic flows should be inherent part of ecosystem services classification. *Ecosystem Services*, 19:1–5, 6 2016.
14. U.S. EPA. Final ecosystem goods and services classification system (FEGS-CS). *EPA/600/R-13/ORD-004914*, 2013.
15. D. Landers. National ecosystem services classification system (NESCS): Framework design and policy application. *U.S. Environmental Protection Agency*, EPA/800/R-15/002, 2015.
16. Sandra Díaz, Sebsebe Demissew, Julia Carabias, Carlos Joly, Mark Lonsdale, Neville Ash, Anne Larigauderie, Jay Ram Adhikari, Salvatore Arico, András Báldi, Ann Bartuska, Ivar Andreas Baste, Adem Bilgin, Eduardo Brondizio, Kai M.A. Chan, Viviana Elsa Figueroa, Anantha Duraiappah, Markus Fischer, Rosemary Hill, Thomas Koetz, Paul Leadley, Philip Lyver, Georgina M. Mace, Berta Martin-Lopez, Michiko Okumura, Diego Pacheco, Unai Pascual, Edgar Selvin Pérez, Belinda Reyers, Eva Roth, Osamu Saito, Robert John Scholes, Nalini Sharma, Heather Tallis, Randolph Thaman, Robert Watson, Tetsukazu Yahara, Zakri Abdul Hamid, Callistus Akosim, Yousef Al-Hafedh, Rashad Allahverdiyev, Edward Amankwah, T. Stanley Asah, Zemede Asfaw, Gabor Bartus, Anathea L. Brooks, Jorge Caillaux, Gemedo Dalle, Dedy Darnaedi, Amanda Driver, Gunay Erpul, Pablo Escobar-Eyzaguirre, Pierre Failler, Ali Moustafa Mokhtar Fouda, Bojie Fu, Haripriya Gundimeda, Shizuka Hashimoto, Floyd Homer, Sandra Lavorel, Gabriela Lichtenstein, William Armand Mala, Wadzanayi Mandivenyi, Piotr Matczak, Carmel Mbizvo, Mehrasa Mehrdadi, Jean Paul Metzger, Jean Bruno Mikissa, Henrik Moller, Harold A. Mooney, Peter Mumby, Harini Nagendra, Carsten Nesshover, Alfred Apau Oteng-Yeboah, György Pataki, Marie Roué, Jennifer Rubis, Maria Schultz, Peggy Smith, Rashid Sumaila, Kazuhiko Takeuchi, Spencer Thomas, Madhu Verma, Youn Yeo-Chang, and Diana Zlatanova. The IPBES conceptual framework — connecting nature and people. *Current Opinion in Environmental Sustainability*, 14:1–16, 6 2015.
17. Sandra Díaz, Unai Pascual, Marie Stenseke, Berta Martín-López, Robert T. Watson, Zsolt Molnár, Rosemary Hill, Kai M.A. Chan, Ivar A. Baste, Kate A. Brauman, Stephen Polasky, Andrew Church, Mark Lonsdale, Anne Larigauderie, Paul W. Leadley, Alexander P.E. Van Oudenhoven, Felice Van Der Plaat, Matthias Schröter, Sandra Lavorel, Yildiz Aumeeruddy-Thomas, Elena Bukvareva, Kirsten Davies, Sebsebe Demissew, Gunay Erpul, Pierre Failler, Carlos A. Guerra, Chad L. Hewitt, Hans Keune, Sarah Lindley, and Yoshihisa Shirayama. Assessing nature's contributions to people: Recognizing culture, and diverse sources of knowledge, can improve assessments. *Science*, 359:270–272, 1 2018.

18. Sharachchandra Lele, Oliver Springate-Baginski, Roan Lakerveld, Debal Deb, and Prasad Dash. Ecosystem services: Origins, contributions, pitfalls, and alternatives. *Conservation and Society*, 11:343, 2013.
19. Harvey Locke, Johan Rockström, Peter Bakker, Manish Bapna, Mark Gough, Jodi Hilty, Marco Lambertini, Jennifer Morris, Paul Polman, Carlos M Rodriguez, et al. A nature-positive world: the global goal for nature. 2021.
20. State of the planet presented at the un secretary general's address. https://www.un.org/en/climatechange/un-secretary-general-speaks-state-planet. Accessed: 2022-01-08.
21. World Economic Forum. Nature risk rising: Why the crisis engulfing nature matters for business and the economy, 2020.
22. David Tilman, Forest Isbell, and Jane M Cowles. Biodiversity and ecosystem functioning. *Annual review of ecology, evolution, and systematics*, 45:471–493, 2014.
23. Stefan Bringezu, Anu Ramaswami, Heinz Schandl, Meghan O'Brien, Rylie Pelton, Jean Acquatella, E Ayuk, A Chiu, Robert Flanegin, Jacob Fry, et al. Assessing global resource use: A system approach to resource efficiency and pollution reduction. *IRP Reports*, 2017.
24. Joseph R Burger, Craig D Allen, James H Brown, William R Burnside, Ana D Davidson, Trevor S Fristoe, Marcus J Hamilton, Norman Mercado-Silva, Jeffrey C Nekola, Jordan G Okie, et al. The macroecology of sustainability. *PLoS biology*, 10(6):e1001345, 2012.
25. Food Supply crops and livestock products. https://www.fao.org/faostat/en/#data/QCL. Accessed: 2022-03-13.
26. Our World in Data water use and stress. https://ourworldindata.org/water-use-stress. Accessed: 2022-03-13.
27. David S Woodruff. Populations, species, and conservation genetics. *Encyclopedia of biodiversity*, page 811, 2001.
28. B Looney. Statistical review of world energy globally consistent data on world energy markets. *and authoritative publications in the field of energy. Review of World Energy Data*, 70:8–20, 2021.
29. Our World in Data primary resource consumption. https://ourworldindata.org/global-energy-200-years. Accessed: 2022-03-13.
30. Rachel M Hoesly, Steven J Smith, Leyang Feng, Zbigniew Klimont, Greet Janssens-Maenhout, Tyler Pitkanen, Jonathan J Seibert, Linh Vu, Robert J Andres, Ryan M Bolt, et al. Historical (1750–2014) anthropogenic emissions of reactive gases and aerosols from the community emissions data system (CEDS). *Geoscientific Model Development*, 11(1):369–408, 2018.
31. Wenche Aas, Augustin Mortier, Van Bowersox, Ribu Cherian, Greg Faluvegi, Hilde Fagerli, Jenny Hand, Zbigniew Klimont, Corinne Galy-Lacaux, Christopher Lehmann, et al. Global and regional trends of atmospheric sulfur. *Scientific reports*, 9(1):1–11, 2019.
32. Vigdis Vestreng, Gunnar Myhre, Hilde Fagerli, Stefan Reis, and Leonor Tarrasón. Twenty-five years of continuous sulphur dioxide emission reduction in Europe. *Atmospheric chemistry and physics*, 7(13):3663–3681, 2007.
33. Nickolay A Krotkov, Chris A McLinden, Can Li, Lok N Lamsal, Edward A Celarier, Sergey V Marchenko, William H Swartz, Eric J Bucsela, Joanna Joiner, Bryan N Duncan, et al. Aura OMI observations of regional SO 2 and NO 2 pollution changes from 2005 to 2015. *Atmospheric Chemistry and Physics*, 16(7):4605–4629, 2016.
34. USEPA national air quality. https://www.epa.gov/air-trends. Accessed: 2022-03-13.
35. Bhavik R Bakshi. *Sustainable Engineering: Principles and Practice*. Cambridge University Press, 2019.
36. Nat Fonder and Tom Headley. The taxonomy of treatment wetlands: a proposed classification and nomenclature system. *Ecological Engineering*, 51:203–211, 2013.
37. Global Biodiversity Outlook. Progress towards the Aichi biodiversity targets: An assessment of biodiversity trends, policy scenarios and key actions. 2014.
38. NOAA national oceanic and atmospheric administration. https://www.noaa.gov/. Accessed: 2022-03-13.

39. AGAGE advanced global atmospheric gases experiment. http://agage.mit.edu/. Accessed: 2022-03-13.
40. Johan Rockström, Will Steffen, Kevin Noone, Åsa Persson, F Stuart Chapin III, Eric Lambin, Timothy M Lenton, Marten Scheffer, Carl Folke, Hans Joachim Schellnhuber, et al. Planetary boundaries: exploring the safe operating space for humanity. *Ecology and society*, 14(2), 2009.
41. Johan Rockström, Will Steffen, Kevin Noone, Åsa Persson, F Stuart Chapin, Eric F Lambin, Timothy M Lenton, Marten Scheffer, Carl Folke, Hans Joachim Schellnhuber, et al. A safe operating space for humanity. *nature*, 461(7263):472–475, 2009.
42. Will Steffen, Katherine Richardson, Johan Rockström, Sarah E Cornell, Ingo Fetzer, Elena M Bennett, Reinette Biggs, Stephen R Carpenter, Wim De Vries, Cynthia A De Wit, et al. Planetary boundaries: Guiding human development on a changing planet. *Science*, 347(6223), 2015.
43. Kate Raworth. *A safe and just space for humanity: can we live within the doughnut?* Oxfam, 2012.
44. John A Dearing, Rong Wang, Ke Zhang, James G Dyke, Helmut Haberl, Md Sarwar Hossain, Peter G Langdon, Timothy M Lenton, Kate Raworth, Sally Brown, et al. Safe and just operating spaces for regional social-ecological systems. *Global Environmental Change*, 28:227–238, 2014.

Part II
Engineering's Demand for Ecosystem Services

Chapter 3
Quantifying the Direct and Indirect Demand for Ecosystem Services

Kyuha Lee and Bhavik R. Bakshi

3.1 Demand for Ecosystem Services

Sustainability assessment methods such as life cycle assessment (LCA) and footprint analysis calculate environmental impacts of products and processes due to emissions to the environment such as greenhouse gases and other pollutants and inputs from the environment such as water and minerals. In this book, those environmental impacts are referred to as the demand for ecosystem services. The notation D_k is used to represent the demand for the k-th ecosystem service. For instance, CO_2 emissions from a natural gas-burning power plant impose a demand on ecosystems to sequester this CO_2 and provide the carbon sequestration service to the facility. Likewise, water consumption at the facility is the demand for the water provisioning service from the watershed ecosystem. In this sense, the demand for ecosystem services corresponds to environmental impacts from processes and economic activities. Table 3.1 shows some examples of demand and supply for ecosystem services [1, 2].

The demand for ecosystem services may be imposed directly or indirectly. For example, in the process shown in Fig. 3.1, the power plant directly demands the carbon sequestration ecosystem service due to its emission of carbon dioxide that is formed by the combustion of natural gas. The household has no CO_2 emissions,

---

K. Lee
Lowrie Department of Chemical and Biomolecular Engineering, The Ohio State University, Columbus, OH, USA

B. R. Bakshi (✉)
School for Engineering of Matter, Transport and Energy, School of Sustainability, School of Complex Adaptive Systems, Arizona State University, Tempe, AZ, USA

Lowrie Department of Chemical and Biomolecular Engineering, The Ohio State University, Columbus, OH, USA
e-mail: bhavik.bakshi@asu.edu

B. R. Bakshi (ed.), *Engineering and Ecosystems*,
https://doi.org/10.1007/978-3-031-35692-6_3

Table 3.1 Examples of the demand and supply for ecosystem services

Ecosystem service (k)	Demand for ecosystem service (D_k)	Supply of ecosystem service (S_k)	Typical ecosystem contributor
Climate regulation	Greenhouse gas emissions	Carbon sequestration	Forest, Grassland
Air quality regulation	Air pollutant emissions	Air pollutant removal	Forest, Grassland
Nutrient retention	Nutrient runoff	Nutrient removal	Wetland
Water quality regulation	Water pollutant emissions	Water pollutant removal	Wetland
Water provisioning	Freshwater consumption	Freshwater supply	Watershed
Fossil energy source provisioning	Fossil resource consumption	Fossil resource supply	Fossilization processes
Soil retention	Soil erosion	Soil formation	Soil
Pollination	Pollinators needed	Pollinators available	Pollinators

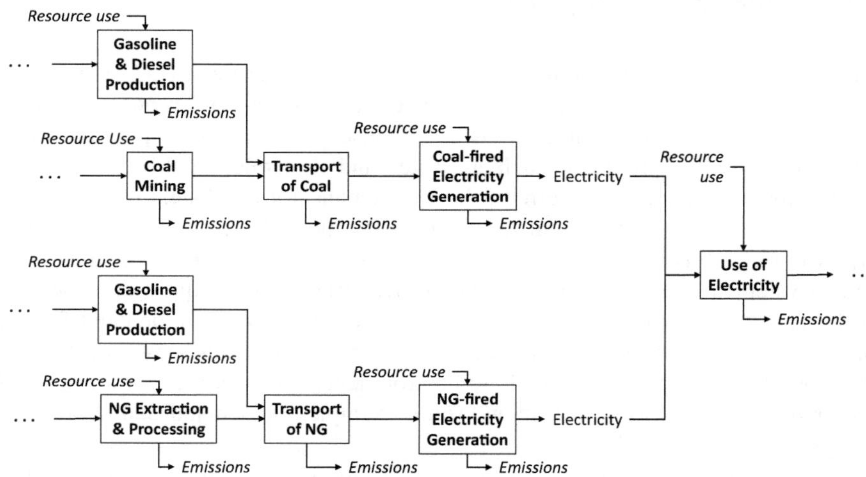

Fig. 3.1 System boundary of an example LCA study on the comparison between coal-fired electricity and natural gas (NG)-fired electricity. Bold boxes represent processes and italics represent environmental intervention flows

so does not demand carbon sequestration directly. However, since the household consumes electricity that is generated by the power plant, the household indirectly emits CO_2 and demands the carbon sequestration ecosystem service. Accounting for the direct and indirect flows is essential for assessing sustainability to prevent the shifting of impacts outside a narrow system boundary. For example, considering electric cars to be zero-pollution due to no direct emissions can be misleading since the emissions usually shift up the supply chain to the power plant. In this chapter, we describe methods to quantify the direct and indirect demand for ecosystem services by methods such as life cycle assessment and footprint analysis.

In sustainability assessment methods, environmental impacts of processes and economic activities can be quantified by mathematical calculations that are addressed in Sect. 3.2. That is, quantifying the demand for ecosystem services relies on the existing methods of LCA and footprint analysis. The rest of this section introduces the basics of LCA with an example of electricity generation.

LCA has been developed for decades to quantify environmental impacts of products or processes throughout their life cycle that ranges from the extraction phase of upstream resources (e.g., fossil resources) to the end-of-life phase (e.g., waste disposal and recycling). Therefore, LCA is often called a cradle-to-grave analysis. Figure 3.1 shows the system boundary of an example LCA study: the LCA of electricity generation. Left of the electricity generation process corresponds to the extraction of upstream resources for electricity generation, while the righthand side corresponds to the downstream processes of electricity use.

LCA has been standardized by ISO [3] and follows four steps as shown below.

1. **Goal and scope definition**: In conducting an LCA study, the goal of LCA study, a functional unit, and a system boundary need to be defined first. With respect to the electricity example shown in Fig. 3.1, the goal is to compare life cycle greenhouse gas emissions (i.e., global warming potential [GWP]) of two product systems for electricity: coal-fired electricity and NG-fired electricity. Many LCA works are comparative studies between more than two product systems since the results obtained from the LCA represent relative indicators, not absolute ones. Therefore, one of the main purposes for conducting LCA studies is to recommend practices that are less bad between the options. A functional unit needs to be defined properly to be common between the options for the goal. In this example, the functional unit can be kilowatt-hour (kWh) of electricity. Also, the system boundary is defined based on the goal of the LCA study and data availability. The system boundary needs to include upstream and downstream processes of a product to avoid shifting of impacts across the life cycle because each process in the life cycle has its own environmental impacts. If all options in the study share the same downstream phases as shown in Fig. 3.1, the use and end-of-life phases can be excluded from the analysis. Such an LCA study is called a cradle-to-gate analysis.
2. **Life cycle inventory analysis**: In this step, every required life cycle inventory (LCI) data is collected from a variety of data sources. Some typical LCI data sources are introduced in Sect. 3.3. The type of data includes, but not limited to the amounts of product inputs, main product, by-products, coproducts, resource use, and emissions.
3. **Life cycle impact assessment**: Life cycle resource use and emissions are calculated based on the LCI data collected in the previous step. Life cycle impact indicators, such as global warming and eutrophication potentials, are calculated using life cycle impact assessment (LCIA) characterization factors. The resulting life cycle impacts can be normalized and aggregated using normalization factors and weighting factors, respectively, depending on the goal

that is defined in the first step. The details for mathematical calculations are addressed in Sect. 3.2.

4. **Interpretation of results**: In the last step, LCIA results are interpreted to make recommendations to reduce life cycle impacts. Hotspot inventories with respect to life cycle impact indicators can be identified as well. This step helps make decisions to change practices less bad to the environment.

3.2 Methods to Quantify Direct and Indirect Demand

LCA can calculate direct and indirect environmental flows (i.e., direct and indirect demands for ecosystem services) of products and processes. The direct demand means on-site resource consumption and emissions from an immediate process that produces the desired product. The indirect demand refers to the resource consumption and emissions from upstream and downstream processes. For example, the direct CO_2 emissions of coal-fired electricity are the on-site CO_2 emissions from the coal-fired electricity generation. On the other hand, the indirect CO_2 emissions for coal-fired electricity include CO_2 emissions from the upstream processes such as coal mining and transportation of coal. The sum of direct and indirect emissions corresponds to the life cycle emission. For specific flows, the life cycle emission is also called footprint: for greenhouse gases, it is the carbon footprint.

An LCA model consists of two equations: the product transaction equation and the environmental intervention equation. The product transaction equation contains data about the transaction of products between processes to produce the desired amount of a final product (i.e., final demand of a product). The environmental intervention equation calculates direct and indirect resource use and emissions for the final demand. In terms of the electricity example in Fig. 3.1, the transaction equation calculates how many coal and NG products are needed to produce 1 kWh of electricity, which is the final demand in this example. Also, the intervention equation calculates the amounts of direct and indirect emissions and resource use to produce 1 kWh of electricity.

Depending on the goal and scope of LCA study, a different way of formulating LCA models is required. For example, if it is expected that detailed LCI data are easily available from LCI databases, process-based LCA model could be appropriate since the model contains a lot of process details. If it is too demanding to collect numerous LCI data along the life cycle, environmentally extended input–output (EEIO) model could be suitable because the model accounts for the entire economy. The EEIO model covers the entire life cycle activities in return for the details of process data. Table 3.2 compares the pros and cons of various LCA models. In the following section, the mathematical formulation for those LCA models is introduced. Underbar (e.g., $\underline{A}$) and overbar (e.g., $\overline{A}$) notations in the mathematical formulation refer to value chain process scale for the process-based LCA model and economy scale for the EEIO model, respectively.

Table 3.2 Pros and cons of various LCA models

Pros	Cons
Process-based LCA model	
+ The model has a lot of process details. + Free LCI databases are available.	– Collecting LCI data is time-consuming work. – Commercial LCI databases are expensive. – It is technically impossible to cover the entire life cycle network.
Environmentally extended input–output (EEIO) model	
+ The model covers the entire life cycle network of a given region. + The U.S. model is available for free.	– The model is based on highly aggregated economy sectors (i.e., lack of details in data).
Multiregional input–output (MRIO) model	
+ Region-specific analysis can be performed.	– Regional data are expensive and challenging to collect.
Integrated hybrid LCA model	
+ The model not only covers the entire life cycle network of a given region but also contains a lot of process details.	– Upstream and downstream cutoff flows between value chain process and economy scales need to be identified. – Price for every product needs to be known to connect process data in physical units to economy data in monetary units.

3.2.1 Process-Based LCA Model

A process-based LCA model is based on process data in physical units (e.g., kg, m^3, and MJ). The product transaction equation of this model can be formulated using physical process data as follows.

$$\underline{A}\underline{s} = \underline{f},$$

$$\text{where } \underline{A} = \begin{bmatrix} \underline{a}_{11} & \underline{a}_{12} & \cdots & \underline{a}_{1n} \\ \underline{a}_{21} & \underline{a}_{22} & \cdots & \underline{a}_{2n} \\ \vdots & \vdots & \ddots & \vdots \\ \underline{a}_{m1} & \underline{a}_{m2} & \cdots & \underline{a}_{mn} \end{bmatrix} = \{\underline{a}_{ij}\} \in \mathbb{R}^{m\times n},\ \underline{s} = \begin{bmatrix} \underline{s}_1 \\ \underline{s}_2 \\ \vdots \\ \underline{s}_n \end{bmatrix}, \text{ and } \underline{f} = \begin{bmatrix} \underline{f}_1 \\ \underline{f}_2 \\ \vdots \\ \underline{f}_m \end{bmatrix}.$$

The matrix $\underline{A}$ is called a technology matrix that contains data about product input and output flows between processes. The rows ($i = 1, 2, \cdots, m$) and columns ($j =$

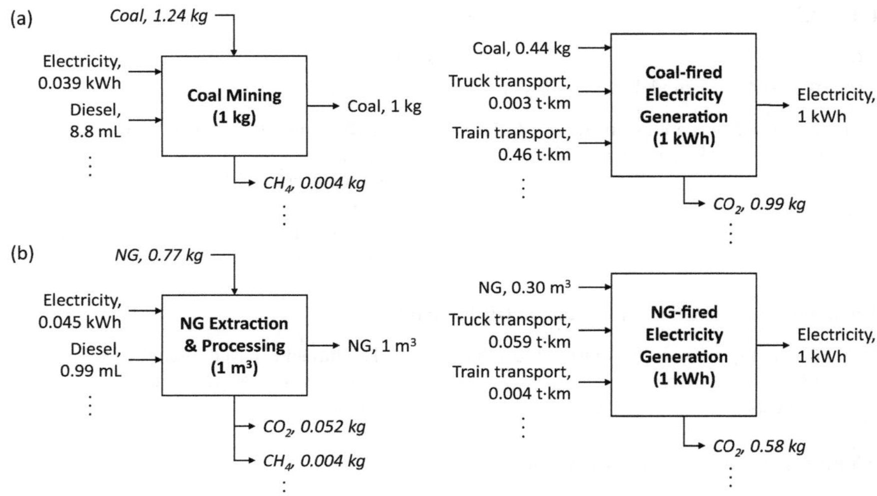

Fig. 3.2 Process-based LCA model for coal-fired and NG-fired electricity

$1, 2, \cdots, n$) of $\underline{A}$ correspond to products and processes, respectively. In most cases, the number of products is equal to the number of processes (i.e., $m = n$). Vectors $\underline{s}$ and $\underline{f}$ represent a scaling vector for each process (j) and a final demand vector for each product (i), respectively. Vector $\underline{s}$ is determined by $\underline{s} = \underline{A}^{-1}\underline{f}$.

With respect to the electricity example, Fig. 3.2 shows partial LCI data that are collected for the process-based LCA model. Matrices $\underline{A}$ for coal electricity (A^{coal}) and NG electricity (A^{NG}) can be formulated as follows:

$$A^{coal} = \begin{bmatrix} 1\,\frac{\text{kWh}}{\text{kWh}} & -0.039\,\frac{\text{kWh}}{\text{kg}} & \cdots \\ -0.44\,\frac{\text{kg}}{\text{kWh}} & 1\,\frac{\text{kg}}{\text{kg}} & \cdots \\ \vdots & \vdots & \ddots \end{bmatrix} \quad \text{and}$$

$$A^{NG} = \begin{bmatrix} 1\,\frac{\text{kWh}}{\text{kWh}} & -0.045\,\frac{\text{kWh}}{\text{m}^3} & \cdots \\ -0.30\,\frac{\text{m}^3}{\text{kWh}} & 1\,\frac{\text{m}^3}{\text{m}^3} & \cdots \\ \vdots & \vdots & \ddots \end{bmatrix}.$$

For instance, the first and second rows of $\underline{A}^{coal}$ represent an electricity product and a coal product, respectively. Also, the first and second columns of $\underline{A}^{coal}$ correspond to a coal-fired electricity generation process and a coal mining process, respectively. Then, a_{11} in $\underline{A}^{coal}$ refers to the amount of electricity generated from the coal-fired electricity generation process. a_{11} has a positive value indicating the generation of electricity. Also, a_{21} shows the amount of coal product that is used to produce the a_{11}

amount of electricity. a_{21} has a negative value indicating the consumption of coal for electricity generation. Vectors $\underline{f}^{coal}$ and $\underline{f}^{NG}$ are defined as $\underline{f}^{coal} = [1, 0, \cdots, 0]^T$ and $\underline{f}^{NG} = [1, 0, \cdots, 0]^T$, respectively.

Also, the intervention equation of process-based LCA model is formulated as follows:

$$\underline{B}\underline{s} = \underline{r},$$

$$\text{where } \underline{B} = \begin{bmatrix} \underline{b}_{11} & \underline{b}_{12} & \cdots & \underline{b}_{1n} \\ \underline{b}_{21} & \underline{b}_{22} & \cdots & \underline{b}_{2n} \\ \vdots & \vdots & \ddots & \vdots \\ \underline{b}_{o1} & \underline{b}_{o2} & \cdots & \underline{b}_{on} \end{bmatrix} = \{\underline{b}_{kj}\} \in \mathbb{R}^{o \times n}, \underline{s} = \begin{bmatrix} \underline{s}_1 \\ \underline{s}_2 \\ \vdots \\ \underline{s}_n \end{bmatrix}, \text{ and } \underline{r} = \begin{bmatrix} \underline{r}_1 \\ \underline{r}_2 \\ \vdots \\ \underline{r}_o \end{bmatrix}.$$

Matrix $\underline{B}$ is referred to by the intervention matrix that includes data about each of the resource use and emissions ($k = 1, 2, \cdots, o$) for each process (j). $\underline{r}$ represents life cycle interventions which are calculated by $\underline{r} = \underline{B}\underline{s} = \underline{B}\underline{A}^{-1}\underline{f}$. For the electricity example, matrices $\underline{B}^{coal}$ and $\underline{B}^{NG}$ can be formulated as follows:

$$\underline{B}^{coal} = \begin{bmatrix} 0.99\,\frac{\text{kgCO}_2}{\text{kWh}} & 0\,\frac{\text{kgCO}_2}{\text{kg}} & \cdots \\ 0\,\frac{\text{kgCH}_4}{\text{kWh}} & 0.004\,\frac{\text{kgCH}_4}{\text{kg}} & \cdots \\ 0\,\frac{\text{kg coal}}{\text{kWh}} & -1.24\,\frac{\text{kg coal}}{\text{kg}} & \cdots \\ 0\,\frac{\text{m}^3\text{ NG}}{\text{kWh}} & 0\,\frac{\text{m}^3\text{ NG}}{\text{kg}} & \cdots \\ \vdots & \vdots & \ddots \end{bmatrix} \quad \text{and}$$

$$\underline{B}^{NG} = \begin{bmatrix} 0.58\,\frac{\text{kgCO}_2}{\text{kWh}} & 0.052\,\frac{\text{kgCO}_2}{\text{m}^3} & \cdots \\ 0\,\frac{\text{kgCH}_4}{\text{kWh}} & 0.004\,\frac{\text{kgCH}_4}{\text{m}^3} & \cdots \\ 0\,\frac{\text{kg coal}}{\text{kWh}} & 0\,\frac{\text{kg coal}}{\text{kg}} & \cdots \\ 0\,\frac{\text{m}^3\text{ NG}}{\text{kWh}} & -0.77\,\frac{\text{m}^3\text{ NG}}{\text{kg}} & \cdots \\ \vdots & \vdots & \ddots \end{bmatrix}.$$

In these matrices, $k = 1$ and $k = 2$ correspond to CO_2 emissions and CH_4 emissions, respectively. Therefore, $\underline{b}_{11}$ and $\underline{b}_{21}$ in $\underline{B}^{coal}$ indicate CO_2 emissions and CH_4 emissions, respectively, of the coal-fired electricity generation process ($j = 1$). Also, $k = 3$ and $k = 4$ correspond to coal and NG resource use, respectively. Since these resource use flows are inputs to the processes, $\underline{b}_{32}$ in $\underline{B}^{coal}$ and $\underline{b}_{42}$ in $\underline{B}^{NG}$ have negative signs.

To calculate life cycle impact indicators (i.e., midpoint indicators), LCIA characterization factors are multiplied with life cycle interventions as follows:

$$Q\underline{r} = \underline{h},$$

$$\text{where } Q = \begin{bmatrix} q_{11} & q_{12} & \cdots & q_{1o} \\ q_{21} & q_{22} & \cdots & q_{2o} \\ \vdots & \vdots & \ddots & \vdots \\ q_{p1} & q_{p2} & \cdots & q_{po} \end{bmatrix} = \{q_{lk}\} \in \mathbb{R}^{p \times o}, \underline{r} = \begin{bmatrix} \underline{r}_1 \\ \underline{r}_2 \\ \vdots \\ \underline{r}_o \end{bmatrix}, \text{ and } \underline{h} = \begin{bmatrix} \underline{h}_1 \\ \underline{h}_2 \\ \vdots \\ \underline{h}_p \end{bmatrix}.$$

Matrix Q is the LCIA characterization factor matrix that contains characterization factors for each intervention flow (k) to calculate midpoint indicators ($l = 1, 2, \cdots, p$). The midpoint indicators are calculated by $\underline{h} = Q\underline{r} = Q\underline{B}\underline{A}^{-1}\underline{f}$. For the electricity example, if $l = 1$ represents global warming potential (GWP), q_{11} and q_{12} correspond to the characterization factors for CO_2 emissions ($k = 1$) and CH_4 emissions ($k = 2$), respectively, to calculate the GWP ($\underline{h}_1$). The GWP has a mass unit of CO_2 equivalent (e.g., $kgCO_2eq$). According to the LCIA characterization factors provided by the EPA [4], $q_{11} = 1$ $kgCO_2eq/kgCO_2$ and $q_{12} = 25$ $kgCO_2eq/kgCH_4$. Therefore, Q can be formulated by

$$Q = \begin{bmatrix} 1\,\frac{\text{kgCO}_2\text{eq}}{\text{kgCO}_2} & 25\,\frac{\text{kgCO}_2\text{eq}}{\text{kgCH}_4} & 0\,\frac{\text{kgCO}_2\text{eq}}{\text{kg coal}} & 0\,\frac{\text{kgCO}_2\text{eq}}{\text{m}^3\text{ NG}} & \cdots \\ \vdots & \vdots & \vdots & \vdots & \ddots \end{bmatrix}.$$

Thus, the GWPs for coal-fired electricity generation and NG-fired electricity generation can be calculated as follows:

$$\begin{aligned} \text{GWP}^{coal} &= \underline{h}_1^{coal} = Q\underline{B}^{coal}\underline{A}^{coal^{-1}}\underline{f}^{coal} = 1.08 \text{ kgCO}_2\text{eq} \\ \text{GWP}^{NG} &= \underline{h}_1^{NG} = Q\underline{B}^{NG}\underline{A}^{NG^{-1}}\underline{f}^{NG} = 0.68 \text{ kgCO}_2\text{eq}. \end{aligned}$$

One of the strengths for performing the process-based LCA model is that the model includes detailed process data. Therefore, sustainability assessment can be performed on a variety of products and processes if process data along the life cycle are easily available. Sources of LCI data for the process-based LCA model are introduced in Sect. 3.3. The process-based LCA model, however, does not account for the entire life cycle network since it is technically impossible to collect the tremendous amounts of process data along the entire life cycle network.

3.2.2 Environmentally Extended Input–Output (EEIO) Model

Environmentally extended input–output (EEIO) model has been developed to account for the entire life cycle network within the economy of a given region. The EEIO model is the environmentally extended version of the economic input–output (IO) model. The IO model is based on commodity transaction data between

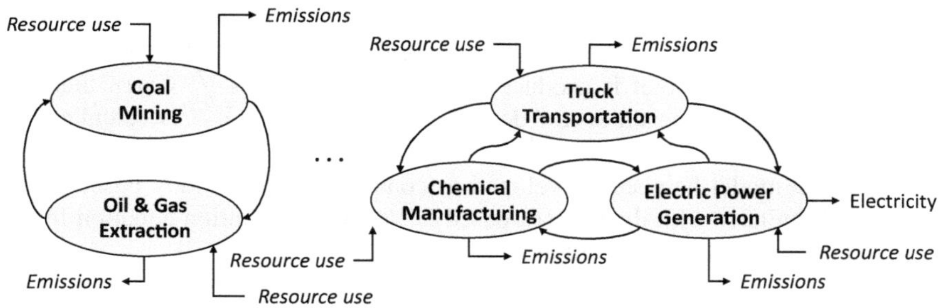

Fig. 3.3 EEIO model for electricity generation. Both coal-fired and NG-fired electricity generation technologies are assigned to an electric power generation sector in the EEIO model

economy sectors in monetary units. Unlike the processes in the process-based LCA model, economy sectors are highly aggregated. For example, coal-fired electricity generation and NG-fired electricity generation are two different technologies. In the IO model, however, all electricity generation technologies are assigned to a single economy sector, which is the electric power generation sector as shown in Fig. 3.3. This sector can also be further aggregated into the utility sector that includes a water supply system and a NG distribution system, as well as the electric power generation. In Fig. 3.3, ellipses, curved arrows, and angled arrows represent economy sectors, commodity flows, and intervention flows, respectively. The commodity transaction between sectors in the IO model is represented by the direct requirement matrix ($\overline{A}$) which consists of coefficients about the direct requirement of an input commodity to produce one-dollar worth of the output commodity. The matrix $\overline{A}$ can be obtained from make ($\overline{V}$) and use ($\overline{U}$) matrices. Matrices $\overline{V}$ and $\overline{U}$ contain data about commodity outputs from each sector and commodity inputs to each sector, respectively. For example, the aggregated electricity sector in the 2012 U.S. economy supplies 327,938 million dollars of electricity commodity while using 14,900 million dollars of coal mining commodity and 12,825 million dollars of oil and gas extraction commodity. Matrices $\overline{V}$ and $\overline{U}$ are combined into the direct requirement matrix ($\overline{A}$) using $\overline{A} = \overline{U}(\overline{V}^T)^{-1}$. Thus, $\overline{A}$ is dimensionless.

The transaction equation of the IO and EEIO model is formulated as follows:

$$(I - \overline{A})\overline{x} = \overline{f},$$

$$\text{where } \overline{A} = \begin{bmatrix} \overline{a}_{11} & \overline{a}_{12} & \cdots & \overline{a}_{1n'} \\ \overline{a}_{21} & \overline{a}_{22} & \cdots & \overline{a}_{2n'} \\ \vdots & \vdots & \ddots & \vdots \\ \overline{a}_{m'1} & \overline{a}_{m'2} & \cdots & \overline{a}_{m'n'} \end{bmatrix} = \{\overline{a}_{i'j'}\} \in \mathbb{R}^{m' \times n'}, \overline{x} = \begin{bmatrix} \overline{x}_1 \\ \overline{x}_2 \\ \vdots \\ \overline{x}_{n'} \end{bmatrix}, \text{ and } \overline{f} = \begin{bmatrix} \overline{f}_1 \\ \overline{f}_2 \\ \vdots \\ \overline{f}_{m'} \end{bmatrix}.$$

i''s ($= 1, 2, \cdots, m'$) and j''s ($= 1, 2, \cdots, n'$) refer to commodities and sectors. Vector $\overline{x}$ represents the total commodity output (i.e., economic throughput) from each sector, j'. This vector is equal to the sum of $\overline{A}\overline{x}$ (the monetary value of every commodity consumed by sectors to produce the final demand) and $\overline{f}$ (the monetary value of final demand that is produced). That is, $\overline{x} = \overline{A}\overline{x} + \overline{f}$. Given that $\overline{f}$ is known (products from each sector demanded by consumers), $\overline{x}$ is calculated by $\overline{x} = (I - \overline{A})^{-1}\overline{f}$.

The EEIO model has been developed to conduct the LCA study based on the IO model. Similarly with the process-based LCA, the intervention equation for the EEIO model is formulated as follows:

$$\overline{B}\overline{x} = \overline{r},$$

$$\text{where } \overline{B} = \begin{bmatrix} \overline{b}_{11} & \overline{b}_{12} & \cdots & \overline{b}_{1n'} \\ \overline{b}_{21} & \overline{b}_{22} & \cdots & \overline{b}_{2n'} \\ \vdots & \vdots & \ddots & \vdots \\ \overline{b}_{o1} & \overline{b}_{o2} & \cdots & \overline{b}_{on'} \end{bmatrix} = \{\overline{b}_{kj'}\} \in \mathbb{R}^{o \times n'}, \overline{x} = \begin{bmatrix} \overline{x}_1 \\ \overline{x}_2 \\ \vdots \\ \overline{x}_{n'} \end{bmatrix}, \text{ and } \overline{r} = \begin{bmatrix} \overline{r}_1 \\ \overline{r}_2 \\ \vdots \\ \overline{r}_o \end{bmatrix}.$$

Matrix $\overline{B}$ is the economy scale intervention matrix that represents emissions and resource use ($k = 1, 2, \cdots, o$) to produce one dollar amount of commodities from each sector (j'). For instance, 6.09 kg of CO_2 is emitted to produce \$1.0 amount of commodities from the electricity sector. If $\overline{B}$ is unknown, it can be obtained by $\overline{B} = \overline{M}\hat{\overline{x}}$. Matrix $\overline{M}$ represents total interventions from each sector (e.g., total CO_2 emissions from the electricity sector). $\overline{r}$ is calculated by $\overline{r} = \overline{B}(I - \overline{A})^{-1}\overline{f}$ and represents life cycle interventions for producing the economy scale final demand ($\overline{f}$).

Using Q, the life cycle impact indicators ($\overline{h}$) are calculated by $\overline{h} = Q\overline{r} = Q\overline{B}(I - \overline{A})^{-1}\overline{f}$. For example, the GWP for producing \$1.0 of electricity in the United States is calculated to be 6.48 $kgCO_2eq$.

Although the EEIO model includes the entire economy of a given region as a system boundary, the model is based on the aggregated economy sectors. For example, the United States EEIO (USEEIO) model by the U.S. EPA has been developed for 388 economy sectors [5], while the United States LCI (USLCI) by National Renewable Energy Laboratory (NREL) for the process-based LCA model contains inventory data for more than 27,000 processes [6]. Therefore, the EEIO model lacks details in data.

3.2.3 Multiregional Input–Output (MRIO) Model

If analysis of multiple regions needs to be performed, a multiregional model that accounts for the regional heterogeneity must be developed. As shown in Fig. 3.4, for instance, region 1 requires more inputs from the coal mining sector for the electric power generation sector than region 2. Also, the electric power generation sector in region 1 needs an interregional coal input flow from the coal mining

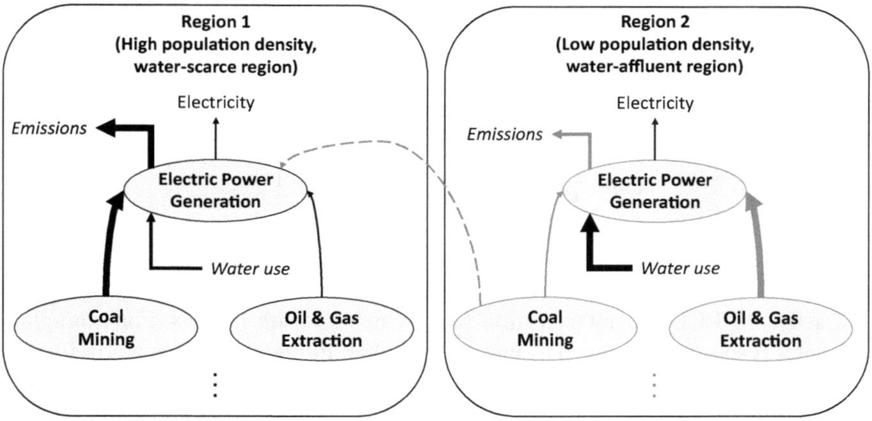

Fig. 3.4 MRIO model for electricity generated in two regions

sector in region 2. Moreover, the electric power generation sector in region 1 has larger emissions, as indicted by the thickness of the arrows, but requires a smaller water resource to generate electricity than region 2. In this case, the electric power generation in region 1 has different commodity inputs and interventions from region 2. The impacts from emissions and water use in region 1 could be different from the impacts from the same amounts of emissions and water use in region 2 because population density and resource availability are not the same between the regions [7]. In this context, the multiregional input–output (MRIO) model has been developed to address regional heterogeneity.

The transaction equation for the MRIO model can be formulated as follows:

$$(I - \overline{A}_{MR})\overline{x}_{MR} = \overline{f}_{MR},$$

$$\text{where } \overline{A}_{MR} = \begin{bmatrix} \overline{A}^{11} & \overline{A}^{12} & \cdots & \overline{A}^{1r} \\ \overline{A}^{21} & \overline{A}^{22} & \cdots & \overline{A}^{2r} \\ \vdots & \vdots & \ddots & \vdots \\ \overline{A}^{r1} & \overline{A}^{r2} & \cdots & \overline{A}^{rr} \end{bmatrix}, \ \overline{x}_{MR} = \begin{bmatrix} \overline{x}^{1} \\ \overline{x}^{2} \\ \vdots \\ \overline{x}^{r} \end{bmatrix}, \text{ and } \overline{f}_{MR} = \begin{bmatrix} \overline{f}^{1} \\ \overline{f}^{2} \\ \vdots \\ \overline{f}^{r} \end{bmatrix}$$

The subscript MR refers to the multiregional matrix. Superscripts, such as $1, 2, \cdots, r$, represent each region. For example, diagonal elements of matrix $\overline{A}_{MR}$ (i.e., $\overline{A}^{11}, \overline{A}^{22}, \cdots, \overline{A}^{rr}$) represent direct requirement matrices for regions $1, 2, \cdots, r$, respectively. Non-diagonal elements of matrix $\overline{A}_{MR}$ correspond to interregional commodity flow matrices. For instance, $\overline{A}^{12}$ is the matrix for commodity flows from region 1 to region 2. Similarly, $\overline{x}_{MR}$ and $\overline{f}_{MR}$ are, respectively, throughput and final demand vectors for each region.

Also, the multiregional intervention matrix can be formulated as follows:

$$\overline{B}_{MR}\overline{x}_{MR} = \overline{r}_{MR},$$

$$\text{where } \overline{B}_{MR} = \begin{bmatrix} \overline{B}^{11} & \overline{B}^{12} & \cdots & \overline{B}^{1r} \\ \overline{B}^{21} & \overline{B}^{22} & \cdots & \overline{B}^{2r} \\ \vdots & \vdots & \ddots & \vdots \\ \overline{B}^{r1} & \overline{B}^{r2} & \cdots & \overline{B}^{rr} \end{bmatrix}, \ \overline{x}_{MR} = \begin{bmatrix} \overline{x}^{1} \\ \overline{x}^{2} \\ \vdots \\ \overline{x}^{r} \end{bmatrix}, \text{ and } \overline{r}_{MR} = \begin{bmatrix} \overline{r}^{1} \\ \overline{r}^{2} \\ \vdots \\ \overline{r}^{r} \end{bmatrix}.$$

$\overline{B}_{MR}$ is a multiregional intervention matrix. Diagonal elements of matrix $\overline{B}_{MR}$ correspond to intervention matrices for each region (i.e., region $1, 2, \cdots, r$).

Lastly, the LCIA characterization factor can vary with regions. For example, if the water resource in region 1 is more scarce than in region 2 as shown in Fig. 3.4, the impacts from the same amount of water resource consumption are worse in region 1 than region 2. In such case, the characterization factor for water resource use in region 1 needs to be larger than region 2 [8]. Accordingly, the resulting life cycle impact indicators ($\overline{h}_{MR}$) vary with regions as shown below:

$$Q_{MR}\overline{r}_{MR} = \overline{h}_{MR},$$

$$\text{where } Q_{MR} = \begin{bmatrix} Q^{1} & 0 & \cdots & 0 \\ 0 & Q^{2} & \cdots & 0 \\ \vdots & \vdots & \ddots & \vdots \\ 0 & 0 & \cdots & Q^{r} \end{bmatrix}, \ \overline{r}_{MR} = \begin{bmatrix} \overline{r}^{1} \\ \overline{r}^{2} \\ \vdots \\ \overline{r}^{r} \end{bmatrix}, \text{ and } \overline{h}_{MR} = \begin{bmatrix} \overline{h}^{1} \\ \overline{h}^{2} \\ \vdots \\ \overline{h}^{r} \end{bmatrix}.$$

3.2.4 Integrated Hybrid LCA Model

As shown in Table 3.2, the characteristics of the process-based LCA model and the EEIO model are complementary with respect to the life cycle analysis boundary and details in their LCI data. The integrated hybrid LCA model has been developed to account for the entire economy while employing detailed process data [9]. Activities that are excluded from the process-based LCA model are included in the hybrid LCA model by connecting corresponding economic activities from the EEIO model to the process-based model. Figure 3.5 shows one example of the hybrid LCA model for NG-fired electricity. In this example, a NG transportation process is excluded from the process-based LCA model. The corresponding economy sector in the EEIO model to the NG transportation process is pipeline transport sector. In the hybrid model, the NG transportation process is substituted by the pipeline transport sector from the EEIO model. In this sense, the hybrid LCA model accounts for the entire life cycle network while keeping the details of process data. However, since the EEIO model covers every economic activity, activities in the process-based model often overlap with the corresponding activities in the EEIO model. As shown in Fig. 3.5, processes for NG-fired electricity generation and NG extraction

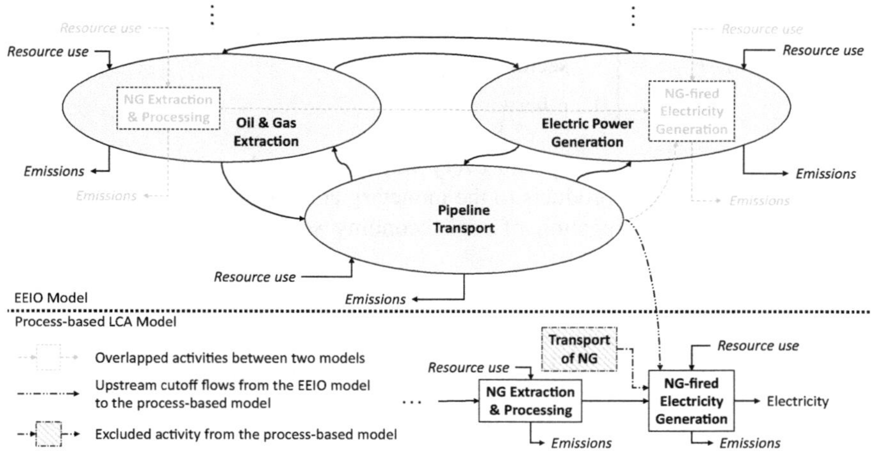

Fig. 3.5 Integrated hybrid LCA model for NG-fired electricity

and processing in the process-based model are included in sectors for electric power generation and oil & gas extraction in the EEIO model, respectively. To avoid double-counting of those activities, therefore, the EEIO model needs to be disaggregated from the process-based model. In other words, product transaction flows between processes and value chain scale intervention flows from processes need to be removed from the corresponding economy commodity transaction flows and economy scale intervention flows.

Disaggregation of the direct requirement matrix ($\overline{A}$) needs to be performed for make ($\overline{V}$) and use ($\overline{U}$) matrices. For the disaggregation of matrices $\overline{V}$ and $\overline{U}$, value chain scale technology matrix ($\underline{A}$) first needs to be separated into value chain scale make ($\underline{V}$) and use ($\underline{U}$) matrices by $\underline{A} = \underline{V}^T - \underline{U}$. In general, positive and negative elements in $\underline{A}$ are assigned to $\underline{V}^T$ and $\underline{U}$, respectively. Also, the disaggregation of the economy scale intervention matrix ($\overline{B}$) needs to be performed for economy scale total intervention matrix ($\overline{M}$). Disaggregation procedures for $\overline{V}$, $\overline{U}$, and $\overline{M}$ are as follows [10]:

1. Construct product-commodity (P_F) and process-sector (P_P) permutation matrices by matching value chain scale products and processes with economy scale commodities and sectors, respectively, as follows:

$$P_F = \{p_{F_{i',i}}\} = \begin{cases} 1 & \text{if value chain product } i \text{ corresponds to the economy} \\ & \text{commodity } i' \\ 0 & \text{otherwise} \end{cases}$$

$$P_P = \{p_{P_{j,j'}}\} = \begin{cases} 1 & \text{if value chain process } j \text{ corresponds to the economy sector } j' \\ 0 & \text{otherwise} \end{cases}$$

2. Construct a price vector (p) for every value chain scale product to convert the physical amounts of products to the monetary amounts of commodities.
3. Perform the disaggregation of each economy scale matrix by the following equations:

$$\overline{V}^* = \overline{V} - (P_P)^T \underline{V} \hat{p} (P_F)^T$$
$$\overline{U}^* = \overline{U} - P_F \hat{p} \underline{U} P_P - \underline{X}_u P_P - P_F \hat{p} \underline{X}_d$$
$$\overline{M}^* = \overline{M} - \underline{B} P_P$$

The superscript asterisk sign indicates disaggregated economy scale matrices. Matrices $\underline{X}_u$ and $\underline{X}_d$ represent the matrices for upstream cutoff flows of economy commodities to the value chain processes and downstream cutoff flows of value chain products to the economy sectors, respectively. Matrices $\underline{X}_u$ and $\underline{X}_d$ have $(i' \times j)$ and $(i \times j')$ dimensions, respectively. $\underline{X}_u$ has monetary units while $\underline{X}_d$ has physical units.

The disaggregated direct requirement matrix ($\overline{A}^*$) and economy scale intervention matrix ($\overline{B}^*$) are obtained by $\overline{A}^* = \overline{U}^* \{\overline{V}^{*T}\}^{-1}$ and $\overline{B}^* = \overline{M}^* \hat{\overline{x}}^*$, respectively.

Accordingly, the transaction equation for the integrated hybrid LCA model is formulated as follows:

$$\begin{bmatrix} I - \overline{A}^* & -\underline{X}_u \\ -\underline{A}_d & \underline{X} \end{bmatrix} \begin{bmatrix} \overline{s} \\ \underline{s} \end{bmatrix} = \begin{bmatrix} \overline{y} \\ \underline{y} \end{bmatrix}.$$

Vector $\overline{x}$ in the EEIO model can be represented by an economy scale scaling vector ($\overline{s}$). Matrix $\underline{X}_u$ corresponds to the matrix for upstream cutoff flows from the EEIO model to the process-based model. This upstream cutoff matrix represents economy commodity input flows to the processes. To construct $\underline{X}_u$, the physical amounts of excluded product flows from the process-based model need to be known. The monetary value of products also needs to be known since the EEIO model is based on monetary units. For example, if 0.35 t·km of NG transportation is needed to generate 1 kWh of electricity from the NG-fired electricity generation process as shown in Fig. 3.5, the monetary amount of cutoff flow for the pipeline transportation is obtained by multiplying 0.35 t·km with the price of NG transportation per t·km. Accordingly, the economy scale commodity transaction equation in the hybrid model is shown by $(I - \overline{A}^*)\overline{s} = \overline{y} + \underline{X}_u \underline{s}$, where $\underline{X}_u \underline{s}$ represents the demand for upstream cutoff commodities that are needed for value chain processes.

Also, matrix $\underline{A}_d$ corresponds to the matrix for downstream cutoff flows from the process-based model to the EEIO model. The downstream flows of products from the process-based model are included in this downstream cutoff matrix. If downstream activities of the main product do not need to be included in the system boundary of LCA study (i.e., if the study is a cradle-to-gate analysis), by-product and coproduct flows in the process-based model can be included in $\underline{A}_d$. To construct the matrix $\underline{A}_d$, economy sectors where downstream products are consumed need to be identified. Then, the physical amounts of downstream cutoff flows need to be normalized by economic throughput from those sectors. That is, $\underline{A}_d$ is obtained by $\underline{A}_d = \underline{X}_d\hat{\overline{x}}$. Accordingly, the product transaction equation in the hybrid model is represented by $\underline{X}\underline{s} = \underline{y} + \underline{A}_d\overline{s}$. $\underline{A}_d\overline{s}$ corresponds to the demand for downstream cutoff products that are consumed by economy sectors.

In the hybrid LCA model, life cycle interventions ($\underline{\overline{r}}$) and midpoint indicators ($\underline{\overline{h}}$) are calculated by the following equations:

$$\begin{bmatrix} \overline{B}^* & \underline{B} \end{bmatrix} \begin{bmatrix} \overline{s} \\ \underline{s} \end{bmatrix} = \underline{\overline{r}} \quad \text{and} \quad Q\underline{\overline{r}} = \underline{\overline{h}}.$$

Double notations on $\underline{\overline{r}}$ and $\underline{\overline{h}}$ indicate multiple scales that are across economy and value chain process scales.

3.3 Sources of Data and Software

In this section, we introduce various public and commercial sources of LCI data and several software programs to perform the LCA study. LCI analysis, which is the second step in conducting LCA, can be very time-consuming work. Table 3.3 shows various LCI data sources. GREET (The Greenhouse Gases, Regulated Emissions, and Energy Use in Transportation Model) is a public process-based LCA model for U.S. transportation-related activities that include various types of power generation technologies [11]. However, its data are limited to transportation and energy-related activities. NREL USLCI is a public U.S. LCI database for the process-based LCA model [6]. The USLCI database covers diverse activities and includes data on upstream cutoff flows for the extension of the model to the hybrid model. Ecoinvent is a commercial LCI database for the process-based model [12]. The LCI data for various regions (mostly Europe) are available in the Ecoinvent LCI database.

With respect to IO-based LCA models such as EEIO and MRIO models, the U.S. EPA has developed both EEIO and MRIO models for the United States. The USEEIO model is a public U.S. EEIO model [5]. This model accounts for the entire U.S. economy in 2013 and includes various environmental intervention data for every economy sector. U.S. state-level MRIO model is a public MRIO model for the 2012 U.S. economy [13]. This model includes MRIO data for 51 states in the United States. If a more detailed regional IO model is needed, RIMS II and IMPLAN models are commercial U.S. regional IO models [15, 16]. Also, regional interventions data are available from various sources. CAIT Climate Data

Table 3.3 Various sources of life cycle inventory data

Source of data	Type of data	Ref.
LCI database for the process-based LCA model		
GREET	U.S. transportation-related process-based LCA model	[11]
NREL	U.S. LCI data for the process-based model	[6]
Ecoinvent	Commercial global LCI data for the process-based model	[12]
LCI database for the EEIO and MRIO models		
EPA	2013 U.S. EEIO model	[5]
EPA	2012 U.S. state-level MRIO model	[13]
BEA	U.S. make and use tables for the IO model	[14]
BEA RIMS II	Commercial U.S. regional IO models	[15]
IMPLAN	Commercial U.S. regional IO models	[16]
EPA	U.S. GHG emissions and sinks data for aggregated sectors	[17]
CAIT	U.S. state-level GHG emissions data for aggregated sectors	[18]
EPA NEI	U.S. county-level air pollutant emissions data for aggregated sectors	[19]
EPA EnviroAtlas	U.S. watershed-level water use and nutrient emissions data for aggregated sectors	[20]

Table 3.4 Various LCA software programs

Program	Features	Ref.
OpenLCA	Open-source LCA software program	[21]
SimaPro	Commercial LCA software program	[22]
GaBi	Commercial LCA software program	[23]
EIO-LCA	Web-based program for the U.S. EEIO model	[24]
Eco-LCA	Web-based program for the U.S. EEIO model with emphasis to ecological impacts	[25]

Explorer has U.S. state-level GHG emissions data for aggregated economy sectors [18]. National Emissions Inventory (NEI) from the EPA has U.S. county-level air pollutant emissions data [19]. EnviroAtlas from the EPA has U.S. watershed scale water use and nutrient emissions data [20].

Also, various LCA software packages that use the LCI data collected from Table 3.3 are available. Table 3.4 shows several LCA software programs. OpenLCA is an open-source LCA software program. The LCI data obtained from the USLCI, Ecoinvent, and USEEIO can be directly imported to OpenLCA. SimaPro and GaBi are commercial LCA software programs. Both EIO-LCA (Economic Input–Output Life Cycle Assessment) and Eco-LCA (Ecologically based Life Cycle Assessment) are web-based software programs for the U.S. EEIO model while the latter emphasizes contributions from ecosystems.

3.4 Conclusions

In assessing sustainability, the demand for ecosystem services can be quantified using various established LCA models. Depending on the goal and scope of sustainability assessment study, the choice of models can vary. Besides the models introduced in this chapter, there have been many advanced sustainability assessment methods developed as well. Most models are based on the LCA approach and the mathematical formulation described in this chapter. Using those models, the demand for most provisioning and regulating services can be quantified. However, it is still challenging to quantify the demand for some ecosystem services, such as supporting and cultural services, since such data are not readily available. Moreover, most LCA models do not consider the supply of ecosystem services, which also needs to be quantified in assessing sustainability. Therefore, systematic approaches and database construction to quantify the demand for ecosystem services are needed. These are described in the next few chapters.

References

1. Roy Haines-Young and Marion B Potschin. Common international classification of ecosystem services (CICES) v5. 1 and guidance on the application of the revised structure. *European Environment Agency (EEA)*, 2018.
2. Bhavik R Bakshi, Guy Ziv, and Michael D Lepech. Techno-ecological synergy: A framework for sustainable engineering. *Environmental science & technology*, 49(3):1752–1760, 2015.
3. International Organization for Standardization. *Environmental Management: Life Cycle Assessment; Principles and Framework*. Number 2006. ISO, 2006.
4. Jane Bare. Traci 2.0: the tool for the reduction and assessment of chemical and other environmental impacts 2.0. *Clean Technologies and Environmental Policy*, 13(5):687–696, 2011.
5. Yi Yang, Wesley W Ingwersen, Troy R Hawkins, Michael Srocka, and David E Meyer. USEEIO: A new and transparent United States environmentally-extended input-output model. *Journal of cleaner production*, 158:308–318, 2017.
6. National Renewable Energy Laboratory (NREL). Us life cycle inventory database. 2012.
7. Yi Yang and Reinout Heijungs. A generalized computational structure for regional life-cycle assessment. *The International Journal of Life Cycle Assessment*, 22(2):213–221, 2017.
8. Anne-Marie Boulay, Jane Bare, Lorenzo Benini, Markus Berger, Michael J Lathuillière, Alessandro Manzardo, Manuele Margni, Masaharu Motoshita, Montserrat Núñez, Amandine Valerie Pastor, et al. The WULCA consensus characterization model for water scarcity footprints: assessing impacts of water consumption based on available water remaining (aware). *The International Journal of Life Cycle Assessment*, 23(2):368–378, 2018.
9. Sangwon Suh. Functions, commodities and environmental impacts in an ecological–economic model. *Ecological Economics*, 48(4):451–467, 2004.
10. Rebecca J Hanes and Bhavik R Bakshi. Process to planet: A multiscale modeling framework toward sustainable engineering. *AIChE Journal*, 61(10):3332–3352, 2015.
11. Michael Wang. The greenhouse gases, regulated emissions, and energy use in transportation (greet) model: Version 1.5. *Center for Transportation Research, Argonne National Laboratory*, 2008.
12. R Frischknecht, N Jungbluth, HJ Althaus, R Hischier, G Doka, R Dones, T Heck, S Hellweg, G Wernet, T Nemecek, et al. Overview and methodology. data v2. 0 (2007). Ecoinvent report no. 2007.
13. Yi Yang, Wesley W Ingwersen, and David E Meyer. Exploring the relevance of spatial scale to life cycle inventory results using environmentally-extended input-output models of the united states. *Environmental modelling & software*, 99:52–57, 2018.

14. U.S. Bureau of Economic Analysis (BEA). Interactive access to industry economic accounts data: Input-output. Available at: https://apps.bea.gov/iTable/itable.cfm?reqid=52&step=1 Accessed September 2019.
15. U.S. Bureau of Economic Analysis (BEA). Regional input-output modeling system (rims ii). Available at: https://apps.bea.gov/regional/rims/ Accessed September 2019.
16. Scott A Lindall, Douglas C Olson, and Gregory S Alward. Deriving multi-regional models using the IMPLAN national trade flows model. *Journal of Regional Analysis and Policy*, 36(1100-2016-89756), 2006.
17. U.S. Environmental Protection Agency (EPA). Inventory of U.S. greenhouse gas emissions and sinks: Åvailable at: https://www.epa.gov/ghgemissions/inventory-us-greenhouse-gas-emissions-and-sinks Accessed September 2019.
18. World Resources Institute (WRI). Cait climate data explorer, 2015.
19. U.S. Environmental Protection Agency (EPA). Air emissions inventories. Available at: https://www.epa.gov/air-emissions-inventories Accessed September 2019.
20. Brian R Pickard, Jessica Daniel, Megan Mehaffey, Laura E Jackson, and Anne Neale. EnviroAtlas: A new geospatial tool to foster ecosystem services science and resource management. *Ecosystem Services*, 14:45–55, 2015.
21. Andreas Ciroth. ICT for environment in life cycle applications openLCA—a new open source software for life cycle assessment. *The International Journal of Life Cycle Assessment*, 12(4):209, 2007.
22. Mark Goedkoop, Anne De Schryver, Michiel Oele, Sipke Durksz, and Douwe de Roest. Introduction to LCA with SimaPro 7. *PRé Consultants, The Netherlands*, 2008.
23. Sabrina Spatari, Michael Betz, Harald Florin, Martin Baitz, and Michael Faltenbacher. Using gabi 3 to perform life cycle assessment and life cycle engineering. *The International Journal of Life Cycle Assessment*, 6(2):81, 2001.
24. Christopher Weber, Deanna Matthews, Aranya Venkatesh, Christine Costello, and H Matthews. The 2002 us benchmark version of the economic input-output life cycle assessment (EIO-LCA) model. *Green Design Institute, Carnegie Mellon University*, 2009.
25. Shweta Singh and Bhavik R Bakshi. Eco-lca: A tool for quantifying the role of ecological resources in lca. In *2009 IEEE International Symposium on Sustainable Systems and Technology*, pages 1–6. IEEE, 2009.

Chapter 4
Water Provisioning Services

Shelly Bogra

4.1 Introduction

According to the Millennium Ecosystems Assessment report and as introduced in Chap. 2, "Ecosystems Services" are the benefits that humans derive from nature both directly and indirectly [1]. They include goods such as food, fiber, timber, water for human consumption and services such as temperature regulation, nutrient availability, soil-fertility to support human activities and well-being. Ecosystems services are broadly categorized as provisioning, regulating, cultural, and supporting. Hydrological ecosystems services focus on services supplied in the context of water in both quantitative and qualitative terms. The water provisioning ecosystem service in the form of appropriate quantity of seasonal rainfall and adequate quantity of environmental flows are necessary to maintain the balance between the quantity withdrawn and used by humans versus the amount needed to sustain ecosystems within a local or regional boundary such as a watershed. That is, the balance between input-flows by nature and withdrawals by nature and humans is the key to maintain the qualitative health of ecological reservoirs such as rivers and other surface water bodies, whose vitality is key to the survival of other ecosystems services and for maintaining appropriate amount of oxygen levels to sustain aquatic animals which are further consumed by humans.

The current rate of consumptive and nonconsumptive uses of water to satisfy various human demands is exceeding the capacity of water-reservoirs at regional scale. For example, increasing groundwater depletion in north-western parts of India

S. Bogra (✉)
Environmental Impacts and Sustainability Department, Norwegian Institute of Air Research, Oslo, Norway

Independent Researcher, Environmental Sustainability Assessments, Faridabad, Haryana, India
e-mail: sbog@nilu.no; shbogr@gmail.com

B. R. Bakshi (ed.), *Engineering and Ecosystems*,
https://doi.org/10.1007/978-3-031-35692-6_4

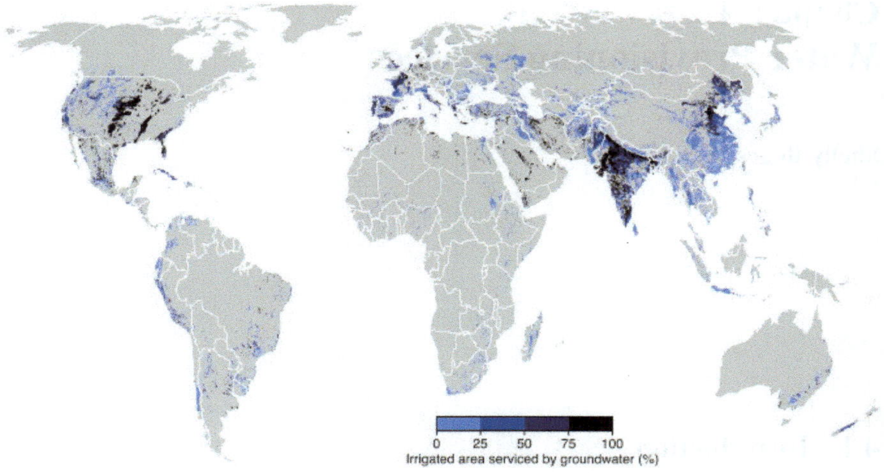

Fig. 4.1 Role of groundwater in irrigation around the globe. (Reproduced with permission from [10])

[2] or falling water level in Ogallala aquifer in the USA [3] is due to extraction beyond the recharging capacity of local precipitation patterns [4], whereas the drying of Aral Sea [5, 6] and lake Urmia in Middle East [7, 8] represents extreme cases of overextractive human demands on surface water bodies. Recent studies indicate that at the global scale, groundwater contributes about 50% or more to global crop production [4], with a recent report [9] indicating that between 6% and 20% of the 39 million groundwater wells across the globe have only few meters of water available in their stock. Such a state of dependence and lack of adequate water in future implies dire threats to food and water security of millions [10] (see Fig. 4.1).

While consumptive water-based studies [9, 4, 11] adequately support addressing of food and water security discussions in the context of local water-use policies, quantitative global assessments related to pollution of water bodies such as lakes or rivers from activities such as mining [12, 13] or industrial processes such as textile-dying, leather-tanning, and chemical-processing are largely lacking in literature.

The importance of water pollution lies in the fact that discharged toxic chemicals require significant volumes of water for dilution. This implies that due to large volumes of water released from many industrial and nonindustrial processes, impact on local water-ecosystems is usually not insignificant [13] and many a times can lead to a complete destruction of water bodies and living biotic that these hydrological ecosystems support. Further, the downstream consequences can become dire for local people whose food, water, and health security are inherently dependent on such local ecosystems and their services.

Owing to exorbitant profits offered by such activities, such polluting industries are usually owned by either large private business entities or government agencies themselves [14, 15]. Such control and bias through subsidies and allocation of public money for private enterprises [16], thereby, also hinders attempts to map such pollution-based impacts. Furthermore, such extractive and environmentally detrimental industrial-activities also satisfy monetary growth objective of policy planners, especially in lesser environmental-friendly regions [17, 18]. Thus, destruction of freshwater bodies and other natural resources such as loss of top-soil or biodiversity is largely ignored for economic-growth considerations [16].

Furthermore, in a largely opaque system of international monetary flows, the huge monetary benefits accrued to few members of global society through such extractive operations [14, 18] hinders policies aimed at maintaining the quality of water and connected abiotic and biotic systems. In other words, pursuance of such activities by stakeholders whose livelihoods do not depend on such local ecosystems plays a crucial role in continued destruction of many such ecological reservoirs.

However, with absolute water scarcity becoming a key reality in many regions, with possible future increase in operational costs of industries themselves due to lack of adequate local freshwater supplies, such impacts are being acknowledged in many regions. For example, rising marine pollution requiring "sea-snot" cleanup in coastal region of Turkey [19, 20] highlights the significance of such indirect water demand for the purpose of dilution of chemical pollution. Thus, future research dealing with sustainable use of water-ecosystems is expected to gain traction on such water pollution accounts.

A key revelation missing in the above discussion is the indirect dependence of global economic system on water-enabled anthropogenic goods and services. The current global economic system transfers huge volumes of food, fiber, materials, and information from one end of the globe to another. This transfer requires secondary economic activities such as processing, packaging, and storage of primary products and informational flows. The infrastructure behind all such secondary activities further requires significant volumes of natural resources, including water. For example, the electricity used for industrial food-processing or for irrigation requirements is usually dependent on large volumes of water-flows from well-fed riverine-systems. As this indirect water-dependence is hidden, its importance is often neglected in key policy areas such as those focusing on urbanization, food, energy growth, or downright economic growth. Further, without accounting for quality of water, especially temperature gradients between input and output water-flows [21], studies on electricity generation report consumptive water-use and not extractive water-use [22]. Such reporting can lead to erroneous idea of water stress, especially for precipitation-less months and storage requirements, when due impetus is being given to hydro-infrastructure as a carbon-less energy source. Further, such indirect resource extraction or destruction (loss of biota due to high steam released from supercritical power-plants) can also be detrimental to local ecosystems if exceeded beyond a certain limit.

Another aspect that would become increasingly important in future is concerned with environmental trade-offs due to technologies implemented for climate-mitigation purposes. The large-scale deployment of key technologies addressing climate action plans such as CCS can further exacerbate the pressures of local water-ecosystems in many regions [23, 24]. Especially, the densely populated and growing economies such as the USA, China, India, South Africa, and Brazil may face disproportionate increase in water-withdrawals since electricity generation from fossil fuel-based infrastructure remains consistently high (see SI, Table 4 [23]).

That said, it is often difficult to quantify the requirements or impacts of such indirect dependence for single products or services, especially when supply chains are disbursed all over the globe. As indicated earlier, such impacts can many a times exceed the provisioning capacity of local hydrological reservoirs. However, assessing such indirect impacts is extremely crucial for maintaining the sanctity of such ecosystems, and further for satisfying both environmental and human paradigms of sustainability. Next, we introduce one of the prominent tools conceptualized to quantify the direct and indirect impacts of production systems or consumptive activities, which can ideally be applied at different scales of interest.

Water footprint (WF) is one of the indicators in the family of footprints, with carbon, ecological, and nitrogen being some of the others, that is capable of measuring the impact of human activities on water-systems. The indicator is capable of measuring both the consumptive and nonconsumptive impacts that originate due to various human activities. Thus, in a way, WF approach is one of the popular methods that measures the demand created by anthropogenic activities in terms of water requirements along with measuring the impact of such withdrawal and use on local water-ecosystems. The following sections provide greater details of this indicator, its uses in measuring impact of human activities, the challenges involved in assessing this indicator at various scales (spatial and temporal), and areas where future research be directed to make this indicator a truly guiding measure that can be used to support various human activities while sustaining the sanctity of water-ecosystems at various scales.

4.2 Water Footprint: The Concept

Water footprint (WF) is an impact assessment indicator that quantifies contribution of water as an ecosystem service for various human activities either directly or indirectly [1]. This indicator has its origins in assessment of water-embodied in food-trade and direct water needs of human consumption. In the context of food, the concept was first pioneered by Allan [25, 26] who defined it as water-embodied in food that is traded across regions. The same is also referred to as virtual-water trade in various assessments. That is, water that is embodied or hidden in a certain product is called virtual water. On the other hand, direct water requirements for human consumption were given due importance in the works of Gleick [27], among others. Both these assessments represent an anthropocentric perspective, since they measure water provisioning for human needs. However, both approaches in a way laid a descriptive foundation of future research and methods which are fundamental to the concept of WF that has evolved today [28, life cycle assessment (LCA)-based WF standard].

However, on the other hand, both ecosystems and human perspectives of water requirements were advocated by Falkenmark [29] who wrote, *Water, through its many different functions, plays multiple roles in the dynamics of ecosystems (citing [30]) and social systems. It has the function of determinant and life elixir of terrestrial ecosystems, as a carrier of nutrients and as a habitat of aquatic ecosystems. In social systems, it has fundamental societal functions for human life-support, food production, energy production, as a transport medium, as a mobile dissolvent, in continuity-related propagation of impacts, as a microclimate moderator, as a global-scale energy carrier, etc.*. Ripl [30] stressed the biophysical viewpoint of water wherein he states water as a key element of biosphere whereas human society is just a subsystem.

The current research on water footprinting encompasses both human and environmental flow requirements, and attempts to capture how withdrawal, use, and consumption by humans are creating pressure on both ecosystems and their functioning along with what that impact finally means for humans. However, to understand both the impacts and the approaches currently used to assess them, it is imperative to understand various types of water, in both flow and stock forms, defined and used in the water-footprinting literature. The concept of water footprint (WF) deals with quantitative consumptive and nonconsumptive measurement of water-related impacts of various human activities. That is, the concept explicitly deals with both direct and indirect water-related impacts of various activities and offers rigorous methodologies and concepts that can uniquely measure various types of water-related impacts on both abiotic and biotic systems. The various concepts and methods are explained in succeeding paragraphs. The next paragraph offers definition for three distinctive types of freshwater that are commonly quantified in WF analysis.

The first type of water is called "blue-water" and refers to water that is part of ecosystem reservoirs such as lakes, rivers, and beneath the top-soil, commonly called as aquifers. Thus, blue-water accounts for freshwater that is primarily withdrawn from surface and groundwater bodies. The blue-water can be considered as a stock when extracted from aquifers that are primarily recharged through local precipitation, whereas it can be considered as flow if retrieved from surface water bodies primarily fed by rivers which in turn are fed by both local and nonlocal precipitation.

The second type is known as "green-water" and refers to the water that originates from precipitation and stays in soil as moisture. This water is usually used in the context of vegetation and is not available for human-use (consumption or withdrawal). Thus, green water can be taken as stock or flow which is available only to vegetation via soil-ecosystem (see Section 2.3 [31]). The last type of water is called "grey-water," and it refers to quantity of freshwater required to dilute the impact of used-water, which constitutes both toxic and nontoxic water, and can be called as indirect impact of degraded water on freshwater. Thus, grey water can also be called as destruction of freshwater to dilute the impact of certain activities that do not consume freshwater in total but releases significant effluents that cause extended impacts on water bodies, which further affect other abiotic and biotic dimensions.

Furthermore, when released from point-source, it represents a flow, which, if stored in green or grey infrastructure, turns into a stock.

Since the above paragraph stressed on various types of water, this paragraph provides details about various types of measurements undertaken to measure the impacts. The in-flow water-measurements are usually performed in the context of withdrawal, use, and consumption. The first term, that is, withdrawal, accounts for water that is removed from the system, which may or may not return via return-flows or run-off, etc., to same system boundary such as a river, an aquifer, watershed, basin, etc. It is easier to understand this term in the context of blue-water that was defined earlier as water that is retrieved from surface and groundwater bodies. So, withdrawal will usually refer to the quantity that is taken out from a water body for a specific human activity, such as water-withdrawn by power-plants from a river for cooling purposes. On the other hand, the use or consumption of water would depend on the system for which water is extracted. Thus, in the case of a power-plant, the water used for cooling would be different from water that is extracted from a reservoir since some water may be returned back to the reservoir after use. More extensive the network, more are the resource requirements. Further, water used for cooling as a fraction of water-withdrawn from a reservoir could serve as an indicator of resource-efficiency (specifically, water-use efficiency) of a power-plant. The last term, namely, consumption, refers explicitly to the quantity of water that is retrieved from the system and does not return to the same system boundary. That is, the quantity of water that is forever removed from the reservoir. Usually, consumption refers to the process of evaporation and is frequently and mostly used with vegetation and the activities that are dependent on various land-vegetation types, such as crop production, hydro-dams, and water evaporated through both surface water bodies and moisture evaporated through soil. This environmentally controlled evaporative process is governed by both solar energy received at the earth's surface and various layers of earth's controlling boundaries that regulate both temperature and hydrological functioning at the planetary level.

The terms defined in the previous paragraphs offer basic definitions for various units which are used for attempting various types of water footprinting assessments. The next paragraph focuses on two distinctive approaches that are widely used in water footprinting assessments to measure the WF of human activities. These two approaches explicitly measure activity-related impacts; however, as they do not offer spatial (region-wise) and temporal impacts explicitly, other methods that are frequently used with these approaches are also briefly discussed in here. That said, the information offered herein is not an exhaustive one and does not comprehensively mention all the lacunae that have been part of such assessments. Hence, the interested reader is encouraged to read the literature for thorough critical analysis of the cited work.

The two approaches that have gained prominence in the water-footprinting literature have their roots in systems thinking. Further, both advocate accounting for both direct and indirect, consumptive and nonconsumptive water-impacts. That said, however, they had different systems as starting point of their progression. The first approach, proposed by the Water Footprinting Network (WFN) called water-footprinting methodology, was pioneered by Arjen Y. Hoekstra [32, 33]. In this methodology, an analytic function incorporating both weather and crop data estimates the ideal water requirements of both land and vegetation evaporation (consumptive) rates [34]. Since it is based on regional weather and crop (vegetation) characteristics, this approach initially exclusively dealt with water-impacts of crop production and has ever since been the most prominent approach to quantify water-impacts of both rain-fed and irrigated agriculture in terms of green, blue, and grey water-impacts.

As the quantitative measurements offered by this approach are based on an analytical function that can explicitly account for crop and regional weather characteristics anywhere on the planet, hence, using the global weather and crop information, this method has been successfully applied across the globe. Further, in this approach, both green and blue water-impacts are measured in consumptive units whereas the grey water accounted in this approach primarily focuses on non-consumptive nitrogen- and phosphorous-based impacts of fertilization processes. Though the WFN approach suggests measuring indirect impacts as well, it usually does not offer extended analysis, in the context of impacts related to biota and abiotic components in an extended boundary. For example, though this approach has been used in the context of consumptive impacts of hydropower-plants which are promoted as cleaner sources of energy generation and brings about another picture of environmental impacts of such large infrastructures on ecosystems [35], it fails to offer consumptive and nonconsumptive impacts associated with fossil fuel-based electricity generation infrastructure. However, at a regional (local) scale, it does have the potential to offer extended assessments if integrated with proper methods that account for both consumptive and nonconsumptive impacts in relation with crop production. For instance, it has been effectively used to assess water footprints of biofuels at regional scale [36].

The second approach, water-use in life cycle assessment (WULCA) [37], is also based on the systems thinking; however, compared to WFN approach, WULCA explicitly considers the impact on the local water-ecosystems. Stated another way, WULCA quantifies the impacts of extraction and pollution on the natural ecosystems which serve as sources for meeting anthropogenic water requirements. As this methodology is based on environmental-LCA approach [38], it initially focused on assessing water-based impacts of industrial products and processes. Realizing the significant consumptive demands of blue-water for agriculture, the goal and scope of LCA has expanded to include vegetation-oriented products with the aim of assessing region-specific quantitative and nonquantitative impacts. The WULCA approach has developed a tool called AWARE (Available WAter Remaining) [39, 40] to specifically account for the role of scarcity and availability; however, degradation due to pollution is still an unsettled category of impact [41] in WULCA.

The main difference that is originating within these two approaches is that within the LCA framework, WULCA explicitly contemplates to account the impact of human activities on water-ecosystems, in the form of scarcity indices and qualitative impacts on both biota and humans [42, 43], whereas WFN's quantitative WF measures per unit requirements across different geographic regimes. Thus, WFN implicitly measures the individual water-impacts of specific products; however, it cannot measure the cumulative impacts on the water-ecosystems (due to different activities) while WULCA proposes to explicitly measure that impact (via regional scarcity indices) (for different water types and distinct impacts—qualitative and quantitative). Additionally, WULCA approach explicitly acknowledges to account for temporal dimension [44]; the database offered by WFN does not offer such information explicitly. The key caveat stated for WULCA approach is that does not include green water [45] (or its scarcity [31]), which is a significant contributor to rain-fed agriculture, grasslands, and forestry, especially in the context of droughts [46]. Nevertheless, both the approaches frequently stress use of latest hydrological and Geographic Information Systems (GIS) data at regional level to account for water-oriented impacts (see Discussion [47, 48]).

4.3 Vulnerability to Water Depletion

The explicit need to assess water-impacts arises primarily due to rising local water scarcity across the globe. Decreasing water levels in major aquifers across the globe (Ogallala aquifer in the USA, Ganges–Brahmaputra, and many others) pose major challenges to the security of food, water, and global trade flows at both local and global scale. This section outlines the vulnerability due to water scarcity (as an indicator of quantitative availability). To highlight the water stress across the subhumid to arid areas, Wada et al. [49] estimated that total global groundwater depletion has increased from 126 km^3 (± 32 km^3) per annum in 1960 to 283 km^3 (± 40 km^3) per annum in 2000. The study further states that this water is a significant contributor to global sea-level rise. In a local context, Rodell et al. [2] estimated groundwater depletion rates of India and ascertained a mean depletion rate of 4.0 ± 1.0 cm per year (equivalent to total volume of about 17.7 ± 4.5 km^3 per year) for the Indian states of Rajasthan, Punjab, and Haryana (including Delhi). The abovementioned studies highlight the regional impact on water supplies due to human activities (especially irrigation for agriculture); hence, these studies do not exactly fall under the purview of WF methodologies described earlier. However, that said, these studies would still fall under the ISO framework for water-impact assessments, namely, ISO 14046, which is also known as ISO standard for water-footprinting assessments [28].

As per ISO 14046, there can be various purposes to conduct water footprint-oriented assessments, with primary among those being quantification of water-oriented impacts of various products and processes. The initial step of consistent and reliable quantification allows opportunities for reduction of impacts by identifying sources of inefficiency at various levels, with the larger objective of successful management of water resources. However, for the WF assessment to be acceptable as per ISO 14046 standard, certain steps are absolutely vital, foremost being that it should follow the systematic framework of the LCA standard, namely, ISO 14040.

Other expected aspects of such assessments include that it should uniquely identify different water-oriented impacts of different stages of life cycle of the products such as quantitative (embodied) and qualitative (polluting). That is, it should successfully differentiate between boundaries of the impacts by being modular in approach. Since availability of water differs in spatial and temporal context due to dependence of water supplies on both the renewable (hydrological cycles) and nonrenewable (deep aquifer based) sources, assessments should incorporate such aspects as well to be part of WF approach.

So, in conclusion, one can say that a WF assessment is the one that successfully identifies and quantifies various water-oriented impacts (in terms of inputs and outputs) of products, processes, or systems that satisfy various human demands along with ascertaining what those water-impacts means to natural environment (other biotic species), human life, and available resources of water. Further, since LCA is suggested as a base framework for such assessments, all the steps applicable to LCA (goal and scope definition, inventory compilation, assessment, and interpretation) are part of comprehensive WF studies.

To highlight different aspects of water-footprinting assessment, herein we discuss few examples at different scales. The purpose of these examples is to offer the readers different perspectives of how water-impacts are distributed across both products (in terms of quantity) and space. However, other studies can also be found in literature that highlight temporal, qualitative, ecological (land, soil, biodiversity, etc.), and human health-oriented impacts, among others.

4.4 Applications

In the context of quantitative impacts, the following example highlights how global food-trade is linked with local water scarcity. Further, the same study highlights how supply chains can make a local resource a global product. The considered example is a study by Dalin et al. [50], wherein regional groundwater depletion (impact on water resources as considered in WF methodology) data (obtained through GIS-based hydrological assessments at spatial scale) is linked with irrigation-based food-trade at the global scale. To be more specific, the study estimates how much of irrigation-based food is dependent on nonrenewable groundwater. Further, the results of this study indicate that the regional boundaries of India, Pakistan, and the USA are the local hotspots that are exporting their nonrenewable groundwater for supporting global food-trade. Thus, in a way, the results highlight the risks posed to both food and water security of both developed and developing and populated nations such as Iran, Mexico, China, and the USA, since both the production and consumption are dependent on depleting reservoirs. Thus, with such an assessment the researchers link the water-impact (i.e., the decreasing water-availability) to product (groundwater-irrigated food) and process (global food-trade), while integrating a hydrological dataset (spatial groundwater assessment) with a spatial process (using inventory of food-based global supply chains).

The next example considered focuses on water-energy nexus and measures WF in terms of water-consumption (Spang et al. 2014 [51]). This study reports that nearly 52 billion cubic meters (BCM) is consumed for energy production (using renewable sources and fossil fuels while excluding hydropower) in 150 countries across the globe. Thus, in this study, water-consumption is taken as an indicator of WF. However, though consumption is taken as an appropriate measure, the study suggests the need for improvement in both data quality and reporting standards for water-based estimates.

One of the reasons for obtaining such water data can be that water-related measurements are performed via observational (remote sensing and GIS), measurement-based (water-quality measurements of nutrients such as nitrogen, phosphorous), and analytical (WFN methodology measuring ideal requirements) methods, and thereby, it is difficult to use and defend such nonhomogeneous data for consistent assessments and further for governance issues. Further, if assessments and methods report water "quantification" numbers in terms of withdrawal, use, consumption, intake, input at various system boundaries (product level, process level, site-scale, etc.), it becomes difficult to compare the varied estimates as well. That is, if one study reports withdrawals and other one reports consumption, then a global decision to reduce water-use may lead to erroneous results since consumption (i.e., removal by evaporation only) in power-plants is much smaller than withdrawal. Thus, to reduce ambiguity in various assessments, ISO 14046 also suggests to use a qualifier to clarify what is being measured [28].

To stress the point of boundary selection and type of measurement, in the next example we consider the economy as a system [52]. The reported study considers geography of India as spatial scale and combination of all products produced in India as one system. This study follows the methodology of environmentally extended input–output (EEIO) modeling, and thereby, in this assessment the economic system of India is considered as one system. Since different products have different levels and types of water-dependence (one product may consume more while other product may pollute more), their impacts vary across the breadth of WF methodology. Thus, to offer a consistent assessment, the study reports only withdrawal (as one single indicator) which represents dependence in terms of intake. Further, by using an economic input–output system as a base framework, the indirect dependence by industries or sectors on primary producers such as agricultural or electricity that withdraw water directly from the ecosystems is also obtained. That is, in this work, the dependence of industries such as food and beverages on rice or wheat sectors is captured via indirect withdrawal. Or, dependence of manufacturing industries to water-embodied in electricity sector is elucidated explicitly.

The explicit quantification of the dependence of industrial and service sectors on water-withdrawn in the upstream supply chains (although aggregated at sector scale in economic input–output table) offers information of indirect life cycle water-dependence which is usually lacking in LCAs that are attempted at producer scale. That is, this EEIO-based assessment quantifies water-dependence of upstream contributions and thereby assists in revealing obscured dependence which assists in completing the boundary-truncation problem encountered in LCAs of specific

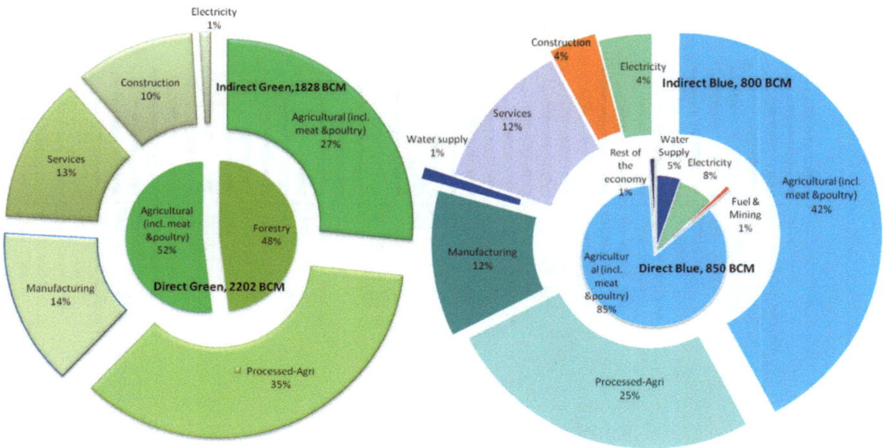

Fig. 4.2 Direct and indirect water-flows within Indian economy. (Reproduced with permission from [52])

products. Thus, albeit aggregated, the water-coefficients of this study offer full WFs of agricultural, services, and industrial supply chains for the Indian economic system (Fig. 4.2).

The study also estimates direct water-withdrawal by products (sector's output) based on product water-intensities. Further, the regional impacts of total direct withdrawal by sector-wise outputs, and especially for the agricultural products (that are the major source of removal of large volumes of water permanently from local water bodies), are also obtained in terms of surface, ground, and scarce-groundwater. Particularly, the scarce-groundwater (determined via estimating water-scarcity indices at regional scale) represents part of the water resources that would not be available for foreseen future and thereby represent the vulnerability for all economic activities that are dependent on this source.

Additionally, by estimating the water-scarcity indicators, this study elucidates the quantitative impact on ecosystems capacity at regional scale. That is, the water-scarcity indices represent the midpoint indicators as suggested by life cycle impact assessment step in the LCA framework [38], wherein the purpose of midpoint indicators is to offer a quantitative description whether the resource use (water in this case) is sustainable at chosen spatial and temporal scale. From a social dependence perspective, it highlights the water-risks at the regional scale associated with food and water security for the population of India. For example, considering groundwater-scarcity indices at the state-level, the study suggests that economic activities based in regions facing groundwater scarcity imply risk to their water supply and downstream supply chains. The nationwide electricity production is also captured in the system (economic input–output) table. Since electricity sector withdraws surface water, this implies surface water-dependence contributes to

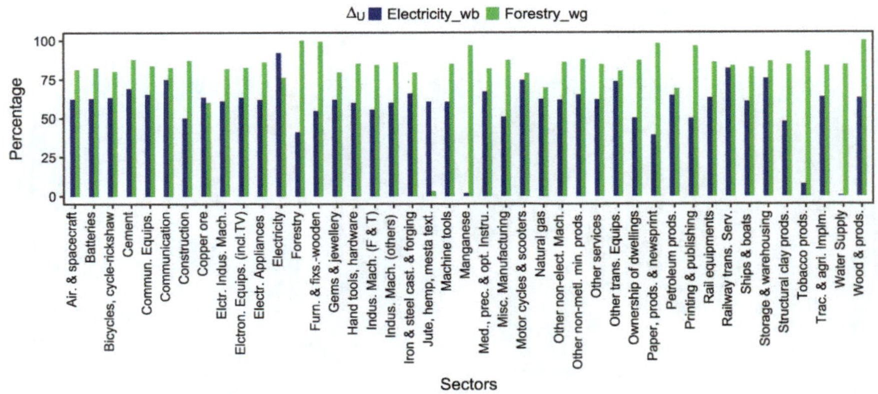

Fig. 4.3 Total water-flows to rest of the economy (RoE), blue-water from electricity sector (Δ_U^{wb} or Δ_U Electricity_wb), and green water for forestry (sector 25) Δ_U^{wg} or Δ_U Forestry_wg in %. (Reproduced with permission from [53])

extraction from environmental flows which are necessary for generation of other ecosystem services in "water-scarce and respective downstream" regions.

Summarizing, the analyses from the study highlighted following key aspects of the Indian economy in terms of water-withdrawal: (i) contribution of green water is larger (72%) than blue-water, with total direct withdrawal of 3052 billion cubic meter (BCM), (ii) green water utilized by forestry sector (1057 BCM) is only slightly lower than 1145 BCM utilized by agricultural sector, (iii) major contribution of blue-water (745 BCM) happens in agricultural sector with electricity and water supply (domestic use) following closely. The scarce water-dependence is higher in north-western states for staple crops, whereas cash crops are primarily grown in central India.

A successive study [53] based on the abovementioned study [52] elucidated the indirect dependence of economic-sectors on water supplies (the indirect assessment is the most prominent feature of the EEIO modeling) (see Fig. 4.3). The mapping of indirect dependence of industries or economic activities is usually missing from product-wise LCA due to boundary-truncation problem and has been stressed as a key area to focus on industry-focused WF assessments [54].

For example, highlighting the indirect dependence of beverages sector on products that have direct embodiment of water or water-use directly in their supply chain, Ercin et al. reported that a hypothetical carbonated product packed in a bottled container has an indirect WF of above 99% in their upstream supply chain [55]. The upstream products considered in this study included were sugar-based crops and packaging, among others. That is, through this study, the major part of the WF of industrial products is shown to lie in the upstream supply chain. This implies that the risks to industry and service sectors are not directly visible to stakeholders. Further, without the tracking of spatial location of the upstream constituents, it is difficult to even visualize the delocalized risks associated with the major constituents. And specifically in the case of water, without regional scarcity information, it may prove difficult to assess sustainability since the industry may be only a small player at regional water-use.

The EEIO-based study [53] mapped the indirect contribution of water-withdrawn by paddy, electricity, and forestry sectors to the other sectors of the economy. The analysis highlighted that major beneficiaries (economic throughput) from these sectors are food-processing, hotels, infrastructural, and construction sectors, among others. Furthermore, the study quantified the green water drives 70% of throughput with rest contributed by blue-water, highlighting the dangers from climate change introduced uncertainties on precipitation or droughts in worst case. The study also mapped scarce surface and groundwater (both forms of blue-water) for electricity and forestry sectors across states (political regions) of India. This analysis highlighted the vulnerability of agriculture-intensive states to future water supply for electricity generation along with offering a key insight that states with large areas under forest-cover have minimum level of groundwater scarcity. These results confirm cropping patterns pursued over past decades (aftermath of green-revolution in India) have become major drivers of blue-water scarcity, whereas forest-cover allows checking groundwater deficits.

To present a comprehensive dimensional analysis of product footprinting in terms of water, the next paragraph discusses local impacts of a single product, namely, biofuels, however, yet the assessments are part of WF methodology. Further, through this example the debate around water as an ecosystem service for environmental and complementary use is sought, since increasing the direct use for anthropogenic activities creates an impact on ecosystems functioning and reserves. That is, "lack" of wateroriented LCA endpoints of environmental impact assessments, namely, human health, ecosystem quality, and resource depletion, are highlighted through these assessments.

Are Biofuels Truly Impact Free?: A Water-Footprinting Perspective

The last set of applications discussed herein deals with water-impacts of low-carbon fuel production in the context of current focus on policy-impetus for a carbon-neutral future. Fossil-fuel consumption (primarily petroleum-based) is identified as one of the primary causes behind rising global CO2 emissions, and biofuels (biodiesel and bioethanol) have been promoted [56, 57, 58] as a panacea for reducing carbon-based impacts or greener fuels. However, is biofuel production truly environment-friendly at the global scale? That is the big question. This section focuses on studies that have assessed water-based impacts of biofuels from local to global scale. Starting with the definition, biofuels are fuels derived from bio-based crops or biomass. Such crops include sugarcane, maize, cassava, rapeseed, and many others. But growing of crops needs water along with land and use of inputs such as fertilizer, pesticides, and fuels. In terms of inputs, this implies direct and indirect contribution of scarce natural resources such as water in some regions and land for food versus fuels production in others. With increasing population, especially in developing and already densely populated nations, the pressures on water sources to satisfy both human and ecosystems needs are bound to increase.

With such a perspective, many biofuel-oriented studies tried to capture aspects from local to global level, while using data obtained from diverse resources.

Using the WFN database, Gerbens-Leenes et al. [59] reported that under the International Energy Agency's (IEA) Advanced Pledges Scenario (APS) of energy increase, global WF of biofuels, including first-generation biofuels such as bioethanol from sugarcane, sugar beet, sweet sorghum, wheat, and maize, and biodiesel from soybean, rapeseed, jatropha and palm oil, is expected to increase more than ten times within the period of 2005–2030. Further, nations such as the USA, China, and Brazil are expected to contribute half of this global WF. Furthermore, the blue WF (due to withdrawals from blue-water sources) is expected to increase blue-water scarcity in many nations, which in turn is expected to increase pressure on water resources. Thus, at a global scale, this study reported that water-costs for biofuel-based transportation may lead to increased water-use in many regions, which is bound to be reflected in future water-availability in water-scarce regions. Focusing on the climate impacts on potential changes on the crop water requirements (CWRs) of corn-based ethanol for transportation in the USA, Dominguez-Faus et al. [60] used a modeling approach integrating a GIS-based model and predicted weather variables to estimate changes in both CWRs and yields of corn-ethanol for a 40-year horizon. Since the weather variables are expected to result in higher evaporation rates for corn, the water requirements are expected to increase by nearly 20%, which in turn would lead to increase in irrigation requirements in the crop-growing regions. Since groundwater from the main reservoir (Ogallala aquifer) is the source of water supply for irrigation, pressure on this source will intensify. Thus, the aforementioned research brings out the climate change impacts on evaporation rates of crops and how any policy that promotes such irrigation-oriented crops may further intensify the human pressure on ecosystems. Further, like Ogallala, many other aquifers support irrigation in different parts of the world, especially in developing and populated countries such as those in South Asia. Thus, the increased crop water requirements will further intensify pressures on such sources and ultimately the debate of fuel versus food would gain prominence in such a scenario.

The study by Zhang et al. [61] highlights the predicament of growing biofuels in populated nations. This study evaluated the water requirement of growing nonfood crops across different regions in the same country, that is, the mainland China. They used the methodology followed by WFN and incorporated crop-type and regional weather data. The results conclude that even nonfood-based crops have significant water requirements and algae-based biofuels have the highest water requirements. Polluted water due to fertilization is also found to be a significant contributor. The study does offer suggestions about which regions would have lowest impacts in case biofuel policy is pursued in China. Though this study offers regional policy suggestions for biofuel production, it did not consider whether overall the regions are facing water-scarcity issues. That is, increase or decrease in water-stress levels due to all crops grown in the region along with direct water-use by humans was not elucidated in this work.

The last study discussed herein highlights the differences in biobased outputs that may serve as fuels to transportation requirements, namely, biodiesel, bioethanol, and bioelectricity. The considered study assessed global WF for 12 different crops that may serve as inputs to bioenergy [62]. The indicated results suggest that compared to using biofuels directly, bioelectricity derived from biomass would offer more efficient conversion per unit of water-embodied, since total biomass could be used in the latter product. Further, the per unit water-dependence of bioethanol seems to be smaller than for biodiesel. Sugar- (or starch-) based crops such as sugar beet, maize, and sugarcane appear to be most suited for bioelectricity and bioethanol, whereas oil-based crops such as soybean and rapeseed appear more suitable for biodiesel production, due to variation in water-embodied per unit quantity produced (different across regions). Such results provide additional information for pursuing renewable energy policy across regions in a resource-scarce world. To reiterate the claim on non-inclusion of end-point-based LCA impacts, none of these studies accounted for impacts on humans or ecosystems from scarcity, pollution, or overall degradation.

4.5 Data, Models, and Software for WF

The literature included in earlier sections indicated few models and data-sources used assessments, and this section formally summarizes few of them; however, for applications of the software especially, the interested researchers should check the details in the cited literature.

The Water Footprint Network (WFN) offers a repository that provides estimates of consumptive water requirements for growing various types of crops and evaporation requirements for hydro-dams [63]. The repository also quantifies grey water requirements (quantity of water required to dilute the impact of pollution) for nitrogen and phosphorous used for various cultivars across regions. Thus, the repository is a unique collection of water-appropriation for products who are major consumers of this ecosystem service. The AWaRe tool, developed by Boulay et al. [39], under the aegis of UNEP-Society for Environmental Toxicology and Chemistry (SETAC) Life Cycle Initiative, serves as a midpoint indicator for water-use and quantifies water remaining in a watershed. Though the tool has a global coverage, it is advocated that context and scale should be considered for product-specific assessments as contrasting results may be obtained due to average and marginal characterization [40].

Following the LCA approach, the Ecoinvent database offers quantification of water-impacts for industrial systems (products or processes) among other categories, This database is frequently used in LCA software such as SIMAPRO [64], OpenLCA [65]p, and GaBi [66]. The Ecoinvent database is a life cycle inventory (LCI) in the context of Europe, LCA Commons [67] is a suit of LCI databases in the context of the USA. The product-specific databases such as those focusing on agricultural products [68, 69] can be screened at LCI repository with software OpenLCA [70]. These LCA-focused databases offer water-impacts in conjunction with other environmental impacts.

4.6 Discussion and Conclusions

Water is an important ecological service that supports various human activities along with supporting numerous ecological functions. Water footprint (WF) is an indicator to quantify both the demand and the impact that human activities create on various hydrological sources, measured only in water-based units. WF assessments can be both consumptive and nonconsumptive. The former usually deals with vegetation growth wherein evaporated water does not return to same spatial and temporal system boundary, whereas the latter deals with deterioration on account of pollution, making further quantity unfit for many human activities and ultimately affects in the form of deprivation.

The purpose of these two types of measurements is to offer total impact that anthropogenic activities create of water resources both directly and indirectly. To make assessments useful for various stakeholders who depend on water directly or directly, the measurements are made in terms of different water types (green water—rainfall and moisture used only by vegetation; blue-water—freshwater from surface and groundwater bodies; and grey water—water that gets polluted or deteriorated). Further, such measurements can be either in withdrawal, in consumptive terms, or in use category, wherein the first two measurements (withdrawal and consumptive) are with respect to water removed from water bodies in gross and net terms, respectively, whereas the use category represent human perspective that excludes material and other losses associated with various anthropogenic systems.

In the context of environmental sustainability- (or water sustainability) oriented assessments, two prominent approaches have evolved to measure water-based impacts, namely, Water Footprint Network's Water Footprint approach and water-use in life cycle assessments (WULCA). The first method uses an analytic function to estimate crop and land-based water evaporation rates, which leads to ideal consumptive water requirements, whereas the second approach is based on LCA framework while focusing only on water-oriented impacts. Since the latter approach is based on systems approach and life cycle thinking, it explicitly stresses to account for the impacts created on water-systems. Whereas in the quantitative measurements of WFNs, the measured quantities differ on account of regional water evaporation rates, that is, regional impacts created by per unit quantity are implicit in consumptive terms. However, the purpose of both types of measurements is to estimate environmental burdens of activities in terms of water as an ecosystems service from local to global scale. That said, both approaches suffer from boundary-truncation issues and fail to capture indirect impacts at larger scales.

To circumvent this issue, role of aggregated methods such as environmentally extended input–output models was discussed using a specific case study. Furthermore, though most assessments report water indicators in a well-defined manner, they are largely based on analytic functions and secondary data. That said, many LCA-based studies do offer primary data; however, LCAs are restricted by regional constraints specifically with respect to water. Thus, nearly all studies stress the need for coherent data formats and assessments such that results and thereby decisions are not based on erroneous data. Further, improvement in methodology and data calibration is also stressed to resolve methodological issues related to regional water assessments and furthermore in global risks to water-embodied trade flows.

To bring out the essence of WF as an environmental sustainability indicator, the discussed case studies highlighted how the locally embodied water in products become a global commodity. Additionally, at another dimension the trade in products also reflects the increasing vulnerability due to rising local water stresses. Further, the local water issues can affect the entire economy that is based on electricity supply. Whereas, the case study of biofuels as a greener source for transportation and other needs reflects the urgency of reducing greenhouse gas levels, satisfying the energy demands of the ever growing human population by using biofuels may shift the burden to water resources. Thus, future cleaner or greener fuels-oriented energy policies should also consider local water-sustainability issues. That is, policies should consider systems approach to assess all types of impacts (including water) and not only carbon-based impacts to have sustainability in true sense.

Last, for the goal of sustenance of water-reservoirs, assessment of socioeconomic drivers is a neglected area in water-footprinting assessment. For example, from a production perspective, when mined water is dumped in lakes or canals, or extremely hot steam is released from supercritical power-plants, or when chemical embedded in plastic bottles mix with extended abiotic and biotic systems in water-ecosystems, the loss of fisheries and drinking water places huge burden (social cost) on local people who do not possess the means to treat such toxic externalities generated by extractive industries. Such a destruction of regional water bodies increases vulnerability of local people who derive their well-being from such ecosystems. The economic driver of profits and growth along with dominant position of stakeholders dependent on such extractive operations in policy-arena becomes a key hindrance and thereby a formidable challenge for pursuance of policies that can lead to socio-environmental sustainability at local scale. Furthermore, from a consumption perspective, increasing water demands due to increasing population and preference for animal-based diets is a strong driver dependent on lifestyle choices, which are still inadequately addressed in WF assessments. Thus, there is urgent need to advocate policies, laws, and ideas, supported by water-sustainability assessments, to seek sustainable production and consumption patterns.

References

1. Millennium Ecosystem Assessment, 2005. Ecosystems and Human Well-being: Synthesis. Island Press, Washington, DC.
2. M. Rodell, I. Isabella Velicogna, and J.S. Famiglietti. Satellite-based estimates of groundwater depletion in India. *Nature*, 460:999–1003, 2009.
3. Y Zhou, H Gholizadeh, GT LaVanchy, and E Hasan. Inspecting the Food–Water Nexus in the Ogallala Aquifer Region Using Satellite Remote Sensing Time Series. *Remote Sensing*, 12(4):2257, 2020.
4. J.S. Famiglietti. The global groundwater crisis. *Nature Climate Change*, 4:945–948, 2014.

5. P. Micklin. Desiccation of the Aral Sea: A Water Management Disaster in the Soviet Union. *Science*, 241(4870):1170–1176, 1988.
6. P. Micklin. The Aral Sea Disaster. *Annual Review of Earth and Planetary Sciences*, 35:47–72, 2007.
7. UNEP. The Drying of Iran's Lake Urmia and its Environmental Consequences, 2012.
8. A. Eimanifar and F. Mohebbi. Urmia Lake (Northwest Iran): a brief review. *Aquat. Biosyst*, 3(5), 2007.
9. S. Jasechko and D. Perrone. Global groundwater wells at risk of running dry. *Science*, 372(6540):418–421, 2021.
10. D. Perrone. Groundwater Overreliance Leaves Farmers and Households High and Dry. *One Earth*, 2:214–217, 2020.
11. B. et al C. Moore. A High Spatiotemporal Assessment of Consumptive Water Use and Water Scarcity in the Conterminous United States. *Water Resources Management*, 29:5185–5200, 2015.
12. J. et al R. Gerson. Artificial lake expansion amplifies mercury pollution from gold mining. *Science Advances*, 6(48):5185–5200, 2020.
13. S. et al Luckeneder. Surge in global metal mining threatens vulnerable ecosystems. *Global Environmental Change*, 69:102303, 2021.
14. R. et al M. S. A. Meira. Brazil's mining code under attack: giant mining companies impose unprecedented risk to biodiversity. *Biodiversity and Conservation*, 25:407–409, 2016.
15. M. Grudnoff. Pouring fuel on the fire: the nature and extent of Federal Government subsidies to the mining industry, 2012.
16. I. et al Stoddard. Three Decades of Climate Mitigation: Why Haven't We Bent the Global Emissions Curve? *Annual Review of Environment and Resources*, 46:653–689, 2021.
17. J. Rehner and S. Rodríguez. Cities built on copper – The impact of mining exports, wages and financial liquidity on urban economies in Chile. *Resources Policy*, 70:101190, 2021.
18. J. C. Orihuela. The environmentalization of mining in Colombia, Chile, and Peru: A comparative analysis of green state formation. *The Extractive Industries and Society*, 8(4):100829, 2021.
19. T. Akdemir and G. Dalgic. The impact of the marine sewage outfalls on the sediment quality: The Black Sea and the Marmara case. *Saudi Journal of Biological Sciences*, 28(1):238–246, 2021.
20. C. Wilcox. Why Turkey's Sea of Marmara Is Full of Marine Snot, 2021.
21. S. Pfister and S. Suh. Environmental impacts of thermal emissions to freshwater: spatially explicit fate and effect modeling for life cycle assessment and water footprinting. *Int J Life Cycle Assess*, 20:927–936, 2015.
22. M. M. Mekonnen, W. P. Gerbens-Leenes, and A. Y. Hoekstraa. The consumptive water footprint of electricity and heat: a global assessment. *Environ. Sci.: Water Res. Technol.*, 1:285–297, 2015.
23. L. et al Rosa. Hydrological limits to carbon capture and storage. *Nat Sustain*, 3:658–666, 2020.
24. W. Schakel, S. Pfister, and A. Ramírez. Exploring the potential impact of implementing carbon capture technologies in fossil fuel power plants on regional European water stress index levels. *Inter J of Greenhouse Gas Control*, 39:318–328, 2015.
25. T. Allan. Economic and political adjustments to scarce water in the Middle East. *Studies in Environmental Science*, 58:375–388, 1994.
26. T. Allan. 'Virtual water': a long term solution for water short Middle Eastern economies? . *Water and Development Session. British Association Festival of Science*, 1997.
27. P. H. Gleick. Basic water requirements for human activities: Meeting basic needs . *Water International*, 21(2):83–92, 1996.
28. International Organization for Standardization (ISO). ISO 14046:2014(en) Environmental management - Water footprint - Principles, requirements and guidelines., 2015. accessed May 19, 2022.

29. M. Falkenmark. Freshwater as shared between society and ecosystems: from divided approaches to integrated challenges. *Philosophical Transactions of the Royal Society of London. Series B: Biological Sciences*, 358(1440):2037–2049, 2003.
30. W. Ripl. Water: the bloodstream of the biosphere. *Philosophical Transactions of the Royal Society of London. Series B: Biological Sciences*, 358 (1440):1921–1934, 2003.
31. J. F. Schyns, A. Y. Hoekstra, and M. J. Booij. Review and classification of indicators of green water availability and scarcity. *Hydrol. Earth Syst. Sci.*, 19:4581–4608, 2015.
32. Water Footprint Network. accessed May 19, 2022.
33. D. Vanham, M. M. Mekonnen, and A. K. Chapagain. Arjen Y. Hoekstra 1967–2019. *Nat Sustain*, 3:80, 2020.
34. A. et al Y. Hoekstra. The Water Footprint Assessment Manual, 2011. accessed May 19, 2022.
35. M. M. Mekonnen and A. Y. Hoekstra. The blue water footprint of electricity from hydropower. *Hydrol. Earth Syst. Sci.*, 16:179–187, 2012.
36. W. Gerbens-Leenes and A. Y. Hoekstra. The water footprint of biofuel-based transport. *Energy Environ. Sci.*, 4:2658–2668, 2011.
37. Water Use in LCA. accessed May 19, 2022.
38. International Organization for Standardization (ISO). ISO 14040:2006(en) Environmental management – Life cycle assessment – Principles and framework, 2006. accessed May 19, 2022.
39. A. et al Boulay. The WULCA consensus characterization model for water scarcity footprints: assessing impacts of water consumption based on available water remaining (AWARE). *Int J Life Cycle Assess*, 23:368–378, 2018.
40. A. Boulay, L. Benini, and S. Sala. Marginal and non-marginal approaches in characterization: how context and scale affect the selection of an adequate characterization model. The AWARE model example. *Int J Life Cycle Assess*, 25:2380–2392, 2021.
41. N. Mikosch, M. Markus Berger, and M. Finkbeiner. Addressing water quality in water footprinting: current status, methods and limitations. *Int J Life Cycle Assess*, 26:157–174, 2021.
42. S. Pfister, A. Koehler, and S. Hellweg. Assessing the Environmental Impacts of Freshwater Consumption in LCA. *Environ. Sci. & Technol.*, 43 (11):4098–4104, 2009.
43. Water footprinting in LCA. accessed May 19, 2022.
44. What is AWARE? accessed May 19, 2022.
45. W. Gerbens-Leenes and Berger. M. Obituary in Remembrance of Tony Allan. *Water*, 13(13):1859, 2021.
46. E. Ercin, T. I. E. Veldkamp, and J. Hunink. Cross-border climate vulnerabilities of the European Union to drought. *Nat Commun*, 12:3322, 2021.
47. A. Boulay, L. Lenoir, and A. Manzardo. Bridging the Data Gap in the Water Scarcity Footprint by Using Crop-Specific AWARE Factors. *Water*, 11(12):2634, 2019.
48. S. Pfister, C. Oberschelp, and T. Sonderegger. Regionalized LCA in practice: the need for a universal shapefile to match LCI and LCIA. *Int J Life Cycle Assess*, 25:1867–1871, 2020.
49. Y. et al Wada. Global depletion of groundwater resources. *Geophysical Research Letters*, 37(20), 2010.
50. C. et al Dalin. Groundwater depletion embedded in international food trade. *Nature*, 543:700–704, 2017.
51. E. et al S. Spang. The water consumption of energy production: an international comparison. *Environ. Res. Lett*, 9(10):105002, 2014.
52. Bogra, S.; Bakshi, B. R.; Mathur, R. A Water-Withdrawal Input–Output Model of the Indian Economy. *Environ. Sci. & Tech.*, 50(3):1313–1321, 2016.
53. Bogra, S. and R. Bakshi, B. . Direct and indirect vulnerability of economic sectors to water scarcity: A hotspot analysis of the Indian economy. *J Ind Ecol.*, 24(6):1323–1337, 2020.
54. A. Y. Hoekstra. *Water Footprint Assessment in Supply Chains*, pages 65–85. Springer International Publishing, Cham, 2017.

55. Ercin, A.E. and Aldaya, M.M. and Hoekstra, A.Y. . Corporate Water Footprint Accounting and Impact Assessment: The Case of the Water Footprint of a Sugar-Containing Carbonated Beverage. *Water Resour Manage*, 25:721–741, 2011.
56. Farigione, F. et al. Land Clearing and the Biofuel Carbon Debt. *Science*, 319 (5867):1235–1238, 2008.
57. B. Fitzherbert, E. et al. How will oil palm expansion affect biodiversity? *Trends in Ecology & Evolution*, 23 (10):538–545, 2008.
58. Achten, W. M. J. and L. V. Verchot . Implications of biodiesel-induced land-use changes for CO2 emissions: case studies in tropical America, Africa, and Southeast Asia. *Ecology and Society*, 16 (4):14, 2011.
59. W. Gerbens-Leenes, P. and R. van Lienden, A. and A.Y.Hoekstra, A. and H.van der Meer, Th. . Biofuel scenarios in a water perspective: The global blue and green water footprint of road transport in 2030. *Global Environmental Change*, 22 (3):764–775, 2012.
60. Dominguez-Faus, R. et al. Climate Change Would Increase the Water Intensity of Irrigated Corn Ethanol. *Environ. Sci. & Tech.*, 47(11):6030–6037, 2013.
61. Zhang, T. and Xie, X. and Huang, Z. Life Cycle Water Footprints of Nonfood Biomass Fuels in China. *Environ. Sci. & Tech.*, 48(7):4137–4144, 2014.
62. Gerbens-Leenes, W. and Y. Hoekstra, A. and H. van der Meer, Th. The water footprint of bioenergy. *PNAS*, 106(25):10219–10223, 2009.
63. M. M. Mekonnen and A. Y. Hoekstra. Review and classification of indicators of green water availability and scarcity. *Hydrol. Earth Syst. Sci.*, 16:179–187, 2012.
64. SimaPro. accessed May 19, 2022.
65. OpenLCA. accessed May 19, 2022.
66. GaBi. accessed May 19, 2022.
67. openLCA Nexus. Federal LCA commons. accessed May 19, 2022.
68. openLCA Nexus. Agri-footprint. accessed May 19, 2022.
69. openLCA Nexus. Agribalyse. accessed May 19, 2022.
70. openLCA Nexus. Databases. accessed May 19, 2022.

Chapter 5
Biogeochemical Cycles: Modeling the Interaction of Carbon and Nitrogen Cycles with Industrial Systems

Shweta Singh

5.1 Introduction to Biogeochemical Cycles

Biogeochemical cycles are the pathways that circulate nutrients across different components of earth systems – both biotic and abiotic. Hence, these cycles perform an essential function of providing the necessary nutrients and materials for sustaining activities on earth. Ecosystems play a crucial role in these material cycles; hence, the support provided by these cycles is accounted as supporting ecosystem service. Accounting for these supporting services from biogeochemical cycles is challenging due to the variety of cycles and understanding the exact service that can be quantified. In this chapter, the focus will be on carbon (C) and nitrogen (N) cycles that provide the two most essential nutrients for production and sustenance of anthropogenic demands. These cycles are much faster and have more active connection with the anthropogenic production systems. Other cycle such as phosphorus (P) cycle is not studied here, as it is much slower and extracted from a much smaller stock of P rock through rock mining. While the biogeochemical cycles have some overlapping compartments and processes with geochemical cycles, the time scale of geochemical cycles that happen in crust and surface of earth is much larger than the biogeochemical cycles. Hence, quantifying dependence on geochemical cycles for decision-making for engineering design is difficult due to the temporal scale variation. If an engineered system is being designed for 25–50 years, the associated nutrients (e.g., metal) provided by geochemical cycles will likely not show any variation in that time scale. However, biogeochemical cycles and related imbalances have a significant impact on the sustainability of ecological systems, availability of nutrients, and associated negative impacts on environment due to

S. Singh (✉)
Agricultural & Biological Engineering, Environmental & Ecological Engineering, and Davidson School of Chemical Engineering (By Courtesy), Purdue University, West Lafayette, IN, USA
e-mail: singh294@purdue.edu

B. R. Bakshi (ed.), *Engineering and Ecosystems*,
https://doi.org/10.1007/978-3-031-35692-6_5

excessive reactive species. Biogeochemistry is an area of active research that cuts across several disciplines including biology, ecology, geology, and chemistry. While the term was first used about 95 years ago (Bashkin 2002, here the author reported the term first being coined 75 years ago, I have calculated it to be around 95 years in 2021 accounting for 19 years since the publication of Bashkin's 2002 book), the field can be mapped to the origin of studies related to natural sciences. Several books provide detailed account of evolution of biogeochemical cycles, present state, and an in-depth understanding of mechanisms that drive these cycles. Hence, I will take the liberty to point the readers to the original texts (Bashkin 2002) for an in-depth study of biogeochemical cycles. For the purposes of detailing the accounting of the role of biogeochemical cycles in engineering activities, the carbon and nitrogen cycles are briefly described in this section.

Carbon (C) Cycle Carbon is one of the most important macroelements for sustenance of the earth system and different production cycles in anthropogenic systems. Biogeochemists mainly focus on describing the carbon cycle by stocks and flows in different compartments of ecosystems, which are further classified into terrestrial, aquatic, or marine ecosystems. Each of these ecosystems' components has distinct compartments where carbon is stored and transformed to different forms at various time scales. The element is continuously distributed between land, water, and atmosphere in different forms of organic and inorganic carbon. The organic form of carbon is mainly present in various living organisms, plants, soil, etc., while the inorganic form is present as exchangeable inorganic ions such as HCO_3-, CO_3^{2-}, and CO_2 in ocean ecosystems. These inorganic forms are also present in aquatic ecosystems. The natural carbon cycle functions at different time scales based on the processes that transform carbon. A shorter time scale carbon cycle recycles carbon from atmosphere (CO_2) to organic carbon via photosynthesis that can be either assimilated into products by anthroposphere or gets decomposed by microorganisms in terrestrial ecosystems. A longer time carbon cycle deposits the organic carbon into deeper sediments of earth systems that eventually become rocks and fossil fuel. This carbon cycle functions over geological time scale. Coal, natural gas, and oil reserves are the fossilized form of carbon that gets converted into fossil fuel over millions of years under earth's extreme environment of high temperature and pressure. This carbon is now being utilized as fuel sources for anthropogenic demands and returned to atmosphere as CO_2, due to which atmospheric concentration of CO_2 has rapidly increased over the past few decades. However, the increase of atmospheric concentration could be larger if ocean and terrestrial ecosystems were not present. Ecologists have provided significant evidence of carbon sequestration services provided by forest and ocean ecosystems due to which the atmospheric CO_2 rise has been lower than the actual CO_2 released from the burning of fossilized carbon. This reliance of anthropogenic systems on carbon cycling provided by ecological systems is a critical ecosystem service provided by the biogeochemical cycle of carbon. The accounting of this service to economic and industrial systems can be done via the Ecologically Based Life Cycle Assessment (Eco-LCA) model, as discussed in Sect. 5.2.

Nitrogen (N) Cycle Nitrogen is the second major macroelement that is necessary for the sustenance of anthropogenic activities. N exists in different forms in various compartments of the biosphere. N is also a critical structural component of proteins, which forms the backbone of DNA; thus, it can be said that N is a key element that forms the foundation of many life forms and is critical for our existence. There are five main processes by which N gets circulated in different ecosystems – (i) fixation: the process by which inert N_2 in air is converted to organic N; (ii) mineralization: the process of converting organic N to inorganic form of N; (iii) nitrification: the process of oxidizing ammonium (NH_4^+) to nitrite and nitrate; (iv) denitrification: the process of converting inorganic N to atmospheric N_2O and N_2; and (v) assimilation: the process of converting inorganic N to organic N. As evident by these processes, the N biogeochemical cycle results in continuous circulation of N from atmosphere to biotic and abiotic compartments in form of inorganic and organic N. However, anthropogenic activities have increased the flow of N from inert N_2 to NH_3, that is, N fixation tremendously by the Haber Bosch process, thus causing accumulation of reactive N in ecosystems, resulting in negative impact on ecological health. The increased fixation of N by the Haber Bosch process has resulted in increased denitrification as well, thus resulting in increased concentration of N_2O in the atmosphere. N_2O also has a greenhouse effect, which is 265–298 times more than that of CO_2 for a 100-year timescale. Hence, anthropogenic activities have also caused imbalance in the N cycle, creating more reactive N species, which has accumulated in the atmosphere resulting in global climate change. Therefore, it is important to quantify the impact and dependence on industrial activities on the flow elements related to N biogeochemical cycle, defined as the ecosystem services provided by N cycle. Similar to the C cycle, the Eco-LCA model provides a framework to account for the dependence and impact of various industrial activities on N biogeochemical cycle as well. The Eco-LCA model and data requirements are described in next section.

5.2 Accounting for the Carbon and Nitrogen Cycle: Eco-LCA Model and Applications

5.2.1 Life Cycle Assessment Methodologies

Life Cycle Assessment (LCA) models provide a robust technique for assessment of environmental impact of industrial activities that take a holistic approach of including indirect impacts of the industrial products/processes. LCA has been around for more than 20 years now and has become a critical assessment tool used by industries to report about their overall emissions and identifying areas of improvement for lowering environmental impacts. The LCA methodology has been standardized by ISO 14040 that describes the best practices to conduct LCA. A typical LCA study consists of four key steps: (i) goal and scope definition, (ii) inventory analysis,

(iii) impact assessment, and (iv) interpretation. The results of LCA can directly be used for product development with lower environmental impacts, strategic planning for research and development toward more environmentally friendly processes and production and informing policy. The goal and scope definition phase of LCA is the distinguishing factor on how LCA results can be used for decision-making. In this phase, practitioners also decide the system boundary for their analysis. LCA can be defined as – Process LCA, Input-Output-based LCA, and Hybrid LCA, depending on the system boundary selected for LCA study. These types of LCA are described in brief below with more details in Chap. 3:

Process LCA: Process LCA originates from the concept of process systems engineering where each industry can be represented as a "unit process" transforming raw materials into products while producing certain emissions/wastes. Thus, in process LCA, the system boundary consists of significant processes that contribute to the manufacturing of a product or support certain process. For example, a process LCA for "plastic bags" may include raw material extraction, raw material processing into monomers, polymer manufacturing, and plastic bag manufacturing as the key processes to be included in the LCA. However, one of the challenges in process LCA is deciding the system boundary, that is, how many processes to include, which causes "cut-off" or "truncation" error. Therefore, process LCA can be supplemented by Input-Output-based LCA.

Input-Output LCA: Input-Output-based LCA or IO-LCA is based on the macroeconomic framework of Input-Output that allows for quantifying the total direct and indirect impacts of economic activity from an economic sector. The economic IO (EIO) based model allows for including all the interconnections between different economic sectors, thus overcoming the challenge of selecting the processes to be included in the system boundary. The traditional EIO model is extended with data on environmental impacts of each sector to enable calculations of total environmental impact of a specific product or economic activity. This is known as Environmentally Extended Input-Output (EEIO) model and provides a method for fast economy-scale LCA of any product/process. However, availability of reliable IO models and appropriate disaggregation level is a challenge. Another disadvantage of using EEIO-based LCA is the aggregation error that is introduced in the analysis as fine scale differences of two production processes belonging to the same economic sector classification can be lost. The reason behind this challenge is that in most IO models, industries with similar products are classified under the same sectoral classification; thus, EIO models are unable to distinguish the impact of production processes that are very similar. To overcome these challenges posed by process LCA and IO LCA, a third approach of hybrid LCA is proposed, described next.

Hybrid LCA: In this approach, the benefits of process LCA and EIO-LCA are combined together to overcome the challenges posed by both approaches. The main processes of the desired LCA system are modeled using the fine details of process information, while any upstream or downstream product streams are modeled using the fast and coarse scale EIO models. This allows for fast calculation of results for upstream/downstream impacts that need to be evaluated but do not necessarily have information for fine scale process scale modeling. If any of the upstream impacts show significant effects on the overall result of LCA study, then the analysts can decide to model these upstream/downstream product streams in fine details. Hauke et al. provide an evaluation criterion to quantify and reduce the truncation error in process LCA.

5.2.2 Ecologically Based Life Cycle Assessment (Eco-LCA) for Carbon and Nitrogen Cycle

The LCA approaches discussed above mainly affect the scale of analysis based on system boundary selection and can also be used to include ecosystem services in any LCA study. This extension to include the role of ecosystem services in LCA studies is called Eco-LCA. As many LCA studies do not include ecosystems in the system boundary of study, the stress on ecosystem caused by overuse of certain ecosystem services get overlooked in decision-making. Hence, the Eco-LCA model provides a valuable method to ensure that the impact of industrial activities on ecosystems through reliance on ecosystem services is properly accounted for in LCA studies. Details of Eco-LCA model are available in literature (Zhang et al. 2010). In summary, Eco-LCA model extends the EIO model with data on ecosystem services as represented in Eq. 5.1.

$$R = \hat{X}^{-1}\left(I - G^{\mathrm{T}}\right)^{-1} V_{\mathrm{ph}}. \tag{5.1}$$

In Eq. 5.1, G matrix represents the supply-side economic model that allows for resource allocation from ith sector to jth sector for total input in ith sector. V_{ph} represents the vector of flows from ecosystems to economy or from economy to ecosystems. To account for the dependence or impact of industrial systems on biogeochemical cycles, the Eco-LCA inventory for carbon and nitrogen flows form the vector V_{ph}. Then, using Eq. 5.1, the dependence of different industries on the biogeochemical cycles is calculated by using the Eco-LCA inventory developed. The details of carbon and nitrogen Eco-LCA inventory are described below.

Carbon Inventory for Eco-LCA Figure 5.1 shows the interaction of technosphere that include the industrial systems with the carbon biogeochemical cycle being mediated by different components of ecosystems such as atmosphere, biosphere, and terrestrial and aquatic ecosystems. The main driver of interactions between the technosphere and the C biogeochemical cycle is the consumption of biomass being produced by terrestrial ecosystems by assimilation of carbon into different products and emissions being generated by industrial systems that gets back to the atmospheric, soil, and aquatic carbon pool. Please note that the depiction is not comprehensive and is meant to only be representative of the interactions. The

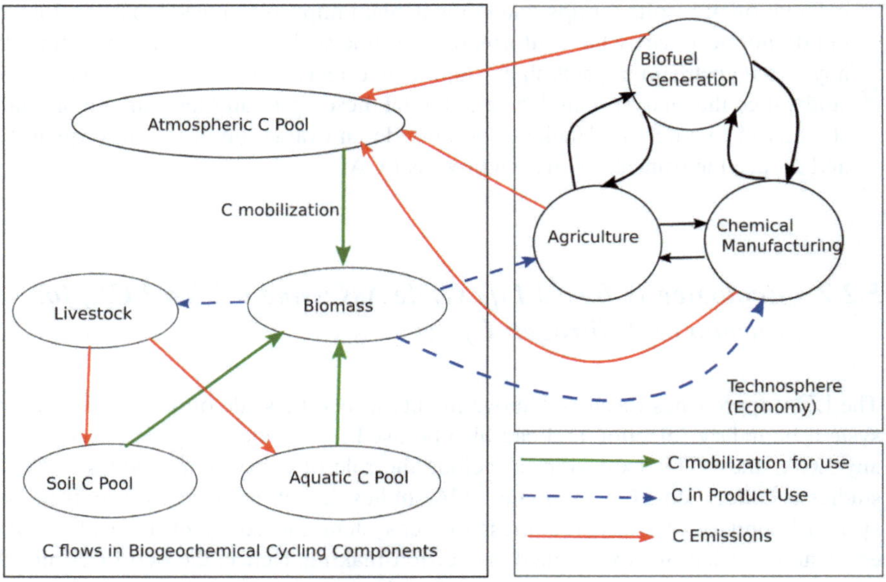

Fig. 5.1 Eco-LCA carbon flow model

consumption of fossil fuels (such as coal, oil, etc.) is also related to anthropogenic interaction with the C biogeochemical cycle that results in carbon emissions. In the Eco-LCA model, C flows that account for interaction of industrial activities with the carbon cycle are mapped to specific economic sectors based on NAICS code. The Eco-LCA C inventory flows are based on the EPA GHG inventory that quantifies the carbon flows – emissions and sinks. These flows form the vector V_{ph} in the Eco-LCA model. Hence, using Eq. 5.1, the total direct and indirect dependence of different economic sectors on various carbon flows is quantified. The list of carbon emissions and sinks included in Eco-LCA carbon inventory is presented in Table 5.1 along with mapping to the peripheral sectors. These peripheral sectors relate the dependence and impact of economy on the C biogeochemical cycle in the form of these flows. Further discussions on Eco-LCA C inventory and US economy dependence on this cycle are given in Singh and Bakshi (2010, 2013).

Nitrogen Inventory for Eco-LCA The Eco-LCA nitrogen inventory consists of N flows in three categories according to anthropogenic interactions with the nitrogen cycle by converting nitrogen into various forms. The three main categories of flows in Eco-LCA N inventory are as follows:

(i) *Reactive N Mobilization*: This Eco-LCA N inventory component consists of all the N flows that convert inert N from atmosphere to a form that can be directly utilized for product formation. In the Eco-LCA N inventory, this is termed as “Green N.” N exists in the atmosphere as inert N_2 gas, which is generally not usable for any product formation except being used as inert gas

Table 5.1 Eco-LCA carbon inventory components

	Eco-LCA inventory component	Peripheral sector
Carbon sequestration	C sink by farmlands	Farming sectors
	C sink by ranchlands	Cattle ranching and farming
	C sink by forestland	Forest nurseries, forest products, and timber tracts
	Soil C sink by cropland remaining cropland	Farming sectors
	Soil C sink by land converted to grassland	Farming and cattle ranching and farming
	C sink by urban trees	Housing maintenance and construction
	C sink in landfills as yard trimming and food scrap stocks	Waste management and remediation services
	CO_2 sink by grasslands remaining grasslands	Cattle ranching and farming
	Ocean sink	–
Carbon emissions	CO_2 emissions by fuel use	Mapping to respective sectors
	CO_2 emissions by liming	All farming sectors
	CO_2 emissions by urea fertilization	Farming sectors, forestry sector
	CO_2 emissions by land converted to cropland	Farming sectors

in industries. The process of converting "inert" form of N to usable format is called mobilization. This N mobilization process also occurs naturally in the ecosystem by microbial activities; however, this natural process is very slow. The flows in this category are: N input to soil by leguminous plants, N fixed by micro-organisms in soil, N from atmospheric deposition, and N fixation by Ammonia production by Haber-Bosch process.

(ii) *Reactive N in Product Use*: This Eco-LCA N inventory consists of flows primarily associated with converting mobilized N into useful products. The main form of N-based products that are included in the Eco-LCA N inventory are N in fertilizer, Manure, Plastics and Synthetics, Explosives, Animal feed, Harvested crops, and Meat for human consumption. To build the inventory, data for N being consumed as each of these product categories are collected and mapped to the respective sectors that consume these products. For example, N in fertilizer is allocated to each farming sector based on the land area and rate of fertilizer application. This data forms the V_{ph} vector for N in products, then usi ng the IO model total dependence of the economic activities on N in product form can be calculated. Similarly, other forms of N based products are also calculated.

(iii) *Reactive N Losses*: This category of inventory components includes N losses as emissions to water, soil, and air in different forms. The main loss flows include N emissions to air as N_2O and NO_X emissions from fuel burning, land use land use change (LULUC), manure management, field burning of agricultural residues, forest fires, nitric acid production, adipic acid production, composting, and waste water treatment. N losses to water include N runoff or leaching mechanisms that consist of N as ammonia, nitrates, nitrites, and organic N due to waste disposal. Sewage sludge is included as N waste to land. Once these reactive N loss flows are quantified for the economy, these are allocated to specific sectors that lead to these losses. For example, N losses due to runoff as ammonia, nitrates, or nitrites are allocated to the farming sectors. The quantification method is given in Singh and Bakshi (2014).

5.2.3 *Carbon-Nitrogen Nexus for Transportation Fuels Using Eco-LCA Carbon and Nitrogen Inventory*

The Eco-LCA carbon and nitrogen inventory was used to evaluate the life cycle impact on carbon and nitrogen flows by using transportation fuels derived from multiple types of feedstock. The types of feedstock compared were fossil-based, bio-based, and waste feedstock. Each of these feedstock affects different carbon and nitrogen flows in the Eco-LCA C & N inventory; thus, the total life cycle impact or dependence on the biogeochemical cycles is different for each of these fuels. The study was performed using tiered hybrid LCA approach. In this study, for each transportation fuel, the process scale data is used for the most important process stage in the fuel life cycle, while all other upstream inputs are modeled using the Eco-LCA carbon and nitrogen inventory to provide a more comprehensive analysis on the impact on carbon and nitrogen footprints. Details of the study are provided in Liu et al. (2018).

Figure 5.2 shows the C-N trade-offs based on LCA using Eco-LCA C & N inventory for various types of transportation fuels. The x axis shows total Nr emissions, while y-axis shows the net C emissions by accounting for C sinks as well using the Eco-LCA C inventory. Fossil-based fuels have highest net C emissions, whereas second-generation fuels show high Nr emissions but lower net C emissions. Fuels in the left lower quadrant are most desired to be used as these show both lower net C emissions and lower Nr emissions.

5.3 Modeling the Anthropogenic Material/Biogeochemical Cycles: A Physical-Input-Output Table (PIOT) Based Approach

While natural biogeochemical cycles have existed for thousands of years, a more recent phenomenon is the anthropogenic material cycles due to flow of materials

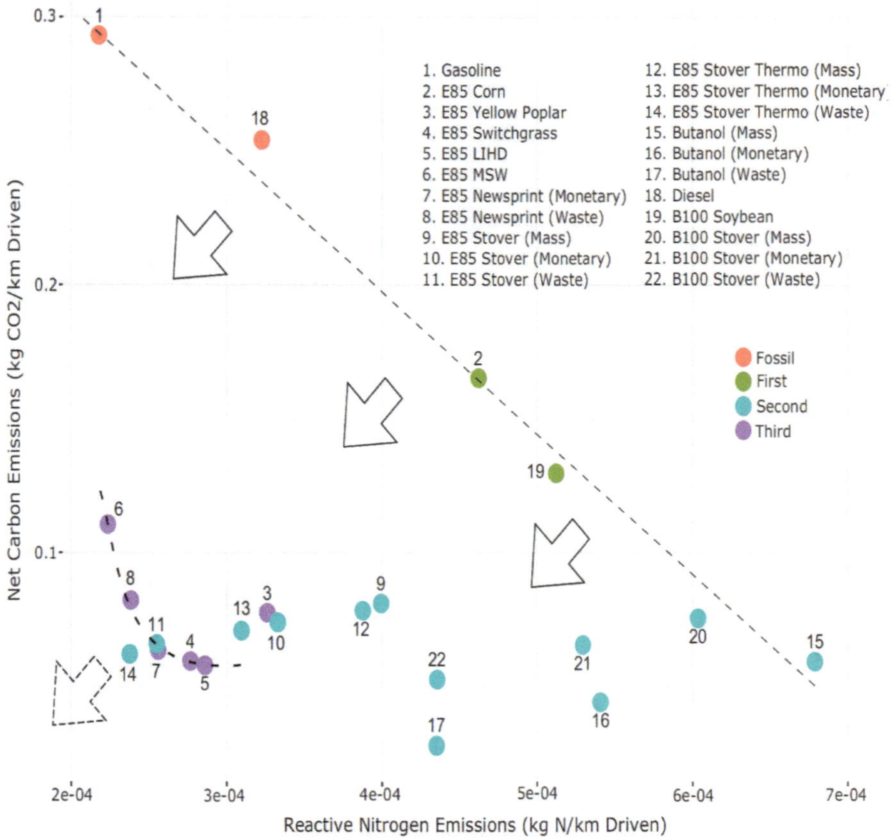

Fig. 5.2 C-N Nexus of transportation fuels. First: "Biofuels from Corn and Soybean"; Second: "Biofuels from Corn Stover"; Third: "Biofuels from low input farming and waste". (Reprinted with Permissions from Journal)

in different products and wastes. Through the Eco-LCA approach as demonstrated in the previous section, one can account for the interaction of industrial activities or human consumption with different biogeochemical flows. However, this does not allow us to understand the flow of different species of these materials within anthropogenic systems such as industrial ecosystems. Hence, it is crucial to also develop models for anthropogenic material cycles such as carbon and nitrogen cycles. This will allow us to fully understand the coupling of biogeochemical cycles with industrial systems, which is necessary for sustainability of biogeochemical cycles and associated ecosystem services.

In order to develop models for anthropogenic material cycles, the physical analogue of macroeconomic Input-Output (IO) model is extremely useful. In contrast to the macroeconomic Input-Output (IO) models that capture the interdependency of

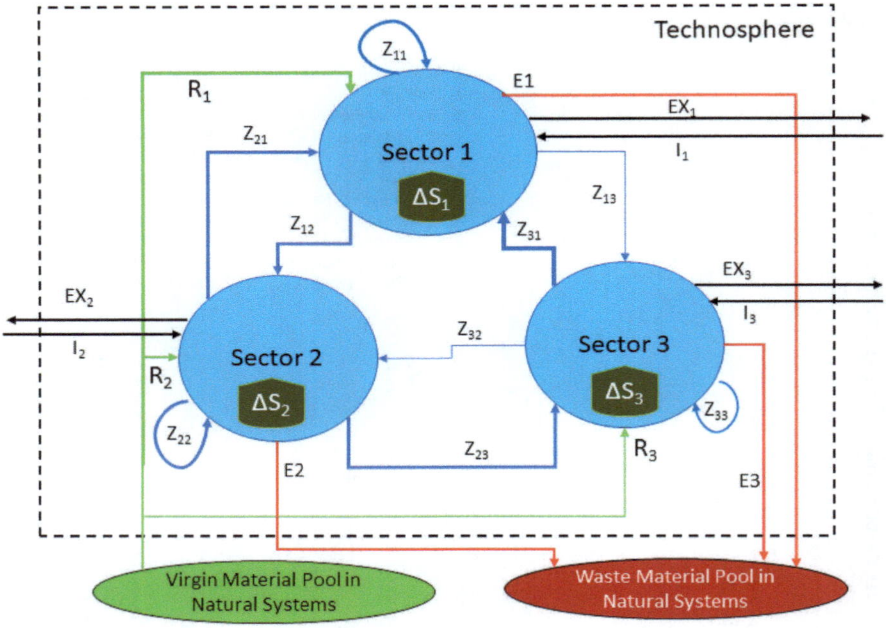

Fig. 5.3 Material flow cycles within technosphere and interactions with natural systems

economic sectors (or industries) in monetary flows, the Physical Input Output Tables (PIOTs) provide the interdependency in terms of material flows. Thus, the PIOTs provide insights into the physical economy and structure of material flows between different industrial sectors. Similar to the biogeochemical cycles, this provides detailed information about different forms of material species and the flow from one key component to another. Figure 5.3 shows a proxy anthropogenic material cycle and depiction of this cycle in a tabular format as PIOT is shown in Table 5.1. The tabular PIOT framework is inspired by the original structure from Hoekstra and van den Bergh (2006). Similar to the models for biogeochemical cycles, the anthropogenic material flow cycle shown in Fig. 5.3 has different subsystems interacting with each other through flow of materials in different forms (species), which can be a compound of various materials and contains embedded quantities of a particular material of interest such as carbon and nitrogen. The subsystems in the case of anthropogenic material cycles are "sectors" or "industries" that are actively utilizing the materials for converting into other useful format. The system boundary of "Technosphere" in Fig. 5.3 represents the geographical region where the material flow cycles are being studied. The scale of these technospheres can be at state scale, national scale, or global scale as per the modeling requirement and data availability. The flows between all the sectors – $Z_{ij}{}^{s}$ – are the flows transferring materials between different subsystems. Interaction of the regional technospheres with other regions is captured by Imports (I_j^s) and Exports (EX_j^s). In case the anthropogenic material cycle is modeled at global scale, it will be a closed system for material flow similar to the biogeochemical cycles. Finally, the interaction of technospheres with the natural systems is modeled as extraction of virgin materials from pool in natural systems and deposition of wastes in the waste material pool. These pools are the common systems where the natural and anthropogenic material cycles interact. Engineered systems are mainly focused on design and operations of material cycles within the technosphere, however overall interaction with natural systems provides the quantification for dependence and impact on biogeochemical cycles in physical units. Currently, the knowledge of anthropogenic material cycles is very scarce as development of these cycles requires a large amount of systematic data collection. The PIOT model shown in Table 5.1 provides a standardized framework for data collection and comparison of these anthropogenic material cycles. Since the PIOT framework is derived from EIO model, it also helps in utilizing the standard IO based method to quantify impact of different economic demands driving production in sectors on anthropogenic material cycles and flows from/to natural systems.

As PIOTs track material flows in the economy, a mass balance is applied for each sector to ensure that all material flows are accounted in modeling the anthropogenic material cycles.

Utilizing this framework, different models for material cycles in the economy can be developed. In order to study detailed dependence on the biogeochemical cycles of C & N, an anthropogenic material flow cycle of C & N needs to be developed. The PIOT framework that tracks the material flow in between different industrial sectors can be used to standardize developing these anthropogenic material flow cycles. The scale of dependence can be regional, national, or global. However, developing this model at global scale will require significant amount of data. An example of using PIOT framework for modeling Nitrogen cycle in Illinois, USA, for the year of 2002 is discussed later based on a published study (Singh et al. 2017).

While the conceptual framework discussed above provides a standardization method for modeling anthropogenic material cycles such as for carbon and nitrogen, developing these tables and network diagrams for material flows is tedious. Hence, the next section of this chapter discusses a bottom-up-approach to develop these PIOT models to map the anthropogenic material cycles for interactions with the natural biogeochemical cycles.

Bottom-Up Approach for Developing PIOTs to Map Anthropogenic Material Cycles A "bottom-up" approach can help develop these models with fine granularity (See Fig. 5.4). This can be compared to the similar approach in the study of biogeochemical cycles, where understanding mechanisms of interaction between different subsystems for driving the flows is necessary. In this bottom-up approach, first identification of major flows associated with chosen materials is done. This can be based on empirical observation. Once the major flows and associated commodities have been identified, a Material Flow Analysis (MFA) diagram is developed with mapping each flow to the corresponding economic sectors in the economy. The system boundary for the MFA of these commodities is "cradle to gate" and traces the transformation of raw materials through intermediate steps

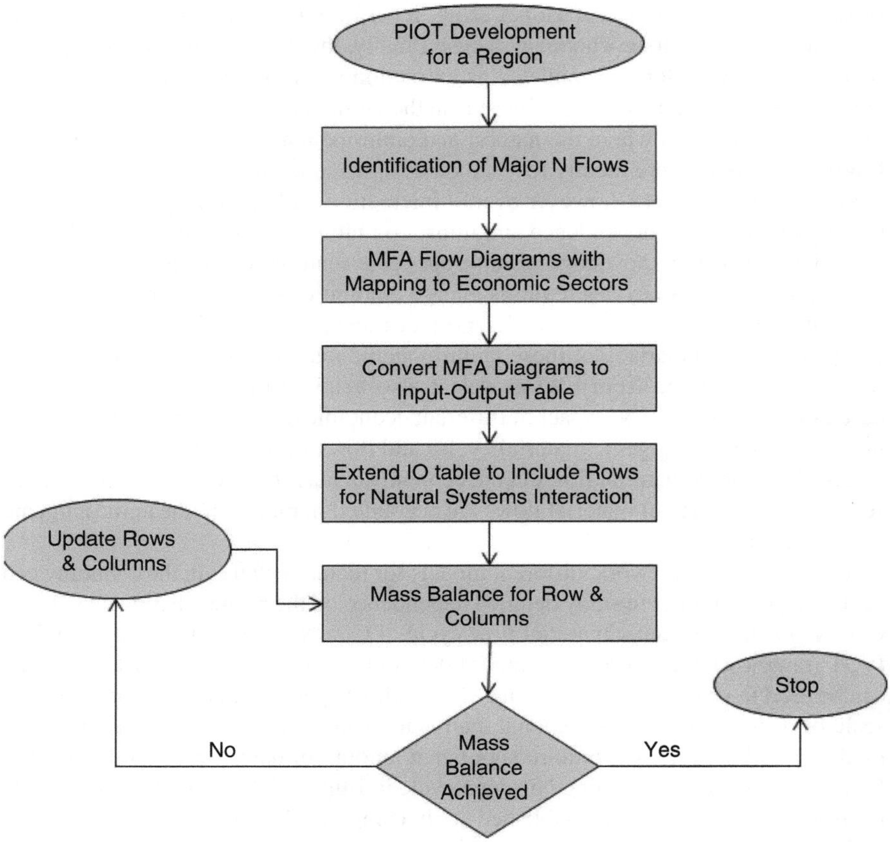

Fig. 5.4 Bottom-up approach to develop model for anthropogenic material cycle using input-output structure. (With permission from Ecological Modelling, Singh et al. 2017. Reprinted with Permissions from Journal)

into final products. The MFA diagrams are finally supplemented with available data on these flows for the selected region to quantify the anthropogenic material flows. Once these data are collected, the MFA diagrams are converted to the Physical Input-Output Table (PIOT) using the structure in Table 5.2. To enable the anthropogenic material flow cycle's interaction with natural biogeochemical cycles of these materials, the PIOT is enhanced by including rows that capture the natural system interactions. These rows are: "raw material from nature" and "emissions" to nature. Similar to the modeling in natural system biogeochemical cycles, mass balances of each material need to be ensured in anthropogenic systems as well.So, at the last, mass balance for each sector is checked by using, *Total Mass Input* = *Total Mass Output* for each sector. Mass balance can be calculated at "commodity" scale or "elemental" scale. Both of these scales are advised for ensuring that the material

Table 5.2 PIOT framework for tracking material flows in industrial systems

		To processing sectors						Total outputs from sector
		1	2	3	Consu. (C)	Export (E)	Import (I)	
From processing sectors	1	Z_{11}	Z_{12}	Z_{13}	C_1	EX_1	I_1	TO_1
	2	Z_{21}	Z_{22}	Z_{23}	C_2	EX_2	I_2	TO_2
	3	Z_{31}	Z_{32}	Z_{33}	C_3	EX_3	I_3	TO_3
Raw materials	Input from Nature	R_1	R_2	R_3				
Residuals production		RP_1	RP_2	RP_3				
Residuals use		RU_1	RU_2	RU_3				
Stock changes		ΔS_1	ΔS_2	ΔS_3				
Emissions		E_1	E_2	E_3				
Total inputs to sector		TI_1	TI_2	TI_3				

Table 5.3 Key drivers of anthropogenic nitrogen flows – Major crop areas in Illinois (2002) USDA NASS

Major crop	Area (Acres)
Corn for grain	10,742,787
Corn for silage	109,847
Wheat for grain	581,084
Soybean for beans	10,505,989
Alfalfa (hay)	416,997
Total of above	22,356,704
Total cropland in Illinois: 24,171,260 Acres	
Harvested cropland in Illinois: 22,562,904 Acres	
Major crops (corn, soybean, wheat, & alfalfa) form 99% of harvested cropland.	

cycles developed for anthropogenic systems do not miss any significant flows. In case, there is an imbalance in the mass flows, either there is some missing data or there are errors in the flow values calculated for each stream in the MFAs put together for PIOT. A simple approach to handle this is to introduce a "slack variable" by adding new rows and columns to ensure sector level mass balances.

Case Study: Developing the Anthropogenic Nitrogen Cycle in Industrial Sectors for Illinois, USA

The strength of PIOT framework to develop anthropogenic material cycles for industrial sectors is shown here via a case study for agro-based economic sectors in the state of Illinois, USA. In this case study, an N cycle is developed using the PIOT framework. Following the approach described above, first important commodities in the agro-based economy that drive the anthropogenic N cycle in Illinois economy were selected. In Illinois, for year of 2002, four major crops, namely, corn (grain, silage), wheat, soybean, and alfalfa, accounted for 99% of harvested crop land. These crops form the backbone of energy and food-based industries in Illinois and provide the connection between ecosystem and economy for nitrogen flows. Table 5.3 shows the total activity in Illinois for growing these four crops in terms of acres that are dedicated to these four crops. These crops drive the N cycle in Illinois by use of fertilizer (major use of active N in economy), fixing nitrogen from air (adding more N to economy), and embodied N in the products that are harvested to make products for industrial and consumer use.

Following the methodology in Fig. 5.4, in next step, MFA diagrams were developed to trace the material flows for these crops. An MFA for Soybean is shown in Fig. 5.5. Three different types of flows can be distinguished from the MFA that will help map the anthropogenic N cycle driven by soybean: (1) intersectoral flows such as flow of one commodity to another sector for transformation into consumable products (soybean flowing from farms to oilseed processing industry), (2) flows from the nature driven by soybean farming (nitrogen fixation in Fertilizer manufacturing sector or directly by the *Nitrosomonas* bacteria in Soybean crops), and (3) flows to the nature (emissions, not shown in MFA). All of these flows are eventually

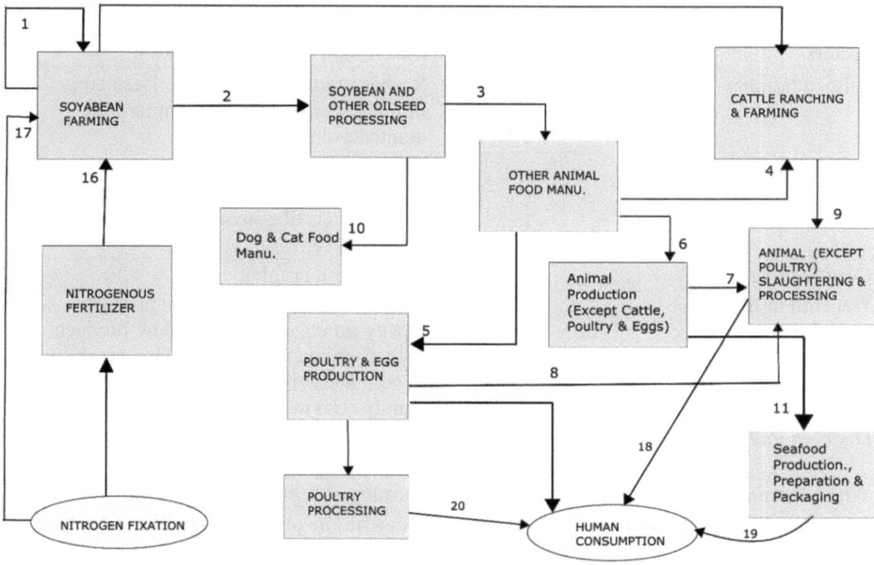

Fig. 5.5 Material flow analysis for soybean in agro-based industries. (Reprinted with Permissions from Journal)

driven by the demand for consumption of different products produced from soybean. This MFA provides the structural information for mapping anthropogenic N cycle. Similarly, MFAs for all other commodities can be drawn. For the major agro-based commodities of Illinois, the MFAs are available in Singh et al. (2017).

After developing MFAs, a mapping to industrial NAICS code was done to combine all the flows associated with different commodities into a PIOT format. The sectors that were important for the anthropogenic N cycle are shown in Table 5.4. These sectors are analogous to the processing compartments for nitrogen flows and some of these sectors involve agroecosystems, that is, natural systems, for example, "corn farming." Hence, developing anthropogenic N cycle also provides an implicit connection to the ecosystems and natural biogeochemical cycles.

Based on all the MFAs and total mass tracking, an overall mass balance of anthropogenic N cycle of Illinois was performed as shown in Fig. 5.6.

An overall mass balance error of 23% was found, which is in line with many biogeochemical cycle error estimation on mass balance flows. This provides a good cross-check for improving confidence in the overall material flows that have been captured for the region. While the overall mass balance provides a good overview, in order to clearly understand the role of different anthropogenic activities (such as industrial production) on flows related to the cycle, a finer scale flow map is required. Hence, MFAs associated with different commodities and industrial production were converted into PIOTs. PIOTs are a matrix as shown earlier (Table 5.2) where each element of the matrix captures the flow from one industry to another

Table 5.4 Sectors driving the anthropogenic nitrogen cycle in Illinois, USA

Sectors	NAICS	Description of sector activities
Oilseed farming	11111	Soybean farming and other oilseed crop farming. For Illinois, soybean farming dominates in this sector
Soybean and other oil seed processing		Industrial activity involved in processing soybean and other oilseed for conversion into products like soymeal, soyoil, animal feed, etc.
Corn farming	11115	Mostly corn farming
Wet corn milling	311221	Industrial establishments that produce mostly starch, syrup, oil, and by-products such as gluten feed and meal by wet milling of corn and sorghum. In Illinois, it was mainly corn wet milling
Dry corn milling	325193	Dry corn milling is mainly used to produce ethanol
Wheat farming	11114	Farming activities growing wheat
Flour milling & malt manu.	311211	Industries involved in processing wheat for conversion to other products or sale to food manufacturing industry
Other animal food manu.	311119	Industries involved in food manufacturing for cattle, hogs, etc.
Dog & cat food manu.	311111	Industries involved in food manufacturing for pets
Cattle ranching & farming	1121	Livestock farming industry
Animal production except cattle & poultry eggs	1122	Hog, pig, sheep, goat farming industry
Poultry & egg production	1123	Poultry farming industry
Poultry processing	311615	Industry engaged in poultry slaughtering and preparing processed poultry and small game meat/meat by-products
Animal (except poultry) slaughtering, & processing	311611	Industry engaged in slaughtering and preparing processed meat from hog, pig, cow, etc.
Nitrogenous fertilizer manu.	325311	Fertilizer manufacturing industry
Bread, bakery, and product manu.	31181	Food manufacturing industry of bread, etc.
Cookie, cracker, and pasta manu.	31182	Food manu.
Snack food manu.	311919	Snack food
Tortilla manu.	311830	Tortilla manufacturing from wheat flour, corn flour, etc.
Breakfast cereal manu.	311230	Cereal manufacturing industry
Frozen food manu.	311411	Industry involved in freezing food such as sweet corn, meat, etc.
Vegetable and fruit canning & drying	311421	Industries involved in preparing canned and dried food for distribution

Adapted from Singh et al. (2017)

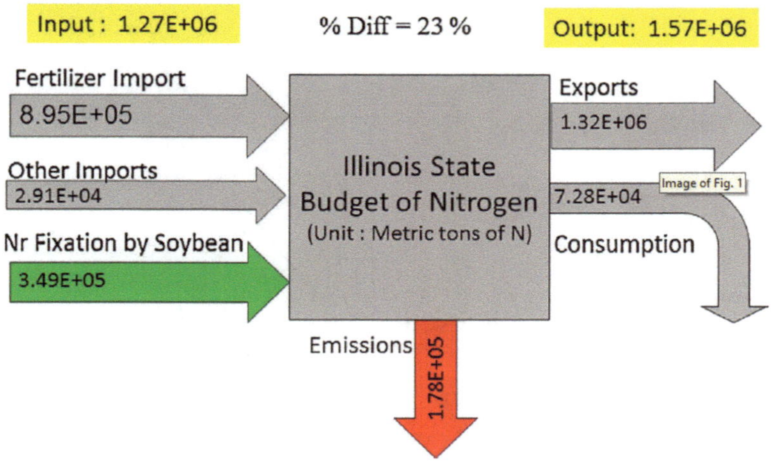

Fig. 5.6 N Balance for anthropogenic N cycle in Illinois. (Reprint with permission from Ecological Indicators. Reprinted with Permissions from Journal)

in physical units. The original PIOT for Nitrogen flows in Illinois is given in Singh et al. 2017. The network representation of these flows is drawn by considering each industry as a "node" and each flow (values in matrix cell) as "edges" of the network. Since the values of the flow depict the amount of nitrogen flowing from one industry to another, the weight of the edges depicts the amount of nitrogen being transferred from one industry to another. For Illinois, the nitrogen flow among the selected industries in Table 5.4 is shown in Fig. 5.7. As shown in Fig. 5.7, PIOT-based network diagram for material flows provides a clear insight into exact industrial activity that is driving the changes in biogeochemical cycles through the anthropogenic cycles. The key process of converting "inert" nitrogen to active nitrogen occurs in two sectors – fertilizer manufacturing (Haber Bosch process) and soybean farming (natural N fixation). However, several downstream sectors including farming, food manufacturing, animal farming, and food processing depend on these flows, thus clearly indicating the dependence of human activities on the biogeochemical cycles or availability of nitrogen through the cycles for converting into active products. Further, several activities are also contributing to the "reactive N" pool of the ecosystems (aquatic ecosystem or air), which impacts the biogeochemical cycles itself. For example, corn farming, oilseed farming, and wheat farming can be seen to contribute to the reactive N pool. The value here is based on the EPIC model (See Singh et al. 2017) and mainly tracks the run-off of nutrient to the land and water systems. In order to track the air emissions due to energy usage, additional data would be required, which was not captured in the original study. These insights at regional scale are particularly useful to understand the dependence and impact of a region on nutrient biogeochemical cycles, thus addressing the accounting of role of supporting ecosystem services such as biogeochemical cycles.

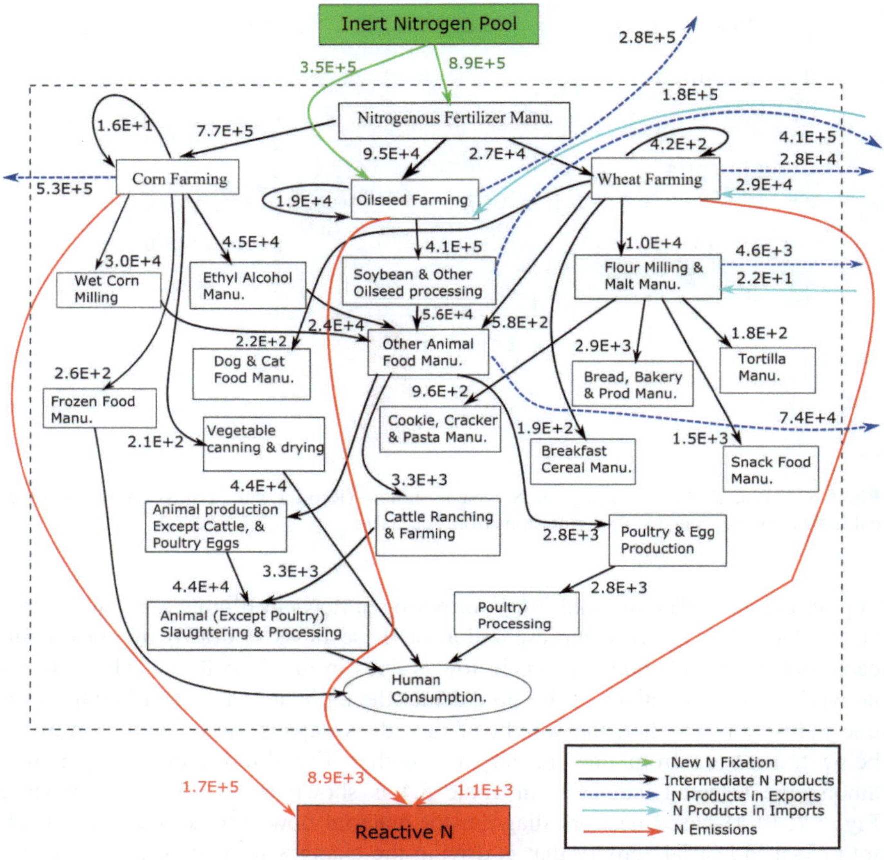

Fig. 5.7 Structure of N cycle in Illinois economy based on PIOT (N Flows linked to arrows in Mt-N/yr). (Reprinted with Permissions from Journal)

Life Cycle Nitrogen Use Efficiency and Total Nitrogen Throughflow in Economic Production

As PIOTs help track the use of nitrogen in different forms in the region by various industries, similar to the ecological systems, each subsystem has its own utilization efficiency of nitrogen based on the total direct and indirect dependence. Using PIOT model, this efficiency is defined in Eq. 5.2:

$$\text{NUE} = \frac{\text{Total useful nitrogen product}}{\text{Total nitrogen throughflow} \in \text{Sector}} \times 100 \tag{5.2}$$

The numerator represents the total useful product from each sector in the economy and denominator represents the total flows (direct and indirect inputs) to support the production. This metric represents how much nitrogen is flowing

Table 5.5 Nitrogen use efficiency for different industrial sectors in Illinois in 2012

Sectors	Total nitrogen throughflow in economy per unit N demand for this sector	NUE (%)
11111 – Oilseed farming	1.22	81.74
311222 – Soyabean and other oil seed processing	2.09	47.84
11115 – Corn farming	2.28	43.86
311221 – Wet corn milling	3.85	25.94
325193 – Ethyl alcohol manu. (dry corn milling)	3.39	29.54
11114 – Wheat farming	2.06	48.56
311211 – Flour milling and malt manu.	3.06	32.69
311119 – Other animal food manu	3.88	25.74
311111 – Dog and cat food manu	3.06	32.69
1121 – Cattle ranching and farming	4.88	20.47
1122 – Animal production except cattle and poultry eggs	4.88	20.47
1123 – Poultry & egg production	4.88	20.47
311615 – Poultry processing	5.88	16.99
311611 – Animal (except poultry) slaughtering and processing	5.88	17.00
325311 – Nitrogenous fertilizer manu.	1.00	100.00
31181 – Bread, bakery, and product manu	4.06	24.63
31182 – Cookie, cracker, and pasta manu.	4.06	24.63
311919 – Snack food manu.	4.06	24.63
311830 – Tortilla manu.	4.06	24.63
311230 – Breakfast cereal manu.	3.06	32.69
311411 – Frozen food manu.	3.28	30.49
311421 – Vegetable and fruit canning and drying	3.28	30.49

through the economy to produce useful nitrogen product from each sector. A sector with low NUE is very nitrogen-intensive; that is, it creates larger impact on nitrogen biogeochemical cycle or has higher dependence on biogeochemical cycle's functioning. Table 5.5 shows the NUE for each of the sectors processing N in Illinois that was calculated using Eq. 5.2.

Further, from the PIOT, the total throughflow of nitrogen in each sector is calculated using Eq. 5.3. Final demand variable in Eq. 5.3 gives the value for "total nitrogen useful product" and the variable (X_{ph}) gives the total nitrogen throughflow. The matrix A_{ph} is calculated from PIOT developed for N flow as $A_{ij} = \frac{Z_{ij}}{\text{TO}_j}$. In this, Z_{ij} is the flow of nitrogen from ith sector to jth sector and TO_j is the total output of N from sector j. Hence, A_{ij} calculates the "direct requirements coefficient"

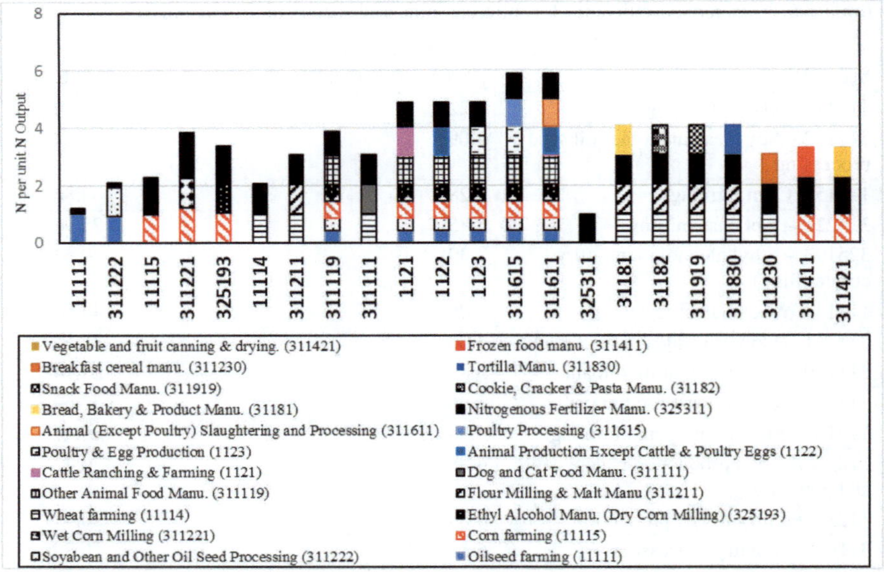

Fig. 5.8 Life cycle N throughput per unit N demand. (Reprinted with Permissions from Journal)

that captures the flow from *i*th sector to *j*th per unit output from sector *j*. Based on the PIOT developed for Illinois, nitrogen throughflow for "1 unit of useful nitrogen product" is shown in Fig. 5.8.

$$X_{\text{ph}} = \left(I - A_{\text{ph}}\right)^{-1} F \tag{5.3}$$

5.4 Summary, Challenges, and Future Work

There has been significant advancement on quantifying the reliance of industrial activities on ecosystem services through the development of Eco-LCA model. There has also been significant improvement in LCA modeling tools itself such as providing the IO models at much finer time and spatial scale. However, the challenges in properly accounting for ecosystems services remain. Specifically, defining the correct metrics to measure supporting ecosystem services is still an unsolved challenge. Some of the supporting services like climate regulation services are hardly being quantified in any LCA study, however becoming much more critical to be formulated as climate mitigation becomes a critical goal for humanity to maintain its existence. Another crucial challenge to be solved is developing spatial databases for ecosystem services as the ecological systems show significant spatial variations; however, they are hardly quantified. A systematic method to develop and maintain these spatial datasets for ecosystem services will be important to expedite the accounting of role of these ecosystem services in supporting various industrial/economic activities.

For the biogeochemical cycles, developing anthropogenic material cycles for different regions is rarely done and there needs to be a coherent effort in understanding the human role in altering the large-scale biogeochemical cycles through these integrated models to be able to correctly quantify industrial reliance on the ecosystem services provided by biogeochemical cycles. To be able to develop these anthropogenic cycles, large-scale efforts on development of PIOTs for different material cycles need to be done. However, this is gigantic task. While some recent developments in automated generation of PIOTs using bottom-up mechanistic engineering models (Vunnava and Singh 2021b; Wachs and Singh 2018) provide an opportunity to create these anthropogenic material cycles, it is still a gigantic task and need collaborative efforts. Aligned with this collaborative possibility, a prototype cloud-based collaborative tool has also been recently developed (Vunnava et al. 2021a, b); however, it needs scaling up to accommodate for large-scale database and computation requirement.

Nonetheless, there has been significant advancement in methods and tools that can finally allow us to appropriately model the industrial system's impact and dependence on biogeochemical cycles, thus improving the quantification of ecosystem services provided to industrial systems by these cycles. The field has certainly made progress in past decade, with potential to expedite the much-needed science to help meet the goal of functioning within the planetary limits. Much work remains to be done in understanding the role of engineering in managing biogeochemical cycles and related ecosystem services along with the role of these natural systems in sustaining our humanity.

Acknowledgments Author would like to acknowledge funding support from NRC postdoctoral fellowship to the author, NSF CBET 1805741, for support on advancing PIOT method and NSF CBET support for the research on Eco-LCA, all of which has formed the foundation for this chapter.

References

Bashkin, Vladimir N., ed. "Biogeochemical Cycling of Macroelements." In *Modern Biogeochemistry*, 73–160. Dordrecht: Springer Netherlands, 2002. https://doi.org/10.1007/0-306-48103-0_3.

Hoekstra, Rutger & van den Bergh, Jeroen C.J.M., 2006. "*Constructing physical input-output tables for environmental modeling and accounting: Framework and illustrations,*" Ecological Economics, Elsevier, vol. 59(3), pages 375-393, September.

Liu, Xinyu, Shweta Singh, Erin L. Gibbemeyer, Brooke E. Tam, Robert A. Urban, and Bhavik R. Bakshi. "The Carbon-Nitrogen Nexus of Transportation Fuels." *Journal of Cleaner Production* 180 (April 10, 2018): 790–803. https://doi.org/10.1016/j.jclepro.2018.01.090.

Singh, Shweta, and Bhavik R. Bakshi. "Accounting for Emissions and Sinks from the Biogeochemical Cycle of Carbon in the U.S. Economic Input-Output Model." *Journal of Industrial Ecology* 18, no. 6 (2014): 818–28. https://doi.org/10.1111/jiec.12156.

Singh, Shweta, and Bhavik R. Bakshi. "Accounting for the Biogeochemical Cycle of Nitrogen in Input-Output Life Cycle Assessment." *Environmental Science & Technology* 47, no. 16 (August 20, 2013): 9388–96. https://doi.org/10.1021/es4009757.

Singh, Shweta, and Bhavik R. Bakshi. "Enhancing the Reliability of C and N Accounting in Economic Activities: Integration of Bio-Geochemical Cycle with Eco-LCA." In *Proceedings of the 2010 IEEE International Symposium on Sustainable Systems and Technology*, 1–6, 2010. https://doi.org/10.1109/ISSST.2010.5507712.

Singh, Shweta, Jana E. Compton, Troy R. Hawkins, Daniel J. Sobota, and Ellen J. Cooter. "A Nitrogen Physical Input-Output Table (PIOT) Model for Illinois." *Ecological Modelling* 360 (September 24, 2017): 194–203. https://doi.org/10.1016/j.ecolmodel.2017.06.015.

Vunnava, Venkata Sai Gargeya, Jaewoo Shin, Lan Zhao, and Shweta Singh. "PIOT-Hub - A Collaborative Cloud Tool for Generation of Physical Input–Output Tables Using Mechanistic Engineering Models." *Journal of Industrial Ecology* n/a, no. n/a. Accessed November 3, 2021a. https://doi.org/10.1111/jiec.13204.

Vunnava, Venkata Sai Gargeya, and Shweta Singh. "Integrated Mechanistic Engineering Models and Macroeconomic Input–Output Approach to Model Physical Economy for Evaluating the Impact of Transition to a Circular Economy." *Energy & Environmental Science* 14, no. 9 (2021b): 5017–34. https://doi.org/10.1039/D1EE00544H.

Wachs, Liz, and Shweta Singh. "A Modular Bottom-up Approach for Constructing Physical Input–Output Tables (PIOTs) Based on Process Engineering Models." *Journal of Economic Structures* 7, no. 1 (October 10, 2018): 26. https://doi.org/10.1186/s40008-018-0123-1.

Zhang Yi, Baral Anil and Bakshi, B.R. "Accounting for Ecosystem Services in Life Cycle Assessment, Part II: Toward an Ecologically Based LCA." *Environmental Science & Technology, 44, 7, 2624-2631*, 2010.

Chapter 6
The Significance of Insect Pollinators: Opportunities and Challenges

Alex Jordan, Mason Unger, and Vikas Khanna

6.1 Introduction

Pollinators serve a crucial role in ecosystems, human nutrition, and the economy. While some pollination is performed by wind, water, or auto-pollination, pollinators are essential for most pollination-dependent plants to move pollen from male to female structures, resulting in fertilization and production of seed and or fruit. Over one-third of the world's food supply and about 75% of angiosperms (flowering plants) depend upon animal-mediated pollination service. While there are many birds, bats, and other larger mammals responsible for pollination, most pollination service is performed by both wild and managed populations of insects including ants, wasps, thrips, flies, honey bees, bumble bees, solitary bees, butterflies, and moths [1, 2]. Pollination-dependent crops include many of the most nutrient-dense foods such as nuts, seeds, oils, fruits, and vegetables (Table 6.1) [1, 3]. Pollination service provided by pollinating organisms has value to human nutrition, and through aiding the production of commodities with material benefits, the service has economic value to the agricultural industry responsible for the growth, harvest, and distribution of these nutritive crops.

In addition to the value of pollination services to the agricultural sector, there are many complex economic linkages in pollination-dependent crops that affect nonagricultural sectors, products, and processes [6]. These are sectors both upstream and downstream of the agricultural sectors, which do not depend upon pollinators directly, but indirectly, they have inputs to or from agriculture that is dependent upon pollination service. Existing research has quantified the extent of this dependence of nonagricultural sectors on pollinators [6]. As an example, Fig. 6.1

A. Jordan · M. Unger · V. Khanna (✉)
Department of Civil and Environmental Engineering, University of Pittsburgh, Pittsburgh, PA, USA
e-mail: khannav@pitt.edu

B. R. Bakshi (ed.), *Engineering and Ecosystems*,
https://doi.org/10.1007/978-3-031-35692-6_6

Table 6.1 Crops dependent upon insect-mediated pollination service [3–5]

Almonds	Citrus (other)	Lettuce	Persimmons	Tangerines
Apples	Coffee	Limes	Plums	Tomatoes
Apricots	Cranberries	Macadamias	Plums & Prunes	Turnips
Avocados	Cucumbers	Mangoes	Pomegranates	Watermelon
Beans	Currants	Melons	Prunes	
Beets	Eggplant	Nectarines	Pumpkins	*Alfalfa*
Berries, other	Figs	Okra	Rapeseed	*Broccoli*
Blueberries	Flaxseed	Olives	Raspberries	*Carrots*
Boysenberries	Grape	Oranges	Safflower	*Cauliflower*
Brussels Sprouts	Grapefruit	Papayas	Sesame	*Celery*
Buckwheat	Guar	Passion fruit	Soybeans	*Clover*
Cabbage	Guavas	Peaches	Squash	*Cotton*
Canola	Kiwifruit	Peanuts	Strawberries	*Onions*
Cherries	Kumquats	Pears	Sunflower	*Sugarbeets*
Chestnuts	Lemons	Peas	Sweet potatoes	
Chicory	Lentils	Peppers	Tangelos	

Lower-right italicized list of crops is indirectly dependent upon pollination service; all others are directly dependent upon pollination service

shows the dependence of top 15 industrial sectors of the US economy on animal pollinators. These economic dependences arise because of both the direct linkages of nonagricultural sectors with agricultural sectors (e.g., *fertilizer manufacturing*, *cereal manufacturing*) and also the indirect linkages of nonagricultural sectors with agricultural sectors (e.g., *oil and gas extraction* providing inputs to *fertilizer manufacturing*, which in turn serves as input to agricultural sectors).

It is because of these upstream and downstream linkages that engineers should be aware of pollinators and the crucial environmental services they provide. Economic sectors that rely on pollination services directly or indirectly include but are not limited to supporting activities for agriculture, agricultural chemical and fertilizer manufacturing, battery manufacturing, and oil and gas extraction. As such, these economic sectors are vulnerable to loss or deterioration of pollination services.

Human activities, especially engineering activities, can negatively and positively impact pollinator health, habitat, and resources. Throughout design processes, engineers often consider trade-offs where the impact on common ecosystem services like clean water and air must be mitigated [7–9]; however, ecosystem services like pollination services are rarely included. Design choices can have a direct effect on both the quantity and diversity of pollinators present in an ecosystem (both markers of an ecosystem's resilience and specialization). Land use, and in particular, natural diversity are significant predictors of pollinator health and abundance. Typically, human and industrial development threatens to fragment habitats, reduce available forage, and overall reduce natural capital, but this does not need to be the case. Engineers and planners must incorporate pollinators and the valuable services they provide into project plans.

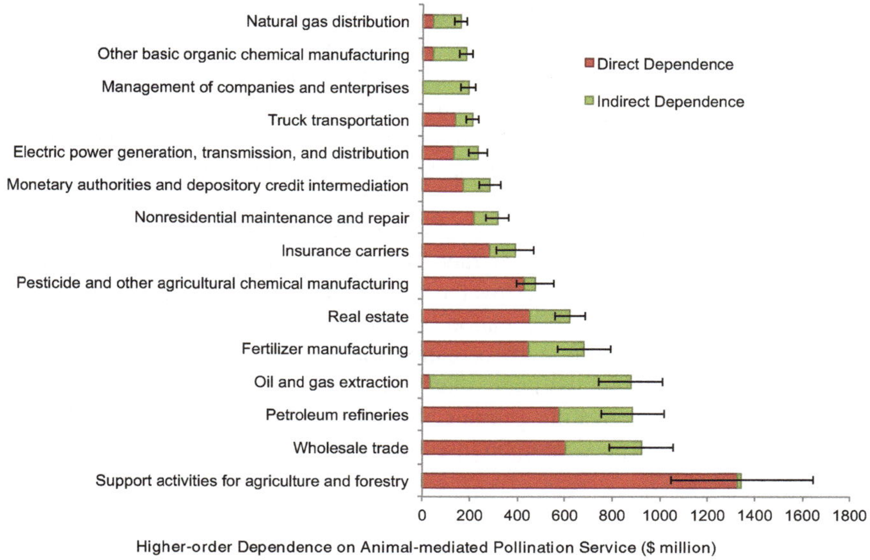

Fig. 6.1 Top 15 industrial sectors in the US economy that are dependent on animal-mediated pollination service. The solid bars represent the mean value for economic dependencies of sectors, while the error bars represent 10th and 90th percentiles obtained via Monte Carlo analysis. (Reprinted with permission from Chopra et al. (2015). Copyright 2015 American Chemical Society)

Evaluating the benefit of pollination services is critical to ensuring that pollination services are conserved. Most design projects require or consider an offset or mitigation of carbon, water, or toxic effluents. Pollinators support diverse ecosystems, which are more resilient and provide the benefits of clean air, carbon cycling/sequestration, water capture and storage, and prevent soil erosion. Without pollinators, ecosystems are highly disconnected and more vulnerable to pests and pathogens, disease, and extreme weather and the ability of these disconnected ecosystems to provide critical ecosystem services is hindered. Engineering projects and designs can receive net benefit from including pollinator habitat and forage in project planning. In rural settings, it is easy to see how pollinators can provide benefit to surrounding landscapes and crops because of their spatial interconnectedness. In urban environments, the foraging distance of pollinators must be considered when planning green space in a city. Creating pollination "islands" (where there is no new forage with the foraging range of a species) establishes fragile networks, which are easily disrupted. Instead, planning for many interconnected, nearby habitats would create a more resilient network of plants and pollinators in a city, which provide numerous benefits to ecosystems, industries, and human well-being.

Pollinators also support the habitat and nutritional resources for many other organisms [10]. As with many ecosystem services, pollination provides many nonmaterial benefits for which value is either not easily calculated or is not possible to enumerate, but these intangible benefits can have much more significance to humans than the material benefits [11]. There is aesthetic value to a meadow of flowers or a diverse landscape of angiosperms that pollinators create and sustain; however, pollination-dependent plants are important cultural and social assets that go beyond aesthetic benefits. Pollination service provides these assets that contribute to human spiritual and heritage values, their sense of place, and cultural identity [12].

Pollinator diversity also means a diverse diet for other wildlife [10, 13]. Insectivores gain a portion of their diet from the pollinating insects, which can be a part of their diet. Many freshwater fish depend upon pollinators as a stable part of their diet. Indirectly, pollination underpins other ecosystem services with their own benefits (e.g., fishing as recreation/sport, culture, tradition, subsistence) [11]. Herbivores rely upon the many plants that are only able to reproduce due to pollination service. Seed-eating animals also depend upon pollinators as the seed set of many plants is dependent upon insect-mediated pollination service [3, 4, 13].

6.2 Quantifying Pollination Services

It is difficult to assess the total value of pollination services, because both ecological and industrial systems depend on these services, which are complex with many aspects lacking simple or even tangible metrics [12, 14]. How does one quantify the aesthetic value of seeing a field of flowers, or breathing in its captivating scent? Despite the difficulty in doing so, valuation of pollinators through valuation of the services they provide is a key method to motivating and stimulating conservation efforts that benefit pollinators [13]. Since intangible effects are hard to visualize and quantify, it is useful to put the value of any ecosystem service into monetary terms, as economic values give us a scale against which to measure immeasurable services [12]. However, monetary terms are difficult to assign to nonmaterial benefits of pollination service [10, 12, 14]. One aspect of value, which can be more readily measured in economic terms, and therefore quantified, is the economic value of pollination service to agricultural and nonagricultural industrial sectors.

6.2.1 Challenges to Quantification

Economic methods require valuation of the dependence of pollination-dependent plants on the services provided by pollinators. Crops can be dependent upon pollination service for yield and/or quality of seed and/or fruit, but the level to which each is dependent varies relative to plant-specific botanical characteristics [3, 4, 14]. A crop can be dependent upon pollinators for either fruit or seed set; thus, dependence is also based on which part of the plant is the commodity of the crop because crops can be directly or indirectly dependent upon pollination service. If a crop is dependent upon pollination service for seed set, but the commodity of that plant is not the seed (e.g., onion), that crop is said to be indirectly dependent upon pollination service [3, 4]. Crops are directly dependent upon pollination service if the commodity of that crop is developed with aid from pollination service (e.g., apple, almond, and blueberry). Some crops have an essential relationship with insect pollinators where fruit or seed set cannot occur without pollination. This can occur when pollen structure, consistency, and fertility are prohibited by other means. This is the case in highbush blueberry cultivars, where pollen is too heavy and sticky to be carried by wind, necessitating pollination by insect pollinators [15, 16]. Some plants are somewhat self- or wind-pollinated and so can still produce some of the commodity of the plant without pollinator assistance. Sometimes referred to as dependency or a dependence-ratio, the dependence of a given crop is a proportion of the commodity that is dependent upon pollination service [3, 4, 14].

Dependence estimates are complex and can be determined by various methods of field study in which researchers compare the plant under conditions of pollinator presence in varying pollinator density and under conditions of pollinator absence [3, 14]. This is normally accomplished by employing some sort of insect exclusion netting either around the whole areas, single plants, or even single flowers. Researchers can introduce or document (depending on which method of exclusion is used) which species is responsible for pollination and their impact on yield, fruit set, and fruit quality. A determination can be made by using this comparison to analyze the improvement or deterioration of commodity yield or quality between the presence or absence of certain pollinators. However, there is a great complexity reflected in an estimate of dependence determined by these studies [3]. Characteristics of the crop such as floral morphology across a variety of commercial cultivars, the capacity of those cultivars to self-pollinate or the necessity to cross-pollinate for production, and the structure and composition of the landscape can all affect the efficiency of pollination service and the dependence of that plant on pollination service [14]. In addition, the composition of the pollinator community where these crops are grown, their relationship with the specific crop, pollinator abundance and density, and the characteristics of those pollinators such as body size and pollen load, as well as environmental conditions such as temperature and precipitation affect the measurement of factors that contribute to calculating dependence of a specific crop [3, 14].

Information on pollination dependence has been collected and summarized by several widely accepted sources for dependence estimates (Table 6.2). This work has had great value in the field of valuation; however, due to overall lack of field study data when considering the complexities of the relationship between pollinators and pollination-dependent crops, estimates have been derived from crop production knowledge, some field data of production, and expert opinion [3]. The estimates are either point estimates that do not recognize a bound of variance or uncertainty [4, 5, 17], or wide, categorical range estimates [18] that do not reflect the nuances of the variety of crops that the estimates represent. Many studies aim to make wide claims about regions, countries, and global populations of bees, which can vary widely over single acres due to available forage, distance to natural habitat, local

Table 6.2 Summary of literature estimating crop pollination dependence coefficients

Pollinator inclusion	Estimate type	Source, year
Honey bees	Point	[5], 1992
		[17], 2000
		[35], 1989
All insects		[4], 2012
	Categorical range	[3], 2007

farming practices, and the presence or absence of managed and wild bees [19–21]. While these population studies can be useful for assessing overall declines of bees, pollinator species interactions must be understood beyond quantification of abundance if more meaningful locally impactful conclusions are to be drawn for the availability of pollination service provided by insects. There is also some discrepancy between sources, and reconciliation is necessary to aptly determine the value of pollination service to the agricultural and subsequently, nonagricultural sectors [14]. Reconciliation can only be achieved with the acquisition of proper field study data that considers cultivar, landscape, pollinator, and climate variation [14]. This lack of consensus or detail in estimates of dependence can pose a challenge to subsequent valuation of pollination service.

Another challenge can be determining the scope and limits of valuation [14]. Where do you draw the line with the boundaries of valuation? As discussed in Sect. 6.2.3, it can be difficult to quantify many aspects of the value of pollination services. Nonuse, option use, or indirect use values often do not have simple means to estimation and direct use value can be determined using terms that are not readily measured or that lack consensus. Much of the valuation can be time and resource expensive, and there may simply not be enough data to draw meaningful conclusions making a complete analysis challenging. It is up to the analyst to draw the boundaries of where to end quantification, which aspects of value to include, and which approach to use—however, such boundary drawing must be done with caution. Because pollination services can be linked to so many processes, products, and other services, a boundary-less analysis of pollination services would show their infinite benefit. The boundaries of the valuation will ultimately determine the communication and impression of the value of pollination services to other members of the field, agencies, and ultimately the general public or those with power to influence the welfare of pollinators [14].

6.2.2 Direct Use Value

Direct use value refers to consumptive and productive use of pollination service. Although other industries (e.g., floral industry, recreational industries) also benefit directly from pollination service, the direct use value of pollination services is by far dominated by the agricultural value [12]. The direct value of pollination service is evident through the well-established beekeeping industry through which farmers can buy or rent colonies of bee species or pay to have them maintained as a standard agricultural production process [14]. There are three methods that have been used to quantify the economic value of pollination services to agriculture: consumer surplus, production value, and replacement cost. Each of these is built upon an initial determination of dependence coefficients. Although there are merits to each of these methods, the production value method lends itself well to subsequently determining the economic value of pollination services to other nonagricultural industry sectors [6, 12]. Any of the described methods could be used to determine the value of a subset of pollinators (honey bees, all managed species of insect pollinators, wild pollinators, mammals) or all pollinators given enough available data. Likewise, the methods are scalable at varying levels of flexibility and can be adjusted to provide estimates for a range of interests (local, multistate, national, and international). Locality-specific valuation analyses can be performed to motivate pollinator protection policy (e.g., Pennsylvania Pollinator Protection Plan [22]).

The consumer surplus method determines the economic value of pollination service attributable to managed honey bees in terms of the change in consumers' and producers' surpluses of pollination-dependent crops in the presence of pollination services [5, 23, 24]. In 1992, Southwick and Southwick introduced this method to determine the value of honey bees in the United States to the agricultural sector by assessing the surplus of pollination-dependent crops gained from pollination service by honey bees [5]. In this method, the long-term supply curve is assumed elastic such that no constraint of land availability or increased production cost is incurred on the behalf of farmers to switch to production of a different crop. A variation of this method utilizes constant price elasticity in the calculation of demand for all crops and instead estimates value based on the loss of agricultural production for each crop. This loss is transformed into a consumer surplus loss, which results in an estimate of the social cost of pollinator decline [23].

The production value method determines the economic value of total crop production that is attributable to pollinators [12]. This method is more relevant for agricultural sectors and has been utilized by many studies on a range of economic scales. This method is especially useful for multisector (including nonagricultural sectors) or large-scale economic analysis that is built upon input-output framework discussed in Sect. 6.2.3 [6, 12].

The replacement cost method determines the economic value of pollination service in terms of the cost to replace the service by alternate means of pollination such as hand or mechanized pollination [25]. These methods are costly, labor intensive, and often times a last option for producers who have neglected or lost pollinators for some other reason. While not ideal, hand and mechanical pollination is already a necessity for some plants especially in remote areas. This cost of replacement is incurred by beekeepers or producers.

In addition to agricultural production sectors, there are many other industrial sectors both relating to agriculture (pesticides, fertilizers) and not relating to agriculture (pharmaceuticals, recreation) that benefit from the production use of pollination service through use of pollination-dependent crops. There are many industrial sectors that rely upon output from these secondary sectors and so on

[6]. These higher-order economic relationships can be quantified through several methods (Table 6.3).

Input-output analysis is an economic modeling technique that has found extensive use in the field of industrial ecology to provide a systems-level framework to describe interactions between industries. The model simplifies the economy based on production in each industrial sector organized as transactional data between sectors [26]. This model allows for higher-order value of pollination service to be valued. In addition to direct use value, it is important to consider other value attributed to pollination service through indirect usage and nonuse.

6.2.3 Indirect Use and Nonuse Value

Indirect use describes the function of pollination service to support other species and the ecosystem and society as a whole or other uses including nonconsumptive uses, option value, or nonuse value. This value surpasses the direct use value as it reflects a whole picture of ecology and society [12]. Option value is ascribed to pollination service for the option of benefiting from pollination service in the future or to retain benefits that may not be currently understood. Nonuse value refers to the intrinsic value ascribed to pollination service and pollinators for simply existing (existence), for benefiting others even though it may not benefit the party assigning value (altruism), or for being available for future generations (bequest) [12].

Indirect use value has been evaluated by willingness-to-pay methods for ecosystem services including pollinator protection policy [50] and through work that seeks to incorporate ecosystem services into life cycle assessment studies [51]. Willingness to pay can be assessed directly (e.g., by survey), or consequentially through assessing costs to finding substitutes for the benefits provided by pollination service either physically (a parallel that can be measured in the market) or behaviorally (as costs to obtain other service or for not having the benefits of the service) [12, 14, 50]. Physically, this might be the cost of hand or mechanized pollination, or it may be the cost of something like artificial flowers as an aesthetic substitute for those provided for by pollination service [12, 25]. Behaviorally, this could be the cost of gaining nutrients typically provided by pollination-dependent foods (vitamin supplements), or the cost of increasing land, water, fertilizer, and pesticide usage to achieve the same level of production provided when pollinators are present. There is also inherent value to pollination service as well as value from the cultural, social, and aesthetic activities provided, enhanced, and maintained by pollination services. Estimations of these can vary depending on the stakeholders involved. Lastly, in addition to pollination value in the aforementioned facets, there are also benefits that are either not well understood or may be completely unknown. The ecosystem is a complex web of interdependence between biotic and abiotic components. Other ecosystem services depend upon pollination service including disease control, pest management, disturbance regulation, erosion control, and nutrient cycling [10].

Table 6.3 Summary of available literature with estimates of pollination service value by various methods [4–6, 10, 17, 23, 25, 27–48]

Valuation method			Author(s)	Year of study	Differentiation of wild and managed Pollinators	Scale	Estimates (millions)	Note
Production value	Economic value		Butler	1943	No			
			Metcalf et al.	1962				
			Martin	1973				
			Levin	1984				
			Winston and Scott-Dupree	1984				
			Fluri and Frick	2005				
	Total economic value of insect pollination	GEM[a]	Bauer and Wing	2010	All	Global	127,000–152,000 (CGE)	Captures variation in pollination benefits across crops, but may generalize between cultivars. Applicable at all scales. Only estimates producer benefits. Assumes maximum efficiency of pollination or maximum density of pollinators. Omits other inputs such as chemicals, labor, and capital. Assumes perfectly elastic demand. Assumes no price change due to reduction in crop supply.

(continued)

Table 6.3 (continued)

Valuation method			Author(s)	Year of study	Differentiation of wild and managed Pollinators	Scale	Estimates (millions)	Note
			Bauer and Wing	2016	All	Global	138,300 (PE) 334,100 (GE)	
		Dependence ratio	Robinson et al.	1989	Honey bees	National [US]	8300[b]	
			Morse and Calderone	2000	Honey bees	National [US]	14,600[c]	
			Losey and Vaughan	2006	Wild	National [US, fruit and vegetable]	3070[d]	
			Chopra, et al.	2007	All	National [US]	14,200–23,800 [19.0[e]]	
			Jordan, et al.	2012	All	National [US]	31,800–36,200 [34.0 [49]]	
			Winfree et al.	2011	Managed and wild	Local [Multi-state, US]		
			Calderone	2012	All	National [US]	19,200[f]	
			Gallai et al.	2009	All	Global	200,000[g] [153,000[h]]	

		Consumer & Producer Surplus			All	Global	265,000-425,000[i]	With regard to pollinator decline, estimates social cost to consumer. Considers market price fluctuation.
			Southwick and Southwick	1992	Honey bees		1600-5200[j]	
			Kasina et al.	2009	Wild bees	Local [Western Kenya]		
			Chacoff et al.	2010	All	National [Argentina]	19,910	
			Lautenbach et al.	2012	Total	Global	290,700[k]	
			O'Grady	1987				
			Ashworth et al.	2009				

(continued)

Table 6.3 (continued)

Valuation method		Author(s)	Year of study	Differentiation of wild and managed Pollinators	Scale	Estimates (millions)	Note
Replacement cost	Replacement of all pollinators by labor or wild pollinators by managed bees	Muth and Thurman	1995	Honey bees	National [US]	419	Applicable at all scales. Does not overestimate pollination benefits. No reliance on crop prices. Assumes producer willingness and ability to pay. Relies upon input and labor prices. Most appropriate in circumstances where replacements have or will be made. Replacement options may not fully replace all benefits or be as effective, because this model does not represent all benefits.

			Allsopp et al.	2008	All	Local [Western Cape, South Africa[l]]	77.0–433.8	
					Honey bees		28.0–122.8	
					Wild insects		49.1–310.9	
			Mouton	2011	Yes	Local [Western Cape, South Africa]		
Contingent valuation method	Willingness to pay for wild pollinator protection		Mwebaze et al.	2010	Honey bees	National [UK]	1770[m]	Captures nonuse in monetary terms Dependent on order of valuation No reliance on market prices Reflects public opinion Requires a full understanding of pollination service benefits Can overestimate as payment is not actually required Expensive

(continued)

Table 6.3 (continued)

Valuation method		Author(s)	Year of study	Differentiation of wild and managed Pollinators	Scale	Estimates (millions)	Note
Landscape service flows	Relate landscape patterns to bee diversity and abundance and crop yields	Chaplin-Kramer et al.	2011	Yes	Local [California]	29,000	
		Morandin and Winston	2006				
		Olschewski et al.	2006				
		Ricketts et al.	2004				

[a]General Equilibrium Model, Computable General Equilibrium model
[b]Dependence ratio multiplied by total crop value
[c]Total crop production, managed honey bees, 2000 data
[d]Native pollinators, fruits and vegetables, US
[e]Mean
[f]Total crop production, managed honey bees, 2010 data
[g]USD, contribution of insect pollination for global agriculture
[h]Euro, € 2005 contribution of insect pollination for global agriculture
[i]USD, consumer surplus loss
[j]Surplus gain, Agriculture, US, honeybees
[k]Calculated as 1.9× increase from Gallai et al. as value was reported
[l]Western Cape deciduous fruit industry, 2005
[m]Pounds, £ 2010 per year

6.3 Pollination in a Life Cycle Assessment Framework

Process and input-output life cycle assessment (LCA) can be used to determine the environmental burdens associated with a product, process, or activity [52]. By assessing the entire life cycle of the object of interest, one can evaluate the environmental impact of released emissions or of materials and energy used during various stages (material extraction, manufacturing, usage, and disposal) of the object's life within boundaries of the system relevant to the scope of the analysis. Traditionally, LCA studies have focused heavily on quantifying the use of nonrenewable resources and emissions. It is vital to include ecosystem goods and services in LCA in order to provoke sustainable development, however currently ecosystem services are not well represented in most life cycle-oriented methods and available tools. Some LCA tools have been developed for the purpose of assessing the role of ecosystem services in process and input-output life cycles (EIO-LCA [53], Eco LCA [54]) [55]. Data on managed species is continually more widely recorded and available, and there are tools being developed for estimating parameters that are useful to determining the role of wild pollinators in LCA (InVEST [56]).

6.3.1 Process LCA

Generally, process life cycle assessment (LCA) does not account for pollination service [57]. Managed pollinators (mostly honey bees with some other species of managed bees) currently have enough data to be included in some process LCAs, but services provided by wild pollinators are presently difficult to include due to a lack of available data on the contribution of wild pollinators to production relative to managed pollinators or all pollinators (with no delineation of wild versus managed) [14, 57].

6.3.2 Economic Input-Output LCA

Economic Input-Output Life Cycle Assessment developed at Carnegie Mellon University builds upon existing economic input-output modeling by combining the economic relationship matrix of EIO with environmental and energy flow [53, 58]. This is done by adding an environmental effects vector to the economic model developed from the work of economist Wassily Leontief. In the 1930s, Leontief formulated an economic input-output table for the US economy showing the transactional relationships between economic sector [58, 59], EIO-LCA takes final demand estimates (Y) and direct/indirect economic requirements (X) from the EIO model and combines them with an environmental impacts sector [58]. This environmental sector (E) is defined by the following equation:

$$E = RX = R[I - A]^{-1}Y \tag{6.1}$$

In Eq. 6.1, R is a matrix with diagonal elements representing the environmental impact per dollar of output in each sector. The R matrix has units of environmental burdens per dollar of output (e.g., kg CO_2/$). This matrix is multiplied by vector X, the output of each sector in dollars. Vector X is defined by $[I - A]^{-1}$ (Eq. 6.2) [60], the total requirements matrix, multiplied by Y, the vector of desired output or final demand [58, 59]. It is called as the total requirements matrix (sometimes the Leontief inverse [59]), because to calculate the term, $[I - A]^{-1}$, all direct and indirect purchases are totaled [58]. In that definition, I is the identity matrix, A is the direct requirements matrix. The total environmental burden, vector E, includes both direct and indirect environmental effects with units of burdens (e.g., kg CO_2) by sector [53, 58].

$$X = [I - A]^{-1}Y \tag{6.2}$$

The EIO-LCA model can include an array of environmental burdens such as air pollutant emissions, global warming potential, ozone depleting substances, or estimates of resource inputs such as fuels, fertilizers, or electricity [53, 58]. As indicated in Eq. 6.1, these burdens are calculated using economic output data from each sector (X) and the R matrix. The values for the R matrix from which these burdens are derived comes from public datasets that report these emissions on a sector-level such as those provided through the United States Environmental Protection Agency (EPA). Sometimes translation, conversion, or reclassification of the reports is necessary to derive the matrices. The EIO-LCA method developed by CMU utilizes the EIO matrix and associated environmental data in US Benchmark Models using the North American Industry Classification System (NAICS) for defining sectors [53, 58].

Benchmark models are created every 5 years in the United States and include more than 400 industry sectors [53, 58]. The data sources and publications used for these models are vast, and much comes from surveys of operating facilities in each industry in addition to reports from the United States Environmental Protection Agency. For example, the 2002 US Benchmark Model uses the US EPA Toxics Release Inventory for updated toxic emission data and the US EPA Inventory of US Greenhouse Gas Emissions and Sinks for estimating greenhouse gas emissions by industrial sectors. Other examples of data sources include the Manufacturing Energy Consumption Survey (MECS), US EPA National Emissions Inventory (NEI), and US EPA National Biannual Resource Conservation and Recovery Act Hazardous Waste Report [53, 58].

The data used to compile many of these surveys and reports has varied quality. It is often self-reported and is subject to measurement error [53, 58]. In addition, reporting requirements vary widely by industrial sector and thus there are gaps in the information available. There are similar international surveys and models, but none are as extensive in the amount of sectors included. Another limitation to this method is that each sector is represented by an aggregated average, and despite having over 400 sectors disaggregated, this can still hinder detailed assessment. For example, there is no distinction in the fruits and nuts sector for the type or quality of product. In addition, there is no distinction between mills and plants with varied efficiency or pollution output that may be specific to or used primarily in a life cycle. Process LCAs can be more specific in this distinction for a particular material or process [53, 58].

Along with inherent uncertainty coming from the original data source, the aggregation of these sources compounds uncertainty, and there is often missing or incomplete data or estimations where data is lacking [53]. There are also many assumptions made for allocation of environmental burden when sectors from the economic data and environmental data are not aggregated in the same way. In addition, the data is only from publicly available sources and not industry-specific such that information that may exist in industry reports is not incorporated into the model simply because it is not available publicly. The model is also based on producer price as opposed to purchaser price, which can differ vastly [53]. In addition, the model is based on constant coefficients, which work for short-term assessments but not in the long term when consumers and industries may adapt to disruptions in sectors including technological changes. Also, the model's linear nature may not aptly represent production processes or flows of economy as they are often nonlinear [53, 59].

Finally, many burdens are not represented for lack of data or because they are not incorporated into the framework of the model. Pollination service, like many ecosystem services, is neglected in LCA, including CMU's EIO-LCA method. Ecosystem services are often considered free and with infinite supply; however, the renewability of these resources has limitations [61, 62]. This limit has been made apparent for many of these resources including pollination as pollinators have faced significant decline due to many factors including industrial use of pesticides [61, 63, 64]. The role of this service in industrial sectors is unexplored compared to finite or nonrenewable sources.

An environmental vector for pollination service does not currently exist in Carnegie Mellon University's online tool for EIO-LCA [53]. In fact, ecosystem goods and services are overlooked in this model. As a resource, "Pollination," exists in EcoLCA, an IO-LCA tool developed by The Ohio State University, however, the vector is not developed and exists as a placeholder in this tool at present [54].

6.3.3 Future Directions

Although managed species of pollinators, especially *Apis mellifera* and *Apis cerana* (European and eastern honey bees), are relatively well studied and data pertaining to managed species is widely recorded and available, there is still much data missing for wild pollinators. In addition, crop-specific field data are available for some crops and cultivars; however, there is generally a lack of systematically collected data across representative crops, cultivars, and landscapes. Creation and implementation of an environmental vector for pollination services for LCA tools (e.g., EIO-LCA, EcoLCA) for managed pollinators and subsequently wild pollinators will allow for an account of the role of pollination services in product and process life cycles.

6.4 Pollination in a Network Analysis Framework

Network analysis is a technique used to quantify characteristics such as dependence and connectivity, magnitude, and importance of interactions on an element-by-element basis. This technique allows for complex systems that are normally analyzed from an outside-in approach (where a system boundary is formed and the practitioner is blind to all processes within) to be instead analyzed on a more specific basis that adds nuance by linking specific members or nodes within the system boundary. Shedding light on otherwise blind processes allows for detailed analysis of the dependent interactions within the system boundary that are directly or indirectly linked to the visible outcomes. This framework is commonly used to analyze interactions in social networks, politics, epidemiology, and electrical circuits, to name a few [65–68].

The ability of network analysis to assess each individual node in a system allows for the establishment of complex linkages that can otherwise be overlooked when only direct inputs and outputs are analyzed. The establishment of a network allows for quantification of the connectivity of each node within the network and for "hotspots" to be identified. These hotspots can be thought of as the nodes, which are the most vital to the network, contribute the greatest benefit, or contain the most critical linkages and would cause the greatest disruption if removed. These analyses can give informational statistics, trace disease outbreaks, and guide conservation efforts to name a few.

6.4.1 Pollination Services as a Network

Traditionally, plant pollinator systems are not assessed from a network perspective. Most studies evaluate pollinators on the basis of a single plant of interest (usually the yield crop) and a few previously established active pollinators of that cultivar [21, 69] In addition to only focusing on one plant in the plant pollinator system, these studies neglect the effects of local flora and fauna to support the local and managed pollinators in the absence of primary crop blooms, which is the majority of the year [21, 70]. Because bloom for some crops may only last for a few weeks, it is important to look at other surrounding forage that may support healthy pollinator communities year-round. This includes providing enough stored food for overwintering and preparing brood for subsequent years, thus ensuring abundant bee populations in the future year. It has also been shown that bee colonies who enter winter with too few bees or food increase the likelihood that the colony will not survive over winter [71]. Commercially available pollen cakes and nectar substitutes are available to supplement the potential lack of forage in high-intensity agricultural areas; however, colonies that feed heavily on these substitutes have been shown to have increased disease and pest susceptibility and decreased overall health [72].

Naturally, pollination services are easily visualized as network model where the simplest model documents the interactions of a community of pollinators on the local flora and fauna in the area. Each pollinator and plant represent an element in the system where the individual pollinators are agents, which act on the nodes (flowering plants) by transferring pollen, harvesting nectar, and aiding in reproduction. In this case, the level of dependence of the nodes on the agents is highly dependent on the physical characteristics of the plant, local forage, and species diversity, especially when plant and pollinator species coevolve to necessitate highly specific interactions that can only be performed by a few community members [73]. This model attempts to answer the questions of who is responsible for pollination and how this interaction occurs (duration and method) in order to better quantify the service provided by local and managed pollinators to specific plants.

Assessing pollinators as a network can allow for species level interactions to be mapped. It has been shown that the specific makeup of local pollination populations can influence the behavior of those species in the community [6]. For instance, honeybees, which are generalists, naturally will move down rows of plants moving from one flower to the next in a linear fashion. While efficient, this does not always increase pollination in species that are self-incompatible and rely on pollen from different varieties of the same plant for reproduction. Here more erratic movement across rows (when different varieties are planted in alternating rows) can increase the amount of compatible pollen transferred and thus increase final fruit set and quality [17]. This phenomenon has been observed in areas where native pollinators are in abundance, honey bees tend to move more across rows due to competition, species signaling, and many other factors [74]. This led to higher pollinator efficiencies on a per visit basis and increased yield. Understanding how species diversity and density impacts species behavior can only be realized through the realization of the specific species level interactions responsible for the observed behavior. This understanding adds nuance otherwise unquantifiable by traditional methods.

In addition to the mapping of species interactions, this type of model provides a framework for which detailed spatiotemporal data can be incorporated to better understand the impact of pollinators on fruit set and quality. Traditional data collected for harvest prediction and assessment including, weather, soil characteristics, fruit yield, and fruit quality can be used to better understand the contribution of pollinators versus other variables known to influence final yield. Increasing this basic model to also include spatial data such as distance from natural habitat, plot size, and spatial location of each interaction can help to better understand the effects of the location of natural flora and fauna and agricultural plots on pollinator species richness, diversity, and abundance.

6.4.2 Challenges in Plant Pollinator Networks

Network analysis is not without its own challenges. The obvious challenge is the sheer amount of data collection that is needed to draw conclusions about local ecosystems services delivered by pollinators to local flora and fauna in addition to the agricultural crop(s) cultivated in the area. In these systems, every flowering plant must be classified and sampled for which pollinators are visiting, in what frequency, for what duration, and at what efficiency. This is highly resource intensive and materializes in many hours of vigorous sampling to ensure that no interactions are omitted, overestimated, or underrepresented in the network.

Timing is also crucial in plant pollinator systems, because over different timescales, the local ecosystems can change drastically. These studies involve gathering data about a seasonally variable system. Would it be fair to assess the seasonal network on a yearly, monthly, seasonally, or even weekly basis? It is unclear where to draw the timescales on pollinator plant network studies due to the seasonality of the data. Changing weather conditions (temperature, rain, and wind), available forage, and pollinator behavior in these networks may vary considerably over long timescales, impacting the connectivity of the network. This can result in the overestimation or underrepresentation of pollinator species and their value. In addition to the general population variability in plant pollinator systems, the actual timing of sampling can affect results. In some systems, some pollinators may be "early risers" or only feed at certain times not captured by human sampling due to the limited amount of time that one individual can sample. Analyzing the networks in a reasonable timeframe can be challenging when assessing management strategies and conservation efforts. This becomes extremely challenging when comparing pollinator networks during bloom for fruit quality and fruit set and year-round models for overall pollinator community health, but it can show the dependencies of primary crop pollinators on the local ecosystem that is available outside of primary crop bloom.

6.4.3 Future Direction

The smallest bottleneck in plant pollinator network analysis is the amount of labor required to draw meaningful conclusions about the network. It is often impractical to document all flowering plants and the interactions of both managed bees and native bees year-round. Also, a researcher must possess an immense level of expertise in plant and pollinator identification. New methods of quantification involving noninvasive sampling are needed to gather the amount of data required to draw conclusions about whole plant pollinator networks. Currently, bees must be observed while pollinating, caught, and identified by experts, which is time consuming and labor intensive. Alternative strategies that could aid in the sampling and identification of the bees would increase the amount of data that could be gathered and help to facilitate better policy and conservation decisions.

In addition to better documenting the interactions of pollinators and plants, better analysis of spatiotemporal variables is needed to understand the nuance associated with plant pollinator networks. The end goal of many agriculturally focused studies is to increase crop yield through better understanding of plant pollinator networks. A fundamental problem is determining the amount of yield directly attributed to pollinators because this involves deducing the differences between the effects of pollinators and spatiotemporal, which can vary over small distances and times.

Once dependable networks are established which draw links between species interactions and yield, while controlling for non–pollinator-dependent variables like local landscape, soil pH, rainfall, nutrients, and overall plant health, these models can be used to better influence farming decisions and may even serve as predictive tools for final yield, assessing the need for managed pollinators, and future pollination service needs.

6.5 Effect of Loss of Pollinators

Decline or loss of pollinators has widespread impacts on agricultural and nonagricultural industries, cultural and social institutions, and various ecosystems and biodiversity. Significant declines in both managed and wild populations of pollinators have been documented [63, 64], and continued loss would be devastating to human nutrition, culture, and industrial activity as well as ecosystems. Obviously, the types of pollinators lost, the extent to which they are lost, and the resulting composition of the pollinator community would influence the type and magnitude of the impacts associated with the loss. However, one can pursue a thought exercise to assess potential impacts by loss of pollinators. Without proper valuation and understanding of the material and nonmaterial benefits that pollinators provide, it is not possible to fully understand the effects of the loss of pollinators.

6.5.1 Agricultural Impact

In agriculture, there is a direct impact on the production of food crops with pollinator loss. In a realistic scenario, the agricultural industry would mitigate losses through various strategies [14]. Speaking only in the interest of preserving the value of capital (weak sustainability), elastic prices of many pollination-dependent crops would respond to the decrease in production and increase accordingly, likely narrowing the gap between current production value and after any pollination service loss. Elasticity in supply and demand allows for substitution on the part of both producers and consumers. Production losses could also be mitigated by increasing acreage of pollination-dependent crop. This would require greater resources for the additional crop acreage and associated costs and impacts. It also can take many years for some pollinator dependent crops to become established (e.g., apples, berries, cherries) increasing the lag time in producer response. For some crops, there would be total production loss without pollination and the industry would have to respond with technological substitution through manual pollination performed by hand or through mechanical pollination [12, 25]. There may also be a greater increase in demand for managed pollinator species. The managed pollinator industry has already faced significant losses over the last 50 years and labors substantially to meet current demand given current seasonal losses [75]. Given the current increased mortality of managed honeybee colonies due to colony collapse disorder and other environmental factors coupled with the aging beekeeping population make it unlikely that the managed pollinator industry could keep up with widespread loss of pollinator loss.

If the extreme case where all pollinators were removed from the system were to happen, how would human nutrition suffer? Surprisingly, many staple crops (crops that support the majority of nutrition like potatoes, rice, and corn) are not dependent on pollinators. This means that while our dinner plates may start to look bland and unappealing, the majority of required nutrients could still be provided. The largest impacts from loss of pollinators come in the form of micronutrient deficiencies. Hidden hunger, where individuals are not starving from lack of food, but instead lack of micronutrients, is a worldwide issue, which impacts 1 in 4 people globally [76]. Fat-soluble vitamins in particular are supplied primarily by pollinator-dependent crops. Particularly of note, vitamin A and provitamin A are nearly 70% and 98% reliant on pollination services [77]. Vitamin A deficiency is one of the most common micronutrient deficiencies worldwide, causing over 500,000 cases especially in underprivileged areas [78]. Some estimates state that a total loss of pollination services would cause increased risk of hidden hunger in up to 50% of the global population [77, 79]. Supplementation would seem like an easy solution to solve hidden hunger, but the plants that are used to make vitamin A supplements are reliant on pollinators. This means that in the absence of pollinators, the supply of micronutrients for supplementation would also decrease, making supplementation only available to those who could afford it. The decline in micronutrient availability due to pollinator loss would most adversely affect poorer, underprivileged populations who are already under various other forms of oppression.

It is very important to understand that preserving pollination service is not a complete solution. As mentioned, pollination service can be provided, albeit at an efficiency deficit, by augmenting current managed pollinator trends or substitution by means of manual or mechanical pollination [12, 24]. However, the preservation of pollinator diversity is essential for long-term ecosystem fitness [10]. The importance of pollinators extends far beyond human endeavors.

6.5.2 Nonagricultural Impact

The impact of pollinator loss would extend beyond agricultural production and has effects upon the production of other related, nonagricultural industries [6]. Related economic sectors relating to fibers, materials extraction, pharmaceuticals, and construction rely on production output from agricultural sectors and indirectly rely on the benefits pollinators provide. These indirect, complex linkages are not well quantified or understood, but would cause unquantified effects cascading through the economy. Moreover, loss of pollinators would be a detriment to ecosystem function and biodiversity [10]. In fact, many pollinator species with no agricultural production value are necessary to ecosystem function [13].

6.5.3 Ecosystem Impact

Through the delivery of pollen to plants, pollinators ensure set and enhance quality of wild fruits and seeds. Many of these wild plant pollinators have no influence on agricultural production value despite their crucial role in nature. This provides and maintains diverse habitat for other organisms and supplies nutrition through trophic webs (including consumption of pollination-dependent plants, pollinators, and secondary or tertiary consumption) [13]. A loss of pollinators also has the potential to disrupt existing cultural behaviors and practices [11, 13]. Recreational activities, tourism, apple- and strawberry-picking, activities relating to identity, and celebration of heritage or self would all be interrupted with potentially no suitable substitute. Other important services like disease and erosion control, ecosystem resilience, biological diversity, would suffer for lack of pollinators [10, 12]. These services are incredibly valuable and connected to the health and welfare of humans as well as greater ecosystems.

6.6 Summary

The preservation and restoration of pollinators are critical to the welfare of humans and ecosystems. Already significant declines in abundance and biodiversity of pollinators have underscored a need for valuation of the benefits they provide. There are tools available for useful valuation of pollinators and pollination services from an economic perspective. However, the pollinators serve ecosystems indirectly through many pathways that ensure ecological and environmental resilience that cannot be accounted for in current economic methods. In addition, many cultural, nonmaterial benefits derived from pollinators may not ever be entirely captured by economics. As with any valuation, it is critical to consider material and nonmaterial benefits and choose the scope of a valuation of pollinators and pollination service with care and always with clarification of limitations as the influence of a valuation can be far-reaching and highly influential.

Current life cycle methods do not adequately account for ecosystem goods and services, and pollination service is especially not represented in process or product LCA frameworks, although placeholders and intuitive avenues for implementation exist. Loss or decline of pollinators will be significant and extensive, having impacts directly in agriculture, indirectly in nonagricultural industries and human culture, and finally in essential ecosystem function. Future directions include creation and integration of this environmental impact category into life cycle analyses and systematic collection of data on both wild and managed pollinators to better understand their role in production as well as the cultural and social benefits attributed to pollinators. More acute valuation of pollination services will motivate and guide conservation, revitalization efforts, and policy decision-making.

References

1. IPBES (2016): Summary for policymakers of the assessment report of the Intergovernmental Science-Policy Platform on Biodiversity and Ecosystem Services on pollinators, pollination and food production. S.G. Potts, V. L. Imperatriz-Fonseca, H. T. Ngo, J. C. Biesmeijer, T. D. Breeze, L. V. Dicks, L. A. Garibaldi, R. Hill, J. Settele, A. J. Vanbergen, M. A. Aizen, S. A. Cunningham, C. Eardley, B. M. Freitas, N. Gallai, P. G. Kevan, A. Kovács-Hostyánszki, P. K. Kwapong, J. Li, X. Li, D. J. Martins, G. Nates-Parra, J. S. Pettis, R. Rader, and B. F. Viana (eds.). Secretariat of the Intergovernmental Science-Policy Platform on Biodiversity and Ecosystem Services, Bonn, Germany. 36 pages. In.
2. Rader, R.; Bartomeus, I.; Garibaldi, L. A.; Garratt, M. P. D.; Howlett, B. G.; Winfree, R.; Cunningham, S. A.; Mayfield, M. M.; Arthur, A. D.; Andersson, G. K. S.; Bommarco, R.; Brittain, C.; Carvalheiro, L. s. G.; Chacoff, N. P.; Entling, M. H.; Foully, B.; Freitas, B. M.; Gemmill-Herren, B.; Ghazoul, J.; Griffin, S. R.; Gross, C. L.; Herbertsson, L.; Herzog, F.; Hipólito, J.; Jaggar, S.; Jauker, F.; Klein, A.-M.; Kleijn, D.; Krishnan, S.; Lemos, C. Q.; Lindström, S. A. M.; Mandelik, Y.; Monteiro, V. M.; Nelson, W.; Nilsson, L.; Pattemore, D. E.; de O. Pereira, N.; Pisanty, G.; Potts, S. G.; Reemer, M.; Rundlöf, M.; Sheffield, C. S.; Scheper, J.; Schüepp, C.; Smith, H. G.; Stanley, D. A.; Stout, J. C.; Szentgyörgyi, H.; Taki, H.; Vergara, C. H.; Viana, B. F.; Woyciechowski, M., Non-bee insects are important contributors to global crop pollination. *Proceedings of the National Academy of Sciences* **2016,***113*, (1), 146–151.
3. Klein, A.-M.; Vaissière, B. E.; Cane, J. H.; Steffan-Dewenter, I.; Cunningham, S. A.; Kremen, C.; Tscharntke, T., Importance of pollinators in changing landscapes for world crops. *Proceedings of the Royal Society B: Biological Sciences* **2007,***274*, (1608), 303–313.
4. Calderone, N. W., Insect pollinated crops, insect pollinators and US agriculture: Trend analysis of aggregate data for the period 1992–2009. *PLoS ONE* **2012,***7*, (5), e37235.
5. Southwick, E. E.; Southwick, J. L., Estimating the Economic Value of Honey Bees (*Hymenoptera: Apidae*) as Agricultural Pollinators in the United States. *Journal of Economic Entomology* **1992,***85*, (3), 621–633.
6. Chopra, S. S.; Bakshi, B. R.; Khanna, V., Economic Dependence of U.S. Industrial Sectors on Animal-Mediated Pollination Service. *Environmental Science and Technology* **2015,***49*, 14441–14451.
7. Charles, M.; Bakshi, B. R., Designing Industrial Landscapes for Mitigating Air Pollution with Spatially-Explicit Techno-Ecological Synergy. *AIChE Journal*, e17347.

8. Charles, M.; Ziv, G.; Bohrer, G.; Bakshi, B. R., Connecting air quality regulating ecosystem services with beneficiaries through quantitative serviceshed analysis. *Ecosystem Services* **2020,***41*, 101057.
9. Ghosh, T.; Bakshi, B. R., Designing biofuel supply chains while mitigating harmful algal blooms with treatment wetlands. *Computers & Chemical Engineering* **2019,***126*, 113–127.
10. Losey, J. E.; Vaughan, M., The economic value of ecological services provided by insects. *BioScience* **2006,***56*, (4), 311–323.
11. Chan, K. M. A.; Guerry, A. D.; Balvanera, P.; Klain, S.; Satterfield, T.; Basurto, X.; Bostrom, A.; Chuenpagdee, R.; Gould, R.; Halpern, B. S.; Hannahs, N.; Levine, J.; Norton, B.; Ruckelshaus, M.; Russel, R.; Tam, J.; Woodside, U., Where are *Cultural* and *Social* in Ecosystem Services? A Framework for Constructive Engagement. *BioScience* **2012,***62*, (8), 744–756.
12. Mburu, J.; Hein, L. G.; Gemmill, B.; Collette, L. *Economic Valuation of Pollination Services: Review of Methods*; Food and Agriculture Organization of the United Nations: Rome, Italy, 2006.
13. Senapathi, D.; Biesmeijer, J. C.; Breeze, T. D.; Kleijn, D.; Potts, S. G.; Carvalheiro, L. G., Pollinator conservation — the difference between managing for pollination services and preserving pollinator diversity. *Current Opinion in Insect Science* **2015,***12*, 93–101.
14. Melathopoulos, A. P.; Cutler, C. G.; Tyedmers, P., Where is the value in valuing pollination ecosystem services to agriculture? *Ecological Economics* **2015,***109*, 59–70.
15. Kirk, A. K.; Isaacs, R., Predicting Flower Phenology and Viability of Highbush Blueberry. *HortScience* **2012,***47*, (9), 1291–1296.
16. Drummond, F., Reproductive Biology of Wild Blueberry (Vaccinium angustifolium Aiton). *Agriculture* **2019,***9*, (4), 69.
17. Morse, R. A.; Calderone, N. W., The value of honey bee pollination in the United States. *Bee Culture* **2000,***128*, 1–15.
18. Klein, A. M.; Vaissiere, B. E.; Cane, J. H.; Steffan-Dewenter, I.; Cunningham, S. A.; Kremen, C.; Tscharntke, T., Importance of pollinators in changing landscapes for world crops. *Proceedings of the Royal Society B* **2007,***274*, 303–313.
19. Garibaldi, L. A.; Steffan-Dewenter, I.; Winfree, R.; Aizen, M. A.; Bommarco, R.; Cunningham, S. A.; Kremen, C.; Carvalheiro, L. G.; Harder, L. D.; Afik, O.; Bartomeus, I.; Benjamin, F.; Boreux, V.; Cariveau, D.; Chacoff, N. P.; Dudenhöffer, J. H.; Freitas, B. M.; Ghazoul, J.; Greenleaf, S.; Hipólito, J.; Holzschuh, A.; Howlett, B.; Isaacs, R.; Javorek, S. K.; Kennedy, C. M.; Krewenka, K. M.; Krishnan, S.; Mandelik, Y.; Mayfield, M. M.; Motzke, I.; Munyuli, T.; Nault, B. A.; Otieno, M.; Petersen, J.; Pisanty, G.; Potts, S. G.; Rader, R.; Ricketts, T. H.; Rundlöf, M.; Seymour, C. L.; Schüepp, C.; Szentgyörgyi, H.; Taki, H.; Tscharntke, T.; Vergara, C. H.; Viana, B. F.; Wanger, T. C.; Westphal, C.; Williams, N.; Klein, A. M., Wild pollinators enhance fruit set of crops regardless of honey bee abundance. *Science* **2013,***339*, (6127), 1608–1611.
20. Kleijn, D.; Winfree, R.; Bartomeus, I.; Carvalheiro, L. G.; Henry, M.; Isaacs, R.; Klein, A. M.; Kremen, C.; M'Gonigle, L. K.; Rader, R.; Ricketts, T. H.; Williams, N. M.; Lee Adamson, N.; Ascher, J. S.; Baldi, A.; Batary, P.; Benjamin, F.; Biesmeijer, J. C.; Blitzer, E. J.; Bommarco, R.; Brand, M. R.; Bretagnolle, V.; Button, L.; Cariveau, D. P.; Chifflet, R.; Colville, J. F.; Danforth, B. N.; Elle, E.; Garratt, M. P. D.; Herzog, F.; Holzschuh, A.; Howlett, B. G.; Jauker, F.; Jha, S.; Knop, E.; Krewenka, K. M.; Le Feon, V.; Mandelik, Y.; May, E. A.; Park, M. G.; Pisanty, G.; Reemer, M.; Riedinger, V.; Rollin, O.; Rundlof, M.; Sardinas, H. S.; Scheper, J.; Sciligo, A. R.; Smith, H. G.; Steffan-Dewenter, I.; Thorp, R.; Tscharntke, T.; Verhulst, J.; Viana, B. F.; Vaissiere, B. E.; Veldtman, R.; Ward, K. L.; Westphal, C.; Potts, S. G., Delivery of crop pollination services is an insufficient argument for wild pollinator conservation. *Nat Commun* **2015,***6*, 7414.
21. Klein, A. M.; Hendrix, S. D.; Clough, Y.; Scofield, A.; Kremen, C., Interacting effects of pollination, water and nutrients on fruit tree performance. *Plant Biology* **2015,***17*, (1), 201–208.

22. Grozinger, C. M.; Lopez-Uribe, M. M.; Patch, H. M.; Strathmeyer, F.; Welliver, R.; Spichiger, S.; Roccasecca, K.; Libhart, T.; Vorisek, C.; Rapasky, S.; Eggen, D.; Sitch, K.; Leppo, B.; Jones, B.; Brightman, T.; Radebaugh, M.; Pulig, B.; Campbell, B. Pennsylvania Pollinator Protection Plan. https://ento.psu.edu/pollinators/publications/p4-introduction
23. Gallai, N.; Salles, J.-M.; Settele, J.; Vaissière, B. E., Economic valuation of the vulnerability of world agriculture confronted with pollinator decline. *Ecological Economics* **2009,***68*, (3), 810–821.
24. Breeze, T. D.; Gallai, N.; Garibaldi, L. A.; Li, X. S., Economic Measures of Pollination Services: Shortcomings and Future Directions. *Cell Press: Trends in Ecology & Evolution* **2016,***31*, (12), 927–939.
25. Allsopp, M. H.; de Lange, W. J.; Veldtman, R., Valuing insect pollination services with cost of replacement. *PLoS ONE* **2008,***3*, (9), e3128.
26. Leontief, W. W., Quantitative input and output relations in the economic system of the United States. *The Review of Economic Statistics* **1936,***18*, (3), 105–125.
27. Butler, C., The position of honey-bee in the national economy. *Adult bee diseases. Annals Applied Biology* **1943,***30*, 189–191.
28. Metcalf, C. L.; Flint, W. P.; Metcalf, R. L., *Detructive and Useful Inescts: Their Habits and Control*. McGraw-Hill Book Company: 1962.
29. Martin, E. C., The use of bees for crop pollination. *American Bee Journal* **1973,***113*, 422–423.
30. Levin, M. D., Value of bee pollination to US agriculture. *Bulletin of the Entomological Society of America* **1984,***29*, (4), 50–51.
31. Winston, M. L.; Scott-Dupree, C., The value of bee pollination to Canadian agriculture. *Canadian Beekeeper* **1984,***11*, (134).
32. Fluri, P.; Frick, R., L'apiculture en Suisse:état et perspectives. *Revue Suisse d'Agriculture* **2005,***37*, 81–86.
33. Bauer, D.; Wing, I. S., Economic Consequences of Pollinator Declines: A synthesis. *Agricultural and Resource Economics Review* **2010,***39*, (3), 368–383.
34. Bauer, D.; Wing, I. S., The macroeconomic cost of catastrophic pollinator declines. *Ecological Economics* **2016,***126*, 1–13.
35. Robinson, W. S.; Nowogrodzki, R.; Morse, R. A., The value of honey bees as pollinators of U.S. crops. *American Bee Journal* **1989,***129*, 411–423, 477–487.
36. Winfree, R.; Gross, B. J.; Kremen, C., Valuing pollination services to agriculture. *Ecological Economics* **2011,***71*, 80–88.
37. Kasina, J.; Mburu, J.; Kraemer, M.; Holm-Mueller, K., Economic benefit of crop pollination by bees: a case of Kakamega small-holder farming in western Kenya. *Journal of Economic Entomology* **2009,***102*, 467–473.
38. Chacoff, N. P.; Morales, C. L.; Garibaldi, L. A.; Ashworth, L.; Aizen, M. A., Pollinator dependence of Argentinian agriculture: current status and temporal analysis. *The Americas Journal of Plant Science and Biotechnology* **2010,***3*, (1), 106–116.
39. Lautenbach, S.; Seppelt, R.; Liebscher, J.; Dormann, C. F., Spatial and temporal trends of global pollination benefit. *PLoS ONE* **2012,***7*, (4), e35954.
40. O'Grady, J. H., *Market Failure in the Provision of Honeybee Pollination; a Heuristic Investigation*. University of Vermont: 1987.
41. Ashworth, L.; Quesada, M.; Casas, A.; Aguilar, R.; Oyma, K., Pollinator-dependent food production in Mexico. *Biological Conservation* **2009,***142*, (4), 1050–1057.
42. Muth, M. K.; Thurman, W. N., Why support the price of honey? *Choices* **1995,***10*, 19–21.
43. Mouton, M. Significance of Direct and Indirect Pollination Ecosystem Services to the Apple Industry in the Western Cape of South Africa. University of Stellenbosch, Stellenbosch, 2011.
44. Mwebaze, P.; Marris, G. C.; Budge, G. E.; Brown, M.; Potts, S. G.; Breeze, T. D.; MacLeod, A., Quantifying the value of ecosystem services: a case study of honeybee pollination in the UK. In *12th Annual BIOECON Conference 'From the Wealth of Nations to the Wealth of Nature: Rethinking Economic Growth*, Venice, Italy, 2010.
45. Chaplin-Kramer, R.; Tuxen-Bettman, K.; Kremen, C., Value of wildland habitat for supplying pollination services to Californian agriculture. *Rangelands* **2011,***33*, 33–41.

46. Morandin, L. A.; Winston, M. L., Pollinators provide economic incentive to preserve natural land in agroecosystems. *Agriculture, Ecosystems & Environment* **2006**,*116*, (3–4), 289–292.
47. Olschewski, R.; Tscharntke, T.; Benítez, P. C.; Schwarze, S.; Klein, A.-M., Economic evaluation of pollination services comparing coffee landscapes in Ecuador and Indonesia. *Ecology and Society* **2006**,*11*, (1), article 7.
48. Ricketts, T. H.; Daily, G. C.; Ehrlich, P. R.; Michener, C. D., Economic value of tropical forest to coffee production. *PNAS* **2004**,*101*, (34), 12579–12582.
49. Jordan, A.; Patch, H. M.; Grozinger, C. M.; Khanna, V., Economic dependence and vulnerability of United States agricultural sector on insect-mediated pollination service. *Environmental science & technology* **2021**,*55*, (4), 2243–2253.
50. Breeze, T. D.; Bailey, A.; Potts, S. G.; Balcombe, K. G., A Stated Preference Valuation of the Non-Market Benefits of Pollination Services in the UK. *Ecological Economics* **2015**,*111*, 76–85.
51. Bruel, A.; Troussier, N.; Guillaume, B.; Sirina, N. In *Considering Ecosystem Services in Life Cycle Assessment to Evaluate Environmental Externalities*, 23rd CIRP Conference on Life Cycle Engineering, 2016; 2016; pp 382–287.
52. Zhang, Y.; Singh, S.; Bakshi, B. R., Accounting for Ecosystem Services in Life Cycle Assessment, Part I: A Critical Review. *Environmental Science & Technology* **2010**,*44*, 2232–2242.
53. Carnegie Mellon University Green Design Institiute, Economic Input-Output Life Cycle Assessment (EIO-LCA). In 2008.
54. The Ohio State University Center of Resilience, Ecologically-Based Life Cycle Assessment (Eco-LCA). In 2009.
55. Zhang, Y.; Baral, A.; Bakshi, B. R., Accounting for Ecosystem Services in Life Cycle Assessment Part II: Toward an Ecologically Based LCA. *Environmental Science & Technology* **2010**,*44*, (7).
56. Sharp, R.; Tallis, H. T.; Ricketts, T. H.; Guerry, A. D.; Wood, S. A.; Chaplin-Kramer, R.; Nelson, E.; Ennaanay, D.; Wolny, S.; Olwero, N.; Vigerstol, K.; Pennington, D.; Mendoza, G.; Aukema, J.; Foster, J.; Forrest, J.; Cameron, D.; Arkema, K.; Lonsdorf, E.; Kennedy, C.; Verutes, G.; Kim, C.-K.; Guannel, G.; Papenfus, M.; Toft, J.; Marsik, M.; Bernhardt, J.; Griffin, R.; Glowinski, K.; Chaumont, N.; Perelman, A.; Lacayo, M.; Mandle, L.; Hamel, P.; Vogl, A. L.; Rogers, L.; Bierbower, W.; Denu, D.; Douglass, J., InVEST: Integrated Valuation of Ecosystem Services and Tradeoffs. In Project, T. N. C.; University, S.; Minnesota, U. o.; Conservancy, T. N.; Fund, W. W., Eds. 2020.
57. Crenna, E.; Sala, S.; Polce, C.; Collina, E., Pollinators in life cycle assessment: towards a framework for impact assessment. *Journal of Cleaner Production* **2017**,*140*, 525–536.
58. Matthews, H. S.; Hendrickson, C. T.; Matthews, D. H., *Life Cycle Assessment: Quantitative Approaches for Decisions That Matter*. 2015.
59. Miller, R.; Blair, P., *Input-Output Analysis: Foundations and Extensions*. Prentice-Hall, Inc.: Englewood Cliffs, New Jersey, 1985.
60. Lucas A. Garibaldi, I. S.-D., 2 Rachael Winfree,3 Marcelo A. Aizen, 4 Riccardo Bommarco,5 Saul A. Cunningham,6 Claire Kremen,7 Luísa G. Carvalheiro,8,9 Lawrence D. Harder,10 Ohad Afik,11 Ignasi Bartomeus,12 Faye Benjamin,3 Virginie Boreux,13,14 Daniel Cariveau,3 Natacha P. Chacoff,15 Jan H. Dudenhöffer,16 Breno M. Freitas,17 Jaboury Ghazoul,14 Sarah Greenleaf,7 Juliana Hipólito,18 Andrea Holzschuh,2 Brad Howlett,19 Rufus Isaacs,20 Steven K. Javorek,21 Christina M. Kennedy,22 Kristin M. Krewenka,23 Smitha Krishnan,14 Yael Mandelik,11 Margaret M. Mayfield,24 Iris Motzke,13,23 Theodore Munyuli,25 Brian A. Nault,26 Mark Otieno,27 Jessica Petersen,26 Gideon Pisanty,11 Simon G. Potts,27 Romina Rader,28 Taylor H. Ricketts,29 Maj Rundlöf,5,30 Colleen L. Seymour,31 Christof Schüepp,32,33 Hajnalka Szentgyörgyi,34 Hisatomo Taki,35 Teja Tscharntke,23 Carlos H. Vergara,36 Blandina F. Viana,18 Thomas C. Wanger,23 Catrin Westphal,23 Neal Williams,37 Alexandra M. Klein13, Wild Pollinators Enhance Fruit Set of Crops Regardless of Honey Bee Abundance.
61. *Millennium Ecosystem Assessment*. Island Press: Washington, DC, 2005.

62. Costanza, R.; d'Arge, R.; de Groot, R.; Farber, S.; Grasso, M.; Hannon, B.; Limburg, K.; Naeem, S.; O'Neill, R. V.; Paruelo, J.; Raskin, R. G.; Sutton, P.; van den Belt, M., The value of the world's ecosystem services and natural capital. *Nature* **1997**,*387*, 253–260.
63. Hallmann, C. A.; Sorg, M.; Jongejans, E.; Siepel, H.; Hofland, N.; Schwan, H.; Stenmans, W.; Müller, A.; Sumser, H.; Hörren, T.; Goulson, D.; de Kroon, H., More than 75 percent decline over 27 years in total flying insect biomass in protected areas. *PLoS One* **2017**,*12*, (10), e0185809.
64. Sánchez-Bayo, F.; Wyckhuys, K. A. G., Worldwide decline of the entomofauna: A review of its drivers. *Biological Conservation* **2019**,*232*, 8–27.
65. Christley, R. M.; Pinchbeck, G. L.; Bowers, R. G.; Clancy, D.; French, N. P.; Bennett, R.; Turner, J., Infection in Social Networks: Using Network Analysis to Identify High-Risk Individuals. *American Journal of Epidemiology* **2005**,*162*, (10), 1024–1031.
66. Deng, Y.; Jiang, Y.-H.; Yang, Y.; He, Z.; Luo, F.; Zhou, J., Molecular ecological network analyses. *BMC Bioinformatics* **2012**,*13*, (1), 113.
67. McNally, R. J., Can network analysis transform psychopathology? *Behaviour Research and Therapy* **2016**,*86*, 95–104.
68. Hafner-Burton, E. M.; Kahler, M.; Montgomery, A. H., Network Analysis for International Relations. *International Organization* **2009**,*63*, (3), 559–592.
69. Burkle, L. A.; Alarcon, R., The future of plant-pollinator diversity: Understanding interaction networks across time, space, and global change. *American Journal of Botany* **2011**,*98*, (3), 528–538.
70. Kennedy, C. M.; Lonsdorf, E.; Neel, M. C.; Williams, N. M.; Ricketts, T. H.; Winfree, R.; Bommarco, R.; Brittain, C.; Burley, A. L.; Cariveau, D.; Carvalheiro, L. G.; Chacoff, N. P.; Cunningham, S. A.; Danforth, B. N.; Dudenhöffer, J.-H.; Elle, E.; Gaines, H. R.; Garibaldi, L. A.; Gratton, C.; Holzschuh, A.; Isaacs, R.; Javorek, S. K.; Jha, S.; Klein, A. M.; Krewenka, K.; Mandelik, Y.; Mayfield, M. M.; Morandin, L.; Neame, L. A.; Otieno, M.; Park, M.; Potts, S. G.; Rundlöf, M.; Saez, A.; Steffan-Dewenter, I.; Taki, H.; Viana, B. F.; Westphal, C.; Wilson, J. K.; Greenleaf, S. S.; Kremen, C., A global quantitative synthesis of local and landscape effects on wild bee pollinators in agroecosystems. *Ecology Letters* **2013**,*16*, (5), 584–599.
71. Döke, M. A.; Frazier, M.; Grozinger, C. M., Overwintering honey bees: biology and management. *Current Opinion in Insect Science* **2015**,*10*, 185–193.
72. Watkins de Jong, E.; DeGrandi-Hoffman, G.; Chen, Y.; Graham, H.; Ziolkowski, N., Effects of diets containing different concentrations of pollen and pollen substitutes on physiology, Nosema burden, and virus titers in the honey bee (Apis mellifera L.). *Apidologie* **2019**,*50*, (6), 845–858.
73. Ramos-Jiliberto, R.; Moisset De Espanés, P.; Franco-Cisterna, M.; Petanidou, T.; Vázquez, D. P., Phenology determines the robustness of plant–pollinator networks. *Scientific Reports* **2018**,*8*, (1).
74. Brittain, C.; Williams, N.; Kremen, C.; Klein, A.-M., Synergistic effects of non- Apis bees and honey bees for pollination services. *Proceedings of the Royal Society B: Biological Sciences* **2013**,*280*, (1754), 20122767.
75. Honey Bee Colonies. In National Agricultural Statistics Service, A. S. B., Ed. 2018; p 20 pages.
76. Tulchinsky, T. H., Micronutrient Deficiency Conditions: Global Health Issues. *Public Health Reviews* **2010**,*32*, (1), 243–255.
77. Eilers, E. J.; Kremen, C.; Smith Greenleaf, S.; Garber, A. K.; Klein, A.-M., Contribution of pollinator-mediated crops to nutrients in the human food supply. *PLoS ONE* **2011**,*6*, (6), e21363.
78. DellaPenna, D., Nutritional Genomics: Manipulating Plant Micronutrients to Improve Human Health. *Science* **1999**,*285*, (5426), 375–379.
79. Ellis, A. M.; Myers, S. S.; Ricketts, T. H., Do pollinators contribute to nutritional health? *PLOS ONE* **2015**,*10*, (1), e114805.

Chapter 7
Biodiversity

Francesca Verones and Martin Dorber

7.1 Role and Importance of Biodiversity

Biodiversity can be described as the "variety of life" (Gaston et al. 2003) and is of utmost importance for humanity's physical, mental, and spiritual well-being. Also, the term "biodiversity" is closely connected to so-called ecosystem services. Here, we will explain what these terms encompass and discuss the importance of biodiversity for human well-being.

There are slightly differing definitions of biodiversity available. One of the most comprehensive definitions of biodiversity was set up in the Convention on Biological Diversity (CBD 1992):

> "Biological diversity" means the variability among living organisms from all sources including, inter alia, terrestrial, marine and other aquatic ecosystems and the ecological complexes of which they are part; this includes diversity within species, between species and of ecosystems.

According to this definition, biodiversity is encompassing genetic diversity (within species), different species (between species), and the differences between ecosystems and biomes.

Genetic diversity is showcased not only by the large number of different species but also by the difference between individuals of a single species (Biology Online 2020). One example for genetic diversity are domestic dogs. All domestic dogs belong to one species, the *Canis familiaris*, yet they are the morphologically most variable mammal species (Vilà et al. 1999). Thus, the genetic diversity between the different dog breeds is large (at least the genetic diversity relating to appearance).

F. Verones (✉) · M. Dorber
Industrial Ecology Programme, Norwegian University of Science and Technology, Trondheim, Norway
e-mail: francesca.verones@ntnu.no

B. R. Bakshi (ed.), *Engineering and Ecosystems*,
https://doi.org/10.1007/978-3-031-35692-6_7

On the other hand, the genetic diversity within one breed is usually small, since humans control the population size and by breeding also control the genes that are passed on. In general, the larger the pool of different genetic traits within one population is, the better are the species' long-term chances of survival, because these small genetic differences act as a "life insurance", for example, to ensure that hereditary diseases do not affect all species. Small genetic diversity may make a species more vulnerable for inbreeding and extinction due to future threats, for example, Australian humpback dolphin (*Sousa sahulensis*) (Parra et al. 2018).

The diversity between species can also be grouped into what is often referred to as alpha diversity and beta diversity. Alpha diversity is the variation of species within a community, that is, this can also be referred to as local diversity (Sepkoski 1988). The measures that are used for quantifying this alpha diversity can differ, but very often species richness is used, that is, the count of how many different species are present in a location. As an example, Muthuramkumar et al. (2000) counted 75 liana species in a region in India. That means the alpha diversity of lianas is 75 in this region.

The beta diversity refers to the differences between the communities, that is, it indicates the taxonomic differentiation between habitats (Sepkoski 1988). Assume, for example, that next to the 75 lianas in the tropical evergreen forest patch in India, we also counted lianas in a nearby (hypothetical) plantation. Assume that we found 17 species of liana there, of which 4 were not found in the evergreen forest patch, while the remaining 13 species overlap. Now, the beta diversity compares the difference between these two patches and essentially counts how many species are unique in each of these ecosystems. The beta diversity would thus be 66 (4 unique species in the plantation and 62 unique species in the forest).

Finally, the concept of gamma diversity indicates the total diversity within a landscape (Muthuramkumar et al. 2000) (e.g., an ecoregion, a region that contains distinct species communities and ecosystem types (Olson et al. 2001)) and thus represents regional biodiversity. In our liana example, the gamma diversity would be 79, that is, adding the 75 liana species from the forest and the four unique ones from the hypothetical plantation.

Ecosystem diversity is the combination of biotic diversity (biodiversity) and abiotic properties, such as different soil types, nutrients, and water. Globally, ecosystem diversity is exemplified through the large diversity among the existing ecosystems, such as deserts, rainforests, tundra, or coral reefs. Within one region ecosystem diversity describes then the complexity of an ecosystem, taking into account the number of niches available, the present trophic levels, and the number and complexity of species within the ecosystem. Coral reefs are, for example, ecosystems with incredible diversity and many different niches (e.g., different kinds of soft and hard corals).

It is important to stress that biodiversity consists of genetic, species, *and* ecosystem diversity, because biodiversity is very often understood by people as "species richness" and, as we will see further below, this is also often the means to assess impacts on "biodiversity."

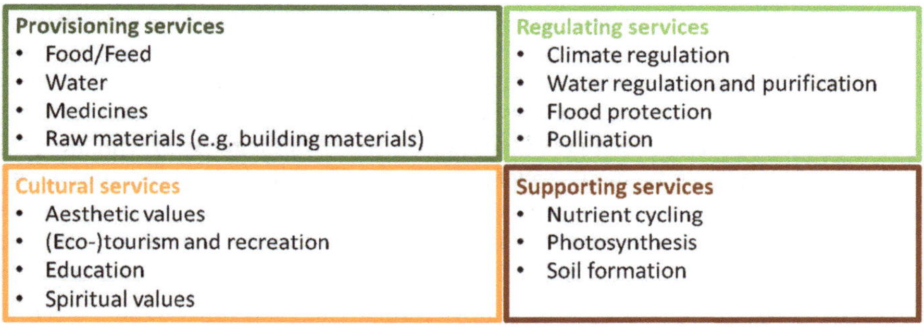

Fig. 7.1 Examples of ecosystem services within the four overarching categories of provisioning, regulating, cultural, and supporting services. (Based on (WWF 2018)

Another topic connected to biodiversity and human well-being is ecosystem services. These are benefits that we humans derive from the ecosystems around us (Millennium Ecosystem Assessment 2003). As described in Chap. 2, the Millennium Ecosystem Assessment differentiated between four categories of ecosystem service, namely, provisioning, regulating, supporting, and cultural ecosystem services (Fig. 7.1).

Provisioning services are also described as ecosystem goods (e.g., Cilliers et al. (2013)) and are usually easiest to assess and value, especially if a market exists for them. They encompass foods like fish and berries, building materials like timber and reed, or energy (hydropower, biomass for burning). Regulating services are a very varied group of services and include aspects such as flood regulation, water purification, carbon sequestration, or disease control. Cultural services include spiritual and aesthetic services, science and education, as well as recreational experiences. Supporting services are underpinning all the other services and are thus sometimes not classified as services themselves (La Notte et al. 2017). Without functioning supporting services, such as nutrient cycling, soil formation, or primary production, other ecosystem services would not work. Therefore, some authors also distinguish between "ecosystem processes" and "final ecosystem services" (e.g., Mace et al. 2012), with final services being the services that deliver goods/welfare directly, such as food and water regulation, while ecosystem processes are resulting from interactions of the ecosystems and organisms within (e.g., pollination, nutrient cycling). It is worth to note that in general, there is still a debate ongoing regarding a final classification and valuation scheme for ecosystem services, since the systems in place depend very much on the purpose of the study (Fisher et al. 2008).

An interesting question is whether biodiversity is an ecosystem service itself. This is called the "conservation perspective" by Mace et al. (2012) and ensures that the intrinsic value of biodiversity (albeit often limited to charismatic species like tigers, elephants, and giant pandas) is recognized. The protection of biodiversity is in this perspective thus considered next to other possible ecosystem services, such as flood protection (Eigenbrod et al. 2009; Nelson et al. 2009). The other perspective is called by Mace et al. (2012) the "ecosystem service perspective" and entails that biodiversity is the same as ecosystem services. However, none of the perspectives encompass the breadth of roles biodiversity can play. As Mace et al. (2012) put it there is a "multilayered" relationship between biodiversity and ecosystem services, with biodiversity being able to take on different roles within an ecosystem, namely, a regulator, a final ecosystem service, or a good. This highlights the complexity and breadth of the biodiversity concept. A regulator is defined by the authors as biodiversity that influences underpinning processes, for example, soil microbes are influencing the cycling of nutrients. Examples for biodiversity as final ecosystem services include medicinal plants or the genetic diversity of wild crops, who act as a safety net for future crop productions and biodiversity "goods" include values like humans caring for the existence of species (Mace et al. 2012).

In general, there is a positive relationship between biodiversity (e.g., species richness as a proxy) and ecosystem services (Costanza et al. 2007; Gamfeldt et al. 2013). More diversity in tree species were found, for example, to be positively correlated with ecosystem services such as game presence and berry production (Gamfeldt et al. 2013). It is very important to stress that biodiversity and their related ecosystem services are crucial for many aspects of our daily lives.

A very basic physical need for us humans is the need for food. Biodiversity is crucial for providing food, since agriculture relies on different crop and livestock species. If a system is richer in biodiversity, this means that functions of species often overlap and therefore provide a buffer against harsh and changing environmental conditions (Di Falco et al. 2010). Today, about two-thirds of the globally consumed calories can be attributed to three crops: rice, maize, and wheat (Gruber 2017). This dependency of our food system on just three species is critical, since the lack in genetic diversity means that the crops are susceptible to pests and stronger and stronger pesticides are needed to avoid yield losses. However, globally, more than 50,000 plant species grow that are edible (Gruber 2017). It is more and more recognized that to adapt to changing environmental conditions (e.g., drought frequencies, increasing temperatures), we should start to use a more diverse set of crops in agriculture. Old crops (with often lower yields) are rediscovered and investigated, such as fonio, an African cereal crop with a large potential to grow in conditions that include more frequent droughts (Abrouk et al. 2020), or wild tomatoes that can help breed new, more pest-resistant tomato varieties (Gruber 2017).

Humans are ecosystems themselves, with microbiota colonizing both the internal and external parts of the human body (Lindley et al. 2019). While some microorganisms can have detrimental effects, a plethora of microorganisms is highly important for our human well-being.

Shelter is another very basic need for humans and animals alike. Plants provide shelter for many different species and also provide building materials for sheltering humans, such as timber or straw. However, the construction sector is one of the most damaging sectors with a substantial impact on biodiversity and ecosystems (Opoku 2019), for example, by the extraction of timber (Fahrig 2001) and other building materials for construction purposes. However, urban systems would greatly benefit from embracong more biodiversity and ecosystem services friendly approaches. This is for example true for cooling and pollutant control (CABE 2006; Elmqvist et al. 2015), water regulation by e.g. trees intercepting rainfall and thus reducing the pressure on drainage systems (Pataki et al. 2011), as well as psychological benefits, such as reduced stress levels (Lee et al. 2009).

Humans need at times medication to get healthy again. The pharmaceutical industry has profited substantially from biodiversity, with numerous drugs and treatments originating from nature, antibiotics being one famous example (WHO et al. 2015). In 2012, still 75% of antibacterials approved by the US Food and Drug Administration have natural origins (Newman et al. 2012) and for future drug discoveries, biodiversity is considered crucial (Neergheen-Bhujun et al. 2017).

Research regarding the influence of biodiversity on the mental well-being of humans is at an early stage. Very often studies focus on the influence of "green spaces" on mental health and well-being, but do not go deeper into investigating the contribution of biodiversity as such (Marselle et al. 2019). These authors report on studies that investigate the role of biodiversity (measured differently per study, e.g., species richness and Shannon Index) regarding topics such as depression, anxiety, emotions, and stress levels. Even though they also found nonsignificant results, 14 of the studies showed positive relationships between biodiversity and mental health and well-being.

Spiritual well-being is also increasingly being recognized as important (Irvine et al. 2019). Spiritual beliefs can induce respect for certain biodiversity. Sacred places, like sacred groves or mountains, are often rich in biodiversity and can contribute to the protection thereof (Irvine et al. 2019). Wilderness recreation (e.g., whale-watching) is based, to a large extent, on the present biodiversity and is described as a secular type of spirituality (Irvine et al. 2019).

Apart from the direct effects on human well-being, of which we mentioned some few examples above, biodiversity is also important for numerous more indirect ecosystem services that are very important for us humans. Due to the plethora of services, it is not possible to give a comprehensive overview, but we hope to be able to give some relevant examples.

Biodiversity plays an important role for water purification (Cardinale 2011) and therefore, wetlands are sometimes used or artificially created to benefit from this service. In addition, wetlands are delivering many other ecosystem services, such as flood regulations, spiritual and aesthetic values, and food (Zedler et al. 2005). Biodiversity is also relevant for other regulating aspects, such as the protection of coastal regions from storms or the protection of the soil through limiting and halting erosion potentials, for example. Pollination is another, widely cited ecosystem service that is reliant on biodiversity. Brittain et al. (2013) found, for example, in almond orchards in California that orchards with a higher diversity in pollinators (honey bees, wild bees, and other wild pollinators like hover flies) had less reduction in pollination when it was very windy than orchards with only honey bees present. Thus, the complementarity of these species is ensuring the provision of the ecosystem service "pollination" even under (some degree) of environmental stress.

These examples show how our lives and biodiversity and ecosystem services are intertwined. However, we lose biodiversity at an alarming rate (Pimm et al. 2014) and it is therefore crucial to understand how we can measure and assess biodiversity. Current species extinction rates are estimated to be 1000 times higher than background extinction rates (Pimm et al. 2014). Rapidly increasing human pressures could further increase the species extinction rate (Sala et al. 2000) and cause a sixth species mass extinction event (Barnosky et al. 2011).

7.2 Quantifying Biodiversity

Biodiversity loss is mainly caused by habitat change, pollution, climate change, invasive species, and overexploitation (Pereira et al. 2012b). To halt this loss, several goals have been set up. There are global goals like the Aichi Targets and the Kunming-Montreal goals as successors thereof under the framework of the Convention on Biological Diversity (CBD), which aim to "*achieve a reduction of the current rate of biodiversity loss at the global, regional and national level*" (Secretariat of the Convention on Biological Diversity 2010; Secretariat of the Convention on Biological Diversity 2019), or the United Nations' Sustainable development goals (SDGs) which had aimed to "*take urgent and significant action to reduce the degradation of natural habitats, halt the loss of biodiversity and, by 2020, protect and prevent the extinction of threatened species*" (United Nations 2015). In addition, there are regional goals, in Europe formulated for example in the European green deal, which aim to "*restore biodiversity and cut pollution*" (European Comission 2020). Independent of the specific goals and spatial coverage, they have one thing in common: fulfillment of these biodiversity targets requires indicators, which can identify a change in biodiversity over time. However, due to the many layers of biodiversity, we cannot monitor all aspects of biodiversity at the same time. In this part, we therefore focus on indicators related to species richness, and due to data availability, these indicators are most commonly applied (Chiarucci et al. 2011). A variety of species richness indicators have been developed, which we will group into four categories here: species richness (number of species), abundance (number of individuals per species), evenness (the relative abundance of the different species in an area), and indicators accounting for how threatened species are. While the first three give each species the same value, the last indicator group gives threatened species higher conservation value than common species, and thus assumes that a threatened species contributes more to regional or national biodiversity than the ubiquitous species (Duelli et al. 2003).

An example for a simple species richness indicator is the so-called potentially disappeared fraction of species (PDF). The Life Cycle Assessment tool (See Sect. 10.3) uses the "potentially disappeared fraction of species" (PDF) as the unit for quantifying ecosystems quality impact. It is calculated by dividing the calculated species loss with the species richness of the corresponding area (Verones et al. 2017a).

The Living Planet Index (LPI) is a good example for an indicator related to species' abundance. The LPI tracks the abundance of 20,811 populations of 4392 species (mammals, birds, fish, reptiles, and amphibians) around the world and includes data for threatened and nonthreatened species. The LPI has its base year in 1970 (for which the value is set to 1) and then tracks if the populations are increasing, declining, or remaining stable. The 2020 global LPI shows an average 68% decline in monitored populations between 1970 and 2016. The LPI indicates the trend for the development of population numbers of selected vertebrate species, but it is not designed to show how many species have been lost or are extinct. Hence, it cannot be used as an indicator for species richness (Almond et al. 2020).

One indicator for "evenness" is the Shannon Index. The proportion of species relative to the total number of species is calculated, and then multiplied by the natural logarithm of this proportion. The resulting product is summed across species and multiplied by -1 (because the natural logarithm is negative) (Pielou 1966).

Finally, an example for an indicator focusing on "threatened species" is the IUCN Red List Index. The Red list Index (RLI) from the *International Union for Conservation of Nature (IUCN) (Butchart* et al. 2004a*)* shows trends in overall extinction risk for species over time. It is based on the IUCN Red List of Threatened Species, which classifies species into seven categories: least concern, non-threatened, vulnerable, endangered, critically endangered, extinct in the wild, and extinct. In addition, species with too little data for a thorough assessment are indicated as data-deficient. Species are assigned to one of these category if they meet a defined threshold in at least one of five criteria (Butchart et al. 2005). These quantitative criteria are based on population size, rate of decline, and area of distribution (Butchart et al. 2010). Since the RLI is monitoring a trend over time, the RLI can only be calculated for a set of species where the IUCN species category has been assessed at least twice (Butchart et al. 2005). To calculate the Red List index, it is counted how many species are within each threat category in each of the two temporally different assessments. In the second step, the differences in species numbers per category between these two time steps are evaluated. The number of species changing categories then shows a genuine improvement or a deterioration status over time (Lusseau et al. 2007). The RLI varies between 1 and 0, where a value of 1 represents that no species is expected to become extinct in the near future and an RLI value of 0 that all species have gone extinct. Hence, an upward trend in the RLI signifies that the rate of biodiversity loss has been reduced since the last assessment.

All these example indicators have in common that they need information about the presence of species in a specific area. In addition, some indicators (e.g., LPI and RLI) also need information on how abundant they are. For a species richness estimate, we only need to find one individual per species and area, while for the abundance, an estimate of the number of all individuals is required. As a result, species abundance estimates are harder to obtain than species richness estimates. One example for the IUCN Red list is the killer whale (*Orcinus orca*), where the geographic range is known, but the population trend, and therefore the abundance, is unknown (Reeves et al. 2017). On the other hand, an abundance estimate automatically confirms that a species is present in a region. Hence, abundance methods can contribute to species richness estimates, but pure species richness estimates methods cannot be used to estimate the abundance. Thus, we need to be familiar with approaches for quantifying the species richness and methods to estimate species abundance. Before presenting specific methods, we would like to point out that there are many different methods to estimate species richness and abundance, and that we are aware that the applied methods can vary between species. The following section therefore only gives examples of methods, with a focus on mammals and is not meant to give a comprehensive overview.

Species richness represents the number of species at a specific site or region and thus relies on species observations. Several options to obtain these species observations exist. One option is to use a human as species observer, either by foot or in combination with means of transportation (e.g., cars, boats, or airplanes) to count all species that can be found in one area. A second option is to use technical helping tools, like camera traps (cameras that are remotely activated via an active or passive sensor), drones, or even satellite images. Camera traps are mainly used to identify ground-dwelling vertebrates (mostly mammals). Due to technological advances and reductions in cost, the use of remote cameras has grown exponentially in the last decade (Steenweg et al. 2017). As a result, there is a growing global camera trap network, which inter alia can be used to monitor grizzly bears (*Ursus arctos*) or mule deer (*Odocoileus hemionus*) (Steenweg et al. 2017).

Since drones nowadays are relatively easy to use and cheaper with increased image quality, they are increasingly being used to monitor different fauna, including birds, elephants, and marine mammals (Lyons et al. 2019). A recent case-study highlighted that with the use of drones, birds can be monitored more accurately than with the traditional human, ground-based method (Hodgson et al. 2018). Furthermore, it has been shown that WorldView-3 satellite imagery can be used to monitor whales from space (Bamford et al. 2020).

The so far described methods looked at observing the species directly. However, as some species can have long flight distances (Møller 2008), which is the distance between an animal and an observer at the moment of flight initiation (Boer et al. 2004), and may be difficult to observe directly (e.g., a species that is night active and lives in dense forest), it is also possible to indirectly identify a species with, for example, excrements (e.g., done for wild boars (Acevedo et al. 2006), feathers, hairs (e.g., grizzly bear hair traps (*Ursus arctos*) (Apps et al. 2004)), sound (Stowell et al. 2019), or other signs that a specific species is present.

During species observations in the field, it is important to keep in mind that most animal species are mobile and not sedentary. Species have different home ranges (Ofstad et al. 2016), different movement patterns (the wolverine (*Gulo gulo*), e.g., may walk more than 20 km per day to search for food) and different species have different migration patterns. For the artic tern (*Sterna paradisaea),* an annual movement of 80,000 km has been reported, as they migrate from boreal and high arctic breeding grounds to the Southern Ocean (Egevang et al. 2010). As a result, the number of different species observations will increase with observation time. It is important to note that the relationship between number of species and observation time is not linear. While in the beginning, it is relatively easy to find a new species, finding the "last unknown" species will take a lot more time. A camera trap study to monitor terrestrial mammal species within Tanzania's Ruaha National Park, for example, needed 650 camera trap days to identify 30 species, while they only discovered 8 more species in the next 2170 camera days (Guralnick et al. 2015). Furthermore, the species activity can be different between seasons or depends on the weather. If the species are, for example, less active, it is harder to observe them, and hence, more time is required to find the same number of species. The study from Guralnick et al. (2015), for example, needed 650 camera trap days to identify 30 species in the dry seasons, while they need 1358 camera trap days in the wet season for the same number of species. In addition, some migrating species will only be observable during a certain time of the year. As monitoring normally only has a limited timeframe, we should keep in mind that our species richness counts are normally only an estimate. As a result, the current number of species on Earth is still unknown, but the number of species has been estimated to be around 8.7 million eukaryote species (Mace et al. 2011). However, as for many species groups only a small group has formally been described, it is important to point out that we can today only identify a small, probably atypical, part of the species richness biodiversity (Purvis et al. 2000).

Does this mean that it is always necessary to set up fieldwork surveys oneself, as described, for example, in Field et al. (2002), to estimate spices richness of a study area? The answer is "No," because there are a growing number of online databases, which collect global species occurrence records (an occurrence is a species observation at a specific location and time, offering evidence of the occurrence of a species (or other taxon) at that particular place and time). An example of such a database is the database of the Global Biodiversity Information Facility (GBIF - www.gbif.org). At the time of writing this chapter, the database contained over 2,300,000,000 species occurrence records, provided by more than 2100 institutions. Previously, these occurrence points have been mainly collected with a research focus, including a systematic assessment, but today, citizen science is playing an increasingly important role in the collection of these occurrence points. Citizen science can be defined as the involvement of nonprofessionals in scientific research and environmental monitoring (Chandler et al. 2017). With webpages and apps like inaturalist (www.inaturalist.org), you can contribute to increase the number of species occurrence points, by simply reporting what species you "discovered" in your free time. However, as this citizen science data is not assessed in a systematic way (compared to monitoring programs), this data can be biased toward some species and regions, which must be considered when using the data (Phillips et al. 2009).

One way to move directly from species occurrence points to a species richness estimate, is to obtain all occurrence points of the target species group (e.g. fish) for one area (e.g. catchment) and then count the number of different species (Dorber et al. 2019b).

The second option is to use these occurrence points in combination with, for example, climate data to run a species distribution model. Species distribution models are numerical tools that combine species occurrence observations with environmental estimates. Main purpose of the models is to predict the distributions of the species across landscapes (Elith et al. 2009). A commonly used software to perform this modeling is Maxent (Elith et al. 2011). The distribution of a species can then be stored as a so called "species range map". The advantage of range maps is that they provide estimates of species richness in areas where no field work has been performed, just based on the underlying environmental conditions.

Range maps can also be drawn based on expert knowledge. Then they are more informed than via Maxent models, but it is very time-consuming to collect the necessary species data and draw these maps. IUCN, for example, provides at the time of writing expert-based range maps for 105,500 species.

The advantage of the range maps is that by stacking all available individual species maps on top of each other, they provide information of species richness in an area.

So far, we looked at methods to measure species richness, but for abundance and threat level indicators we need to know the number of individuals of each species in a certain region. The ideal way to obtain this information is to do a total count of all individuals (also called complete species census) in a certain area (Haig et al. 2005). However, due to complexity of finding and locating wild animals, this method is only applicable to small areas and a limited amount of species in parallel. An example here would be elephants (*Loxodonta africana*) in the Krueger national park (Ferreira et al. 2017).

There are numerous approaches available for estimating abundances and we will just present the examples of the fixed-transect method and the capture-recapture method.

To estimate abundance a transect with a fixed width can be placed in a study area (the shape can vary) (Schwarz et al. 1999). Assume that we have a study area of 20,000 m^2 and we set up 6 transects with a length of 200 m and a diameter of 10 m each. One person walks along these transects and counts all individuals of the targets species that are visible within this transect. Individuals outside the transects are not counted even if observed. Let's assume that we counted 20 individuals of a certain species inside the transects. Now we can say that there is a relationship between counted individuals to total individuals and transect area to study area. As a result, the number of individuals in the study area can be estimated by dividing the study area with the total transect area and multiplying it by the number of counted individuals. This gives us a final estimate of 200 individuals for the example study area mentioned.

For species that might be difficult to detect visually (e.g., fish) the capture-recapture method offers an additional option to estimate species abundance (Schwarz et al. 1999). As the name indicates, the first step is to define a setup to capture species in the desired study area. The species that are captured are marked and afterwards released into the wild. Assume that we captured and marked five individuals in the first capture. After the individuals had time to mix with the natural population a second capture (recapture) is performed (it is important to note that the recapture setup should be the same as in the first capture). In the second capture we captured in total eight individuals, out of which two were already marked. Now there is a relationship between individuals marked in the first capture to total individuals and marked individuals recaptured to total individuals. Using this relationship, we can estimate the number of individuals to 20. For this method it is important that the capture probability between first capture and recapture capture does not change. This could happen due to habitat change (seasonality) or growth of the individuals (from juvenile to adult).

As some species might be physically hard to catch (especially predators), capture and recapture approaches can also be carried out in and indirect way. For some species where specific individuals are distinguishable by distinct features, such as differences in spots, stripes, or fluke shapes, for example, the jaguar (*Panthera onca*), the capture-recapture method can be carried out with camera traps (Jędrzejewski et al. 2016). In addition, the capture-recapture approach can be performed with noninvasive genetic sampling (Lukacs et al. 2005). This is, for example, done for grizzly bears (*Ursus arctos*), by using wire as hair trap .The hairs can then be used to analyze the DNA and to "mark" the individual bears (Apps et al. 2004).

Abundance can also be an indirect number of species numbers estimates. For example, the number of camera trap pictures or number of beaver dams. For beavers, for example, it is common to look for active beaver colonies with planes. A colony is then classified as active based on various signs, typically a lodge and food cache. It can then, for example, be assumed that each colony hosts five beavers to estimate the abundance (Woolf et al. 2003).

In summary, we have shown how we can obtain species richness and abundance information to calculate different categories of indicators: species richness, abundance, evenness, and indicators accounting for how threatened species are. When using and interoperating these indicators, we have to be aware of the inherent uncertainty of these measures and that the entire species richness is still unknown.

Which of the previously described indicator should be used to measure biodiversity? To answer this question, Santini et al. (2017) performed a virtual case study, where they investigated the results of 12 biodiversity indicators when applied to 9 different scenarios of biodiversity change. As their results between the indicators differed, the authors concluded that biodiversity monitoring shall not be done by using only one indicator. Also Duelli et al. (2003) conclude that there is no single best indicator for biodiversity. However, these authors also point out that the choice of indicators depends on the aspect of biodiversity to be evaluated (hence on the research question or biodiversity goal) and that it is guided by a value system based on personal and/or professional motivation.

Hence, there is not one single biodiversity indicator that could be universally used and recommended. However, global targets like the SDGs (United Nations 2015) need indicators to monitor the progress and a decision has to be made which indicators are used. In the next chapter we will show some example of how biodiversity indicators can contribute to decision-making.

7.3 Biodiversity in Decision Making

As described in the sections before, biodiversity is a complex term and there are multiple ways in which biodiversity, or parts thereof, can be quantified. When it comes to decision-making, there are also multiple indicators and methodologies that can be used.

We are fully aware that there is a large variety of methods and approaches available that cannot all be covered within this chapter. We thus focus on some selected methods and indexes that we regard as important tools for sustainability assessments, centering mostly around species diversity and with the largest focus on life cycle assessment (LCA). Tools that routinely account for other aspects of biodiversity are unfortunately scarce within sustainability assessments.

7.3.1 Red List Index

The calculations behind the Red List Index (RLI) have been mentioned in the previous section. The RLI is indicating the projected relative extinction risk of a species and can be calculated for any species that has been assessed at least twice. The RLI can be used, for example, to outline how much progress has been made toward reaching internationally set biodiversity targets, such as the biodiversity targets set by the Convention on Biological Diversity (CBD) (Butchart et al. 2005). The RLI is indicating how much, in between two points in time, a species is moving toward or away from extinction. In order to remark progress toward the goal of halting biodiversity loss, we would need to see a positive trend in the RLI, which would show that the species assemblage assessed is on average moving away from the threat of extinction. Often, however, trends are negative, meaning the species is moving closer to extinction. However, the trend might become less steep and that means that the species is moving closer to extinction at a slower rate. Butchart et al. (2005) report, for example, negative trends for birds (−6.9%) and amphibians (−13.7%), meaning that at that time, the loss of species diversity for these species groups was continuing.

RLIs can be calculated and interpreted at global level (e.g., for mammals (Hoffmann et al. 2010), birds (Butchart et al. 2004b; Butchart 2008), and amphibians (Butchart et al. 2005)), but, based on national or regional Red Lists, they can also be calculated on a national level, as exemplified for 11 taxonomic groups in Finland by Juslén et al. (2013). In Finland, the RLI was on average decreasing by 0.3% between 2000 and 2010. However, as the authors correctly point out it can be misleading to only look at one overall assessment or to focus on just one taxonomic group. Of the 11 taxonomic groups that were assessed, 6 showed an increasing trend and 5 a decreasing one. That means that there are species groups that move away from the extinction threat, but overall, the trend is still negative. While it is very helpful to have global and averaged assessments for a first screening and general, global trends, it is very important to try and increase the spatial and taxonomic resolution for assessing trends with the Red List Index approach, if the data allows for it. One example is the use of the RLI to assess the status of vascular plant on a national *and* subnational level in Spain (Saiz et al. 2015), which shows different rates of decrease in different areas. The authors do point out though, that red listing of species is a very work-intensive task and since at least two complete lists (at different points in time) are needed for applying the RLI approach, this is an approach that cannot be used in all countries or regions (yet).

7.3.2 Living Planet Index

The Living Planet Index (LPI) has also been mentioned in the previous section. It takes mammals, birds, fish, amphibians, and reptiles into account. Since it measures the changes in the abundance of individual species populations, it can be used at very different scales, for example, on global or national levels, to identify for each country individually how species abundance changes. This is, for example, the case in Uganda, which publishes the "State of Uganda's Biodiversity" series, based on the LPI (Pomeroy et al. 2017). The 2017 report shows that while some species have been decimated very strongly (e.g., 90% of the Kampala bats and 60% of Crowned Cranes have disappeared since 1970), others, like Marabou storks, are increasing.

The LPI has also been used to investigate how effective protected areas are (Milligan et al. 2014). The authors calculated the LPI for more than 4000 populations of over 1500 different species in protected areas within 130 countries and could show that in 39% of the countries, wildlife populations were declining within protected areas. Apart from showing globally large differences between protected area types and countries, the study also showed that there is a strong taxonomic bias (Milligan et al. 2014). While fish abundance in protected areas increased globally by 182% since 1970 and mammals and birds also show an increase (10% and 57% respectively), amphibians and reptiles show a strong negative abundance trend within protected areas (−74%). On the other hand, migratory fish have in general (i.e., not restricted to protected areas) seen a reduction of 76% with values being largest in Europe (−93%) and smallest in North America (−28%) (Deinet et al. 2020).

7.3.3 Environmental Impact Assessment (EIA)

Apart from single indicators, more holistic assessment approaches are available. An environmental impact assessment (EIA) is a study that is conducted to investigate the possible environmental impacts of a project. An EIA is carried out prior to the decision-making on that project, in order to weight the environmental impacts against the benefits of a project. If an EIA is carried out early in the planning process, it can help to identify approaches to minimize the identified environmental impacts. Results of an EIA do not need to fulfill a predetermined environmental outcome, but rather they present the decision-makers with the loss of environmental values of the different options investigated. Regulations and legislation regarding EIA vary from country to country, but generally it involves a screening, scoping, the development of the actual report, subsequent decision-making and, if necessary, monitoring during implementation (European Commission 2017).

During the screening it is evaluated whether a project indeed needs a thoroughly carried out EIA. This can be determined either by a case-by-case manner or through predefining certain thresholds. Some types of projects are automatically subject to an EIA and do not need a screening first. In Europe, this includes any type of project involving, for example, the refining of crude-oil, nuclear power plants, chemical production plants, motorways, and roads with more than four lanes, dams storing more than 10 million m^3 of water, pipelines with a certain diameter, quarries, or high-voltage lines (European Parliament et al. 2011). Projects types that need to be screened include, for example, aquaculture projects, wind farms, cement kilns, rubber industry, or industrial estate development (European Parliament et al. 2011).

In the following scoping phase, it is defined what information needs to be provided to make an informed decision about the environmental impacts of the planned project, as well as the decision regarding which approaches should be employed.

The actual report then always contains the baseline scenario and its foreseen environmental impacts, as well as the proposition of several alternatives and their respective environmental impacts. Based on this report, the decision-makers assess whether the project contains significant effects on the environment and whether the project can be carried out or not.

The types of environmental impacts that should be considered depend on the project in question (e.g., an aquaculture project will have a completely different set of impacts than a power line through a forested area), but some factors that are suggested are investigating impacts on fauna and flora, soils, human health, hydrology (including water quality), climate, noise, visual pollution, or air quality. Both the magnitude of the impact and the spatial extent of the impact, as well as the probability of the impact should be assessed and information about the duration, frequency, and whether the impact is reversible or not should be indicated (European Commission 2017).

However, the actual methods in how these impacts should be assessed are not predefined and vary from impact to impact. Assessments can both be qualitative or quantitative, and they can, for example, include map analyses, statistical models, check lists, and, in most cases, involve field surveys to collect data for the EIA. If chemical impacts are to be assessed, for example, environmental risk assessments (ERAs) can be used (Morris et al. 2001). In some cases, also a Life Cycle Assessment (see more details further below) can form part of an EIA, even though the scope of an LCA is usually not site-specific, while an EIA relies on site-specific data. An LCA can, however, showcase the impacts that are occurring throughout the supply chain.

Including biodiversity in EIA is very often done based on the abundance of species (see description in the previous section), while other aspects of biodiversity, such as genetic variation or ecosystem functioning is, similar as in most other tools, often neglected (Atkinson et al. 2000; Gontier et al. 2006; Khera et al. 2010). However, due to the liberty an EIA provides with the choice of methods, it is of course possible to include the other aspects if data can be collected and if these impacts are deemed relevant. In an analysis of 42 EIAs in Southern France seven included impacts on all levels of biodiversity (i.e. genetic variation, species diversity and ecosystem functioning) (Bigard et al. 2017). About half of the EIAs presented in this study used expert advice of fauna and flora to determine potential impacts and while almost all EIAs included field visits to collect actual data on flora and fauna only two thirds of the EIA actually did so in more than one season. Gontier et al. (2006) found in a review of 38 EIAs in Europe that even though all conducted some qualitative form of biodiversity assessment, only eight actually tried to present some quantitative results. The main method used for biodiversity assessments were inventories of species presence, while some more specialized approaches included habitat surveys or the use of indicator species (Gontier et al. 2006). All in all, the authors conclude that biodiversity is still underrepresented in many EIAs, even though many different options of ecological modeling (coupled e.g. with GIS) are available nowadays.

7.3.4 Life Cycle Assessment (LCA)

Life Cycle Assessment (LCA) is a mature environmental management methodology that is routinely used by businesses (Unilever 2020), governments (European Commission et al. 2010; Frischknecht et al. 2013) and other stakeholders for environmental decision-support. It can be used to assess impacts along the entire value chains of products, processes or services. LCA is designed to be used in a comparative manner, comparing the environmental impacts of the assessed product with a product with the same functionality (e.g. comparing a glass bottle and a plastic bottle), in order to show which alternative is preferable from an environmental point of view and to identify where in the value chains the largest impacts and trade-offs between impacts can be found (Hellweg et al. 2014).

LCA consists of four phases: the goal and scope definition, the life cycle inventory collection, the life cycle impact assessment phase, and the interpretation (ISO 2006). The goal and scope phase is about defining the system boundaries, choosing an appropriate functional unit, and making choices about what should be in- and excluded from the assessment. The life cycle inventory is used to collect all the material uses and emissions created throughout the entire life cycle of a product or a process, that is, how many kg of wood, how many liters of water and how many m^2 of land etc. have been used, and how many kg of greenhouse gas or other emissions have been released into the environment.

Life Cycle Impact Assessment (LCIA) then groups the different inventory results into different impact categories. There are several LCIA methods available, some of the most well-known and recent ones are TRACI (Bare 2011), ReCiPe 2016 (Huijbregts et al. 2017), ImpactWorld+ (Bulle et al. 2019), LC-IMPACT (Verones et al. 2020), and LIME (in Japan) (Itsubo et al. 2003). Depending on the LCIA methodology, there are different impact categories included and behind each of these categories, there is a different model. It is important to understand that even though most software is carrying out LCIAs automatically, there are different assumptions and models behind each of them. A good overview of some of the most recent methods is also presented in Rosenbaum (2018).

One aspect that is differentiating some of the methods is also whether impact categories are at midpoint or at damage (also called endpoint) level. Midpoint indicators are traditionally defined as an indicator "*located on the impact pathway at an intermediate position between the LCI results and the ultimate environmental damage*" (Jolliet et al. 2004). Examples of midpoint indicators are kg CO_2-equivalents or kg 1,4DCB-equivalents. These are indicators that are very helpful to compare the impacts within one category (e.g. comparing the climate change impact of carbon dioxide vs. methane), but they do not tell us much about the consequences for biodiversity or human health. Endpoint indicators, on the other hand, model the entire impact pathway up to the damage in either ecosystem quality, resources, or human health. Many newer methods, especially many of those that display impacts on biodiversity, are not modeling the impacts via a midpoint, but go directly to damage (e.g., de Souza et al. 2013; Chaudhary et al. 2015; Verones et al. 2017c; Dorber et al. 2019a). Therefore, the LCIA framework has been updated to reflect this difference (Verones et al. 2017a) and it is now explicitly mentioned that *"[m]odels stopping at midpoint level, or models going directly to damage, or models encompassing both, are equally appropriate"* (Verones et al. 2017a). Impact categories at damage level that are dealing with biodiversity are (mostly) using "potentially disappeared fraction of species" (PDF) as the unit for quantifying the impacts. While this is a more abstract concept, it is very helpful for comparing impacts across impact categories. Since all impact categories are indicated in PDF, we can now compare the impact of, for example, climate change with the impact of eutrophication or land use. This possibility of comparison often makes damage level metrics easier to communicate to stakeholders.

As mentioned earlier, biodiversity is in many models set equal with species richness. This is true for most LCIA models as well, with a few exceptions (e.g., Souza et al. (2013)). Most LCIA methodologies cover impacts of climate change, land use, water consumption, eutrophication, terrestrial acidification, and ecotoxicity on biodiversity (Rosenbaum 2018).

The way biodiversity is modeled is of course specific to each individual approach, but many models are based on some form of species area, species discharge, or species sensitivity distributions.

The species-area relationship (SAR) is a widely used concept in ecology (Rosenzweig 1995). It has to be noted that there are many different forms of SARs (Matthews et al. 2020), but all of them ultimately relate some form of an area to species numbers. We do not aim to comprehensively describe the history and forms of SARs here, but only focus on the SARs relevant for and used in LCIA.

The so-called classical SAR is a power function between the Area A and the species number S, with c and z two parameters that depend on the region, taxonomic group in question, and the sampling regime (Rosenzweig 1995).

$$S = c \cdot A^z$$

In LCIA models, we use this equation to assess the number of species that are lost (S_{lost}), that is, we calculate based on the "original" number of species (S_{org}), the corresponding habitat area (A_{org}), and the new habitat area (A_{new}), that is, the area that is left of a habitat after a land use change.

$$S_{\text{lost}} = S_{\text{org}} \cdot \left(1 - \left(\frac{A_{\text{new}}}{A_{\text{org}}}\right)^z\right)$$

As an example, imagine the original area being a pristine forest with all its species (S_{org}). If a certain area of that forest is now deforested, the new area will be what is left of that forest in question.

Dividing S_{lost} by S_{org} will then give us the potentially disappeared fraction (PDF) of species. The classical SAR has been used early on in LCIA models (Koellner et al. 2008; Schmidt 2008). However, the classical SAR assumes that all nonnatural areas, such as urban or agricultural areas, are completely hostile to biodiversity (Pereira et al. 2012a). We do know that there are species that can survive very well in human-modified landscapes. Foxes, badgers, or raccoons, for example, often thrive in cities, exploiting human food sources and making use of multiple sheltering opportunities (Bateman et al. 2012). On the other hand, there are species that are highly sensitive to habitat change and will perish more easily. The Abbott's duiker (*Cephalophus spadix*), a small, forest-dwelling mammal endemic to Tanzania, is, for example, highly sensitive to habitat fragmentation (Keinath et al. 2017). The classical SAR does not account for these differences in responses.

The matrix-calibrated SAR (Koh et al. 2010) is one further development of the SAR that takes the matrix effects (the habitat that is provided by human-modified land) into account and assesses the response of each taxa separately. This version of the SAR has also been used in LCIA models (de Baan et al. 2013). However, also the matrix-calibrated SAR assumes that all species are lost if no natural habitat remains in the region. Again, some species are capable of surviving in the absence of any natural habitat, for example, the house mouse (*Mus musculus*).

Yet another updated version of the SAR is the countryside SAR (Pereira et al. 2006). That one accounts for the fact that species use habitats differently (i.e., their affinity to certain habitat is different) and predicts that some species are indeed able to survive (albeit perhaps not necessarily to thrive) in the absence of natural habitat.

The countryside SAR has also been used in LCIA models, to model the impacts of land use specifically for individual taxa (mammals, birds, reptiles, amphibians, and plants) and for different land use types (annual crops, permanent crops, intensive and extensive forestry, urban areas, and pasture) (Chaudhary et al. 2015). The countryside SAR does not account for fragmentation effects though (Kuipers et al. 2019b); therefore, new models are being developed for use in LCA, in order to take the spatial configuration of habitats better into account. One example for a model that includes habitat loss and fragmentation in LCIA simultaneously is the species-habitat relationship proposed by Kuipers et al. (2021a, b).

Similar to an SAR, species-discharge relationships (SDRs) exist for freshwater ecosystems and are relating the number of fish species to the amount of discharge in a river (Xenopoulos et al. 2006). The discharge in a river can be reduced due to water consumption. Water consumption describes the abstraction of water (in this case, from a river) that is used and not returned to the watershed of origin. Water that is abstracted from a river and used for irrigation, for example, or for cooling water in cooling towers will be either incorporated into the product or evaporated and thus does not return to the river, which is therefore deprived of some part of its original discharge. Consequently, species numbers will decrease, because the species have less habitat (less volume) available. The SDR has been used in LCIA to assess the impacts of water consumption in several models (e.g., Hanafiah et al. 2011a; Tendall et al. 2014b; Dorber et al. 2019b; Pierrat et al. 2023).

Another frequently used modeling type is the species sensitivity distribution (SSD) that is used, for example, in the impact categories of ecotoxicity, freshwater eutrophication, or terrestrial acidification. An SSD is a cumulative probability distribution showing the investigated impact for multiple species; thus, it shows the sensitivity of an ecosystem to a specific substance (Hauschild et al. 2015). The impact in question can be, for example, death of species or chronic impacts (e.g., growth inhibition). In order to derive an SSD, data for the sensitivity of individual species to a given substance needs to be collected. This data is either derived in own lab tests or collected from the available literature. If possible, the aim should be to collect information for at least three different species from three different trophic levels. In most cases, the information needed would be the EC50 or LC50 of a given substance for a specific species. LC50 and EC50 stand for the lethal and effective concentration at which 50% of the species die or show an impact. Once the data has been collected (uniformly for all species, i.e., either all LC or all EC values), it is sorted from lowest to highest value, that is, the lowest EC50 value, representing the most sensitive species, is listed first. After plotting all the data points in an empirical cumulative distribution function (concentration, the hazardous concentration on the x-axis, the fraction of affected species on the y-axis), a cumulative distribution function is fitted (e.g., normal, lognormal, logistic). The best fit is the model that is subsequently used. In LCIA, the effect factors for the impacts of these substances on ecosystems are then derived by picking the concentration value from the curve where 50% of the species are exposed to concentrations that are higher than their respective EC50s, that is, 50% of the species are affected. Recently, models have started to use EC10 (i.e. 10% of species affected) to derive HC20 values (Owsianiak et al. 2023). This is because most toxicity levels are low and chronic values are thus better able to represent the environmental conditions that EC50, which are often based on acute tests (i.e. death of the species).

7.3.5 Input-Output Models

Another method that can be used to trace the impacts on biodiversity through our complex world of globalized supply chains is input-output analysis, or, more accurately, environmentally extended multiregional input-output models (EE-MRIOs). Basically input-output models describe the economic flows between all different sectors of a country within a year. They essentially describe the trade relations between all countries on an individual level. An environmental extension is then indicating the environmental impact of that sector in a year (Kitzes 2013). Since the trade relations between all countries are traceable in an EE-MRIO, indirect environmental impacts can be traced through the entire system. Marques et al. (2017) list several available EE-MRIO. The environmental extensions available are for some of these models limited to carbon emissions only, while some are also able to assess land-use-related, water-use-related, and pollution-related aspects.

Regarding impacts on biodiversity, Lenzen et al. (2012) were the first to add a biodiversity extension. They linked the threats listed for vulnerable, endangered and critically endangered species from the IUCN Red List to one or several, specific economic sectors with more than 15,000 commodities and found that 30% of the global threats for these species are due to international trade. Other approaches than using the number of threatened species have used mean species abundance (Wilting et al. 2017), potentially disappeared fractions of species (linking it with an LCIA approach) (Verones et al. 2017b; Koslowski et al. 2020), species-yr (Koslowski et al. 2020), or occupied bird ranges and missing individual birds (Kitzes et al. 2017) as indicators for biodiversity impacts.

For most assessment methods, data scarcity is an issue. Knowledge gaps can both be related to unknown species and poorly researched regions. For studies at global level, these issues are even more relevant, since there may be significant bias in the results due to these knowledge gaps.

For MRIOs, apart from the lack of biodiversity data, coarse spatial resolution (country scale at best), lack of details for certain economic sectors, and good information for linking observed threats to specific industries are additional issues (Moran et al. 2016). However, ongoing research is contributing to closing these research gaps.

7.4 Application Example of Biodiversity Impacts Within LCA

In this section, we will use the study from Dorber et al. (2020) to showcase how Life Cycle Impact Assessment (LCIA) can be used to quantify biodiversity impacts of hydropower reservoirs and how the results shall be interpreted. In addition, we will show the kind of biodiversity data that is used in applied LCIA models.

For sustainable development, an increase in the share of renewable energy in the global energy mix is urgently needed. Reservoir-based hydropower plays an important role in future energy supply as its expansion is often attractive from an energy and sustainable development perspective, since it provides, in most cases, cheap electricity with a comparably low carbon footprint (UNEP 2016). So far, there has been a strong focus on the technical development potential of hydropower, meaning that studies like Gernaat et al. (2017) have identified where new hydropower reservoirs could be constructed that are economically feasible. In addition, most focus has been on the climate change mitigation potential of such reservoirs (Pehl et al. 2017). However, hydropower expansion can also have serious biodiversity impacts (Gracey et al. 2016), leading to loss of ecosystems services and therewith ultimately affecting human well-being. It is therefore not enough to focus on technical feasibility and climate change impacts only for site selections of future hydropower reservoirs, but there is a need to also include biodiversity-related impacts into the decision-making process. However, so far, only a few studies have quantified biodiversity impacts of hydropower electricity on a global scale (Gibon et al. 2017; Zarfl et al. 2019) and these studies were either limited by species coverage or by the spatial detail included.

Dorber et al. (2020) built their study on Gernaat et al. (2017), who identified globally 1956 possible new hydropower reservoirs that could produce 3.9 PWh yr^{-1} with a production cost below 0.1 US$ per kWh. Gernaat et al. (2017) focused on technical and economic feasibility only. Dorber et al. (2020) then used newly developed life cycle impact assessment models: to assess potential biodiversity impacts of these possible future hydropower reservoirs. Their aim was to answer the following three research questions:

1. Where can hydropower electricity be produced with the least biodiversity impact?
2. How much biodiversity impact can be avoided by not exploiting the full hydropower potential?
3. How much does site selection require a trade-off between terrestrial and aquatic biodiversity impacts?

Why did the authors choose to use the LCA framework? While the first research question benefits of the strength of LCA to compare two products/processes, the second research question uses the options to perform a hot-spot analysis with LCA. The third research question highlights that it is important to include as many relevant impacts as possible, as required for an LCA.

More specifically, Dorber et al. (2020) quantified the potential terrestrial and freshwater biodiversity impacts of land occupation, water consumption, and methane emissions during a possible future reservoir operation.

To quantify the terrestrial biodiversity impact of land occupation, they used an LCIA model from Dorber et al. (2019a), denoting the PDF per m^2 future land occupation. The LCIA model itself is based on the SAR concept (see previous section) and used IUCN range maps as species richness data input.

To quantify the aquatic biodiversity damage of water consumption, LCIA models using the species-discharge relationship (SDRs, see previous section) from several studies (Hanafiah et al. 2011b; Tendall et al. 2014a) were used. To obtain the SDRs, both studies used species occurrence points from databases as input data (as described in Sect. 7.2). For studying the impacts of methane emissions on aquatic and terrestrial biodiversity, Dorber and colleagues used the LCIA model from LC-Impact (Verones et al. 2020), denoting the potentially disappeared fraction of species per degree temperature increase. The model itself is based on the SAR and a species distribution model.

Since some of the LCIA models in its original version calculated regional PDFs (indicating a fraction of potential regional species extirpations), Dorber et al. (2020) used the global extinction probability (GEP) from Kuipers et al. (2019a) to convert them into global PDFs (indicating a fraction of potential global species extinctions). The GEPs are based on local species range sizes, species threat levels, and species richness. Hence, here, in addition to species richness information also, an indicator accounting for how threatened species are is included.

Dorber et al. (2020) did not find a strong correlation between biodiversity impact and methane emissions per kWh. This in turn means, that, if mitigating climate change is the main motivation for increased hydropower production (and thus only climate change indicators are used), it is likely that a potential biodiversity impact is overlooked. This strengthens our conviction that for sustainable hydropower development, biodiversity impacts have to be one of the decisions layers. However, as also pointed out in the publication, other factors like electricity demand, social aspects, or human health impacts should also be included in the final decision-making.

Overall, the results show that careful selection of future hydropower reservoirs has a large potential to limit biodiversity impacts as, for example, globally, 3.9% of the hydropower potential accounts for 51% of terrestrial biodiversity impact. In other words, already half of the terrestrial biodiversity impact would be avoided, if only the other 96% of the global hydropower potential would be used. These results are triggered by significant impact differences between the reservoirs. For example, the reservoir with highest terrestrial biodiversity impact produces one kWh with 1,475,000 times the impact of the reservoir with lowest terrestrial biodiversity impact (Fig. 7.2).

As dominant explanatory factors of the variance in the quantified aquatic and terrestrial biodiversity impact, Dorber and colleagues identified local environmental factors (river size, ecoregion area, river location, species richness, and global extinction probability) and not the pure amount of water consumed or land occupied per

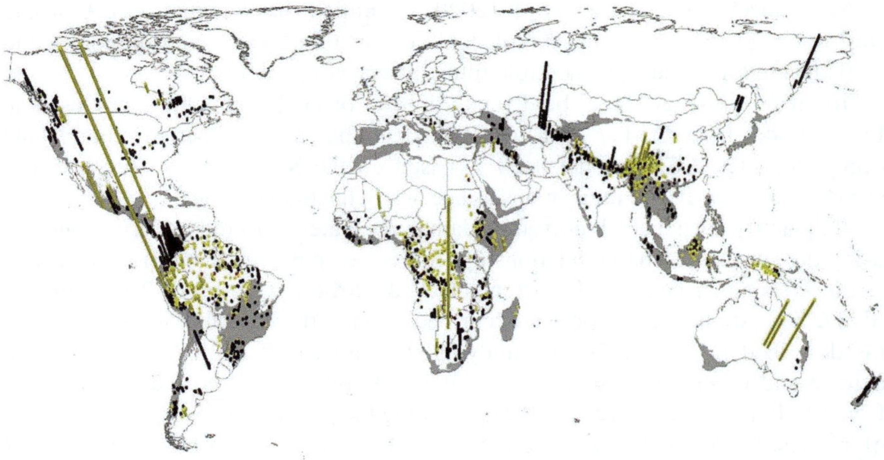

Fig. 7.2 Bars show the terrestrial biodiversity impact in PDF*y per kWh hydropower electricity produced for each possible future hydropower reservoir. The y-axis goes from 3.0×10^{-18} to 4.4×10^{-12} PDF*y/kWh. The color code indicates the yearly electricity production of each reservoir. Black: $\leq$1000 GWh; yellow: between 1000 and 30,000 GWh; red: $\geq$ 30,000 GWh. Reservoirs in a grey area will be located in a biodiversity hotspot (Dorber et al. 2020)

kWh. This finding highlights that we need models, which are based on local species biodiversity data to really reflect local conditions. Furthermore, 906 of the potential hydropower reservoirs could add additional stress to already threatened terrestrial species, as they are located in biodiversity hotspots, which are characterized as areas with high endemic species richness and where biodiversity is already threatened (Myers et al. 2000). As hydropower is only one of several stressors for terrestrial and aquatic biodiversity loss, these results highlight that it is important that we keep monitoring the status of biodiversity and its loss.

As a final result, Dorber et al. (2020) showed that there is a trade-off risk between terrestrial and aquatic biodiversity impacts, as construction of reservoirs with low terrestrial impacts will not automatically be accompanied by a low aquatic biodiversity impact, and vice versa. Hence, it is important that we use tools for decision-making that can assess multiple species groups and impact categories at the same time.

However, Dorber and colleagues also note that the current aquatic biodiversity impact model uses only fish as species biodiversity indicator, while the terrestrial biodiversity impact is based on four species groups (terrestrial mammals, birds, amphibians, and reptiles) and this reminds us that we always assess an incomplete picture of biodiversity. But, at the same time, this highlights the research needs, to for example, develop an SDR for macro-invertebrates.

When interpreting the mentioned results, it is important to note that the study only considered the operation phase of the reservoirs. However, for a final decision, a complete life cycle assessment, including more life cycle stages (e.g., dam construction and deconstruction), is needed as these life cycle stages would add further impacts, leading, among others, also to a higher biodiversity impact. Furthermore, the results could change if LCIA methods for additional biodiversity impact pathways such as impacts of habitat fragmentation from dams become available. Hence, this highlights that more knowledge about how humans are affecting biodiversity is needed.

Overall, a sustainable future electricity mix will depend on a mix of different renewable energy sources (Bogdanov et al. 2019), accompanied by different biodiversity impacts. A comparison of these impacts could also be done with the approach of life cycle assessment. For wind power, for example, the availability of LCIA models is growing (Laranjeiro et al. 2018; May et al. 2020). If we continue to implement knowledge about how humans are impacting biodiversity into impact assessment tools (including more biodiversity data), we may one day find a renewable electricity mix, which balances human well-being and biodiversity. This balance is needed as human well-being ultimately relies on biodiversity and their ecosystem services.

References

Abrouk, M., Ahmed, H. I., Cubry, P., Šimoníková, D., Cauet, S., et al. (2020). "Fonio millet genome unlocks African orphan crop diversity for agriculture in a changing climate." Nature Communications **11**(1): 4488.

Acevedo, P., Vicente, J., Höfle, U., Cassinello, J., Ruiz-Fons, F. and Gortazar, C. (2006). "Estimation of European wild boar relative abundance and aggregation: a novel method in epidemiological risk assessment." Epidemiology and Infection **135**(3): 519-527.

Almond, R., Grooten, M. and Peterson, T. (2020). Living Planet Report 2020-Bending the curve of biodiversity loss, World Wildlife Fund.

Apps, C. D., McLellan, B. N., Woods, J. G., Proctor, M. F. and Gehrt (2004). "Estimating Grizzly Bear Distribution and Abundance Relative to Habitat and Human Influence." Journal of Wildlife Management **68**(1): 138–152.

Atkinson, S. F., Bhatia, S., Schoolmaster, F. A. and Waller, W. T. (2000). "Treatment of biodiversity impacts in a sample of US environmental impact statements." Impact Assessment and Project Appraisal **18**(4): 271–282.

Bamford, C. C. G., Kelly, N., Dalla Rosa, L., Cade, D. E., Fretwell, P. T., Trathan, P. N., Cubaynes, H. C., Mesquita, A. F. C., Gerrish, L., Friedlaender, A. S. and Jackson, J. A. (2020). "A comparison of baleen whale density estimates derived from overlapping satellite imagery and a shipborne survey." Scientific Reports **10**(1).

Bare, J. (2011). "TRACI 2.0: the tool for the reduction and assessment of chemical and other environmental impacts 2.0." Clean Technologies and Environmental Policy **13**(5): 687–696.

Barnosky, A. D., Matzke, N., Tomiya, S., Wogan, G. O. U., Swartz, B., Quental, T. B., Marshall, C., McGuire, J. L., Lindsey, E. L., Maguire, K. C., Mersey, B. and Ferrer, E. A. (2011). "Has the Earth's sixth mass extinction already arrived?" Nature **471**(7336): 51–57.

Bateman, P. W. and Fleming, P. A. (2012). "Big city life: carnivores in urban environments." Journal of Zoology **287**(1): 1–23.

Bigard, C., Pioch, S. and Thompson, J. D. (2017). "The inclusion of biodiversity in environmental impact assessment: Policy-related progress limited by gaps and semantic confusion." Journal of Environmental Management **200**: 35–45.

Biology Online. (2020). "Genetic Diversity." Retrieved 7 Janaury, 2021, from https://www.biologyonline.com/dictionary/genetic-diversity.

Boer, H. Y. d., Breukelen, L. v., Hootsmans, M. J. M. and Wieren, S. E. v. (2004). "Flight distance in roe deer Capreolus capreolus and fallow deer Dama dama as related to hunting and other factors." Wildlife Biology **10**(1): 35–41.

Bogdanov, D., Farfan, J., Sadovskaia, K., Aghahosseini, A., Child, M., Gulagi, A., Oyewo, A. S., de Souza Noel Simas Barbosa, L. and Breyer, C. (2019). "Radical transformation pathway towards sustainable electricity via evolutionary steps." Nature Communications **10**(1): 1077.

Brittain, C., Kremen, C. and Klein, A.-M. (2013). "Biodiversity buffers pollination from changes in environmental conditions." Global Change Biology **19**(2): 540–547.

Bulle, C., Margni, M., Patouillard, L., Boulay, A.-M., Bourgault, G., et al. (2019). "IMPACT World+: A globally regionalized Life Cycle Impact Assessment method." International Journal of LCA.

Butchart, S. H., Stattersfield, A. J., Bennun, L. A., Shutes, S. M., Akçakaya, H. R., Baillie, J. E., Stuart, S. N., Hilton-Taylor, C. and Mace, G. M. (2004a). "Measuring global trends in the status of biodiversity: Red List Indices for birds." PLoS Biol **2**(12): e383.

Butchart, S. H. M. (2008). "Red List Indices to measure the sustainability of species use and impacts of invasive alien species." Bird Conservation International **18**(S1): S245–S262.

Butchart, S. H. M., Stattersfield, A. J., Baillie, J., Bennun, L. A., Stuart, S. N., Akçakaya, H. R., Hilton-Taylor, C. and Mace, G. M. (2005). "Using Red List Indices to measure progress towards the 2010 target and beyond." Philosophical Transactions of the Royal Society B: Biological Sciences **360**(1454): 255–268.

Butchart, S. H. M., Stattersfield, A. J., Bennun, L. A., Shutes, S. M., Akçakaya, H. R., Baillie, J. E. M., Stuart, S. N., Hilton-Taylor, C. and Mace, G. M. (2004b). "Measuring Global Trends in the Status of Biodiversity: Red List Indices for Birds." PLOS Biology **2**(12): e383.

Butchart, S. H. M., Walpole, M., Collen, B., van Strien, A., Scharlemann, J. P. W., et al. (2010). "Global Biodiversity: Indicators of Recent Declines." Science **328**(5982): 1164–1168.

CABE, Commission for Architecture and the Built Environment (2006). Making contracts work for wildlife: how to encourage biodiversity in urban parks.

Cardinale, B. J. (2011). "Biodiversity improves water quality through niche partitioning." Nature **472**(7341): 86–89.

CBD, Convention on Biological Diversity. (1992). "Convention on Biological Diversity." Retrieved 22 July, 2015, from https://www.cbd.int/convention/text/.

Chandler, M., See, L., Buesching, C. D., Cousins, J. A., Gillies, C., Kays, R. W., Newman, C., Pereira, H. M. and Tiago, P. (2017). Involving Citizen Scientists in Biodiversity Observation. The GEO Handbook on Biodiversity Observation Networks. M. Walters and R. J. Scholes. Cham, Springer International Publishing: 211–237.

Chaudhary, A., Verones, F., De Baan, L. and Hellweg, S. (2015). "Quantifying Land Use Impacts on Biodiversity: Combining Species-Area Models and Vulnerability Indicators." Environ. Sci. Technol. **49**(16): 9987–9995.

Chiarucci, A., Bacaro, G. and Scheiner, S. M. (2011). "Old and new challenges in using species diversity for assessing biodiversity." Philosophical Transactions of the Royal Society B: Biological Sciences **366**(1576): 2426–2437.

Cilliers, S., Cilliers, J., Lubbe, R. and Siebert, S. (2013). "Ecosystem services of urban green spaces in African countries—perspectives and challenges." Urban Ecosystems **16**(4): 681–702.

Costanza, R., Fisher, B., Mulder, K., Liu, S. and Christopher, T. (2007). "Biodiversity and ecosystem services: A multi-scale empirical study of the relationship between species richness and net primary production." Ecological Economics **61**(2): 478–491.

de Baan, L., Mutel, C. L., Curran, M., Hellweg, S. and Koellner, T. (2013). "Land Use in Life Cycle Assessment: Global Characterization Factors Based on Regional and Global Species Extinction." Environmental Science and Technology **47**(16): 9281–9290.

de Souza, D., Flynn, D. B., DeClerck, F., Rosenbaum, R., de Melo Lisboa, H. and Koellner, T. (2013). "Land use impacts on biodiversity in LCA: proposal of characterization factors based on functional diversity." The International Journal of Life Cycle Assessment **18**(6): 1231–1242.

Deinet, S., Scott-Gatty, K., Rotton, H., Twardek, W. M., Marconi, V., et al. (2020). The Living Planet Index (LPI) for migratory freshwater fish - Technical Report., World Fish Migration Foundation, The Netherlands.
Di Falco, S., Bezabih, M. and Yesuf, M. (2010). "Seeds for livelihood: Crop biodiversity and food production in Ethiopia." Ecological Economics **69**(8): 1695–1702.
Dorber, M., Arvesen, A., Gernaat, D. and Verones, F. (2020). "Controlling biodiversity impacts of future global hydropower reservoirs by strategic site selection." Scientific Reports **10**(1): 21777.
Dorber, M., Kuipers, K. and Verones, F. (2019a). "Global characterization factors for terrestrial biodiversity impacts of future land inundation in Life Cycle Assessment." Science of The Total Environment: 134582.
Dorber, M., Mattson, K. R., Sandlund, O. T., May, R. and Verones, F. (2019b). "Quantifying net water consumption of Norwegian hydropower reservoirs and related aquatic biodiversity impacts in Life Cycle Assessment." Environmental Impact Assessment Review **76**: 36–46.
Duelli, P. and Obrist, M. K. (2003). "Biodiversity indicators: the choice of values and measures." Agriculture, Ecosystems & Environment **98**(1–3): 87–98.
Egevang, C., Stenhouse, I. J., Phillips, R. A., Petersen, A., Fox, J. W. and Silk, J. R. D. (2010). "Tracking of Arctic terns Sterna paradisaea reveals longest animal migration." Proceedings of the National Academy of Sciences **107**(5): 2078–2081.
Eigenbrod, F., Anderson, B. J., Armsworth, P. R., Heinemeyer, A., Jackson, S. F., Parnell, M., Thomas, C. D. and Gaston, K. J. (2009). "Ecosystem service benefits of contrasting conservation strategies in a human-dominated region." Proceedings of the Royal Society B: Biological Sciences **276**(1669): 2903–2911.
Elith, J. and Leathwick, J. R. (2009). "Species Distribution Models: Ecological Explanation and Prediction Across Space and Time." Annual Review of Ecology, Evolution, and Systematics **40**(1): 677–697.
Elith, J., Phillips, S. J., Hastie, T., Dudík, M., Chee, Y. E. and Yates, C. J. (2011). "A statistical explanation of MaxEnt for ecologists." Diversity and distributions **17**(1): 43–57.
Elmqvist, T., Setälä, H., Handel, S. N., van der Ploeg, S., Aronson, J., Blignaut, J. N., Gómez-Baggethun, E., Nowak, D. J., Kronenberg, J. and de Groot, R. (2015). "Benefits of restoring ecosystem services in urban areas." Current Opinion in Environmental Sustainability **14**: 101–108.
European Comission. (2020). "European Green Deal." from https://ec.europa.eu/info/research-and-innovation/strategy/european-green-deal_en.
European Commission (2017). Environmental Impact Assessment of Projects. Guidance on Scoping (Directive 2011/92/EU as amended by 2014/52/EU).
European Commission and Joint Research Centre and Institute for Environment and Sustainability (2010). ILCD Handbook. International Reference Life Cycle Data System. Framework and requirements for Life Cycle Impact Assessment models and indicators.
European Parliament and Council of the European Union (2011). DIRECTIVE 2011/92/EU OF THE EUROPEAN PARLIAMENT AND OF THE COUNCILof 13 December 2011on the assessment of the effects of certain public and private projects on the environment.
Fahrig, L. (2001). "How much habitat is enough?" Biological Conservation **100**(1): 65–74.
Ferreira, S. M., Greaver, C. and Simms, C. (2017). "Elephant population growth in Kruger National Park, South Africa, under a landscape management approach." 2017 **59**(1).
Field, S. A., Tyre, A. J. and Possingham, H. P. (2002). "Estimating bird species richness: How should repeat surveys be organized in time?" Austral Ecology **27**(6): 624–629.
Fisher, B. and Kerry Turner, R. (2008). "Ecosystem services: Classification for valuation." Biological Conservation **141**(5): 1167–1169.
Frischknecht, R. and Büsser Knöpfel, S. (2013). Swiss Eco-Factors 2013 according to the Ecological Scarcity Method. Methodological fundamentals and their application in Switzerland. Environmental studies no. 1330. , Federal Office for the Environment, Bern, Switzerland.

Gamfeldt, L., Snäll, T., Bagchi, R., Jonsson, M., Gustafsson, L., Kjellander, P., Ruiz-Jaen, M. C., Fröberg, M., Stendahl, J., Philipson, C. D., Mikusiński, G., Andersson, E., Westerlund, B., Andrén, H., Moberg, F., Moen, J. and Bengtsson, J. (2013). "Higher levels of multiple ecosystem services are found in forests with more tree species." Nature Communications **4**(1): 1340.

Gaston, K. J. and Spicer, J. I. (2003). Biodiversity: An Intrduction, 2nd edition, Wiley-Blackwell.

GBIF (The Global Biodiversity Information Facility). "*Search occurrences*." from http://www.gbif.org/occurrence/search.

Gernaat, D. E. H. J., Bogaart, P. W., Vuuren, D. P. v., Biemans, H. and Niessink, R. (2017). "High-resolution assessment of global technical and economic hydropower potential." Nature Energy **2**(10): 821–828.

Gibon, T., Hertwich, E. G., Arvesen, A., Singh, B. and Verones, F. (2017). "Health benefits, ecological threats of low-carbon electricity." Environmental Research Letters **12**(3): 034023.

Gontier, M., Balfors, B. and Mörtberg, U. (2006). "Biodiversity in environmental assessment—current practice and tools for prediction." Environmental Impact Assessment Review **26**(3): 268–286.

Gracey, E. O. and Verones, F. (2016). "Impacts from hydropower production on biodiversity in an LCA framework—review and recommendations." The International Journal of Life Cycle Assessment **21**(3): 412–428.

Gruber, K. (2017). "Agrobiodiversity: The living library." Nature **544**(7651): S8–S10.

Guralnick, R., Cusack, J. J., Dickman, A. J., Rowcliffe, J. M., Carbone, C., Macdonald, D. W. and Coulson, T. (2015). "Random versus Game Trail-Based Camera Trap Placement Strategy for Monitoring Terrestrial Mammal Communities." Plos One **10**(5).

Haig, S. M., Ferland, C. L., Cuthbert, F. J., Dingledine, J., Goossen, J. P., Hecht, A. and McPHILLIPS, N. (2005). "A complete species census and evidence for regional declines in Piping Plovers." The Journal of wildlife management **69**(1): 160–173.

Hanafiah, M. M., Xenopoulos, M. A., Pfister, S., Leuven, R. S. and Huijbregts, M. A. J. (2011a). "Characterization Factors for Water Consumption and Greenhouse Gas Emissions Based on Freshwater Fish Species Extinction." Environ. Sci. Technol. **45**(12): 5572–5278.

Hanafiah, M. M., Xenopoulos, M. A., Pfister, S., Leuven, R. S. E. W. and Huijbregts, M. A. J. (2011b). "Characterization Factors for Water Consumption and Greenhouse Gas Emissions Based on Freshwater Fish Species Extinction." Environmental Science & Technology **45**(12): 5272–5278.

Hauschild, M. Z. and Huijbregts, M. A. J. (2015). Life Cycle Impact Assessment, Springer.

Hellweg, S. and Milà i Canals, L. (2014). "Emerging approaches, challenges and opportunities in life cycle assessment." Science **344**(6188): 1109–1113.

Hodgson, J. C., Mott, R., Baylis, S. M., Pham, T. T., Wotherspoon, S., Kilpatrick, A. D., Raja Segaran, R., Reid, I., Terauds, A., Koh, L. P. and Yoccoz, N. (2018). "Drones count wildlife more accurately and precisely than humans." Methods in Ecology and Evolution **9**(5): 1160–1167.

Hoffmann, M., Hilton-Taylor, C., Angulo, A., Böhm, M., Brooks, T. M., et al. (2010). "The Impact of Conservation on the Status of the World's Vertebrates." Science **330**(6010): 1503–1509.

Huijbregts, M. A. J., Steinmann, Z. J. N., Elshout, P. M. F., Stam, G., Verones, F., Vieira, M., Zijp, M., Hollander, A. and van Zelm, R. (2017). "ReCiPe2016: a harmonised life cycle impact assessment method at midpoint and endpoint level." The International Journal of Life Cycle Assessment **22**(2): 138–147.

inaturalist. "https://www.inaturalist.org/."

Irvine, K. N., Hoesly, D., Bell-Williams, R. and Warber, S. L. (2019). Biodiversity and Health in the Face of Climate Change M. R. Marselle, J. Stadler, H. Korn, K. N. Irvine and A. Bonn, Springer Open.

ISO (2006). Environmental Management - Life Cycle Assessment - Principles and Framework. International Standard ISO 14040. International Organisation for Standardisation. Geneva, Switzerland.

Itsubo, N. and Inaba, A. (2003). "A new LCIA method: LIME has been completed." The International Journal of Life Cycle Assessment **8**(5): 305–305.

Jędrzejewski, W., Puerto, M. F., Goldberg, J. F., Hebblewhite, M., Abarca, M., Gamarra, G., Calderón, L. E., Romero, J. F., Viloria, Á. L., Carreño, R., Robinson, H. S., Lampo, M., Boede, E. O., Biganzoli, A., Stachowicz, I., Velásquez, G. and Schmidt, K. (2016). "Density and population structure of the jaguar (Panthera onca) in a protected area of Los Llanos, Venezuela, from 1 year of camera trap monitoring." Mammal Research **62**(1): 9–19.

Jolliet, O., Müller-Wenk, R., Bare, J., Brent, A., Goedkoop, M., Heijungs, R., Itsubo, N., Peña, C., Pennington, D., Potting, J., Rebitzer, G., Stewart, M., de Haes, H. and Weidema, B. (2004). "The LCIA midpoint-damage framework of the UNEP/SETAC life cycle initiative." The International Journal of Life Cycle Assessment **9**(6): 394–404.

Juslén, A., Hyvärinen, E. and Virtanen, L. K. (2013). "Application of the Red-List Index at a National Level for Multiple Species Groups." Conservation Biology **27**(2): 398–406.

Keinath, D. A., Doak, D. F., Hodges, K. E., Prugh, L. R., Fagan, W., Sekercioglu, C. H., Buchart, S. H. M. and Kauffman, M. (2017). "A global analysis of traits predicting species sensitivity to habitat fragmentation." Global Ecology and Biogeography **26**(1): 115–127.

Khera, N. and Kumar, A. (2010). "Inclusion of biodiversity in environmental impact assessments (EIA): a case study of selected EIA reports in India." Impact Assessment and Project Appraisal **28**(3): 189–200.

Kitzes, J. (2013). "An Introduction to Environmentally-Extended Input-Output Analysis." Resources **2**(4).

Kitzes, J., Berlow, E., Conlisk, E., Erb, K., Iha, K., Martinez, N., Newman, E. A., Plutzar, C., Smith, A. B. and Harte, J. (2017). "Consumption-Based Conservation Targeting: Linking Biodiversity Loss to Upstream Demand through a Global Wildlife Footprint." Conservation Letters **10**(5): 531–538.

Koellner, T. and Scholz, W. R. (2008). "Assessment of land use impacts on the natural environment. Part 2: Generic characterization factors for local species diversity in central europe." Int. J. Life Cycle Assess. **13**: 32–48.

Koh, L. P. and Ghazoul, J. (2010). "A matrix-calibrated species-area model for predicting biodiversity losses due to land-use change." Conservation Biology **24**(4): 994–1001.

Koslowski, M., Moran, D. D., Tisserant, A., Verones, F. and Wood, R. (2020). "Quantifying Europe's biodiversity footprints and the role of urbanization and income." Global Sustainability **3**: e1.

Kuipers, K. J. J., Hellweg, S. and Verones, F. (2019a). "Potential Consequences of Regional Species Loss for Global Species Richness: A Quantitative Approach for Estimating Global Extinction Probabilities." Environmental Science & Technology **53**(9): 4728–4738.

Kuipers, K. J. J., May, R. F., Graae, B. J. and Verones, F. (2019b). "Reviewing the potential for including habitat fragmentation to improve life cycle impact assessments for land use impacts on biodiversity." The International Journal of Life Cycle Assessment.

Kuipers, K. J. J., Hilbers, J. P., Garcia-Ulloa, J., Graae, B. J., May, R., Verones, F., Huijbregts, M. A. J. and Schipper, A. M. (2021a). Habitat fragmentation amplifies threats from habitat loss to mammal diversity across the world's terrestrial ecoregions. One Earth.

Kuipers, K. J. J., May, R. and Verones, F. (2021b). Considering habitat conversion and fragmentation in characterisation factors for land-use impacts on vertebrate species richness. Science of The Total Environment **801**: 149737.

La Notte, A., D'Amato, D., Mäkinen, H., Paracchini, M. L., Liquete, C., Egoh, B., Geneletti, D. and Crossman, N. D. (2017). "Ecosystem services classification: A systems ecology perspective of the cascade framework." Ecological Indicators **74**: 392–402.

Laranjeiro, T., May, R. and Verones, F. (2018). "Impacts of onshore wind energy production on birds and bats: recommendations for future life cycle impact assessment developments." The International Journal of Life Cycle Assessment **23**(10): 2007–2023.

Lee, J., Park, B.-J., Tsunetsugu, Y., Kagawa, T. and Miyazaki, Y. (2009). "Restorative effects of viewing real forest landscapes, based on a comparison with urban landscapes." Scandinavian Journal of Forest Research **24**(3): 227–234.

Lenzen, M., Moran, D., Kanemoto, K., Foran, B., Lobefaro, L. and Geschke, A. (2012). "International Trade Drives Biodiversity Threats in Developing Nations." Nature **486**(7401): 109–112.

Lindley, S. J., Cook, P. A., Dennis, M. and Gilchrist, A. (2019). Biodiversity, Physical Health and Climate Change: A Synthesis of Recent Evidence. Biodiversity and Health in the Face of Climate Change. M. R. Marselle, J. Stadler, H. Korn, K. N. Irvine and A. Bonn.

Lukacs, P. M. and Burnham, K. P. (2005). "Review of capture–recapture methods applicable to noninvasive genetic sampling." Molecular ecology **14**(13): 3909–3919.

Lusseau, D., Butchart, S. H. M., Resit Akçakaya, H., Chanson, J., Baillie, J. E. M., Collen, B., Quader, S., Turner, W. R., Amin, R., Stuart, S. N. and Hilton-Taylor, C. (2007). "Improvements to the Red List Index." PLoS ONE **2**(1).

Lyons, M. B., Brandis, K. J., Murray, N. J., Wilshire, J. H., McCann, J. A., Kingsford, R. T., Callaghan, C. T. and Shepard, E. (2019). "Monitoring large and complex wildlife aggregations with drones." Methods in Ecology and Evolution **10**(7): 1024–1035.

Mace, G. M., Mora, C., Tittensor, D. P., Adl, S., Simpson, A. G. B. and Worm, B. (2011). "How Many Species Are There on Earth and in the Ocean?" PLoS Biology **9**(8).

Mace, G. M., Norris, K. and Fitter, A. H. (2012). "Biodiversity and ecosystem services: a multilayered relationship." Trends in Ecology and Evolution **27**(1): 19–26.

Marques, A., Verones, F., Kok, M. T. J., Huijbregts, M. A. J. and Pereira, H. M. (2017). "How to quantify biodiversity footprints of consumption? A review of multi-regional input–output analysis and life cycle assessment." Current Opinion in Environmental Sustainability **29**: 75–81.

Marselle, M. R., Martens, D., Dallimer, M. and Irvine, K. N. (2019). Biodiversity and Health in the Face of Climate Change. M. R. Marselle, J. Stadler, H. Korn, K. N. Irvine and A. Bonn.

Matthews, T. J., Triantis, K. A. and Whittaker, R. J. (2020). The Species-Area Relationship. Theory and Application, Cambridge University Press.

May, R., Middel, H., Stokke, B. G., Jackson, C. and Verones, F. (2020). "Global life-cycle impacts of onshore wind-power plants on bird richness." Environmental and Sustainability Indicators **8**.

Millennium Ecosystem Assessment (2003). Ecosystems and Human Well-being: A Framework for Assessment. Island Press, Washington DC.

Milligan, H., Deinet, S., McRae, L. and Freeman, R. (2014). Protecting Species: Status and Trends of the Earth's Protected Areas. Preliminary Report, Zoological Society of London, UK.

Møller, A. P. (2008). "Flight distance and population trends in European breeding birds." Behavioral Ecology **19**(6): 1095–1102.

Moran, D., Petersone, M. and Verones, F. (2016). "On the suitability of input–output analysis for calculating product-specific biodiversity footprints." Ecological Indicators **60**: 192–201.

Morris, P. and Therivel, R. (2001). Methods of Environmental Impact Assessment. Second Edition, Spon Press.

Muthuramkumar, S. and Parthasarathy, N. (2000). "Alpha diversity of lianas in a tropical evergreen forest in the Anamalais, Western Ghats, India." Diversity and Distributions **6**(1): 1–14.

Myers, N., Mittermeier, R. A., Mittermeier, C. G., da Fonseca, G. A. B. and Kent, J. (2000). "Biodiversity hotspots for conservation priorities." Nature **403**(6772): 853–858.

Neergheen-Bhujun, V., Awan, A. T., Baran, Y., Bunnefeld, N., Chan, K., Dela Cruz, T. E., Egamberdieva, D., Elsässer, S., Johnson, M.-V. V., Komai, S., Konevega, A. L., Malone, J. H., Mason, P., Nguon, R., Piper, R., Shrestha, U. B., Pešić, M. and Kagansky, A. (2017). "Biodiversity, drug discovery, and the future of global health: Introducing the biodiversity to biomedicine consortium, a call to action." Journal of global health **7**(2): 020304–020304.

Nelson, E., Mendoza, G., Regetz, J., Polasky, S., Tallis, H., Cameron, D., Chan, K. M., Daily, G. C., Goldstein, J., Kareiva, P. M., Lonsdorf, E., Naidoo, R., Ricketts, T. H. and Shaw, M. (2009). "Modeling multiple ecosystem services, biodiversity conservation, commodity production, and tradeoffs at landscape scales." Frontiers in Ecology and the Environment **7**(1): 4–11.

Newman, D. J. and Cragg, G. M. (2012). "Natural Products As Sources of New Drugs over the 30 Years from 1981 to 2010." Journal of Natural Products **75**(3): 311–335.

Ofstad, E. G., Herfindal, I., Solberg, E. J. and Sæther, B.-E. (2016). "Home ranges, habitat and body mass: simple correlates of home range size in ungulates." Proceedings of the Royal Society B: Biological Sciences **283**(1845).

Olson, D. M., Dinerstein, E., Wikramanayake, E. D., Burgess, N., Powell, G. V. N., Underwood, E., D'Amico, J. A., Itoua, I., Strand, H. E., Morrison, J. C., Loucks, C. J., Allnutt, T. F., Ricketts, T. H., Kura, Y., Lamoreux, J. F., Wettengel, W. W., Hedao, P. and Kassem, K. R. (2001). "Terrestrial Ecoregions of the World: A New Map of Life on Earth (Note: data update from 2004 used)." BioScience **51**(11): 933–938.

Opoku, A. (2019). "Biodiversity and the built environment: Implications for the Sustainable Development Goals (SDGs)." Resources, Conservation and Recycling **141**: 1–7.

Owsianiak, M., Hauschild, M. Z., Posthuma, L., Saouter, E., Vijver, M. G., Backhaus, T., Douziech, M., Schlekat, T. and Fantke, P. (2023). Ecotoxicity characterization of chemicals: Global recommendations and implementation in USEtox. Chemosphere **310**: 136807.

Parra, G. J., Cagnazzi, D., Jedensjö, M., Ackermann, C., Frere, C., Seddon, J., Nikolic, N. and Krützen, M. (2018). "Low genetic diversity, limited gene flow and widespread genetic bottleneck effects in a threatened dolphin species, the Australian humpback dolphin." Biological Conservation **220**: 192–200.

Pataki, D. E., Carreiro, M. M., Cherrier, J., Grulke, N. E., Jennings, V., Pincetl, S., Pouyat, R. V., Whitlow, T. H. and Zipperer, W. C. (2011). "Coupling biogeochemical cycles in urban environments: ecosystem services, green solutions, and misconceptions." Frontiers in Ecology and the Environment **9**(1): 27–36.

Pehl, M., Arvesen, A., Humpenöder, F., Popp, A., Hertwich, E. G. and Luderer, G. (2017). "Understanding future emissions from low-carbon power systems by integration of life-cycle assessment and integrated energy modelling." Nature Energy **2**(12): 939–945.

Pereira, H. M., Borda-de-Água, L. and Martins, I. S. (2012a). "Geometry and scale in species–area relationships." Nature **482**(7386): E3–E4.

Pereira, H. M. and Daily, G. C. (2006). "Modeling biodiversity dynamics in countryside landscapes." Ecology **87**(8): 1877–1885.

Pereira, H. M., Navarro, L. M. and Martins, I. S. (2012b). "Global Biodiversity Change: The Bad, the Good, and the Unknown." Annual Review of Environment and Resources **37**(1): 25–50.

Phillips, S. J., Dudík, M., Elith, J., Graham, C. H., Lehmann, A., Leathwick, J. and Ferrier, S. (2009). "Sample selection bias and presence-only distribution models: implications for background and pseudo-absence data." Ecological Applications **19**(1): 181–197.

Pielou, E. C. (1966). "Shannon's formula as a measure of specific diversity: its use and misuse." The American Naturalist **100**(914): 463–465.

Pierrat, E., Barbarossa, V., Núñez, M., Scherer, L., Link, A., Damiani, M., Verones, F. and Dorber, M. (2023). Global water consumption impacts on riverine fish species richness in Life Cycle Assessment. Science of The Total Environment **854**: 158702.

Pimm, S. L., Jenkins, C. N., Abell, R., Brooks, T. M., Gittleman, J. L., Joppa, L. N., Raven, P. H., Roberts, C. M. and Sexton, J. O. (2014). "The biodiversity of species and their rates of extinction, distribution, and protection." Science **344**(6187): 1246752–1246752.

Pomeroy, D., Tushabe, H. and Loh, J. (2017). The State of Uganda's Biodiversity 2017. . NAtional Biodiversity Data Bank, Makerere University, Kampala, Uganda.

Purvis, A. and Hector, A. (2000). "Getting the measure of biodiversity." Nature **405**(6783): 212–219.

Reeves, R., Pitman, R. L. and Ford, J. K. B. (2017). "Orcinus orca. The IUCN Red List of Threatened Species 2017: e.T15421A50368125. https://doi.org/10.2305/IUCN.UK.2017-3.RLTS.T15421A50368125.en." Retrieved 11 February, 2021.

Rosenbaum, R. K. (2018). Overview of Existing LCIA Methods—Annex to Chapter 10. Life Cycle Assessment: Theory and Practice. M. Z. Hauschild, R. K. Rosenbaum and S. I. Olsen. Cham, Springer International Publishing**:** 1147–1183.

Rosenzweig, M. (1995). Species diversity in space and time, Cambridge University Press: Cambridge, United Kingdom.

Saiz, J. C. M., Lozano, F. D., Gómez, M. M. and Baudet, Á. B. (2015). "Application of the Red List Index for conservation assessment of Spanish vascular plants." Conservation Biology **29**(3): 910–919.

Sala, O. E., Chapin, F. S., Armesto, J. J., Berlow, E., Bloomfield, J., Dirzo, R., Huber-Sanwald, E., Huenneke, L. F., Jackson, R. B., Kinzig, A., Leemans, R., Lodge, D. M., Mooney, H. A., Oesterheld, M., Poff, N. L., Sykes, M. T., Walker, B. H., Walker, M. and Wall, D. H. (2000). "Global biodiversity scenarios for the year 2100." Science **287**(5459): 1770–1774.

Santini, L., Belmaker, J., Costello, M. J., Pereira, H. M., Rossberg, A. G., Schipper, A. M., Ceaușu, S., Dornelas, M., Hilbers, J. P., Hortal, J., Huijbregts, M. A. J., Navarro, L. M., Schiffers, K. H., Visconti, P. and Rondinini, C. (2017). "Assessing the suitability of diversity metrics to detect biodiversity change." Biological Conservation **213**: 341–350.

Schmidt, J. (2008). "Development of LCIA characterisation factors for land use impacts on biodiversity." Journal of Cleaner Production **16**: 1929–1942.

Schwarz, C. J. and Seber, G. A. (1999). "Estimating animal abundance: review III." Statistical Science: 427–456.

Secretariat of the Convention on Biological Diversity (2010). Strategic Plan for Biodiversity 2011–2020 and the Aichi Targets. Montreal, Quebec, Canada.

Secretariat of the Convention on Biological Diversity. (2019). "Preparations for the Post-2020 Biodiversity Framework." Retrieved 18.02.2019, from https://www.cbd.int/post2020/.

Sepkoski, J. J. (1988). "Alpha, Beta, or Gamma: Where Does all the Diversity Go?" Paleobiology **14**(3): 221–234.

Souza, D. M., FLynn, D., Rosenbaum, R. K., de Melo Lisboa, H. and Koellner, T. (2013). "Land use impacts on biodiversity: proposal of characterization factors based on functional diversity." int. J. Life Cycle Assess. **18**(6): 1231–1242.

Steenweg, R., Hebblewhite, M., Kays, R., Ahumada, J., Fisher, J. T., Burton, C., Townsend, S. E., Carbone, C., Rowcliffe, J. M., Whittington, J., Brodie, J., Royle, J. A., Switalski, A., Clevenger, A. P., Heim, N. and Rich, L. N. (2017). "Scaling-up camera traps: monitoring the planet's biodiversity with networks of remote sensors." Frontiers in Ecology and the Environment **15**(1): 26–34.

Stowell, D., Wood, M. D., Pamuła, H., Stylianou, Y., Glotin, H. and Orme, D. (2019). "Automatic acoustic detection of birds through deep learning: The first Bird Audio Detection challenge." Methods in Ecology and Evolution **10**(3): 368–380.

Tendall, D. M., Hellweg, S., Pfister, S., Huijbregts, M. A. J. and Gaillard, G. (2014a). "Impacts of River Water Consumption on Aquatic Biodiversity in Life Cycle Assessment—A Proposed Method, and a Case Study for Europe." Environmental Science & Technology **48**(6): 3236–3244.

Tendall, D. M., Hellweg, S., Pfister, S., Huijbregts, M. A. J. and Gaillard, G. (2014b). "Impacts of River Water Consumption on Aquatic Biodiversity in Life Cycle Assessment - a proposed method, and a case study for Europe." Environ. Sci. Technol. **48**(6): 3236–3244.

UNEP (2016). *Green Energy Choices: The benefits, risks and trade-offs of low-carbon technologies for electricity production*. Report of the International Resource Panel. E. G. Hertwich, J. Aloisi de Larderel, A. Arvesen, P. Bayer, J. Bergesen, E. Bouman, T. Gibon, G. Heath, C. Peña, P. Purohit, A. Ramirez, S. Suh, (eds.).

Unilever. (2020). "Lifecycle Assessments." Retrieved 02.12., 2020, from https://www.unilever.com/sustainable-living/reducing-environmental-impact/lifecycle-assessments/index.html.

United Nations (2015). Transforming our world: The 2030 agenda for sustainable development - A/RES/70/1.

Verones, F., Bare, J., Bulle, C., Frischknecht, R., Hauschild, M., et al. (2017a). "LCIA framework and cross-cutting issues guidance within the UNEP-SETAC Life Cycle Initiative." Journal of Cleaner Production **161**: 957–967.

Verones, F., Hellweg, S., Antón, A., Azevedo, L. B., Chaudhary, A., et al. (2020). "LC-IMPACT: a regionalized life cycle damage assessment method " Journal of Industrial Ecology.

Verones, F., Moran, D., Stadler, K., Kanemoto, K. and Wood, R. (2017b). "Resource footprints and their ecosystem consequences." Scientific Reports **7**: 40743.

Verones, F., Pfister, S., van Zelm, R. and Hellweg, S. (2017c). "Biodiversity impacts from water consumption on a global scale for use in life cycle assessment." The International Journal of Life Cycle Assessment **22**(8): 1247–1256.

Vilà, C., Maldonado, J. E. and Wayne, R. K. (1999). "Phylogenetic Relationships, Evolution, andGenetic Diversity of the Domestic Dog." The American Genetic Association **90**: 71–77.

WHO, (World Health Organization) and Secretariat of the Convention on Biological Diversity (2015). Connecting global priorities: biodiversity and human health: a state of knowledge review.

Wilting, H. C., Schipper, A. M., Bakkenes, M., Meijer, J. R. and Huijbregts, M. A. J. (2017). "Quantifying Biodiversity Losses Due to Human Consumption: A Global-Scale Footprint Analysis." Environmental Science & Technology **51**(6): 3298–3306.

Woolf, A., Nelson, T. and Bluett, R. (2003). "Beaver relative abundance in southern Illinois watersheds." Transactions of the Illinois State Academy of Science **96**: 135–144.

WWF (2018). Living Planet Report 2018: Aiming Higher. M. Grooten and R. E. A. Almond, WWF, Gland, Switzerland.

Xenopoulos, M. A. and Lodge, D. M. (2006). "GOING WITH THE FLOW: USING SPECIES–DISCHARGE RELATIONSHIPS TO FORECAST LOSSES IN FISH BIODIVERSITY." Ecology **87**(8): 1907–1914.

Zarfl, C., Berlekamp, J., He, F., Jähnig, S. C., Darwall, W. and Tockner, K. (2019). "Future large hydropower dams impact global freshwater megafauna." Scientific Reports **9**(1): 18531.

Zedler, J. B. and Kercher, S. (2005). "Wetland Resources: Status, Trends, Ecosystem Services, and Restorability." Annual Reviews of Environment and Resources **30**: 39–74.

Part III
Nature's Capacity to Supply Ecosystem Services

Chapter 8
Tools for Mapping and Quantifying Ecosystem Services Supply

Zhenyu Wang, Karen T. Lourdes, Perrine Hamel, Theresa G. Mercer, and Alex M. Lechner

8.1 Introduction

The concept of ecosystem services highlights the contribution of ecosystems to human well-being while bridging ecological and social systems (Daily and Matson 2008; Haines-Young 2009; de Groot et al. 2010; Fisher et al. 2011; Chung and Kang 2013; Bryan et al. 2013). They are classified into four categories, which include provisioning services, regulating services, cultural services, and supporting

Z. Wang
School of Land Science and Technology, China University of Geosciences, Beijing, China

Peking University Shenzhen Graduate School, Shenzhen, China

K. T. Lourdes
Landscape Ecology and Conservation Lab, School of Environmental and Geographical Sciences, University of Nottingham Malaysia, Semenyih, Malaysia

P. Hamel
Asian School of the Environment, Nanyang Technological University, Singapore, Singapore

Earth Observatory of Singapore, Singapore, Singapore

T. G. Mercer
Cranfield Environment Centre, School of Water, Energy and Environment, Cranfield University, Bedfordshire, UK

Biogeography and Planetary Health in the Department of Geography, School of Life and Environmental Sciences, University of Lincoln, Lincoln, UK

A. M. Lechner (✉)
Monash University Indonesia, BSD City, Tangerang Selatan, Banten, Indonesia
e-mail: alex.lechner@monash.edu

B. R. Bakshi (ed.), *Engineering and Ecosystems*,
https://doi.org/10.1007/978-3-031-35692-6_8

services (Millennium Assessment (MA) 2005).[1] Ecosystem services represent the interface for the management of social-ecological systems at different scales (MA 2005; Müller et al. 2010) and quantifying their benefits within the context of socio-ecological systems has increasingly become a focus of research (Raymond et al. 2009; Haines-Young et al. 2012; Schmitt and Brugere 2013; Posner 2015; Reed et al. 2017).

The actual or realized benefits, which society receives from ecosystem services, will depend not only on the supply of ecosystems services but also on the demand from society (Burkhard et al. 2012). Humans have an important role in the delivery of ecosystem services and make critical contributions to the flow of ecosystem services between areas of supply and demand. Therefore, the potential benefits derived from ecosystem services will also depend on stakeholders' management strategies, capacity, access, and need within the context of a range of social, economic, and institutional contexts (Villamagna et al. 2013).

While the importance of differentiating between demand and supply is recognized, existing tools for quantifying ecosystem services primarily focus on ecosystem services supply or implicitly integrate demand without specifically looking at flows and/or treat the beneficiaries as homogenous. Ecosystem services supply focuses on the capacity of natural ecosystems to provide relevant ecosystem goods and services within a given time period (Burkhard et al. 2012; Crossman et al. 2013). However, increasingly more research and tools have been devoted to the quantification of ecosystem services demand and the flows of ecosystem services between supply and demand locations. Ecosystem service supply remains important because it is directly derived from the amount and quality of ecosystems, irrespective of the demand or value assigned to the potential service. It is therefore an essential part of ecosystem services assessments.

The objective of this chapter is to review the research and tools for quantifying ecosystem service supply focusing on commonly used tools. We begin by discussing and reviewing the major types of mapping methods for characterizing single ecosystem services. We then describe how multiple ecosystem services can be considered and ways in which important priority areas for ecosystem services provision can be identified. While the review focuses specifically on ecosystem services supply, the distinction between supply and demand in many modeling papers may not be specifically made. We conclude by discussing the research gaps and future challenges in quantifying ecosystem service supply.

[1] Besides the most widely used MA classification, there are also other classification systems which treat ecosystem services slightly differently, especially the MA's "supporting services" class. For instance, TEEB replaced "supporting service" with "habitat service" (TEEB 2010), while CICES (Common International Classification of Ecosystem Services) does not include "supporting service" leaving only three categories (Haines-Young and Potschin-Young 2018). FEGS-CS (Final Ecosystem Goods and Services Classification System) classify 21 final ecosystem service categories and 358 unique FEGS codes (Landers and Nahlik 2013). More details are provided in Chap. 2.

8.2 Quantifying Ecosystem Services

A wide range of methods have been developed for ecosystem services assessments as discussed in multiple comprehensive reviews (Feld et al. 2009; Seppelt et al. 2011; Hernández-Morcillo et al. 2013; Blattert et al. 2017; Cord et al. 2017). These approaches can be vastly different, even for the same services, and the values quantified can vary from biophysical values to monetary values (La Notte et al. 2012; Reed et al. 2017), biophysical values such as erosion control (Vihervaara et al. 2012), or social values (Raymond et al. 2009; Bryan et al. 2010, 2011; Brown 2013).

Many approaches focus specifically on the spatial characteristics of ecosystem services, such as where services are generated (supply) and where services are received and distributed (demand, Fig. 8.1). From the perspective of ecosystem services supply, many quantification methods exist that are derived from natural sciences. For example, the science of catchment management or ecosystem management provides direct quantifications of ecosystem services provision, even though it was developed long before the concept of ecosystem services was popularized. Thus, in this chapter, we have focused specifically on tools and techniques, which have only been developed from the perspective of quantifying—and in particular mapping of—ecosystem services.

Mapping is a practical and useful tool for integrating and revealing complex spatial information across different scales (Martínez-Harms and Balvanera 2012; Crossman et al. 2013). Given the advantages of mapping approaches, the number of studies on mapping ecosystem services has been growing in recent years. Ecosystem services maps can explicitly reveal the spatial distribution of ecosystem services (Egoh et al. 2008), such as service hotspot areas (Eigenbrod et al. 2010; Leh

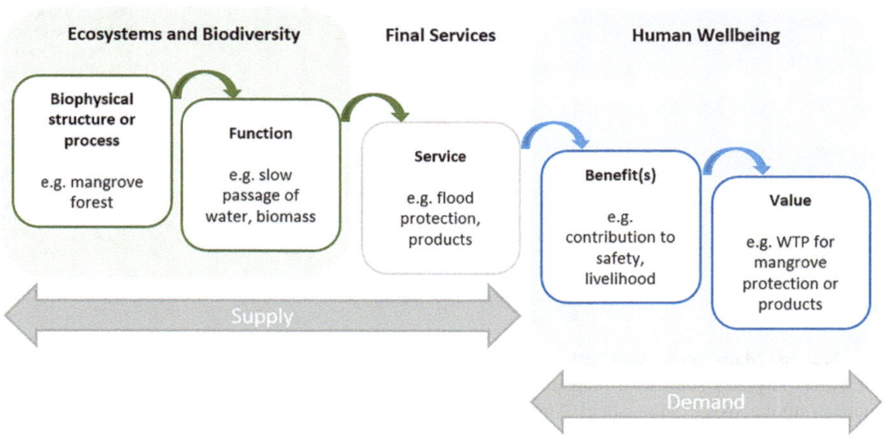

Fig. 8.1 Ecosystem services cascade and relationship between ecosystem services supply and demand. (Adapted from Braat and de Groot (2012))

et al. 2013) and trade-offs and correlations between multiple ecosystem services (Mouchet et al. 2017), which can support decision-making and help communication with stakeholders. With the continued development of ecosystem services mapping methods, many comprehensive off-the-shelf tools have been developed, among which, InVEST (Sharp et al. 2020), ARIES (Villa et al. 2009), and SoIVES (Sherrouse et al. 2011) are widely used. A longer list of tools can be found in a review by de Groot et al. (2018; Table 8) and in the Ecosystems Knowledge Network Tool Assessor (https://ecosystemsknowledge.net/tool). In this chapter we describe the five main types of ecosystem services mapping and quantification methods: (1) Primary data; (2) Causal relationships; (3) Expert knowledge; (4) Participatory mapping; and (5) Biophysical models.

8.2.1 Primary Data

Ecosystem services supply can be mapped directly using primary data. Primary data are derived from field survey or samples (Martínez-Harms and Balvanera 2012) and or remote sensing to represent ecosystem services values. Primary data are most often used in quantifying provisioning services, such as timber (Delphin et al. 2013) and food (Wang et al. 2018a) (Fig. 8.2a). Although primary data offer

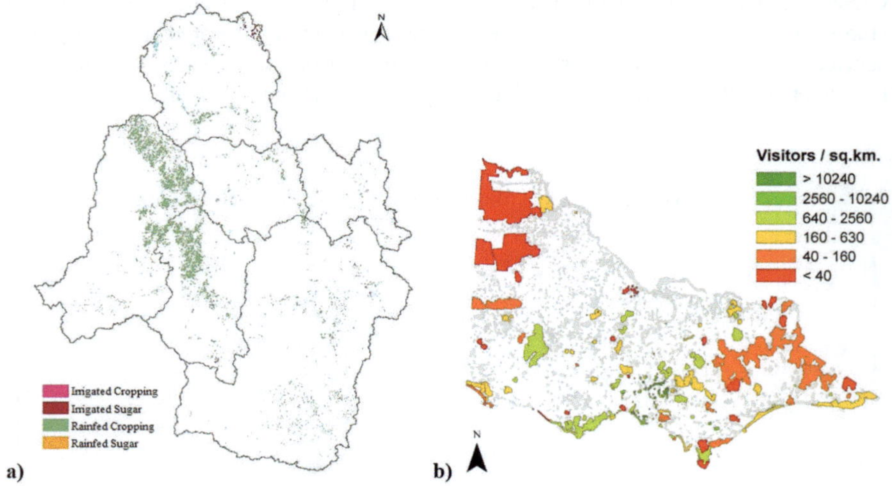

Fig. 8.2 (**a**) Many ecosystem services, especially provisioning services, are commonly produced for other purposes such as a map of agricultural areas. Maps of agricultural land cover classes such as above from Queensland Australia represent primary data, which can be used to map ecosystem services. (Adapted from Wang et al. 2018a). (**b**) Causal relationships use readily available information to characterize ecosystem services. This map of the spatial patterns of Victoria's protected areas popularity was produced using official visitation statistics. (Adapted from Levin et al. 2017)

the most accurate information, such data are not available for all types of ecosystem services. For example, the majority of regulating and supporting services are closely related to complex ecosystem processes and functions, while cultural services are nonmaterial and thus difficult to represent with primary data. Furthermore, availability of primary data is a critical limitation (Martínez-Harms and Balvanera 2012). Primary data cannot always be mapped, particularly if there are issues around data confidentiality or data protection.

8.2.2 *Causal Relationships*

The term causal relationship describes how readily available information can be used to characterize ecosystem processes and services and is one of the most frequently used methods to map different ecosystem services (Martínez-Harms and Balvanera 2012; Schägner et al. 2013). For instance, air quality regulation services of a city could be mapped based on urban greenspace distribution and the vegetation attributes described by remote-sensing-derived leaf area index (Ortolani and Vitale 2016). Recreational services are usually mapped by social and ecological data such as the national parks numbers, tourism statistics, and public access levels (Raudsepp-Hearne et al. 2010; Paracchini et al. 2014) (Fig. 8.2b).

Causal relationships represent a method which utilizes proxies to quantify ecosystem services provision and provide a useful way to estimate ecosystem services when direct ecosystem services indicators are absent. Many of these mapping and quantification techniques implicitly incorporate ecosystem services demand without explicitly modeling supply (i.e., visitor numbers at national parks). However, causal relationship methods need considerable knowledge for understanding the generating processes of ecosystem services (Eigenbrod et al. 2010; Schägner et al. 2013). Uncertainties are produced and the outcomes are not accurate if there are poor causal relationships between the data and the ecosystem service it is meant to represent.

8.2.3 *Expert-Based Model Knowledge*

Expert knowledge is one of the most widely used approaches; it is a simple and effective way to map ecosystem services (Egoh et al. 2008; Müller et al. 2010; Grêt-Regamey et al. 2017). This method incorporates advice from different experts and stakeholders to map ecosystem services. For example, Haines-Young et al. (2012) use "expert" and "literature-driven" methods to establish the multiple links between land cover and use and potential ecosystem service outputs in different geographical contexts across Europe. One of the more popular methods is the application of land cover proxies where each land cover is given a specific value for ecosystem service provision (Jacobs et al. 2015; Burkhard et al. 2012, 2014) to create a

Ecosystem Services

Land cover	Crop production	Water provision	Raw materials	Recreation and tourism	Landscape aesthetics	Spirituality and Cultural	Erosion prevention	Pollination	Carbon sequestration	Local climate and air	Stormwater retention	Habitat for biodiversity
Water bodies	3.4	4.25	3.3	4.3	3.5	2.75	2.1	1.75	3.25	3.5	3.05	3.8
Bare soil	0.6	0.35	1.85	1.25	0.35	0.25	0.7	0.35	0.5	0.4	1.2	1.1
Impervious	0.3	0.25	0.25	1.65	1.65	0.8	2.5	0.45	0.35	0.35	0.35	1.55
Roads	0.25	0.2	0.2	0.35	0.45	0.35	0.4	0.4	0.25	0.25	0.25	0.25
Oil palm	4.5	3.4	4.25	2.25	3.4	2.25	3.9	3.7	4	4.1	3.1	3.75
Rubber	4.5	3.2	4.4	3.2	3.25	3.8	3.9	3.8	4	4	3.4	3.5
Rubber mix	4.4	3.15	4.35	4.1	4.1	4.3	4	3.5	4	4.2	4	4.1
Fruit mix	4.4	3.1	4.3	4.15	4.4	3.6	4	4.35	3.8	4.1	4	4.3
Other ag.	4.3	3.1	4.05	3.65	3.9	4	4.05	4.15	4	4.2	3.7	3.9
Forests	0.8	0.9	4.1	4.4	4.35	4.4	4.3	4.05	4.5	4.2	4.4	4.45

Fig. 8.3 Matrix model/look-up-table for mapping ecosystem services supply with a land cover proxy for Kuala Lumpur, Malaysia. Dark red represents the highest potential to supply ES, while yellow shows the least. These values are then mapped based on their corresponding land cover classes. (Data supplied by Gangul Nelaka)

matrix/look-up-table of ecosystem services versus land cover (Fig. 8.3), which can then be converted into an ecosystem services map based on this relationship. Other examples of this approach include a study by Swetnam et al. (2011) who developed scenarios integrating local stakeholders and experts to define the extent of changes in land cover classes under different sets of drivers. Palomo et al. (2013) defined ecosystem services according to expert-advice and questionnaires and then mapped ecosystem services flows. However, the high levels of subjectivity and the lack of quantitative assessments for ecosystem services are the main disadvantages of the expert knowledge approaches (Hamel and Bryant 2017).

8.2.4 Participatory Mapping

Public Participation Geographic Information System (PPGIS) is used to map ecosystem services using quantitative or qualitative social surveys (Brown 2013; Shoyama and Yamagata 2016). A common approach is to ask participants to identify ecosystem service locations using a point or polygon on a web-based or paper map (Fig. 8.4). These points are then aggregated or interpolated to create a raster surface representing ecosystem service provision. Social Values for Ecosystem Services (SoIVES) is a very popular ecosystem services mapping tool that also uses quantitative social surveys of both point locations and preferences for difference locations and land covers, in conjunction with Maxent create raster surfaces of ecosystem services (Sherrouse et al. 2011). Both approaches map ecosystem services from the perspective of a specific stakeholder group.

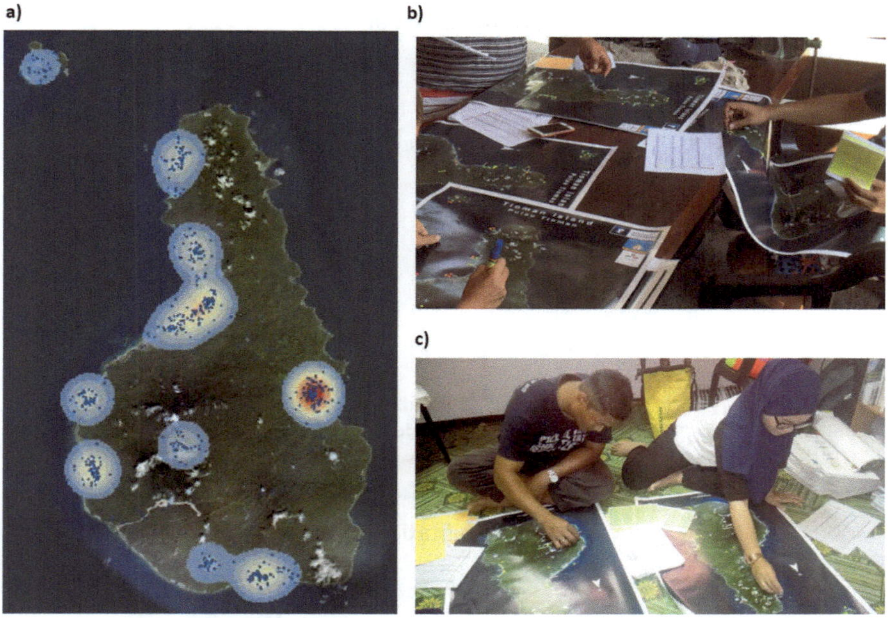

Fig. 8.4 PPGIS mapping with a quantitative social survey carried out on Tioman Island Malaysia. (Figures and data adapted from Lechner et al. (2020)). (**a**) Map of recreational landscape values. Multiple survey participants identified recreational landscape values with sticker dots (**b**, **c**), which were first digitized, then combined, and finally interpolated to produce a recreational landscape value surface

8.2.5 Biophysical Models

The most data intensive mapping approach uses biophysical models to describe the biophysical processes and functions of ecosystems (Kareiva et al. 2011; Petz 2014; Running 2017). Various models from different disciplines and theories are utilized for ecosystem services assessments and are integrated with GIS. Commonly, these biophysical models are based on process or mechanistic models, which are composed of multiple equations, which approximate real world biophysical processes such as erosion or hydrological flows. However, machine learning, or other statistical approaches that mimic biophysical processes, can also be used. These models are commonly used for mapping regulating services and supporting services (Martínez-Harms and Balvanera 2012; Baral et al. 2013a; Pulighe et al. 2016). For instance, the Universal Soil Loss Equation (USLE) can be used to simulate the mechanisms associated with the interaction between soil, precipitation, and vegetation to assess the soil loss and retention for the soil conservation service (Sánchez-Canales et al. 2015; Grafius et al. 2016), while the Carnegie-Ames-Stanford approach (CASA) model can be used to simulate photosynthesis processes to estimate net primary productivity for carbon sequestration service (Dai et al.

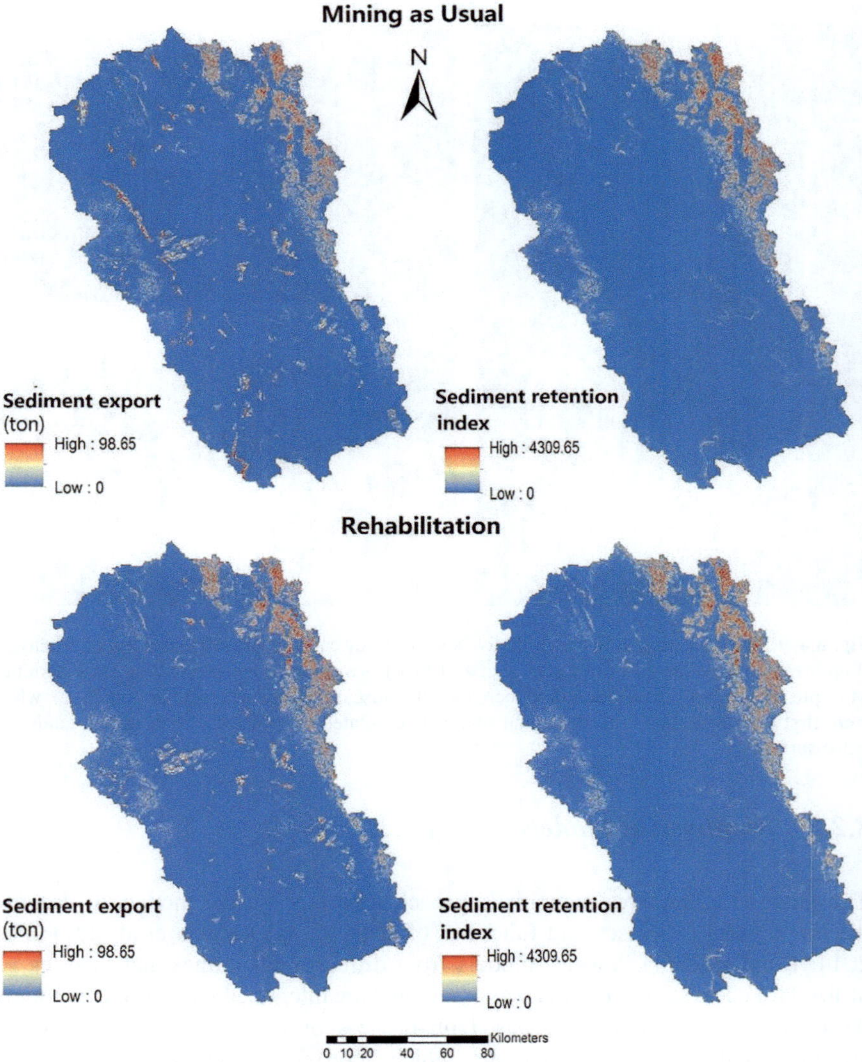

Fig. 8.5 Sediment retention service mapping under two different scenarios: rehabilitation and mining as usual scenarios in an Australian mining region (Wang et al. 2018b) using process-based Sediment Delivery Ratio model in InVEST

2017). However, there are still challenges associated with how well these models characterize a biophysical process (Surfleet and Tullos 2013; Sharp et al. 2020) and/or models selection (Lavorel et al. 2011; Petz and van Oudenhoven 2012) due to a lack of understanding ecosystem processes, subjectivity, and data availability. Integrated Valuation of Ecosystem Services and Trade-offs (InVEST) is one of the most widely applied ecosystem services mapping tools (Sharp et al. 2020) and it includes several biophysical models to characterize a range of ecosystem services such as sediment retention, urban cooling, and flood risk (Fig. 8.5).

8.3 Comparing Ecosystem Service Supply

Ecosystem services assessments typically assess the spatial or temporal change in services to identify optimal land use or management strategies. Several approaches are used to compare multiple ecosystems and identify their spatial trends. In this section, we describe approaches for (1) comparing multiple ecosystem services and (2) identifying priority areas.

8.3.1 Comparison of Multiple Services

A single ecosystem can provide multiple ecosystem services and these services can interact resulting in services trade-offs and synergies (Fig. 8.6). Trade-offs occur when one service decreases as another service increases (Rodríguez et al. 2006; Grêt-Regamey et al. 2013). Synergies are defined as the situation when the changes are positive for both (or many) ecosystem services and trade-offs describe the opposing situation (Rodríguez et al. 2006; Haase et al. 2012; Crossman et al. 2013). Haase et al. (2012) proposed an evaluation matrix of ecosystem services correlations, which describe ecosystem services synergies, trade-offs, losses, and other single-aspect changes.

Ecosystem services trade-offs and synergies have been mapped and assessed in various studies. Among the four types of ecosystem services, provisioning and regulating services are most frequently assessed. For instance, the interactions between water provision and sediment retention services have been assessed in multiple fields of different countries (Williams and Hedlund 2014; Früh-Müller et al. 2016; Fernandez-Campo et al. 2017; Hao et al. 2017; Hamel et al. 2019). There are also many studies focusing on cultural services interactions (Turner et al. 2014; Ament et al. 2017). Different agricultural practices can lead to different correlations among crop yields, soil carbon, and nutrient retention (Qiu and Turner 2013; Kragt and Robertson 2014; Nelson et al. 2009). Correlations between ecosystem services are also commonly assessed in response to future change scenarios. For instance, climate changes can cause changes in hydrological ecosystem services and their interactions with other services (Agropolis et al. 2013; Jiang et al. 2017; Mandle et al. 2017). Analyses of ecosystem services current and future trade-offs and synergies are commonly integrated into land use planning and natural resources management (Castro et al. 2014; Witt et al. 2014; Keith et al. 2017) to help decision-makers balance the protection and promotion of different ecosystem services.

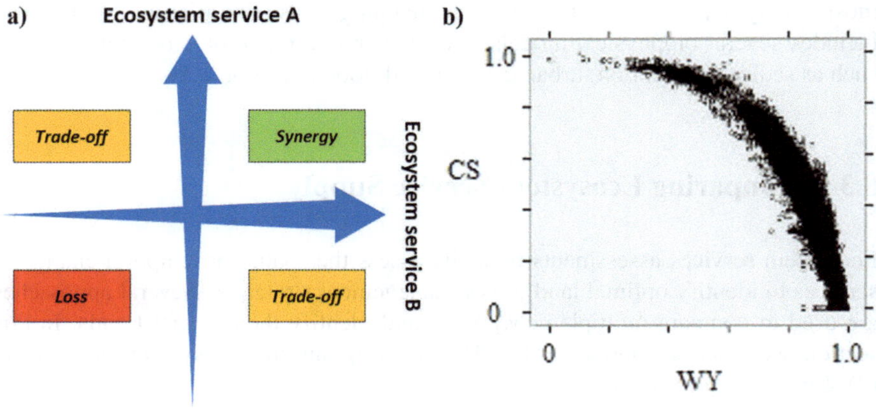

Fig. 8.6 (**a**) Evaluation matrix of ecosystem services trade-offs and synergies. (Adapted from Haase et al. 2012). (**b**) The pixel-scale correlation between carbon sequestration (CS) and water yield (WY) within the whole mining lease of Curragh mine in Queensland, Australia. The chart shows a trade-off between carbon sequestration (CS) and water yield (WY) with a Spearman's rank correlation coefficient of −0.88 ($P = 0.01$). (Adapted from Wang et al. 2020)

Quantitative approaches for measuring ecosystem services trade-offs and synergies include aspatial and spatial methods. Aspatial methods range from aggregated metrics such as correlation coefficients to graphical tools such as radar plots, parallel coordinate plots, or scatterplots (Weil 2017). Correlation coefficients such as the Pearson and Spearman's rank coefficients have been widely used to assess the ecosystem services interactions (Agropolis et al. 2013; Castro et al. 2014; Oñatibia et al. 2015; Staes et al. 2017). Regression models have also been applied to characterize pair-wise relationships between ecosystem services. For instance, Jia et al. (2014) utilized a logistical regression model to analyze ecosystem services trade-offs and synergies for a Grain-for-Green area in China. Maes et al. (2012b) applied multinomial logistic regression models to assess the correlations between ecosystem service supply, biodiversity, and habitat conservation status in Europe.

In addition to aggregated statistical methods for assessing trade-offs, a range of useful tools or indicators have been developed to map these trade-offs. Beyond traditional mapping methods such as hotspot or chloropleth maps (Weil 2017; Burkhard and Maes 2017), interactive tools have become popular to visualize trade-offs and synergies (Natural Capital Project 2020; Fredriksson et al. 2020). For instance, Pang et al. (2017) developed the Landscape simulation and Ecological Assessment (LEcA) tool to analyze synergies and trade-offs among five ecosystem services in Sweden. Trodahl et al. (2017) utilized Land Utilization and Capability Indicator (LUCI) to evaluate the trade-offs between water quality and agricultural productivity in New Zealand. Other useful examples can be found on the Natural Capital Project's visualization website (Natural Capital Project 2020). Where many ecosystem services are assessed, the concept of ecosystem services bundles has been used to describe services that always concurrently appear together; these are commonly identified by cluster analysis methods (Raudsepp-Hearne et al. 2010; Turner et al. 2014; Ament et al. 2017; Mouchet et al. 2017).

8.3.2 *Identifying Priority Areas*

Assessing the spatial patterns in ecosystem service distribution can be valuable in identifying priority – areas for management (Lourdes et al. 2022). Such assessments have been applied to individual and multiple ecosystem services, although variation in the spatial distribution between multiple ecosystem services can be high (Schröter and Remme 2016). In order to meaningfully assess the spatial patterns of multiple ecosystem services, individual services are converted to a common scale (i.e., rescaled) for standardization such as a minimum-maximum normalization (e.g., 0–1) (Dou et al. 2020; Lavorel et al. 2011; Maes et al. 2012a; Mokondoko et al. 2018), or z-score normalization or z-standardization (Jopke et al. 2015; Weil 2017). The type of rescaling method applied is tied strongly to the objective of the assessment, taking note that the absolute or initial values assigned to ecosystem service will change when standardized.

Hotspot mapping is a common cluster analysis method used to distinguish the abundance and distribution of ecosystem services across a landscape. The terms "hotspot" and "coldspot" respectively denote areas of high service provision and low service provision for a single ecosystem service (Egoh et al. 2008). Methods for delineating ecosystem service hotspots are diverse; Schröter and Remme (2016) provide a comprehensive review on the methods available. Popular methods include the top richest cells method and Getis-Ord Gi* statistic. Although both methods delineate ecosystem service hotspots, the two methods highlight unique approaches to identifying spatial patterns in ecosystem services. The top richest cells (quantile) method divides grid cells, ranked from high to low service value, into classes with an equal number of cells (Bai et al. 2011; Dou et al. 2020; Eigenbrod et al. 2010; Orsi et al. 2020). The class with the highest values is defined as a hotspot, with class sizes for hotspots ranging from 5% to 30% of the total cells. For example, Orsi et al. (2020) delineated hotspots as the highest 20% of cells supplying an ecosystem service. While the Getis-Ord Gi* statistic (Getis and Ord 1992) utilizes a spatial clustering method to delineate hotspots. This method identifies areas where high value cells are highly concentrated within a specified distance/neighbourhood (i.e. high values within a neighborhood of low values or vice versa) (Bagstad et al. 2016; Li et al. 2017; Sylla et al. 2020), distinguishing hotspots and coldspots with varying degrees of clustering (i.e. significance). The differences in these two approaches are further detailed by Bagstad et al. (2017).

Areas of overlap between multiple ecosystem services, or 'multiple service hotspots', can be identified by summing hotspots produced through a range of methods, for individual services. The applications of both hotspot mapping and aggregation and comparison of multiple services are diverse (Anderson et al. 2009; Dou et al. 2020; Maes et al. 2012a; Mokondoko et al. 2018). Pan et al. (2020)

assessed areas of overlap between several hydrological ecosystem services for the integrated management of a river basin, while Bogdan et al. (2019) mapped social values, delineating multiple service hotspots for cultural ecosystem services. Bagstad et al. (2016), on the other hand, combined biophysical ecosystem service values and social values, delineating multiple service hotspots for more inclusive management of the Southern Rocky Mountains.

8.4 Data Sources and Uncertainty

One of the big challenges for ecosystem services mapping is the requirement for data and the varieties of sources and principles by which they were created (Crossman 2017). The types of data sources can be distinguished into two categories, primary and secondary data (Egoh et al. 2012; Martínez-Harms and Balvanera 2012; Crossman et al. 2013). Primary data are those derived from sampling in the field, such as field measurements, surveys, and interviews, while secondary data are defined as those derived from readily available information not typically verified in the field including literature-based or modeled data (Martínez-Harms and Balvanera 2012). Along with the types of data sources, data can be classified as biophysical or socioeconomic. Biophysical data are related to the natural and biophysical systems, such as hydrological data, remote sensing, topographical, and land cover data. Socioeconomic data are the data related to social and human activities, such as crop production, population, road lines, and economic data (Martínez-Harms and Balvanera 2012).

Among different type of datasets, land cover data are the most widely used. Land cover change is one of the greatest drivers of changes in ecosystems and their services and, as noted in Sect. 8.3, is commonly used as proxy for mapping ecosystem services (Petter et al. 2013; Baral et al. 2013b; Nahuelhual et al. 2014; Abram et al. 2014; Tolessa et al. 2016). Land cover spatial data can be acquired through different ways. There are many well-established land use and land cover database in many countries, which are acquired through particular land use and cover mapping projects. For instance, the CORINE land cover database of Europe (Haines-Young 2009; Burkhard et al. 2012) and the national land cover dataset (NLCD) of USA (Lawler et al. 2014; Yoo et al. 2014) are widely used for ecosystem services assessments. Where existing data is unavailable, land cover changes can be mapped using remote sensing (Krishnaswamy et al. 2009; Tolessa et al. 2016; Zaehringer et al. 2017). Besides land cover spatial data, other data such as hydrological (Vigerstol and Aukema 2011; Terrado et al. 2014), topographic (Sherrouse et al. 2011; Fernandez-Campo et al. 2017), and climate data (Bangash et al. 2013; Jiang et al. 2017) are also are frequently utilized for ecosystem services mapping.

Scale is always a critical issue in landscape ecology and geographic research (Lechner et al. 2012a) and has a significant effect on ecosystem services mapping (Nemec and Raudsepp-Hearne 2012; Di Sabatino et al. 2013). Some ecological processes are scale dependent (i.e., species environment relationships), while other processes occur at multiple scales (Lechner et al. 2012b; Grêt-Regamey et al. 2014). Ecosystem services supply can also be mapped at different grain sizes and extents, which include pixel, local, regional, national, and global scales (Martínez-Harms and Balvanera 2012). Among them, local and regional scales are the most frequently assessed in which the extents of a study are commonly defined by the boundaries of a biogeographic or hydrologic system such as a mountain range (Grêt-regamey et al. 2012), watershed (Band et al. 2012), forest (Pohjanmies et al. 2017), or urban area (McPhearson et al. 2013).

Mapping scale can influence the results of ecosystem service assessment. For instance, Grafius et al. (2016) mapped ecosystem services at two different scales in three urban areas of the UK, and found sensitivity to scale was dependent on the type of service. Hou et al. (2017) assessed ecosystem service interactions at the pixel and town scales through a case study in the central Loess Plateau of China, which revealed that scale could apparently affect ecosystem services synergies and interactions. Grêt-Regamey et al. (2014) estimated the effects of scale on ecosystem services mapping through four case studies of different countries and suggested a four-step approach to address the scale issues. There are also many other studies focusing on scales of ecosystem services mapping (Larondelle and Lauf 2016; Raudsepp-Hearne and Peterson 2016; Calderón-Contreras and Quiroz-Rosas 2017; Xu et al. 2017), which all demonstrate the scale dependency issue and emphasize the importance of considering scale effects when mapping ecosystem services.

Beyond input data and spatial scale, several other sources of uncertainty affect the quantification and mapping of ecosystem services, including the context and framing of the assessment and, in the case of modeled data, the model structure, parameters, and technical implementation (Hamel and Bryant 2017). Managing these uncertainties involves understanding the potential use of ecosystem services information, potentially through codevelopment approaches, and applying proved analytical methods developed in the field of integrated environmental modeling (Pappenberger and Beven 2006; Petersen et al. 2013; Hamilton et al. 2019; Hamel and Bryant 2017),

8.5 Challenges for Quantifying Ecosystem Services

One of the primary goals for quantifying and mapping ecosystem service is for integration into planning and management (Egoh et al. 2008; Raymond et al. 2009; Potschin and Haines-Young 2012; Grêt-Regamey et al. 2017). There are many examples of mapping ecosystem services for urban planning focusing on the whole urban area (Lauf et al. 2014; Kaczorowska et al. 2015; Albert et al. 2016; Larondelle and Lauf 2016; Pickard et al. 2017; see Lourdes et al. 2021 for a regional review) or particular parts such as urban green spaces (Pulighe et al. 2016; Engström and Gren 2017). Also, ecosystem services have been mapped for conservation planning and natural resource management (Tallis and Polasky 2009; Bottalico et al. 2015;

Gunton et al. 2017) commonly together with biodiversity (Guerry et al. 2012; Sumarga and Hein 2014).

Although studies on ecosystem services mapping and quantification are growing, there are still a number of challenges. Some significant review papers have summarized the multiple challenges and bottlenecks associated with ecosystem services mapping (Crossman et al. 2013; Malinga et al. 2015; Brown and Fagerholm 2015) and characterized the types of ecosystem services mapped for multiple purposes (Willemen et al. 2015; Klein and Celio 2015; Drakou et al. 2015). Building on these existing reviews, we outline four key research challenges and gaps, which need to be considered and should be a focus for future research.

1. Gaps in Data Availability

Data availability can affect the quality of ecosystem services maps and the types of ecosystem services mapping tools available. Where primary data is not used, the application of proxy data is used, which can lead to uncertainties for mapping outputs (Eigenbrod et al. 2010; Schägner et al. 2013). Little is known about how the errors associated with proxy-based methods might affect the inferences drawn from analyses because quantifying the impacts of such errors is difficult without comparisons to primary data (Vrebos et al. 2015). A major challenge for ecosystem services mapping is to develop approaches, which adequately characterize ecosystems when using limited available data. This is especially important for data poor regions in the Global South where both primary data such as land cover maps may be of poor quality or unavailable and basic biophysical information such as the properties of soil may have never been measured or are highly uncertain, thus restricting the use of process-based biophysical models.

2. Inconsistency in Mapping Approaches

Although there are various approaches applied to mapping ecosystem services, there is still a need to understand the uncertainties and biases introduced by different mapping methods (Crossman et al. 2013; Crossman 2017). Different indicators and approaches have been used to map the same service, which can lead to apparent differences in the outputs (Schulp et al. 2014). Also, the same service may be mapped differently according to different research objectives such as mapping only ecosystem supply versus quantifying flows and potential values of ecosystem services. Additional guidance for ecosystem service mapping for decision-making and interpreting outputs will help understand the differences between methods and the implications for the development of decision-support tools (Bagstad et al. 2013; Hamel et al. 2020). Because different actors have different mandates and motivations, understanding information needs will ensure that ecosystem services information is used effectively (Bremer et al. 2020). It is important to note; there is not one optimal method, but the approach taken should be adapted to the decision-making context and uncertainty understood.

3. Assessing Uncertainties in Ecosystem Services Mapping

Uncertainty assessments are commonly conducted in many disciplines from hydrology (Benke et al. 2008), economics (Gilboa et al. 2008) to landscape ecology (Lechner et al. 2012a) as uncertainty can seriously affect the outcome of an analysis. Currently, few studies focus explicitly on analyzing uncertainty in ecosystem services mapping (Grêt-regamey et al. 2012, 2014; Schulp et al. 2014; Hamel and Guswa 2015; Wang et al. 2018a, b). A recent special issue in the journal *Ecosystem Services* in 2018 demonstrated best practices and challenges for "transparent, feasible and useful uncertainty assessment in ecosystem services modelling" (Bryant et al. 2018). Uncertainty was characterized for different models (Maldonado et al. 2018) and different scenarios (Ashley et al. 2018; Monge et al. 2018) and new methods were introduced for uncertainty assessment such as through the application of machine learning (Willcock et al. 2018). However, there are still a lot of challenges for successfully assessing uncertainties of ecosystem services mapping. As ecosystem services mapping draws on methods from a range of disciplines each with their own methods for assessing uncertainty (e.g., social sciences to hydrology), these methods should be incorporated into ecosystem services quantification approaches. It is important that any approach which addresses uncertainty is effective without being so time-consuming that it would be impractical to apply (Hamel and Bryant 2017).

As ecosystem service modeling methods become more complex, incorporating and assessing multiple ecosystem services in more complex ways, can cause errors and uncertainty to propogate and magnify. Approaches that incorporate multiple ecosystem services remain limited in several ways. Many approaches identify only high or low values of ecosystem services provision relative to a study area (i.e., top pixel values) or neighborhood (i.e., Getis-Ord Gi). However, the normalization process ignores the absolute values of these ecosystem services from the societal perspective. Not all ecosystem services are equivalent in their value to society and thus not all high valued ecosystem services should be considered equal when it comes to combining multiple ecosystem services.

4. Mapping Across Temporal Scale

While ecosystem services are commonly mapped across space, there are still relatively few studies, which have mapped historical changes in ecosystem services which have included high temporal resolution (i.e., numerous timesteps), even though abiotic (i.e., rainfall) and biotic (i.e., phenology) systems are dynamic. Although mapping future ecosystem service scenarios is relatively common, ecosystem services hotspots, trade-offs, and priority areas will change in both time and space. Mapping these temporally can add to the predictive capability of the outcomes. Such an approach is especially important in highly dynamic landscapes such as mining sites and agricultural and urban landscapes, which develop very quickly, especially in the Global South.

8.6 Conclusions

This chapter provided an introduction to the current tools for mapping and quantifying ecosystem services supply. While ecosystem service approaches have progressed rapidly in recent years, there are still many challenges. This is especially the case in the Global South where there is rapid land use change, resulting in the loss of biodiversity and ecosystem services. In urban landscapes, modelling aproaches still need further development. For example, InVEST, one of the most widely used ecosystem services modeling package, only recently released a suite of tools for modeling urban ecosystem services. In addition, moving beyond the realm of quantifying and mapping ecosystem services, consideration needs to be given for how the outcomes will be used and by whom. This also poses challenges in terms of how the outcomes should be represented including ways to clearly communicate uncertainties of both the input and output data and the ways in which the models have been validated. Knowledge gaps between practitioners and stakeholders could be reduced by building collaborative connections and including stakeholders early on in the mapping and decision-making process. This ensures that the needs of the end users are met and the underlying questions to be mapped are understood. Co-production of maps could result in output maps that are fit for purpose, easily understood by relevant stakeholders or end users, and could lead to better results around translating mapping outcomes into effective policy, planning, and management.

References

Abram NK, Meijaard, E, Ancrenaz, M, Runting RK, Wells, JA, Gaveau, D, Pellier, A-S, Mengersen, K. 2014. Spatially explicit perceptions of ecosystem services and land cover change in forested regions of Borneo. Ecosyst Serv. 7:116–127.

Agropolis A, Cifor J, Program GC, Rica C. 2013. Synergies and trade-offs between ecosystem services in Costa Rica. The journal for Agropolis is Environmental Conservation (Cambridge University Press) 41:27–36.

Albert C, Galler, C, Hermes, J, Neuendorf, F, Von Haaren, C, Lovett, A. 2016. Applying ecosystem services indicators in landscape planning and management: The ES-in-Planning framework. Ecol Indic. 61:100–113.

Ament JM, Moore, CA, Herbst M, Cumming, GS. 2017. Cultural Ecosystem Services in Protected Areas: Understanding Bundles, Trade-Offs, and Synergies. Conserv Lett. 10:439–449.

Anderson, BJ, Armsworth, PR, Eigenbrod, F, Thomas, CD, Gillings, S, Heinemeyer, A, Roy, DB, Gaston, KJ 2009. Spatial covariance between biodiversity and other ecosystem service priorities. Journal of Applied Ecology, 46(4), 888–896.

Ashley, R, Gersonius, B, Digman, C, Horton, B, Smith, B, & Shaffer, P. 2018. Including uncertainty in valuing blue and green infrastructure for stormwater management. Ecosyst Serv. 33: 237–246.

Bagstad, KJ, Reed, JM, Semmens, DJ, Sherrouse, BC, & Troy, A. 2016. Linking biophysical models and public preferences for ecosystem service assessments: a case study for the Southern Rocky Mountains. Regional Environmental Change, 16(7), 2005–2018.

Band LE, Hwang T, Hales TC, Vose J, Ford C. 2012. Ecosystem processes at the watershed scale: Mapping and modeling ecohydrological controls of landslides. Geomorphology. 137:159–167.

Bagstad, KJ, Semmens, DJ, Ancona, ZH, & Sherrouse, BC. 2017. Evaluating alternative methods for biophysical and cultural ecosystem services hotspot mapping in natural resource planning. Landscape Ecology, 32(1), 77–97.

Bagstad, KJ, Semmens, DJ, Waage, S, Winthrop, R. 2013. A comparative assessment of decision-support tools for ecosystem services quantification and valuation. Ecosyst. Serv. 5: 27–39.

Bai, Y, Zhuang, C, Ouyang, Z, Zheng, H, & Jiang, B 2011. Spatial characteristics between biodiversity and ecosystem services in a human-dominated watershed. Ecological Complexity, 8(2), 177–183.

Bangash RF, Passuello, A, Sanchez-Canales, M, Terrado, M, López, A, Elorza, FJ, Ziv G, Acuña, V, Schuhmacher, M. 2013. Ecosystem services in Mediterranean river basin: Climate change impact on water provisioning and erosion control. Sci Total Environ. 458–460:246–255.

Baral H, Keenan, RJ, Fox, JC, Stork, NE, Kasel, S. 2013a. Spatial assessment of ecosystem goods and services in complex production landscapes: A case study from south-eastern Australia. Ecol Complex. 13:35–45.

Baral H, Keenan, RJ, Stork, NE, Kasel, S. 2013b. Measuring and managing ecosystem goods and services in changing landscapes: a south-east Australian perspective. J Environ Plan Manag. 57:961–983.

Bremer LL, Hamel, P, Ponette-González, AG, Pompeu, PV, Saad, SI, Brauman, KA. 2020. Who Are we Measuring and Modeling for? Supporting Multilevel Decision-Making in Watershed Management. Water Resour Res. 56(1):1–17.

Benke KK, Lowell KE, Hamilton AJ. 2008. Parameter uncertainty, sensitivity analysis and prediction error in a water-balance hydrological model. Math Comput Model. 47:1134–1149.

Blattert C, Lemm R, Thees O, Lexer MJ, Hanewinkel M. 2017. Management of ecosystem services in mountain forests: Review of indicators and value functions for model based multi-criteria decision analysis. Ecol Indic. 79:391–409.

Bogdan, SM, Stupariu, I, Andra-Topârceanu, A, & Năstase, II. 2019. Mapping social values for cultural ecosystem services in a mountain landscape in the Romanian Carpathians. Carpathian Journal of Earth and Environmental Sciences, 14(1): 199–208. https://doi.org/10.26471/cjees/2019/014/072

Bottalico F, Pesola L, Vizzarri M, Antonello L, Barbati A, Chirici G, Corona P, Cullotta S, Garfì V, Giannico V, et al. 2015. Modeling the influence of alternative forest management scenarios on wood production and carbon storage: A case study in the Mediterranean region. Environ Res. 144:72–87.

Braat, L.C., de Groot, R., 2012. The ecosystem services agenda: bridging the worlds of natural science and economics, conservation and development, and public and private policy. Ecosyst. Serv. 1(1): 4–15.

Brown G, Fagerholm N. 2015. Empirical PPGIS / PGIS mapping of ecosystem services: A review and evaluation. Ecosyst Serv. 13:119–133.

Brown G. 2013. The relationship between social values for ecosystem services and global land cover: An empirical analysis. Ecosyst Serv. 5:58–68.

Bryan BA, Higgins A, Overton IC, Holland K, Lester RE, King D, Nolan M, MacDonald DH, Connor JD, Bjornsson T, Kirby M. 2013. Ecohydrological and socioeconomic integration for the operational management of environmental flows. Ecol Appl. 23:999–1016.

Bryan BA, Raymond CM, Crossman ND, King D. 2011. Comparing spatially explicit ecological and social values for natural areas to identify effective conservation strategies. Conserv. Biol. 25 (1): 172–181.

Bryan BA, Raymond CM, Neville DC, MacDonald, DH. 2010. Targeting the management of ecosystem services based on social values: where, what and how? Landscape Urban Plan. 97: 111–122.

Bryant, BP, Borsuk, ME, Hamel, P, Oleson, KLL, Schulp, CJE. 2018. Transparent and feasible uncertainty assessment adds value to applied ecosystem services modeling. Ecosyst. Serv. 33: 103–109.

Burkhard, B, & Maes, J. 2017. Mapping Ecosystem Services. In B. Burkhard & J. Maes (Eds.), Advanced Books (Vol. 1). Pensoft Publishers. https://doi.org/10.3897/ab.e12837

Burkhard, B, Kandziora, M, Hou, Y, Müller, F, 2014. Ecosystem Service Potentials, Flows and Demands-Concepts for Spatial Localisation, Indication and Quantification. Landsc. Online, 34: 1–32.

Burkhard B, Kroll F, Nedkov S, Müller F. 2012. Mapping ecosystem service supply, demand and budgets. Ecol Indic. 21:17–29.

Calderón-Contreras R, Quiroz-Rosas LE. 2017. Analysing scale, quality and diversity of green infrastructure and the provision of Urban Ecosystem Services: A case from Mexico City. Ecosyst Serv. 23:127–137.

Castro AJ, Verburg PH, Martín-lópez B, Garcia-llorente M, Cabello J, Vaughn CC, López E. 2014. Landscape and Urban Planning Ecosystem service trade-offs from supply to social demand : A landscape-scale spatial analysis. Landsc Urban Plan. 132:102–110.

Chung MG, Kang H. 2013. A Review of Ecosystem Service Studies: Concept, Approach and Future Work in Korea. J Ecol Environ. 36:1–9.

Cord AF, Bartkowski B, Beckmann M, Dittrich A, Hermans-Neumann K, Kaim A, Lienhoop N, Locher-Krause K, Priess J, Schröter-Schlaack C, et al. 2017. Towards systematic analyses of ecosystem service trade-offs and synergies: Main concepts, methods and the road ahead. Ecosyst Serv. 28:264–272.

Crossman, ND. 2017. Uncertainties of ecosystem services mapping. In: Burkhard B, Maes J (Eds.) Mapping Ecosystem Services. Advanced Books (Vol. 1). Pensoft Publishers.

Crossman ND, Burkhard B, Nedkov S, Willemen L, Petz K, Palomo I, Drakou EG, Martín-Lopez B, McPhearson T, Boyanova K, et al. 2013. A blueprint for mapping and modelling ecosystem services. Ecosyst Serv. 4:4–14.

Dai EF, Wang XL, Zhu JJ, Xi WM. 2017. Quantifying ecosystem service trade-offs for plantation forest management to benefit provisioning and regulating services. Ecol Evol. 7:7807–7821.

Daily GC, Matson PA. 2008. Ecosystem Services: From Theory to Implementation. Proc Natl Acad Sci. 105:9455–9456.

De Groot, R., Moolenaar, S., van Weelden, M., Konovska, I., de Vente, J. 2018. The ESP Guidelines in a Nutshell. FSD Working Paper 2018-09 (29 pp), Foundation for Sustainable development, Wageningen, The Netherlands.

de Groot RS, Alkemade R, Braat L, Hein L, Willemen L. 2010. Challenges in integrating the concept of ecosystem services and values in landscape planning, management and decision making. Ecol Complex. 7:260–272.

Delphin S, Escobedo FJ, Abd-Elrahman A, Cropper W. 2013. Mapping potential carbon and timber losses from hurricanes using a decision tree and ecosystem services driver model. J Environ Manage. 129:599–607.

Di Sabatino A, Coscieme L, Vignini P, Cicolani B. 2013. Scale and ecological dependence of ecosystem services evaluation: Spatial extension and economic value of freshwater ecosystems in Italy. Ecol Indic. 32:259–263.

Dou, H, Li, X, Li, S, Dang, D, Li, X, Lyu, X, Li, M & Liu, S. 2020. Mapping ecosystem services bundles for analyzing spatial trade-offs in inner Mongolia, China. J. of Clean. Prod., 256: 120444.

Drakou EG, Crossman, ND, Willemen, L, Burkhard, B, Palomo I, Maes, J. 2015. A visualization and data-sharing tool for ecosystem service maps : Lessons learnt, challenges and the way forward. Ecosyst Serv. 13:134–140.

Egoh, B, Drakou, EG, Dunbar, MB, Maes, J. 2012. Indicators for mapping ecosystem services : a review. Luxembourg: Institute for Environment and Sustainability, Joint Research Centre, European Commission.

Egoh B, Reyers, B, Rouget, M, Richardson, DM, Le Maitre, DC, van Jaarsveld, AS. 2008. Mapping ecosystem services for planning and management. Agric Ecosyst Environ. 127:135–140.

Eigenbrod, F, Armsworth, PR, Anderson, BJ, Heinemeyer A, Gillings, S, Roy DB, Thomas CD, Gaston, KJ. 2010. The impact of proxy-based methods on mapping the distribution of ecosystem services. J Appl Ecol. 47:377–385.

Engström G, Gren A. 2017. Capturing the value of green space in urban parks in a sustainable urban planning and design context: Pros and cons of hedonic pricing. Ecol Soc. 22(2): 21.
Feld CK, Da Silva PM, Sousa JP, De Bello F, Bugter R, Grandin U, Hering D, Lavorel S, Mountford O, Pardo I, et al. 2009. Indicators of biodiversity and ecosystem services: A synthesis across ecosystems and spatial scales. Oikos. 118:1862–1871.
Fernandez-Campo M, Rodríguez-Morales B, Dramstad WE, Fjellstad W, Diaz-Varela ER. 2017. Ecosystem services mapping for detection of bundles, synergies and trade-offs: Examples from two Norwegian municipalities. Ecosyst Serv. 28:283–297.
Fisher B, Turner RK, Burgess, ND, Swetnam, RD, Green, J, Green RE, Kajembe, G, Kulindwa, K, Lewis, SL, Marchant R, et al. 2011. Measuring, modeling and mapping ecosystem services in the Eastern Arc Mountains of Tanzania. Prog Phys Geogr. 35:595–611.
Fredriksson, HA, Weil, C, Rönnberg, N. 2020. On the usefulness of map-based dashboards for decision making. TechRxiv. Preprint. https://doi.org/10.36227/techrxiv.12738683.v1
Früh-Müller, A, Hotes, S, Breuer, L, Wolters, V, Koellner, T. 2016. Regional Patterns of Ecosystem Services in Cultural Landscapes. Land. 5:17.
Getis, A & Ord, JK. 1992. The Analysis of Spatial Association by Use of Distance Statistics. Geographical Analysis, 24(3), 189–206.
Gilboa I, Postlewaite AW, Schmeidler D. 2008. Probability and Uncertainty in Economic Modeling. J Econ Perspect.22 (3):173–188.
Grafius DR, Corstanje R, Warren PH, Evans KL, Hancock S, Harris JA. 2016. The impact of land use/land cover scale on modelling urban ecosystem services. Landsc Ecol. 31: 1509–1522.
Grêt-Regamey A, Altwegg J, Sirén EA, van Strien MJ, Weibel B. 2017. Integrating ecosystem services into spatial planning—A spatial decision support tool. Landsc Urban Plan. 165:206–219.
Grêt-regamey A, Brunner SH, Kienast F. 2012. Mountain Ecosystem Services: Who Cares ? Mt Res Dev. 32:S23–34.
Grêt-Regamey A, Celio E, Klein TM, Wissen Hayek U. 2013. Understanding ecosystem services trade-offs with interactive procedural modeling for sustainable urban planning. Landsc Urban Plan. 109:107–116.
Grêt-Regamey A, Weibel B, Bagstad KJ, Ferrari M, Geneletti D, Klug H, Schirpke U, Tappeiner U. 2014. On the effects of scale for ecosystem services mapping. PLoS One. 9:1–26.
Guerry AD, Ruckelshaus MH, Arkema KK, Bernhardt JR, Guannel G, Kim C-K, Marsik M, Papenfus M, Toft JE, Verutes G, et al. 2012. Modeling benefits from nature: using ecosystem services to inform coastal and marine spatial planning. Int J Biodivers Sci Ecosyst Serv Manag. 8:107–121.
Gunton RM, Marsh CJ, Moulherat S, Malchow AK, Bocedi G, Klenke RA, Kunin WE. 2017. Multicriterion trade-offs and synergies for spatial conservation planning. J Appl Ecol. 54:903–913.
Haase D, Schwarz N, Strohbach M, Kroll F, Seppelt R. 2012. Synergies, Trade-offs, and Losses of Ecosystem Services in Urban Regions: an Integrated Multiscale Framework Applied to the Leipzig- Halle Region, Germany. Ecol Soc. 17.
Haines-Young R, Potschin-Young M. 2018. Revision of the Common International Classification for Ecosystem Services (CICES V5.1): A Policy Brief. One Ecosystem 3: e27108.
Haines-Young R, Potschin M, Kienast F. 2012. Indicators of ecosystem service potential at European scales: Mapping marginal changes and trade-offs. Ecol Indic. 21:39–53.
Haines-Young R. 2009. Land use and biodiversity relationships. Land use policy. 26: S178–S186.
Hamel, P, Bremer, LL, Ponette-González, AG, Acosta, E, Fisher, JRB, Steele, B, Cavassani, AT, Klemz, C, Blainski, E, Brauman, KA. 2020. The value of hydrologic information for watershed management programs: The case of Camboriú, Brazil. Sci. Total Environ. 705.
Hamel, P, Blundo-Canto, G, Kowal, V, Bryant, B. P, Hawthorne, P L, & Chaplin-Kramer, R. 2019. Where should livestock graze? Integrated modeling and optimization to guide grazing management in the Cañete basin, Peru. Socio-Env. Sys. Model. 1: 16125.
Hamel P, Bryant BP. 2017. Uncertainty assessment in ecosystem services analyses: Common challenges and practical responses. Ecosyst Serv J. 24:1–33.

Hamel P, Guswa AJ. 2015. Uncertainty analysis of a spatially explicit annual water-balance model : case study of the Cape Fear basin, North Carolina. Hydrol Earth Syst Sci. 19:839–853.
Hamilton SH, Fu B, Guillaume JHA, et al. 2019. A framework for characterising and evaluating the effectiveness of environmental modelling. Environ Model Softw. 118(August 2018):83–98.
Hao R, Yu D, Wu J. 2017. Relationship between paired ecosystem services in the grassland and agro-pastoral transitional zone of China using the constraint line method. Agric Ecosyst Environ. 240:171–181.
Hernández-Morcillo M, Plieninger T, Bieling C. 2013. An empirical review of cultural ecosystem service indicators. Ecol Indic. 29:434–444.
Hou Y, Lü Y, Chen W, Fu B. 2017. Temporal variation and spatial scale dependency of ecosystem service interactions: a case study on the central Loess Plateau of China. Landsc Ecol. 32:1201–1217.
Jia X, Fu B, Feng X, Hou G, Liu Y, Wang X. 2014. The tradeoff and synergy between ecosystem services in the Grain-for-Green areas in Northern Shaanxi, China. Ecol Indic. 43:103–111.
Jiang C, Zhang H, Tang Z, Lbzovskii L. 2017. Evaluating the coupling effects of climate variability and vegetation restoration on ecosystems of the Loess Plateau, China. Land use policy. 69:134–148.
Jacobs S, Burkhard B, Van Daele T, Staes J, Schneiders A. 2015. "The Matrix Reloaded": A review of expert knowledge use for mapping ecosystem services. Ecol Modell. 295:21–30.
Jopke, C, Kreyling, J, Maes, J & Koellner, T 2015. Interactions among ecosystem services across Europe: Bagplots and cumulative correlation coefficients reveal synergies, trade-offs, and regional patterns. Ecological Indicators, 49, 46–52.
Kaczorowska A, Kain J-H, Kronenberg J, Haase D. 2015. Ecosystem services in urban land use planning: Integration challenges in complex urban settings—Case of Stockholm. Ecosyst Serv. 22:1–9.
Kareiva P, Tallis H, Ricketts T, Daily G, Polasky S, editors. 2011. Natural capital: theory and practice of mapping ecosystem services. Cambridge (UK): Oxford University Press.
Keith H, Vardon M, Stein JA, Stein JL, Lindenmayer D. 2017. Ecosystem accounts define explicit and spatial trade-offs for managing natural resources. Nat Ecol Evol. 1:1683–1692.
Klein TM, Celio E. 2015. Ecosystem services visualization and communication : A demand analysis approach for designing information and conceptualizing decision support systems. Ecosyst Serv. 13:173–183.
Kragt ME, Robertson MJ. 2014. Quantifying ecosystem services trade-offs from agricultural practices. Ecol Econ. 102:147–157.
Krishnaswamy J, Bawa KS, Ganeshaiah KN, Kiran MC. 2009. Quantifying and mapping biodiversity and ecosystem services: Utility of a multi-season NDVI based Mahalanobis distance surrogate. Remote Sens Environ. 113:857–867.
La Notte A, Maes J, Grizzetti B, Bouraoui F, Zulian G. 2012. Spatially explicit monetary valuation of water purification services in the Mediterranean bio-geographical region. Int J Biodivers Sci Ecosyst Serv Manag. 8:26–34.
Landers DH and Nahlik AM. 2013. Final Ecosystem Goods and Services Classification System (FEGS-CS). EPA/600/R-13/ORD-004914. U.S. Environmental Protection Agency, Office of Research and Development, Washington, D.C.
Larondelle N, Lauf S. 2016. Balancing demand and supply of multiple urban ecosystem services on different spatial scales. Ecosyst Serv. 22:18–31.
Lauf S, Haase D, Kleinschmit B. 2014. Linkages between ecosystem services provisioning, urban growth and shrinkage - A modeling approach assessing ecosystem service trade-offs. Ecol Indic. 42:73–94.
Lavorel S, Grigulis K, Lamarque P, Colace MP, Garden D, Girel J, Pellet G, Douzet R. 2011. Using plant functional traits to understand the landscape distribution of multiple ecosystem services. J Ecol. 99:135–147.
Lawler JJ, Lewis DJ, Nelson E, Plantinga AJ, Polasky S, Withey JC, Helmers DP, Martinuzzi S, Pennington D, Radeloff VC. 2014. Projected land-use change impacts on ecosystem services in the United States. Proc Natl Acad Sci U S A. 111:7492–7.

Lechner AM, Verbrugge, LNH, Chelliah, A, Ang, MLE, Raymond, CM. 2020. Rethinking tourism conflict potential within and between groups using participatory mapping. Landsc. Urban Plan. 203: 1–11.
Lechner AM, Langford WT, Bekessy SA, Jones SD. 2012a. Are landscape ecologists addressing uncertainty in their remote sensing data? Landsc Ecol. 27:1249–1261.
Lechner, AM, Langford, WT, Jones, SD, Bekessy, SA, & Gordon A. 2012b. Investigating species–environment relationships at multiple scales: Differentiating between intrinsic scale and the modifiable areal unit problem. Ecological Complexity, 11, 91–102.
Leh MDK, Matlock, MD, Cummings, EC, Nalley, LL. 2013. Quantifying and mapping multiple ecosystem services change in West Africa. Agric Ecosyst Environ. 165:6–18.
Levin, N, Lechner, AM, Brown, G. 2017. An evaluation of crowdsourced information for assessing the visitation and perceived importance of protected areas. App. Geogr. 79 :115–126.
Li, Y, Zhang, L, Yan, J, Wang, P, Hu, N, Cheng, W & Fu, B 2017. Mapping the hotspots and coldspots of ecosystem services in conservation priority setting. Journal of Geographical Sciences, 27(6), 681–696.
Lourdes, KT, Gibbins, CN, Hamel, P, Sanusi, R, Azhar, B, & Lechner, AM. 2021. A Review of Urban Ecosystem Services Research in Southeast Asia. Land. 10(1): 40.
Lourdes, K. T., Hamel, P., Gibbins, C. N., Sanusi, R., Azhar, B., & Lechner, A. M. 2022. Planning for green infrastructure using multiple urban ecosystem service models and multicriteria analysis. Landsc Urban Plan. 226, 104500.
MA (Millennium Ecosystem Assessment). 2005. Ecosystems and human well-being: synthesis. Washington (DC): Island Press.
Maes J, Egoh B, Willemen L, Liquete C, Vihervaara P, Schägner JP, Grizzetti B, Drakou EG, Notte A La, Zulian G, et al. 2012a. Mapping ecosystem services for policy support and decision making in the European Union. Ecosyst Serv. 1:31–39.
Maes J, Paracchini ML, Zulian G, Dunbar MB, Alkemade R. 2012b. Synergies and trade-offs between ecosystem service supply, biodiversity, and habitat conservation status in Europe. Biol Conserv. 155:1–12.
Maldonado, AD, Aguilera, PA, Salmerón, A, & Nicholson, AE. (2018). Probabilistic modeling of the relationship between socioeconomy and ecosystem services in cultural landscapes. Ecosyst. Serv. 33: 146–164.
Malinga R, Gordon LJ, Jewitt G, Lindborg R. 2015. Mapping ecosystem services across scales and continents – A review. Ecosyst Serv. 13:57–63.
Mandle, L., Wolny, S, Bhagabati, N, Helsingen, H, Hamel, P, Bartlett, R, Dixon, A, Horton, R, Lesk, C, Manley, D, Mel, M. De, Bader, D, Myint, SNW, Myint, W, Mon, MS. 2017. Assessing ecosystem service provision under climate change to support conservation and development planning in Myanmar. PLOS ONE, 12(9): e0184951.
Martínez-Harms MJ, Balvanera P. 2012. Methods for mapping ecosystem service supply: a review. Int J Biodivers Sci Ecosyst Serv Manag. 8:17–25.
McPhearson T, Kremer P, Hamstead ZA. 2013. Mapping ecosystem services in New York City: Applying a social–ecological approach in urban vacant land. Ecosyst Serv. 5:11–26.
Mokondoko, P, Manson, RH, Ricketts, TH, & Geissert, D 2018. Spatial analysis of ecosystem service relationships to improve targeting of payments for hydrological services. PLOS ONE, 13(2), e0192560.
Monge, JJ, Daigneault, AJ, Dowling, LJ, Harrison, DR, Awatere, S, Ausseil, AG. 2018. Implications of future climatic uncertainty on payments for forest ecosystem services: The case of the East Coast of New Zealand. Ecosyst Serv. 33: 199–212.
Mouchet MA, Paracchini ML, Schulp CJE, Stürck J, Verkerk PJ, Verburg PH, Lavorel S. 2017. Bundles of ecosystem (dis)services and multifunctionality across European landscapes. Ecol Indic. 73:23–28.
Müller F, de Groot R, Willemen L. 2010. Ecosystem services at the landscape scale: The need for integrative approaches. Landsc Online. 23:1–11.

Nahuelhual L, Carmona A, Aguayo M, Echeverria C. 2014. Land use change and ecosystem services provision: A case study of recreation and ecotourism opportunities in southern Chile. Landsc Ecol. 29:329–344.

Natural Capital Project. 2020. Natural Capital Project. https://viz.naturalcapitalproject.org/. Accessed 26 Aug 2020

Nelson, E, Mendoza, G, Regetz, J, Polasky, S, Tallis, H, Cameron, Dr, Chan, KM, Daily, GC, Goldstein, J, Kareiva, PM, Lonsdorf, E, Naidoo, R, Ricketts, TH, & Shaw, Mr. 2009. Modeling multiple ecosystem services, biodiversity conservation, commodity production, and tradeoffs at landscape scales. Front. Ecol. Environ. 7(1): 4–11.

Nemec KT, Raudsepp-Hearne C. 2012. The use of geographic information systems to map and assess ecosystem services. Biodivers Conserv. 22:1–15.

Oñatibia GR, Aguiar MR, Semmartin M. 2015. Are there any trade-offs between forage provision and the ecosystem service of C and N storage in arid rangelands ? Ecol Eng. 77:26–32.

Orsi, F, Ciolli, M, Primmer, E, Varumo, L & Geneletti, D 2020. Mapping hotspots and bundles of forest ecosystem services across the European Union. Land Use Policy, 99.

Ortolani C, Vitale M. 2016. The importance of local scale for assessing, monitoring and predicting of air quality in urban areas. Sustain Cities Soc. 26:150–160.

Palomo I, Martín-López B, Potschin M, Haines-Young R, Montes C. 2013. National Parks, buffer zones and surrounding lands: Mapping ecosystem service flows. Ecosyst Serv. 4:104–116.

Pan, J, Wei, S & Li, Z. 2020. Spatiotemporal pattern of trade-offs and synergistic relationships among multiple ecosystem services in an arid inland river basin in NW China. Ecological Indicators, 114, 106345.

Pang X, Nordström EM, Böttcher H, Trubins R, Mörtberg U. 2017. Trade-offs and synergies among ecosystem services under different forest management scenarios – The LEcA tool. Ecosyst Serv. 28:67–79.

Pappenberger F, Beven. 2006. Ignorance is bliss: Or seven reasons not to use uncertainty analysis. 42 (December 2005): 1–8 https://doi.org/10.1029/2005WR004820.

Paracchini ML, Zulian G, Kopperoinen L, Maes J, Schägner JP, Termansen M, Zandersen M, Perez-Soba M, Scholefield PA, Bidoglio G. 2014. Mapping cultural ecosystem services: A framework to assess the potential for outdoor recreation across the EU. Ecol Indic. 45:371–385.

Petersen, A.C., Janssen, P.H.M., van der Sluijs, J.P., Risbey, J.S., Ravetz, J.R., Wardekker, J.A., Hughes Martinson, H. 2013. Guidance for Uncertainty Assessment and Communication, 2nd edition.

Petter M, Mooney S, Maynard SM, Davidson A, Cox M, Horosak I. 2013. A Methodology to Map Ecosystem Functions to Support Ecosystem. Ecol Soc. 18:31.

Petz K, van Oudenhoven APE. 2012. Modelling land management effect on ecosystem functions and services: a study in The Netherlands. Int J Biodivers Sci Ecosyst Serv Manag. 8:135–155.

Petz K. 2014. Mapping and modelling the effects of land use and land management change on ecosystem services from local ecosystems and landscapes to global biomes. Wageningen: Wageningen University.

Pickard BR, Van Berkel D, Petrasova A, Meentemeyer RK. 2017. Forecasts of urbanization scenarios reveal trade-offs between landscape change and ecosystem services. Landsc Ecol. 32:617–634.

Pohjanmies T, Eyvindson K, Triviño M, Mönkkönen M. 2017. More is more? Forest management allocation at different spatial scales to mitigate conflicts between ecosystem services. Landsc Ecol. 32:2337–2349.

Posner SM. 2015. The impact of ecosystem services knowledge on decisions. The University of Vermont.

Potschin M, Haines-Young R. 2012. Landscapes, sustainability and the place-based analysis of ecosystem services. Landsc Ecol. 28:1053–1065.

Pulighe G, Fava F, Lupia F. 2016. Insights and opportunities from mapping ecosystem services of urban green spaces and potentials in planning. Ecosyst Serv. 22:1–10

Qiu J, Turner MG. 2013. Spatial interactions among ecosystem services in an urbanizing agricultural watershed. Proc Natl Acad Sci U S A. 110:12149–54.

Raudsepp-Hearne C, Peterson GD, Bennett EM. 2010. Ecosystem service bundles for analyzing tradeoffs in diverse landscapes. Proc Natl Acad Sci U S A. 107:5242–7.

Raudsepp-Hearne C, Peterson GD. 2016. Scale and ecosystem services: How do observation, management, and analysis shift with scale–lessons from Québec. Ecol Soc. 21.

Raymond CM, Bryan B, MacDonald DH, Cast A, Strathearn S, Grandgirard A, Kalivas T. 2009. Mapping community values for natural capital and ecosystem services. Ecol Econ. 68:1301–1315.

Reed MS, Allen K, Attlee A, Dougill AJ, Evans KL, Kenter JO, Hoy J, McNab D, Stead SM, Twyman C, et al. 2017. A place-based approach to payments for ecosystem services. Glob Environ Chang. 43:92–106.

Rodríguez JP, Beard TD, Bennett EM, Cumming GS, Cork SJ, Agard J, Dobson AP, Peterson GD. 2006. Trade-offs across Space, Time, and Ecosystem Services. Ecol Soc. 11:28.

Runting RK. 2017. Managing natural capital assets and ecosystem services under global change. Brisbane, Australia: The University of Queensland.

Sánchez-Canales M, López-Benito A, Acuña V, Ziv G, Hamel P, Chaplin-Kramer B, Elorza FJ. 2015. Sensitivity analysis of the sediment dynamics in a Mediterranean river basin: global change and management implications. Sci Total Environ. 502:602–610.

Schägner JP, Brander L, Maes J, Hartje V. 2013. Mapping ecosystem services' values: Current practice and future prospects. Ecosyst Serv. 4:33–46.

Schmitt LHM, Brugere C. 2013. Capturing Ecosystem Services, Stakeholders' Preferences and Trade-Offs in Coastal Aquaculture Decisions: A Bayesian Belief Network Application. PLoS One. 8(10): e75956.

Schröter, M & Remme, RP 2016. Spatial prioritisation for conserving ecosystem services: comparing hotspots with heuristic optimisation. Landscape Ecology, 31(2), 431–450.

Schulp CJE, Burkhard B, Maes J, Vliet J Van, Verburg PH. 2014. Uncertainties in Ecosystem Service Maps : A Comparison on the European Scale. PLoS One. 9:e109643.

Seppelt R, Dormann CF, Eppink F V., Lautenbach S, Schmidt S. 2011. A quantitative review of ecosystem service studies: approaches, shortcomings and the road ahead. J Appl Ecol. 48:630–636.

Sharp, R., Tallis, H.T., Ricketts, T., Guerry, A.D., Wood, S.A., Chaplin-Kramer, R., Nelson, E., Ennaanay, D., Wolny, S., Olwero, N., Vigerstol, K., Pennington, D., Mendoza, G., Aukema, J., Foster, J., Forrest, J., Cameron, D., Arkema, K., Lonsdorf, E., Kennedy, C., Verutes, G., Kim, C.K., Guannel, G., Papenfus, M., Toft, J., Marsik, M., Bernhardt, J., Griffin, R., Glowinski, K., Chaumont, N., Perelman, A., Lacayo, M. Mandle, L., Hamel, P., Vogl, A.L., Rogers, L., Bierbower, W., Denu, D., and Douglass, J. 2020, InVEST 3.8.0 User's Guide. The Natural Capital Project, Stanford University, University of Minnesota, The Nature Conservancy, and World Wildlife Fund.

Sherrouse BC, Clement JM, Semmens DJ. 2011. A GIS application for assessing, mapping, and quantifying the social values of ecosystem services. Appl Geogr. 31:748–760.

Shoyama K, Yamagata Y. 2016. Local perception of ecosystem service bundles in the Kushiro watershed, Northern Japan - Application of a public participation GIS tool. Ecosyst Serv. 22:139–149.

Staes J, Broekx S, Van Der Biest K, Vrebos D, Olivier B, De Nocker L, Liekens I, Poelmans L, Verheyen K, Jeroen P, Meire P. 2017. Quantification of the potential impact of nature conservation on ecosystem services supply in the Flemish Region: A cascade modelling approach. Ecosyst Serv. 24:124–137.

Sumarga E, Hein L. 2014. Mapping ecosystem services for land use planning, the case of central kalimantan. Environ Manage. 54:84–97.

Surfleet CG, Tullos D. 2013. Uncertainty in hydrologic modelling for estimating hydrologic response due to climate change (Santiam River, Oregon). Hydrol Process. 27:3560–3576.

Swetnam RD, Fisher B, Mbilinyi BP, Munishi PKT, Willcock S, Ricketts T, Mwakalila S, Balmford A, Burgess ND, Marshall AR, Lewis SL. 2011. Mapping socio-economic scenarios of land

cover change: a GIS method to enable ecosystem service modelling. J Environ Manage. 92:563–74.
Sylla, M, Hagemann, N & Szewrański, S 2020. Mapping trade-offs and synergies among peri-urban ecosystem services to address spatial policy. Environmental Science & Policy, 112, 79–90.
Tallis H, Polasky S. 2009. Mapping and valuing ecosystem services as an approach for conservation and natural-resource management. Ann N Y Acad Sci. 1162:265–83.
TEEB. 2010. The Economics of Ecosystems and Biodiversity: Mainstreaming the Economics of Nature: A synthesis of the approach, conclusions and recommendations of TEEB.
Terrado M, Acuña V, Ennaanay D, Tallis H, Sabater S. 2014. Impact of climate extremes on hydrological ecosystem services in a heavily humanized Mediterranean basin. Ecol Indic. 37:199–209.
Tolessa T, Senbeta F, Abebe T. 2016. Land use/land cover analysis and ecosystem services valuation in the central highlands of Ethiopia. For Trees Livelihoods. 8028:1–13.
Trodahl MI, Jackson BM, Deslippe JR, Metherell AK. 2017. Investigating trade-offs between water quality and agricultural productivity using the Land Utilisation and Capability Indicator (LUCI)–A New Zealand application. Ecosyst Serv. 26:388–399.
Turner KG, Odgaard MV, Bøcher PK, Dalgaard T, Svenning JC. 2014. Bundling ecosystem services in Denmark: Trade-offs and synergies in a cultural landscape. Landsc Urban Plan. 125:89–104.
Vigerstol KL, Aukema JE. 2011. A comparison of tools for modeling freshwater ecosystem services. J Environ Manage. 92:2403–9.
Vihervaara P, Kumpula T, Ruokolainen A, Tanskanen A, Burkhard B. 2012. The use of detailed biotope data for linking biodiversity with ecosystem services in Finland. Int J Biodivers Sci Ecosyst Serv Manag. 8:169–185.
Villamagna, AM, Angermeier, PL, Bennett, EM. 2013. Capacity, pressure, demand, and flow: A conceptual framework for analyzing ecosystem service provision and delivery. Ecol. Complex. 15:114–121.
Villa F, Ceroni M, Bagstad K, Johnson G, Krivov S. 2009. ARIES (ARtificial Intelligence for Ecosystem Services): a new tool for ecosystem services assessment, planning, and valuation. Venice, Italy: 11th Annual BIOECON Conference on "Economic Instruments to Enhance the Conservation and Sustainable Use of Biodiversity"; 2009 Sep 21–22.
Vrebos D, Staes J, Vandenbroucke T, D′Haeyer T, Johnston R, Muhumuza M, Kasabeke C, Meire P. 2015. Mapping ecosystem service flows with land cover scoring maps for data-scarce regions. Ecosyst Serv. 13:28–40.
Weil CG. 2017. http://www.charlotteweil.fr/masterthesis/docs/toolbox.html. Accessed 12 Aug 2020
Wang Z, Lechner AM, Baumgartl T.2018a. Ecosystem Services Mapping Uncertainty Assessment : A Case Study in the Fitzroy Basin Mining Region. Water. 10 (1): 88.
Wang Z, Lechner AM, Baumgartl T. 2018b. Mapping cumulative impacts of mining on sediment retention ecosystem service in an Australian mining region. Int J Sustain Dev World Ecol. 25: 69–80.
Wang Z, Lechner AM, Yang Y, Baumgartl T, Wu J. 2020. Mapping the cumulative impacts of long-term mining disturbance and progressive rehabilitation on ecosystem services. Sci Total Environ. 717:137214.
Willcock, S, Martínez-López, J, Hooftman, DAP, Bagstad, KJ, Balbi, S, Marzo, A, Prato, C, Sciandrello, S, Signorello, G, Voigt, B, Villa, F, Bullock, JM, Athanasiadis, IN. 2018. Machine learning for ecosystem services. Ecosyst Serv. 33: 165–174.
Willemen L, Burkhard B, Crossman N, Drakou EG, Palomo I. 2015. Editorial : Best practices for mapping ecosystem services. Ecosyst Serv. 13:1–5.
Williams A, Hedlund K. 2014. Indicators and trade-offs of ecosystem services in agricultural soils along a landscape heterogeneity gradient. Appl Soil Ecol. 77:1–8.
Witt A, Frank S, Fu C, Koschke L, Makeschin F. 2014. Making use of the ecosystem services concept in regional planning — trade-offs from reducing water erosion. : 1377–1391.

Xu S, Liu Y, Wang X, Zhang G. 2017. Scale effect on spatial patterns of ecosystem services and associations among them in semi-arid area: A case study in Ningxia Hui Autonomous Region, China. Sci Total Environ. 598:297–306.

Yoo J, Simonit S, Connors JP, Kinzig AP, Perrings C. 2014. The valuation of off-site ecosystem service flows: Deforestation, erosion and the amenity value of lakes in Prescott, Arizona. Ecol Econ. 97:74–83.

Zaehringer JG, Schwilch G, Andriamihaja OR, Ramamonjisoa B, Messerli P. 2017. Remote sensing combined with social-ecological data: The importance of diverse land uses for ecosystem service provision in north-eastern Madagascar. Ecosyst Serv. 25:140–152.

Chapter 9
Designing with Nature: Incorporating Hydrologic Services in Engineering Projects

Perrine Hamel and Andrew J. Guswa

9.1 Introduction

Today's water challenges can be summarized pretty simply: too much water or too little water of sufficient quality. Extreme flooding events touch millions of people annually, with a human or economic cost higher than any other natural disaster (Jha et al. 2012). At the same time, droughts or poor water management leave some regions short of the necessary resources for domestic, industrial, or environmental uses. Climate change is expected to exacerbate this reality, and water managers seek solutions to overcome the shortfalls of conventional engineering approaches. Integrated water resource management (IWRM) was developed with this goal, using a systemic approach to understand how social, technical, and environmental factors can increase the resilience and sustainability of water resources (see Schoeman et al. 2014).

Important components of the solutions to water challenges are nature-based solutions (NBSs). They are defined as "actions to protect, sustainably manage and restore natural and modified ecosystems in ways that address societal challenges effectively and adaptively, to provide both human well-being and biodiversity benefits" (IUCN 2016). NBSs are getting increased attention in both academic and policy realms, as they hold the promise of meeting both environmental and development goals.

P. Hamel (✉)
Asian School of the Environment, Nanyang Technological University, Singapore, Singapore

Earth Observatory of Singapore, Singapore, Singapore
e-mail: perrine.hamel@ntu.edu.sg

A. J. Guswa
Picker Engineering Program, Smith College, Northampton, MA, USA

B. R. Bakshi (ed.), *Engineering and Ecosystems*,
https://doi.org/10.1007/978-3-031-35692-6_9

With regard to IWRM, there are three main services provided by NBSs: flood risk mitigation, water quality improvement, and water supply. For these services to be incorporated into engineering design, key questions need to be answered related to NBSs' efficacy: To what extent can NBSs reduce the amount of runoff and peak flows? To what extent are they able to remove contaminants and purify waters? To what extent can they increase streamflows during low-flow periods and/or increase groundwater recharge?

The answers to these questions and the potential for nature-based designs to address water-resources challenges depend on the social, technical, and environmental context (Keeler et al. 2019). Due to the natural processes NBSs rely on, climate and geography will influence their behavior. For example, more intense precipitation or steeper slopes will result in more runoff, reducing the capacity of natural vegetation to infiltrate precipitation. In addition, social and technical contexts affect water challenges themselves, and hence the likelihood that NBSs will address them. For instance, water demand management or the construction of a desalination plant will impact water resources management and the place of NBSs in the strategy.

The complex interactions between social, technical, and environmental factors mean that the potential of NBSs will require cooperation between scientists in various disciplines as well as water engineers. The science of ecosystem services (ES) – the benefits people derive from nature – has developed over the past decades to improve our understanding of the interdependence of nature and people and to quantify the value of natural capital in providing key benefits to people (Guerry et al. 2015). In this chapter, we illustrate how the science of ES, and all the disciplines it draws on, may support IWRM in several ways: by producing information on ecological functions to support engineering design; by developing new approaches to incorporate people in the design phase, as beneficiaries and contributors of knowledge; and by facilitating the communication on the value of nature to a broad range of stakeholders.

The following sections describe how NBSs are becoming part of the water engineering discussion by reviewing the potential of these solutions, highlighting their cobenefits and trade-offs, and finally discussing the opportunities offered by ES science to support IWRM.

9.2 Potential of Nature-Based Solutions for Water Services

This section provides an overview of the functions performed by NBSs with regard to the three main water services: flood risk mitigation, water supply, and water quality management. Common types of NBSs include street trees, parks and open space, engineered stormwater management devices (bioswales, raingardens, etc.), green roofs, waterways and wetlands, upland forests or grasslands, and community/allotment gardens (Table 9.1 and Fig. 9.1). We review the factors moderating the level of service and practical implications for engineering design. Of note,

Table 9.1 Suite of nature-based solutions and their cobenefits

	Nature-based solutions						
Urban ecosystem services	Street trees	Parks & open space	Engineered devices (bioswales, raingardens, etc.)	Green roofs	Waterways & wetlands	Restoration or protection of upland forests or grasslands	Community/allotment gardens
Flood risk reduction		X	X		X	X	
Water quality management	X	X	X	X	X	X	X
Water supply	X	X	X			X	X
Air quality improvement	X	X					
Carbon sequestration	X	X					
Coastal protection						X	
Urban heat reduction	X	X		X	X		X
Recreation		X			X	X	
Mental health	X	X	X		X		X
Urban agriculture		X					X

Adapted from Keeler et al. (2019)

Fig. 9.1 Examples of NBSs that provide hydrologic services in urban or rural environments: (from left to right, top and bottom rows) urban parks, community gardens, afforestation or forest protection, street trees, wetlands, and green roofs

we focus here on the ecological and technical factors, while acknowledging that socioeconomic factors affect the level of risk associated with each service and therefore the risk mitigation service provided by natural infrastructure. For example, low-quality housing may be more vulnerable to flooding, making the service provided by NBSs (or traditional infrastructure) more valuable (see Keeler et al. 2019, for a review of socioeconomic factors affecting water services).

9.2.1 Flood Risk Mitigation

Flooding occurs for multiple reasons: when river flow cannot be contained within the natural or man-made channel (riverine flooding); when rainfall intensity exceeds infiltration capacity over an area (pluvial flooding, with the particular case of stormwater flooding in urban areas); and when large storm systems or rising sea levels affect coastal areas (coastal flooding). We focus here on the first two, associated with freshwater rather than coastal water, while acknowledging that sea-level rise or storm surge may interact with freshwater flooding in coastal environments.

Following are the main functions of NBSs with regard to flood-risk mitigation:

- Reduce runoff production
- Slow surface flows
- Create space for water (in floodplains or basins)

The functions are distinct and natural infrastructure may perform one or several of them to a different extent. This partially explains the inconsistencies in the literature, with some authors claiming that the role of natural infrastructure in flood risk reduction is overestimated (Calder and Aylward 2006). In fact, several ecological and technical factors moderate the effect of natural infrastructure, meaning that the relevance of a given type of natural infrastructure varies widely with context. Starting with ecological factors, the characteristics of a storm event (in particular intensity and duration), type of soil, and location and type of natural infrastructure all influence the risk-mitigation effect (Keeler et al. 2019). For example, landscape interventions in the UK were found to reduce peak flow for moderate rainfall events, but their effect in large basins for extreme events is limited (Dadson et al. 2017). Soils with low infiltration capacity, either naturally or due to compaction, will also generate more runoff and therefore reduce the performance of NBSs with regard to flood risk.

In addition, the type and location of built infrastructure will affect flood-hazard reduction. For example, the presence of natural infrastructure (recreation or protected areas) in a flood plain will not only reduce flood risk downstream but also reduce exposure, since it restricts housing and built infrastructure in flood-prone areas. Another example of built infrastructure affecting NBSs is the presence of a dam, which makes natural flood control less valuable. In urban environments, the density and quality of the stormwater sewer network, if present, will also affect the value of NBSs with respect to volume reduction (but generally not undermining the effect on stormwater quality, see Sect. 9.2.3).

Because of these multiple interactions, evidence for the effect of NBSs may seem inconsistent. However, some facts emerge from the literature. First, for smaller events, the reduction in runoff production from most types of NBSs is uncontested. Second, engineered systems such as vegetated retention basins have the capacity to reduce floodwaters. Third, large vegetated areas such as forest or riparian vegetation can reduce risk by preventing development (which might otherwise create impervious areas or compact soil, thereby increasing runoff production), and by reducing exposure in the case of floodplains. Finally, NBSs have cobenefits related to sediment retention, which are also relevant to flood risk: sediment not only reduces flood storage capacity in reservoirs, but also changes river morphology in floodplains, with sediment build-up reducing the capacity to accommodate flood waters downstream.

From an engineering standpoint, the variability in performance due to ecological or technical factors calls for designing flood-risk reduction projects with a mix of green and gray infrastructure. Depending on the project, whether it addresses riverine or pluvial flood risk, and for prevention or risk reduction, several tools can support the design process.

Stormwater flood risk reduction A number of urban hydrology models now allow users to represent the effect of NBSs on stormwater flow (e.g., SWMM, MUSIC; Elliott and Trowsdale 2007). These models can be used to assess a single storm event and quantify peak flow reduction associated with NBSs. The increased interest in NBSs for stormwater management also prompted the development of dedicated tools (e.g., SUSTAIN, Gwang Lee et al. 2012; InVEST flood risk reduction tool, Sharp et al. 2019) that typically require less hydrologic skill and little calibration. These tools can support siting or preliminary design for engineering projects.

Preventing flood risk The effect of protecting or restoring forests on peak flow or runoff volume can be assessed through simple approaches like the NRCS Curve Number method or the rational method (for small urban watersheds). If greater accuracy or spatial differentiation is needed, semidistributed hydrologic models like SWMM, MIKE-SHE (e.g., in Dadson et al. 2017), or distributed models like TOPMODEL (Beven and Kirkby 1979), LISFLOOD (Van Der Knijff et al. 2010), and CADDIES (Guidolin et al. 2016) can be used. One caveat for the use of these models is that they require extensive calibration or, in the case of global models, they may not focus on vegetated land use (Ward et al. 2015). To facilitate project assessment (in particular, comparison among management options), analytical methods are being developed to quantify the effect of existing natural assets such as wetlands (Watson et al. 2016).

Reducing existing flood risk In addition to estimating the effect of peak flow reduction by NBSs, hydraulic models like HEC-RAS (Brunner 2001) or LISFLOOD-FP (Bates et al. 2010) can be used to understand the effect of floodplain reconnection – for example, the Yolo by-pass project in California (Opperman et al. 2009). The more complex models cited above (fully distributed models) can produce flood extent maps that can help assess the extent of the flood reduction, with the caveats related to model calibration and poor representation of NBSs.

9.2.2 Water Supply

With respect to water supply, that is, the availability of liquid water for human use (domestic, industrial, irrigation, hydropower, cooling), the landscape performs three functions:

- Concentrates water in space; precipitation that falls over an expansive area is collected in streams and funneled to large rivers and, eventually, the oceans.
- Disperses water in time; precipitation that occurs at punctuated moments is spread out through time as it makes its way through the landscape to rivers and oceans.
- Converts solid and liquid water to water vapor; some of the precipitation that falls on the landscape is evaporated and transpired and is no longer available for local use – although that vapor will subsequently precipitate somewhere else (Ellison et al. 2012).

To a large extent, topography and geology govern the first function, and this chapter will focus on the latter two functions. In the presence of a large reservoir (e.g., greater than 10% of mean-annual streamflow, Guswa et al. 2017), the dispersal of water in time provided by the landscape is irrelevant to water supply, and the effects of the landscape are straightforward: more evapotranspiration means less water. Multiple reviews indicate that reduction in forest cover results in less evapotranspiration and more available water (Andréassian 2004; Bosch and Hewlett 1982; Brown et al. 2013; Bruijnzeel 2004). Similarly, Filoso et al. (2017) synthesized results from 167 papers that reported the effects of forest restoration on water yield from 308 sites globally; 80% of the sites reported a decline in water yield following restoration.

When reservoir storage is not available (or only modestly available), however, the timing of water availability, not just the total amount, becomes important. In such cases, the interaction of two functions – the loss of water to evapotranspiration and the dispersion of water in time – leads to complexity in the system and prevents the development of simple rules of thumb for the effects on water supply. Some investigators have found that increased forest cover leads to increased low flows (e.g., Ogden et al. 2013; Price 2011), whereas others have found that forests lead to both lower average yield and lower low-flows (e.g., Brown et al. 2013; Scott and Lesch 1997). The ambiguity is consistent with the synthesis by Filoso et al. (2017) who found that forest restoration resulted in a reduction of baseflow for 63% of the sites and an increase or no change in baseflow for 37% of sites.

These contradictions are sometimes explained by separating the effects of vegetation from soils (Bruijnzeel 2004). While taller vegetation and increased leaf area (e.g., forest vegetation) results in increased evapotranspiration, uncompacted soils with high organic content and macropores (e.g., forest soils) increase infiltration and extend the residence time of water in the soil. Another hypothesis is based on the seasonality of low flows. If the seasonality of low flows coincides with the seasonality of precipitation, that is, if low flows are due to precipitation drought (as they are in Mediterranean climates), then increases in forest cover that increase infiltration during the wet season may increase low flows (Guswa et al. 2007). However, if the seasonality of low flows coincides with the seasonality of actual evapotranspiration (as it does in the eastern USA), then increases in forest cover may further reduce low flows (Guswa et al. 2017).

The uncertainty of the effect of landscape change on low flows makes simple predictions of the effect of natural infrastructure on water supply challenging. Depending on the decision context and the precision required, a number of models and tools are available to the engineer.

Estimates of water yield with reservoir storage Guswa et al. (2017) provide a methodology for determining the potential impacts of landscape change on water supply as a function of reservoir size. Another example using integrated modeling is proposed by Guo et al. (2000). Large reservoirs obviate the need for the temporal dispersion function of a watershed, and a simple model of annual water yield may suffice. Examples include the InVEST annual water-yield model (Sharp et al. 2019) and others that are based on the Budyko curve (Budyko 1961).

Annual estimates of yield without reservoir storage In this case, estimates of annual water yield need to be supplemented by the separation of that yield into baseflow and stormflow components, and it is the baseflow that provides the steady, reliable supplies. Guswa et al. (2018) developed a simple model based on the NRCS curve-number approach to separate annual streamflow into baseflow and stormflow components.

Monthly estimates of yield without reservoir storage A number of parsimonious hydrologic models operate at the monthly scale, for example, abcd (Thomas et al. 1983), HBV (Bergström 1995), and DWBM (Zhang et al. 2008). These models are simpler than semidistributed and fully distributed daily models, though the connection between landscape changes and effects on model parameters is less direct. Nonetheless, Hamel et al. (2017) demonstrated that the DWBM model provides estimates of the relative changes in minimum monthly flows due to changes to the landscape that are robust with respect to parameter uncertainty. The InVEST seasonal water-yield model is a spatially explicit model of monthly flows that enables the spatial attribution of baseflow generation (Sharp et al. 2019).

Daily estimates of water yield Trading simplicity for sophistication are a set of models that operate at the daily or subdaily timescale and represent space in a semidistributed or fully distributed way. Semidistributed models, such as SWAT (Neitsch et al. 2011), PRMS (Leavesley et al. 1983), VIC (Liang et al. 1994), and TOPMODEL (Beven and Kirkby 1979), separate the landscape into a set of hydrologically similar groups based on topography, soils, and land-cover. Fully distributed models, such as MIKESHE (DHI 1998), GSFLOW (Markstrom et al. 2008), and HSPF (Bicknell et al. 1997), represent the landscape as a grid of connected pixels, each with its own characteristics. All of these models have a significant level of complexity and require a knowledgeable user to implement for a particular site.

9.2.3 Water Quality Management

When it comes to attributes of water quality (sediment, nutrients, and pathogens), the effects of natural versus human-modified landscapes are clearer, though often difficult to quantify. With respect to water quality, the landscape and ecosystem perform a set of functions:

- Generation – in addition to point sources of pollution, landscapes serve as non-point-sources of sediment, nutrients, and pathogens.
- Physical retention and dilution – topography, flowpaths, and land-cover will dictate which parts of the landscape have the potential to retain contaminants from upgradient.
- Transformation – biogeochemical processes operating on the landscape have the potential to transform nutrients and pathogens and remove them from water.

In the United States, the water-supply system for the city of New York is a famous example of the value of these processes to water quality. From the early 1990s through 2017, New York spent over $1.7 billion on natural infrastructure so as to avoid a $10 billion filtration facility with a $100 million/year operational cost (Hu 2018). The cities of Boston, MA; San Francisco, CA; Portland, OR; and Seattle, WA, also avoid the need to filter their water supplies via the benefits of natural landscape processes and watershed management. Globally, McDonald et al. (2016) examined the effects of watershed degradation on water treatment costs for large cities from 1900 to 2005; average pollutant yields for degraded watersheds increased by 40% for sediment, 47% for phosphorus, and 119% for nitrogen. For 29% of cities, watershed degradation led to increased treatment costs: 53% increase in O&M and 44% increase in capital costs (McDonald et al. 2016).

In contrast to low flows (see above), the direction of change of the effects of landscape change on sediment, nutrients, and pathogen concentrations can usually be predicted. However, quantification of the magnitude of the effect can be highly uncertain. Therefore, reliance on natural infrastructure to achieve water quality goals is best suited for sediment and nutrients, that is, those constituents for which the impact results from an aggregate effect and for which the tolerance for variability in performance is higher. For pathogens (e.g., bacteria, viruses, parasites), tolerance for uncertainty is less, as even a low concentration or localized outbreak can have a significant impact (Jasper et al. 2013). While natural landscapes do play a significant role with respect to human health and infectious disease (e.g., Herrera et al. 2017; McFarlane et al. 2013), the ability to design natural-infrastructure solutions with the required level of certainty is still developing.

With respect to sediment, natural landscapes both limit generation and can provide physical retention. In the northeast United States, erosion from forests is quite low, with sediment yields of 25 kg/ha/r to 250 kg/ha/yr (Patric 1976; Patric et al. 1984; Wolman and Schick 1967). Erosion from agricultural land is 10–100 times as much (de la Cretaz and Barten 2007). Yields from urban construction and development, if not properly mitigated, can be even greater, and Wolman and Schick (1967) reported yields of 7000 to 490,000 kg/ha/yr. for sites in Maryland.

When positioned downgradient from sources, both wetlands and riparian buffers have been shown to reduce sediment loads to receiving water bodies. While both forest and grass strips are effective at trapping sediment, grasses are particularly effective due to both the density of vegetation cover at the ground surface and to their tendency to spread water over a large area (de la Cretaz and Barten 2007). Trapping efficiencies range from 50% to nearly 100% and vary with buffer width, vegetation type, and grain-size distribution of the sediment (de la Cretaz and Barten 2007).

In addition to retaining nutrients that are transported with sediment (e.g., phosphorus and ammonium), riparian wetlands and vegetated buffers can also transform nitrate to nitrogen gas via denitrification – an anaerobic process. Vegetation also takes up both nitrogen and phosphorus, though much of those nutrients may be returned as litter fall at another time, and overall removal efficiency is uncertain. Studies of nitrate reduction show removal rates that range from 25% to 95%

(de la Cretaz and Barten 2007); phosphorus removal is even more varied, with some studies showing an increase in phosphorus from best management practices (Schechter et al. 2013).

Because of the significant uncertainty associated with the natural transport and transformation of water contaminants, modeling tools for design are few. For simple assessments of the effects of land-cover and land-management on the generation, transport, and transformation of sediment and nutrients, the InVEST model (Sharp et al. 2019) provides annual estimates of nutrient and sediment loads. Operating at the daily to sub-daily timescale, the Soil-Water Assessment Tool (SWAT) is a semi-distributed model developed for agricultural management that has seen wide application for the simulation of water, nutrients, and sediment transport (Neitsch et al. 2011). For urban environments, the US EPA distributes and maintains the Storm Water Management Model (SWMM). This dynamic hydrologic and hydraulic model simulates water quantity and quality and can incorporate green infrastructure, such as rain gardens, green roofs, and permeable pavement (Rossman 2015). HSPF (Hydrological simulation program – FORTRAN) is a fully-distributed, dynamic model that simulates the transport of both point and non-point sources of pollution at the watershed scale (Bicknell et al. 1997), as does the MIKE series of models (DHI 1998).

9.3 Designing with a Mix of Conventional Infrastructure and Nature-Based Solutions

9.3.1 Systemic Approach: Cobenefits, Disservices, and Beneficiaries

Cobenefits and Disservices

While natural infrastructure is not always superior to gray infrastructure, there are cobenefits associated with natural infrastructure that are not present with gray. Table 9.1 presents the suite of cobenefits, ranging from provisioning services (e.g., food production in community gardens), to regulating services (air quality improvement, carbon sequestration), and cultural services (tourism, mental health). These benefits are now well accepted, and ES scientists have developed methods to analyze and quantify their contributions to society (Haase et al. 2014; Pataki et al. 2013).

On the other side of these benefits, there are potential disservices associated with NBSs. We already mentioned the disservices related to water resources management in Sect. 9.2: the use of water by vegetation, which reduces availability for other uses, and the potential net source of nutrients, which can be detrimental to freshwater ecosystems. In addition to those, NBSs may negatively impact human health and well-being through potential disservices that mirror the cobenefits listed in Table 9.1. For example, street trees and urban vegetation may produce allergenic pollens, thereby reducing air quality. Urban vegetation may provide habitat for unwanted species. Other disservices include potential insecurity (in the case of poorly lit areas and potentially dangerous wildlife in urban parks) or net positive carbon budget associated with construction or maintenance of NBSs (Keeler et al. 2019). These disservices need to be included in assessment of NBSs and alternative management solutions to consider the full impact of engineering decisions.

Beneficiaries

Central to the concept of ecosystem services is the definition of beneficiaries, that is, people benefiting from the implementation of a NBSs. Beneficiaries of hydrologic services are mainly determined from their exposure and vulnerability to water-related risks – flooding and water scarcity. For flood risk, the position on the landscape, for example, in flood plains and low-lying areas, will be a primary determinant, together with metrics of social vulnerability (e.g., age group, language) or vulnerability of built infrastructure (housing quality). For water scarcity, whether it results from a water quality or quantity issue, beneficiaries will depend on the local and regional water resources management: whether people source water from surface or subsurface water, which treatment options are available, etc. Assessing beneficiaries and understanding the potential equity issues associated with management options can be facilitated by ecosystem services tools such as those cited above (e.g., InVEST, ARIES).

9.3.2 Practical Opportunities and Constraints

In addition to ecological factors, there are practical opportunities and constraints associated with NBSs. An important consideration is cost, which is difficult to evaluate in generic terms. Sometimes, NBSs have a clear economic advantage (e.g., New York City, in Sect. 9.2.3, and São Paulo, in Sect. 9.4.1); in other cases, the cost of NBSs may be greater than conventional engineering solutions, especially when accounting for both construction and maintenance costs. Indeed, in a 2017 survey, 26 of 31 US municipalities reported green infrastructure was more challenging than gray infrastructure with respect to developing project operation and maintenance cost estimates (U.S. Government Accountability Office 2017). Uncertain and potentially higher costs may present opportunities for partnerships when ecosystem cobenefits are taken into consideration. For example, a subterranean concrete box for stormwater retention may be cheaper than a bioretention basin; however, the latter may provide an opportunity for a water utility to partner with another public agency (e.g., parks and recreation), a private institution (e.g., golf course), or nonprofit group or neighborhood association.

Additionally, since they rely on ecosystem functions, NBSs are less generalizable across geographies than gray infrastructure (Pataki 2015). And, since the designs are visible (as opposed to buried), natural infrastructure also requires greater attention to community norms and values (Nassauer and Raskin 2014). Similarly, local legal and regulatory frameworks vary in their acceptance of nature-based solutions and may require long-established technologies. Thus, when compared with traditional engineering designs, place and location take on greater significance for natural infrastructures and NBSs require greater collaboration with ecologists, landscape architects, planners, and regulators.

9.3.3 Incorporating Synergies and Trade-Offs into Engineering Projects

The last two sections have illustrated the multiple dimensions defining the performance or feasibility of NBSs. To incorporate these dimensions into engineering projects, a suite of tools is available from the fields of policy analysis, engineering, and integrated environmental modeling (Jakeman et al. 2008). Often, the goals of an assessment are to synthesize multiple objectives, reduce or quantify uncertainty, and facilitate comparison among different solutions. Classical decision-support tools like multicriteria decision analyses, economic valuation and cost-benefit analyses, or robust decision-making are among the most common examples, and ad hoc tools have also been developed to support decisions related to NBSs. For example, RIOS (Vogl et al. 2015) was developed to aggregate biophysical information, costs, and other practical constraints to support the siting of NBSs for a range of water-related objectives. For stormwater management objectives, SUSTAIN supports stormwater engineers to evaluate the cost-effectiveness of NBSs in their watersheds (U.S. EPA 2011).

While the above tools were designed to address specific questions, the eco-engineering decision scaling approach developed by LeRoy Poff et al. (2015) is a holistic framework that aims to support an entire project. It was developed to explicitly address multiple objectives and perspectives on water management (from ecologists and engineers), and deal with hydrologic or future climate uncertainties. The framework comprises five steps: (i) stakeholder engagement to define management options, performance indicators, and failure points; (ii) development of a systems decision model representing the important relationships between hydrologic variables, performance indicators, and external drivers; (iii) vulnerability analysis to assess the response to a change in climate or external drivers; (iv) comparison of available management options and definition of alternatives; and (v) assessment of the feasibility of the solutions. Importantly, step ii) requires a good understanding of the performance of NBSs proposed in the project and is subject to the modeling constraints described in Sect. 9.2. Similar practical frameworks are being developed by engineering companies and multilateral banks, such as the World Bank guidebook "Integrating Green and Gray (Browder et al. 2019)".

9.4 Case Studies

9.4.1 Water Supply: Green-Gray Infrastructure Planning in São Paulo

The São Paulo Metropolitan region faced a major drought in 2014–2015. By February 2015, the production of the Cantareira system, the city's primary water-supply system, had fallen to less than half of its typical production, with reservoir levels reaching a historical low. This crisis had important political and financial implications – the estimated losses for the water utility were around US$470 million (Sabesp 2015) – and it renewed discussions about the resilience of the water system. In 2016, a consortium of organizations (The World Resources Institute, The FEMSA Foundation, The Nature Conservancy, the International Union for Conservation of Nature, Instituto BioAtlântica, and the Boticario Group Foundation) joined forces to assess the value of green and gray infrastructures as a water-supply strategy.

The assessment built on the green-gray infrastructure methodology developed by WRI (World Resources Institute 2013), a framework to compare NBSs and traditional infrastructure in a systematic way. The assessment compared several scenarios of reforestation and conservation in the Cantareira system by assessing their effect on sediment export and sediment treatment costs. The analyses found likely cost savings, through reduction in water-treatment costs, from forest restoration and conservation in target areas (Ozment et al. 2018). These savings increased as initial investments were made, assuming a rate of vegetation growth (and therefore sediment retention service) over 30 years. Location of forest restoration or protection projects within the watershed strongly affected the estimated impact, given the role that near-stream ecosystems play in retaining sediment flows.

Two points are worth reflecting on in this study: first, large uncertainties were noted in the sediment and baseflow modeling analyses. The potential baseflow increase due to increased infiltration was not included in the financial analysis due to large uncertainties. Knowledge gaps in hydrological modeling thus remain one barrier to information. Second, even with this uncertainty, the business case was an opportunity to engage diverse groups of stakeholders (water utilities, investors, NGOs) in a reflection on the value of green infrastructure, and provide a concrete road map for the group (Ozment et al. 2018). It spurred a conversation on the multiple facets of forest protection, in particular with regard to the participation of rural communities whose lands are affected by projects.

9.4.2 Water Quality: Combining Infrastructures to Address Combined-Sewer Overflows in Boston Harbor

The clean-up of Boston Harbor in the late 1990s and early 2000s is one of the great environmental success stories of recent history (Dolin 2008). The construction

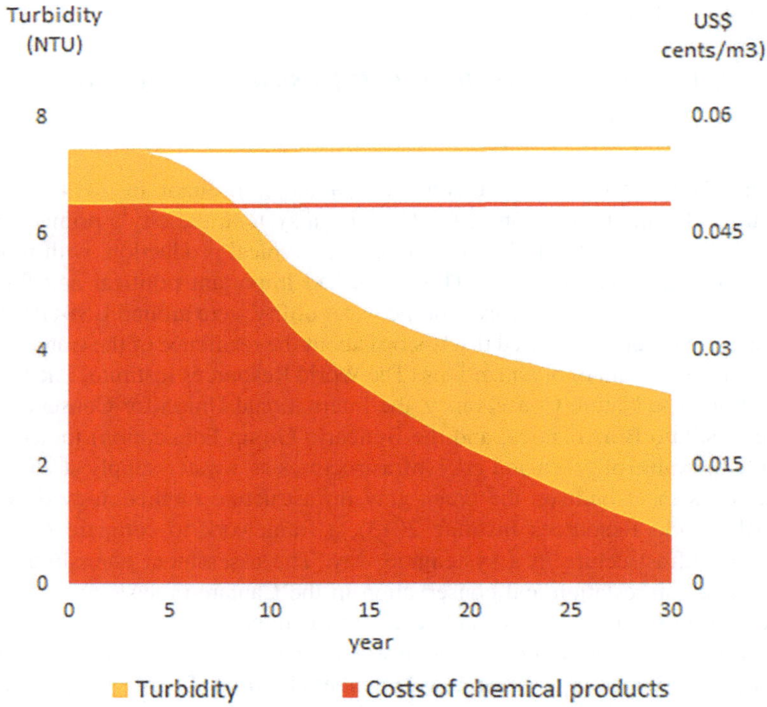

of the massive wastewater treatment facility (peak capacity of 1.2 billion gallons per day) transformed Boston Harbor from "the dirtiest harbor in America" to one that is swimmable in only a few years (MWRA 2014). Despite this success, the historic infrastructure of the Boston area that combines stormwater with sanitary sewage continues to present challenges; during times of heavy rain, some of the combined sewage is discharged directly to Boston Harbor. In August 2012, the U.S. Environmental Protection Agency (U.S. EPA) issued a consent decree that required the Boston Water and Sewer Commission (BWSC) to "minimize the discharge of sewage and other pollutants into the water bodies in and around Boston" (U.S. Department of Justice 2012).

Problems of combined-sewer overflows (CSOs) are not unique to Boston, and the U.S. EPA has articulated that combinations of gray and green infrastructures can provide viable and cost-effective solutions (U.S. EPA 2014). Gray infrastructure solutions include sewer separation and off-line storage (i.e., storage of wet-weather flows in tanks or basins to be treated later). By reducing the quantity and/or rate of stormwater flows into combined sewers, green infrastructures – such as bioretention basins, green roofs, and tree trenches – can reduce the size and need for gray infrastructure (U.S. EPA 2014). Because green infrastructure affects both the quantity and timing of runoff, the integration of hydrologic and hydraulic models improves predictions of the effects on combined-sewer overflows (U.S. EPA 2014).

In Boston, projects to demonstrate the efficacy of integrated gray and green strategies are underway. In October 2017, the BWSC celebrated the completion of a green infrastructure project at the Washington Irving Middle School. That project – a partnership between the BWSC and the Boston Public Schools – comprises replacement of paved areas with green space, the construction of a vegetated swale, and the addition of an outdoor classroom and bioretention area (City of Boston 2014). Additional projects are being designed or implemented for four other public schools, along with other sites, including City Hall Plaza.

9.5 Discussion

9.5.1 Synergies Between Engineering and Ecosystem-Services Science

In the introduction, we proposed that ES science could contribute to IWRM by (i) incorporating ecological functions as opportunities and constraints in engineering design, (ii) developing approaches to better incorporate people into the design phase, and (iii) better communicating the value of nature to a broad range of stakeholders. The case studies illustrated key points related to each of these potential benefits.

First, with regard to ecological functions, the São Paulo case study highlighted the role of ecology and ecohydrology in supporting the implementation of NBSs. Engineering projects will benefit from more knowledge on the behavior of NBSs with regard to sediment retention and baseflow, especially how the type, location, or maintenance of vegetated systems will affect their performance. Second, forest protection and restoration projects in São Paulo spurred reflections on the operational constraints and opportunities, for different beneficiary groups: the water utility, a direct beneficiary of the services, but also rural communities who are key stakeholders in these projects. ES science recommends the use of participatory approaches to better incorporate the knowledge and interests of these communities into project design. Finally, both case studies illustrated that NBSs can serve to raise awareness and educate the public. The NGO consortium in São Paulo helped advance the conversation on green infrastructure by evaluating the economic benefits of NBSs (and their potential cobenefits). The Boston stormwater management demonstration projects contributed to raise awareness on the role of nature among students, who learn about the water cycle, pollution control, and ecological issues.

9.5.2 Implications for Teaching Water-Resources Engineering

The incorporation of natural infrastructure and hydrologic ecosystem services into water-resources designs merits a shift in mindset when it comes to teaching water-resources engineering. Three facets characterize the shift from traditional engineering thought.

Borrowing from medicine, the first shift is a reorientation to first look for opportunities for prevention over treatment. That is, as our cities and urban areas grow and expand, look first to the preservation of landscape characteristics that benefit water resources, such as infiltration. The ability to recognize such features and to create designs that retain such features will be important skills for engineers of the future.

The second shift is to complement reductionist approaches with integrative thinking. The ability to break a complex system into simpler component parts is a powerful skill in engineering. At the same time, cobenefits that reach across multiple sectors are a primary strength of natural infrastructure. Design engineers must be able to articulate to clients and stakeholders the worth of these multiple benefits along with the achievement of the primary water-resources objectives. This requires an ability to work with experts across multiple disciplines, including ecology, economics, and landscape architecture.

Third is to shift from designing-to-avoid-failure to creating designs that acknowledge and tolerate uncertainty in performance. This shift is necessitated by both the greater degree of uncertainty associated with natural infrastructure and the recognition that our climate is changing. Design approaches that require stationarity and well-understood materials are not suitable for incorporating natural infrastructure into water-resources engineering under a changing climate.

More than new tools or models, engineering education must include this expansion of the engineering mindset – loss prevention, integrative thinking, embracing uncertainty – in order to effectively incorporate natural infrastructure into water-resources designs.

9.6 Summary and Outlook

This chapter presented the state-of-the-art on the role of NBSs for providing three water services: flood risk mitigation, water supply, and water quality management. The key processes through which NBSs provide water services are well accepted, allowing, in theory, an understanding of when NBSs may usefully complement traditional infrastructure. However, uncertainties related to the magnitude of these processes impede the incorporation of NBSs into the engineering toolbox.

To promote the adoption of green infrastructure, hydrologic research needs to further progress to improve process understanding: in particular, to better quantify the magnitude of hydrologic services and disservices provided by NBSs. In addition, the development of new tools and approaches in engineering will facilitate the implementation of NBSs, in replacement of or in combination with traditional infrastructure. Such approaches include participatory approaches to include stakeholders in the design of NBSs, valuation methods to quantify cobenefits and disservices, and multicriteria assessment methods to compare engineering solutions across multiple dimensions. In parallel to the research conducted in each of these directions, engineering education needs to adapt to the new paradigms in IWRM – teaching students to consider the downsides of traditional infrastructure and design solutions that reflect our rapidly changing world.

References

Andréassian, V., 2004. Waters and forests: From historical controversy to scientific debate. J. Hydrol. 291, 1–27. https://doi.org/10.1016/j.jhydrol.2003.12.015

Bates, P.D., Horritt, M.S., Fewtrell, T.J., 2010. A simple inertial formulation of the shallow water equations for efficient two-dimensional flood inundation modelling. J. Hydrol. 387, 33–45. https://doi.org/10.1016/j.jhydrol.2010.03.027

Bergström, S., 1995. The HBV model, in: V.P, S. (Ed.), Computer Models in Watershed Hydrology. Water Resources Publications, Highland Ranch, CO, pp. 443–476.

Beven, K.J., Kirkby, M.J., 1979. A physically-based variable contributing area model of basin hydrology. Hydrol. Sci. Bull. 24, 43–69.

Bicknell, B.R., Imhoff, J.C., Kittle, J.L., J., Donigian, A.S., J., Johanson, R.., 1997. Hydrological Simulation Program–Fortran, User's manual for version 11. EPA/600/R-97/080. Athens, Ga.

Bosch, J.M., Hewlett, J.D., 1982. A review of catchment experiments to determine the effect of vegetation changes on water yield and evapotranspiration. J. Hydrol. 55, 3–23. https://doi.org/10.1016/0022-1694(82)90117-2

Brown, A.E., Western, A.W., McMahon, T.A., Zhang, L., 2013. Impact of forest cover changes on annual streamflow and flow duration curves. J. Hydrol. 483, 39–50. https://doi.org/10.1016/j.jhydrol.2012.12.031

Browder, G., Ozment, S., Rehberger Bescos, I., Gartner, T., Lange, G.-M., 2019. Integrating Green and Gray: Creating Next Generation Infrastructure. Washington, DC: World Bank and World Resources Institute.http://hdl.handle.net/10986/31430

Bruijnzeel, L.A., 2004. Hydrological functions of tropical forests: not seeing the soil for the trees? Agric. Ecosyst. Environ. 104, 185–228. https://doi.org/10.1016/j.agee.2004.01.015

Brunner, G.W., 2001. HEC-RAS River Analysis System: Hydraulic Reference Manual. Davis, CA.

Budyko, M., 1961. The Heat Balance of the Earth's Surface. Washington D.C.

Calder, I.R., Aylward, B., 2006. Forest and floods: Moving to an evidence-based approach to watershed and integrated flood management. Water Int. 31, 87–99.

City of Boston, 2014. Mayor Walsh, students celebrate new green infrastructure at Washington Irving Middle School. Office of the Mayor, Martin J. Walsh, Press Release, 3 October 2017.

Dadson, S., Hall, J.W., Murgatoyd, A., Acreman, M., P, B., K, B., L, H., J, H., Holman, I.P., Lane, S.N., O'Connell, E., Penning-Roswell, E., Reynard, N., Sear, D., Thorne, C., Wilby, R., 2017. A restatement of the natural science evidence concerning catchment-based "natural" flood management in the UK. Proc. R. Soc. 473, 31. https://doi.org/10.1098/rspa.2016.0706

de la Cretaz, A.L., Barten, P.., 2007. Land use effects on streamflow and water quality in the Northeastern United States. CRC Press, New York City.

DHI, 1998. MIKE-SHE Users' guide and Technical Reference Manual.

Dolin, E.J., 2008. Political Waters, The Long, Dirty, Contentious, Incredibly Expensive but Eventually Triumphant History of Boston Harbor – a Unique Environmental Success Story.

Elliott, A.H., Trowsdale, S.A., 2007. A review of models for low impact urban stormwater drainage. Environ. Model. Softw. 22, 394–405.
Ellison, D., N. Futter, M., Bishop, K., 2012. On the forest cover–water yield debate: from demand- to supply-side thinking. Glob. Chang. Biol. 18, 806–820. https://doi.org/10.1111/j.1365-2486.2011.02589.x
EPA, U., 2014. Greening CSO Plans: Planning and Modeling Green Infrastructure for Combined Sewer Overflow (CSO) Control. Publication # 832-R-14-001.
Filoso, S., Bezerra, M.O., Weiss, K.C.B., Palmer, M.A., 2017. Impacts of forest restoration on water yield : A systematic review. PLoS One 1–26. https://doi.org/10.1371/journal.pone.0183210
Guerry, A.D., Polasky, S., Lubchenco, J., Chaplin-Kramer, R., Daily, G.C., Griffin, R., Ruckelshaus, M., Bateman, I.J., Duraiappah, A., Elmqvist, T., Feldman, M.W., Folke, C., Hoekstra, J., Kareiva, P.M., Keeler, B.L., Li, S., McKenzie, E., Ouyang, Z., Reyers, B., Ricketts, T.H., Rockström, J., Tallis, H., Vira, B., 2015. Natural capital and ecosystem services informing decisions: From promise to practice. Proc. Natl. Acad. Sci. 112, 7348 LP-7355.
Guidolin, M., Chen, A.S., Ghimire, B., Keedwell, E.C., Djordjević, S., Savić, D.A., 2016. A weighted cellular automata 2D inundation model for rapid flood analysis. Environ. Model. Softw. 84, 378–394. https://doi.org/10.1016/j.envsoft.2016.07.008
Guo, Z., Xiao, X., Li, D., 2000. AN ASSESSMENT OF ECOSYSTEM SERVICES: WATER FLOW REGULATION AND HYDROELECTRIC POWER PRODUCTION. Ecol. Appl. 10, 925–936. https://doi.org/10.1890/1051-0761(2000)010[0925:AAOESW]2.0.CO;2
Guswa, A.J., Hamel, P., Dennedy-Frank, P.J., 2017. Potential effects of landscape change on water supplies in the presence of reservoir storage. Water Resour. Res. 53, 2679–2692. https://doi.org/10.1002/2016WR019691
Guswa, A.J., Hamel, P., Meyer, K., 2018. Curve number approach to estimate monthly and annual direct runoff. J. Hydrol. Eng. 23, 1–10.
Guswa, A.J., Rhodes, A.L., Newell, S.E., 2007. Importance of orographic precipitation to the water resources of Monteverde, Costa Rica. Adv. Water Resour. 30, 2098–2112. https://doi.org/10.1016/j.advwatres.2006.07.008
Gwang Lee, J., Selvakumar, A., Alvi, K., Riverson, J., Zhen, J.X., Shoemaker, L., Lai, F., 2012. A watershed-scale design optimization model for stormwater best management practices. Environ. Model. Softw. 37, 6–18. https://doi.org/10.1016/j.envsoft.2012.04.011
Haase, D., Larondelle, N., Andersson, E., Artmann, M., Borgström, S., Breuste, J., Gomez-Baggethun, E., Gren, Å., Hamstead, Z., Hansen, R., Kabisch, N., Kremer, P., Langemeyer, J., Rall, E.L., McPhearson, T., Pauleit, S., Qureshi, S., Schwarz, N., Voigt, A., Wurster, D., Elmqvist, T., 2014. A quantitative review of urban ecosystem service assessments: concepts, models, and implementation. Ambio 43, 413–33. https://doi.org/10.1007/s13280-014-0504-0
Hamel, P., Guswa, A., Sahl, J., Zhang, L., 2017. Predicting dry-season flows with a monthly rainfall-runoff model: performance for gauged and ungauged catchments. Hydrol. Process. 1–15. https://doi.org/10.1002/hyp.11298
Herrera, D., Ellis, A., Fisher, B., Golden, C.D., Johnson, K., Mulligan, M., Pfaff, A., Treuer, T., Ricketts, T.H., 2017. Upstream watershed condition predicts rural children's health across 35 developing countries. Nat. Commun. 8, 811. https://doi.org/10.1038/s41467-017-00775-2
Hu, W., 2018. A billion-dollar investment in New York's water. New York Times.
IUCN. 2016. World Conservation Congress resolution: WCC-2016-Res-069-EN Annex 1: IUCN Definitional Framework on Nature-based Solutions. Available at: https://portals.iucn.org/library/sites/library/files/resrecfiles/WCC_2016_RES_069_EN.pdf. Accessed 31 Jul 2023
Jakeman, A.J., Chen, S.H., Rizzoli, A.E., Voinov, A.A., 2008. Environmental Modelling, Software and Decision Support, in: Developments in Integrated Environmental Assessment, Developments in Integrated Environmental Assessment. Elsevier, pp. 1–13. https://doi.org/10.1016/S1574-101X(08)00601-7
Jasper, J.T., Nguyen, M.T., Jones, Z.L., Ismail, N.S., Sedlak, D.L., Sharp, J.O., Luthy, R.G., Horne, A.J., Nelson, K.L., 2013. Unit Process Wetlands for Removal of Trace Organic Contaminants

and Pathogens from Municipal Wastewater Effluents. Environ. Eng. Sci. 30, 421–436. https://doi.org/10.1089/ees.2012.0239
Jha, A.K., Bloch, R., Lamond, J., 2012. Cities and Flooding: A Guide to Integrated Urban Flood Risk Management for the 21st Century. Washington D.C.
Keeler, B.L., Hamel, P., Donahue, M., ..., 2019. Social-ecological and technological factors moderate the value of urban nature. Nat. Sustain. 2, 29–38.
Leavesley, G.H., Lichty, R.W., Troutman, B.M., Saindon, L.G., 1983. Precipitation-runoff modeling system; user's manual, Water-Resources Investigations Report.
Liang, X., Lettenmaier, D.P., Wood, E.F., Burges, S.J., 1994. A simple hydrologically based model of land surface water and energy fluxes for general circulation models. J. Geophys. Res. Atmos. 99, 14415–14428. https://doi.org/10.1029/94JD00483
Markstrom, S.L., Niswonger, R.G., Regan, R.S., Prudic, D.E., Barlow, P.M., 2008. GSFLOW - Coupled Ground-water and Surface-water FLOW model based on the integration of the Precipitation-Runoff Modeling System (PRMS) and the Modular Ground-Water Flow Model (MODFLOW-2005).
Mcdonald, R.I., Weber, K.F., Padowski, J., Boucher, T., Shemie, D., 2016. Estimating watershed degradation over the last century and its impact on water-treatment costs for the world ' s large cities. Proc. Natl. Acad. Sci. https://doi.org/10.1073/pnas.1605354113
McFarlane, R.A., Sleigh, A.C., McMichael, A.J., 2013. Land-Use Change and Emerging Infectious Disease on an Island Continent. Int. J. Environ. Res. Public Health 10, 2699–2719. https://doi.org/10.3390/ijerph10072699
MWRA, 2014. Massachusetts Water Resources Authority [WWW Document]. URL http://www.mwra.state.ma.us/01news/2008/bhpenvironentalsuccess/bhpenvsuccess.htm (accessed 1.19.18).
Nassauer, J.I., Raskin, J., 2014. Urban vacancy and land use legacies: A frontier for urban ecological research, design, and planning. Landsc. Urban Plan. 125, 245–253. https://doi.org/10.1016/j.landurbplan.2013.10.008
Neitsch, S.L., Arnold, J.G., Kininy, J.R., Williams, J.R., 2011. Soil and Water Assessment Tool Theoretical Documentation v2009. Texas, USA.
Ogden, F.L., Crouch, T.D., Stallard, R.F., Hall, J.S., 2013. Effect of land cover and use on dry season river runoff, runoff efficiency, and peak storm runoff in the seasonal tropics of Central Panama. Water Resour. Res. 49, 8443–8462. https://doi.org/10.1002/2013WR013956
Opperman, J.J., Galloway, G.E., Fargione, J., Mount, J.F., Richter, B.D., Secchi, S., 2009. Sustainable Floodplains Through Large-Scale Reconnection to Rivers. Science (80-.). 326, 1487 LP-1488.
Ozment, S., Feltran-Barbieri, R., Hamel, P., Gray, E., ..., Piazzetta Valente, T., 2018. Natural infrastructure in São Paulo's water system. World Resources Institute.
Pataki, D.E., 2015. Grand challenges in urban ecology. Front. Ecol. Evol.
Pataki, D.E., McCarthy, H.R., Gillespie, T., Jenerette, G.D., Pincetl, S., 2013. A trait-based ecology of the Los Angeles urban forest. Ecosphere 4, 1–20. https://doi.org/10.1890/ES13-00017.1
Patric, J.H., 1976. Soil Erosion in the Eastern Forest. J. For. 74, 671–677.
Patric, J.H., Evans, J.O., Helvey, J.D., 1984. Summary of Sediment Yield Data from Forested Land in the United States. J. For. 82, 101–104.
Poff, N.L., Brown, C.M., Grantham, T.E., Matthews, J.H., Palmer, M.A., Spence, C.M., Wilby, R.L., Haasnoot, M., Mendoza, G.F., Dominique, K.C., Baeza, A., 2015. Sustainable water management under future uncertainty with eco-engineering decision scaling. Nat. Clim. Chang. advance on.
Price, K., 2011. Effects of watershed topography, soils, land use, and climate on baseflow hydrology in humid regions: A review. Prog. Phys. Geogr. 35, 465–492. https://doi.org/10.1177/0309133311402714
Rossman, L.A., 2015. Storm Water Management Model User's Manual v5.1, EPA/600/R-14/413b, U.S. Environmental Protection Agency, Cincinnati, OH.
Sabesp. 2015. Crise hídrica, estratégias e soluções da Sabesp: Para a região metropolitana de São Paulo. São Paulo: Sabesp

Schechter, S.P., Canfield, T.J., Mayer, P.M., 2013. A meta-analysis of phosphorus attenuation in best management practices (BMP) and low impact development (LID) practices in urban and agricultural areas. US EPA Report EPA/600/R-13/208.

Schoeman, J., Allan, C., Finlayson, C.M., 2014. A new paradigm for water ? A comparative review of integrated, adaptive and ecosystem-based water management in the ... comparative review of integrated, management in the Anthropocene. Int. J. Water Resour. Dev. 30, 377–390. https://doi.org/10.1080/07900627.2014.907087

Scott, D.F., Lesch, W., 1997. Streamflow responses to afforestation with Eucalyptus grandis and Pinus patula and to felling in the Mokobulaan experimental catchments, South Africa. J. Hydrol. 199, 360–377. https://doi.org/10.1016/S0022-1694(96)03336-7

Sharp, R.S., ..., Hamel, P., et al., 2019. InVEST 3.5 User's Guide. Available at: http://data.naturalcapitalproject.org/nightly-build/invest-users-guide/html/.

Thomas, H.A., Marin, C.M., Brown, M.J., Fiering, M.., 1983. Methodology for water resource assessment, report to U.S. Geological Survey. Rep. NTIS 84-124163. Springfield, Va.

U.S. EPA, 2011. Enhanced Framework (SUSTAIN) and Field Applications for Placement of BMPs in Urban Watersheds. Publication No. EPA/600/R-11/144.

U.S. Government Accountability Office, 2017. Stormwater Management: EPA Pilot Project to Increase Use of Green Infrastructure Could Benefit from Documenting Collaborative Agreements, Report to the Honorable Sheldon Whitehouse.

US Department of Justice, 2012. Press release 12-1039 [WWW Document]. URL https://www.justice.gov/opa/pr/settlement-requires-boston-water-and-sewer-commission-remedy-sewer-and-stormwater-discharges (accessed 1.19.18).

Van Der Knijff, J.M., Younis, J., De Roo, a.P.J., 2010. LISFLOOD: a GIS-based distributed model for river basin scale water balance and flood simulation. Int. J. Geogr. Inf. Sci. 24, 189–212. https://doi.org/10.1080/13658810802549154

Vogl, A., Tallis, H., Douglass, J., Sharp, R., Veiga, F., Benitez, S., Le6n, J., Game, E., Petry, P., Guimerães, J., Lozano, J.S., 2015. Resource Investment Optimization System: Introduction & Theoretical Documentation.

Ward, P.J., Jongman, B., Salamon, P., Simpson, A., Bates, P., Groeve, T. De, Muis, S., Perez, E.C. De, Rudari, R., Trigg, M.A., Winsemius, H.C., Cross, G.R., Red, R.C., 2015. Usefulness and limitations of global flood risk models. Nat. Clim. Chang. 5, 712–715. https://doi.org/10.1038/nclimate2742

Watson, K.B., Ricketts, T., Galford, G., Polasky, S., O'Niel-Dunne, J., 2016. Quantifying flood mitigation services: The economic value of Otter Creek wetlands and floodplains to Middlebury, VT. Ecol. Econ. 130, 16–24. https://doi.org/10.1016/j.ecolecon.2016.05.015

Wolman, M.G., Schick, A.P., 1967. Effects of construction on fluvial sediment, urban and suburban areas of Maryland. Water Resour. Res. 3, 451–464. https://doi.org/10.1029/WR003i002p00451

World Resources Institute, 2013. Natural Infrastructure: Investing in Forested Landscapes for Source Water Protection in the United States.

Zhang, L., Potter, N., Hickel, K., Zhang, Y., Shao, Q., 2008. Water balance modeling over variable time scales based on the Budyko framework – Model development and testing. J. Hydrol. 360, 117–131. https://doi.org/10.1016/j.jhydrol.2008.07.021

Chapter 10
Improved Air Quality and Other Services from Urban Trees and Forests

David J. Nowak

10.1 Introduction

Worldwide, there are an estimated 3.0 trillion trees (Crowther et al. 2015) and 7.7 billion people (http://www.worldometers.info/world-population/). These trees produce numerous benefits to society but also create various costs. Trees can improve human health and well-being by moderating climate, reducing building energy use and atmospheric carbon dioxide (CO_2), improving air quality, mitigating rainfall runoff and flooding, providing protection from ultraviolet radiation and soil erosion, lowering noise levels and providing food, lumber, medicines, aesthetic environments, and recreational opportunities (e.g., Millennium Ecosystem Assessment 2005; Nowak and Dwyer 2007; Costanza et al. 2014). However, trees can also produce environmental and monetary costs associated with allergies from pollen, volatile organic compound emissions, potential increased energy use due to trees near buildings, invasive plants that alter local biodiversity, higher taxes from increased property values, and tree maintenance. Both the positive and negative aspects of trees and forest must be considered when designing landscapes to improve human health and well-being.

As trees can be a dominant element in a landscape, understanding the magnitude and the means of how trees affect the environment can lead to better vegetation management and designs to optimize environmental quality and human health for current and future generations. In urban areas, the impacts of trees become more important due to the relatively high concentrations of humans and impervious surfaces that alter the environment.

D. J. Nowak (✉)
USDA Forest Service, Syracuse, NY, USA
e-mail: david.nowak@usda.gov

B. R. Bakshi (ed.), *Engineering and Ecosystems*,
https://doi.org/10.1007/978-3-031-35692-6_10

In the United States, tree cover in urban areas averages 39.4%, but ranges from a low of 10.1% in North Dakota to a high of 61.6% in Connecticut (Nowak and Greenfield 2018). Tree cover within cities varies depending on the region, with cities developed in forests having greater tree cover than cities developed in grassland and desert areas (Nowak et al. 1996). These regional differences are due to water availability and local tree seed sources, which affect natural regeneration and growth of trees. In US urban areas, only about one in three trees come from tree planting, the rest are due to natural regeneration. The proportion of trees planted varies by region (it tends to increase in drier regions due to limited regeneration), land use (more intensely managed (e.g., residential) areas tend to have a higher proportion of planted trees), and population density (as population density increases, so does the proportion of planted trees) (Nowak 2012). Although forest structure will inherently vary by region and water availability, natural regeneration can be used to sustain urban forests in many areas.

Another dominant element in urban landscapes is impervious cover (e.g., buildings, roads, parking lots). These surfaces provide essential services, but limit water infiltration into soils and tend to increase air temperatures. Impervious cover in US urban areas average 26.6%, but ranges from a low of 16.3% in New Hampshire to a high of 46.4% in Nevada (Nowak and Greenfield 2018). Percent impervious cover among states had less variation than percent tree cover.

At the global scale, tree cover in developed areas averages 26.5%, while impervious cover averages 25.9%. The tree cover in developed land varies by ecoregion, similar to the United States, with tree cover averaging 30.4% in forested regions, 18.2% in grasslands, and 12.0% in desert areas (Nowak and Greenfield 2020). Thus, both trees and impervious surfaces are important landscape components that interact within urban landscapes. Understanding how trees function to affect the local environment can lead to improved engineering solutions with trees to improve environmental quality and human health.

The purpose of this chapter is to summarize how trees affect their local environment and the general benefits provided by trees, with a specific focus on air quality. By understanding how and what trees impact, better designs can be engineered using trees to improve the environment. The chapter concludes with a discussion of a modeling system designed to aid in assessing the environmental impacts and values of trees.

10.2 Tree Processes That Affect the Local Environment

Trees across a landscape vary in species composition, abundance, size, health, and location. These five factors affect two key structural attributes that affect the environment: (1) total leaf area and (2) total woody biomass. Both of these attributes provide physical mass that affects wind flow and solar radiation (e.g., tree shade), but leaf area is likely the most important attribute as it typically provides the greatest surface area and gas exchange with the environment.

Leaf area is the most important component of a tree as it affects not only tree health, but also numerous benefits provided by trees (Table 10.1). The leaf area provides a large visual component related to tree aesthetics and also blocks wind and solar radiation, deflects and masks sounds, and intercepts precipitation, all of which affect the local physical environment. The leaves also provide habitat and food for numerous creatures. More importantly, leaves exchange chemicals with the surrounding environment via leaf stomata.

Leaf stomata are tiny pores on leaves that regulate gas exchange between the leaf interior and exterior environment. Depending upon local moisture conditions, these stomata typically open during the daytime to exchange carbon dioxide, oxygen, and water through processes of photosynthesis and respiration (Salisbury and Ross 1978):

$$\text{Photosynthesis}:\ n\,(CO_2) + n\,(H_2O) + \text{light} \rightarrow (CH_2O)_n + nO_2$$
$$\text{Respiration}:\ (CH_2O)_n + nO_2 \rightarrow n\,(CO_2) + n\,(H_2O) + \text{energy}$$

Growing, healthy trees take in carbon dioxide, storing carbon within its biomass (i.e., carbon sequestration), and release oxygen. As a tree dies and decomposes, carbon from the biomass releases carbon dioxide and consumes oxygen.

When the stomata are open, air enters the leaves via gaseous diffusion. This air contains carbon dioxide and air pollutants that can be removed by the leaf interior water and surfaces. Water vapor from the leaf interior also diffuses into the atmosphere via transpiration. Through the transpiration process (evaporating of leaf water), some of the net radiation that would otherwise warm air temperature is directed to evaporating water (latent heat). Further, warm air passes its heat to the evaporating water, which also reduces the temperature of the air (sensible heat). In addition to water being released when stomata are open, some plant volatile organic compounds (e.g., isoprene) are also released by some species. These compounds can affect the formation of air pollution and may also be useful in attracting pollinators or repelling predators (Kramer and Kozlowski 1979).

While leaves provide important functions, so does the woody above and below ground plant biomass. This biomass, which is on the order of several tonnes per mature tree, also provides physical mass to block wind and solar radiation, deflect and mask sounds, intercept precipitation, and provide habitat and food for numerous creatures. The woody biomass is also the main storage vessel for sequestered atmospheric carbon, which can be lost when the tree dies and decomposes. Tree and leaf biomass vary by species and through time as trees grow and eventually die.

In addition to species and size, tree location is also an important attribute that affects tree benefits. Tree location relative to problem sources (e.g., air pollution from automobile) and the receiver of the impact (e.g., human breathing the pollution) need to be considered when determining tree locations and in designing forests to help combat specific issues. Of utmost importance in selecting tree species and location is ensuring that the tree can survive and thrive at that location. However, designs also need to consider the intent of the design and intended impact from trees. For energy conservation, location around buildings is important; for health effects,

Table 10.1 Tree attributes that most directly affect various tree benefits (see Sects. 10.3 and 10.4 for more detailed information on tree benefits)

Tree attribute	Aesthetics/human physiology	Air quality	Air temperature	Building energy use	Carbon sequestration	Noise	Oxygen production	Property values	Ultraviolet radiation	Water flow and quality	Wildlife/biodiversity	Wood products
Aboveground biomass	√			√	√	√		√	√	√	√	√
Belowground biomass					√					√	√	√
Leaf area	√	√	√	√	√	√	√	√	√	√	√	
Leaf gas exchange		√	√	√	√		√			√		

Note: all tree attributes are important to sustain tree health and affect most benefits to some degree, but only major and most direct impacts are given locations around people are important; for water effect, locations near streams or stream pollutant loading sources are important; etc. Forest designs and plantings need to consider best designs that combat the most important local environmental or social issues.

Through proper design and management, these forest attributes can be stewarded to sustain optimal outcomes even though forest populations change through time. Understanding tree biological, chemical, physical, and social impacts on the local environment and human population can lead to better forest design and management to sustain environmental and human health through time.

10.3 Services from Urban Trees

Vegetation provides numerous benefits to society. In general, due to their large mass and leaf area, trees provide more benefits than shrubs, which generally provide more benefits than herbaceous plants. However, each plant type can have specific benefits that may outweigh other vegetation types (e.g., a colorful flower bed providing an aesthetically pleasing carpet of color). Many of the services and costs provided by vegetation and their management affect human health. Thus, designing and managing natural processes to maximize these benefits and minimize the costs can help improve human health. In addition to tree effects on air temperatures, air quality, and building energy use, which are discussed in more detail in Sect. 10.4, the following are some of the other general benefits derived from trees.

10.3.1 Aesthetics and Human Physiological Responses/Well-Being

Trees and urban green space can provide aesthetic environments for residents, but close association with these green areas also affects human physiology and well-being. The evidence on health effects of trees is increasing, with consistent negative associations between urban green space exposure and mortality, heart rate, and violence, and positive associations with attention, mood, and physical activity. Associations are mixed, with some studies finding associations and other studies finding no association between urban green space exposure and general health, weight status, depression, and stress (via cortisol concentration). The number of studies is too low to generalize about birth outcomes, blood pressure, heart rate variability, cancer, diabetes, or respiratory symptoms (Kondo et al. 2018). Several pathways or mechanisms for these health effects have been suggested, such as increased physical activity (Sallis et al. 2016), social interactions (de Vries et al. 2013), and reduced stress levels (Egorov et al. 2017).

10.3.2 Carbon Storage and Sequestration

Urban trees and forests affect climate change by altering the level of greenhouse gases such as carbon dioxide (CO_2). Trees act as a sink for CO_2 by fixing carbon during photosynthesis and storing carbon within tree biomass. The net long-term CO_2 source/sink dynamics of forests changes through time as trees grow, die, and decay. Human influences on forests (e.g., management) can further affect CO_2 source/sink dynamics of forests through factors such as fossil fuel emissions and harvesting/utilization of biomass (Nowak et al. 2002b). Unlike some other services, the carbon sequestration from trees is temporary as much of the stored carbon will revert back to atmospheric carbon through tree death and decomposition (though some carbon can be retained in soils). However, secondary tree effects such as reduced building energy use can reduce carbon emissions due to lower space condition fuel usage (Nowak et al. 2002b). In the United States, urban forests store 919 million tonnes of carbon valued at $119 billion (Nowak and Greenfield 2018). This storage value will vary through time and can be lost if the urban forest population is not sustained. The US urban forest currently annually sequesters about 37 million tonnes of carbon, valued at $4.8 billion (Table 10.3).

10.3.3 Noise

Field tests have shown that properly designed plantings of trees and shrubs can significantly reduce noise. Leaves and stems reduce transmitted sound primarily by scattering it, while the ground absorbs sound (Aylor 1972). For optimum noise reduction, trees and shrubs should be planted close to the noise source rather than the receptor area (Cook and Van Haverbeke 1971). Wide belts (30 m) of tall dense trees combined with soft ground surfaces can reduce apparent loudness by 50% or more (6–10 decibels) (Cook 1978). For narrow planting spaces (less than 3 m wide), reductions of 3–5 decibels can be achieved with dense belts of vegetation, that is, one row of shrubs along the road and one row of trees behind it (Reethof and McDaniel 1978). Buffer plantings in these circumstances typically are more effective in screening views than in reducing noise.

Vegetation also can mask sounds by generating its own noise as wind moves tree leaves or as birds sing in the tree canopy. These sounds may make individuals less aware of offensive noises, because people are able to filter unwanted noise while concentrating on more desirable sounds (Robinette 1972). The perception of sounds by humans also is important. By visually blocking the sound source, vegetation can reduce individuals perceptions of the amount of noise they actually hear (Anderson et al. 1984). The ultimate effectiveness of plants in moderating noise is determined by the sound itself, the planting configuration used, the proximity of the sound source, receiver, and vegetation, as well as climatic conditions (Nowak and Dwyer 2007).

10.3.4 Oxygen Production

Oxygen production is directly tied to carbon sequestration, but in an inverse fashion. When a tree has a net carbon sequestration, it gives off oxygen. When tree biomass decomposes or is burned, it gives off carbon and consumes oxygen. Urban forests in the coterminous United States are estimated to produce about 61 million metric tons (67 million tons) of oxygen annually, enough oxygen to offset the annual oxygen consumption of about 2/3 of the US population. Although oxygen production is often cited as a significant benefit of trees, this benefit is relatively insignificant and of negligible value due to the large oxygen content of the atmosphere (Nowak et al. 2007).

10.3.5 Property Values

One of the more commonly cited benefits of urban trees relates to increased property values. Effects on property value vary from a slight overall increase in value with locally mixed positive and negative effects (e.g., Saphores and Li 2012) up to around 15% (Morales 1980; Thompson et al. 1999). Increases in property values due to trees are an indication of willingness to pay for various benefits associated with trees that the homeowner receives. However, these value transactions only occur at a point sale (e.g., adding a tree to a property may increase property values, but that value is only realized when the property is sold); however, the trees are providing other values (e.g., cooler air temperature) annually. While increases in property values may be considered a benefit, they are also a cost to the homeowner via higher annual taxes paid due to higher home prices.

10.3.6 Ultraviolet Radiation

Tree leaves absorb 90–95% of ultraviolet (UV) radiation and thereby affect the amount of UV radiation received by people under or near tree canopies (Na et al. 2014). This reduction in UV exposure affects incidence of skin cancer, cataracts, and other ailments related to UV radiation exposure (Heisler and Grant 2000).

10.3.7 Water Cycles and Quality

Trees impact surface stormwater runoff, soil moisture, stream flow, groundwater recharge, and water quality by intercepting precipitation (rain and snow), enhancing soil water infiltration, absorbing soil moisture and chemicals, shading surfaces, and evapotranspiring water. While these processes generally reduce runoff, increase baseflow in streams, and reduce peak stream flow events (e.g., flooding), unmanaged trees can also increase flooding if branches or leaves clog drains or dam streams. However, the relationship between trees and groundwater recharge is complex and can be either positive or negative. Water use by trees can outweigh water availability, thus depleting streamflow and groundwater recharge in certain areas (Albaugh et al. 2013). In dry regions, groundwater recharge is maximized at an intermediate tree density. Below this optimal tree density, the benefits from any additional trees on water percolation exceed their extra water use, leading to increased groundwater recharge, while above the optimum, the opposite occurs (Ilstedt et al. 2016).

Trees also affect water quality by generally decreasing the concentration and amount of sediments, nutrients, metals, pesticides, pathogens, microbes, and other pollutants reaching a water body (Nowak et al. 2020). Trees also shade surfaces and reduce air temperatures, which reduces thermal loads on shaded objects and can reduce the heating of river water, thereby mitigating biological activity that can degrade water quality (e.g., eutrophication) (Yang et al. 2008). At a larger scale, urbanization and forests can influence regional precipitation patterns (e.g., Keys et al. 2017). If managed properly, these hydrologic effects can reduce risk to flooding, help recharge aquifers, impact regional precipitation, and improve human health by reducing sediments, chemicals, and pathogens found within waterways.

10.3.8 Wildlife Populations

Tree species composition and structure directly affect wildlife habitat, food, and local biodiversity. Various procedures can be used to estimate the relationship between local forest structure and wildlife species habitat suitability and insect biodiversity (e.g., Tallamy and Shropshire 2009; Lerman et al. 2014).

10.3.9 Wood Products

Though often considered a waste product in urban areas, dead and removed trees can be used for various products such as timber, palettes, fiber, and chemicals (e.g., ethanol). As US urban forests contain about 1.7 billion tonnes of total tree dry-weight biomass (Nowak and Greenfield 2018), assuming a likely conservative annual mortality rate of 2% (Nowak et al. 2004), total above-ground dry-weight biomass removed annually would be around 26 million tonnes per year. This estimate would be slightly higher than a previous estimate of 16–38 million green tons per year (Bratkovich and Fernholz 2010). This biomass could be used to produce wood products and as a potential income source for cities (e.g., Cesa et al. 2003). In addition, leaf drop could be used to provide nutrients (e.g., N, P, and K) and plants' fruits could be used for food (e.g., Clark and Nicholas 2013).

10.3.10 Cumulative Benefits

Trees' effects in numerous cities have been evaluated and reveal benefits typically in the millions of dollars per year, with values varying by tree population size (Table 10.2). At the US national level, urban forest benefits are conservatively estimated at \$18.4 billion per year; \$5.4 billion from air pollution removal, \$5.4 billion from reduced building energy use, \$4.8 billion from carbon sequestration, and \$2.7 billion from avoided pollutant emissions (Table 10.3). This estimate is conservative as it only addresses four benefits out of a myriad of potential benefits from trees.

10.4 Tree Effects on Air Quality

The World Health Organization (2016) states that air pollution is the largest environmental risk factor. Air pollution significantly affects human and ecosystem health (U.S. EPA 2010). Recent research indicates that global deaths directly or indirectly attributable to ambient air pollution reached almost 4.5 million in 2015 (Cohen et al. 2017). Air pollution is the largest environmental cause of disease and premature death in the world (WHO 2014).

Ambient air pollution caused 107.2 million disability adjusted life years (number of years lost due to ill-health, disability or early death) in 2015 (Cohen et al. 2017). Human health problems from air pollution include: aggravation of respiratory and cardiovascular diseases; increased frequency and severity of respiratory symptoms (e.g., difficulty breathing and coughing, chronic obstructive pulmonary disease (COPD), and asthma); increased susceptibility to respiratory infections, lung cancer, and premature death (e.g., Pope et al. 2002; Marino et al. 2015; Vieira 2015). Recent studies also suggest that air pollution can contribute to cognitive and mental disorders (e.g., Calderón-Garcidueñas et al. 2011; Brauer 2015; Annavarapu and Kathi 2016). People with pre-existing conditions (e.g., heart disease, asthma, emphysema, diabetes) and older adults and children are at greater risk for air pollution-related health effects. In the United States, approximately 130,000 deaths were related to particulate matter less than 2.5 microns ($PM_{2.5}$) and 4700 deaths to ozone (O_3) in 2005 (Fann et al. 2012).

Elevated ambient temperatures are associated with increased mortality due to heat stress (Basu and Ostro 2008). Heat exposure increases mortality risk for groups with pre-existing medical conditions, such as cardiovascular, respiratory, and cerebrovascular diseases (Basu 2009). Several high-risk populations have been identified, including the elderly, children, people engaging in outdoor occupations, and people living alone, especially on higher floors of apartment buildings (Basu and Ostro 2008). In July 1995, Chicago sustained a heat wave that resulted in more than 600 deaths, 3300 emergency department visits, and a substantial number of intensive care unit admissions for near-fatal heat stroke (Dematte et al. 1998). A heat wave in Europe in the summer of the 2003 led to more than 70,000 deaths

Table 10.2 City estimates of various tree benefits and values

City	Year	Trees Number (× 10³)	Carbon sequestration[a] t/year	Carbon sequestration[a] $/year (× 10³)	Pollution removal[b] t/year	Pollution removal[b] $/year (× 10³)	Energy savings[c] $/year (× 10³)	Total $/year (× 10³)	References
Houston, TX	2009	33,975	182,900	34,385	2191	21,122[d]	53,900	109,407	Nowak et al. (2017)
Austin, TX	2007	33,843	83,200	15,642	1137	11,035[d]	18,900	45,577	Nowak et al. (2016b)
Los Angeles, CA	2008	5993	69,800	13,122	1792	14,173	10,200	37,495	Nowak et al. (2011)
New York, NY	2013	6977	55,200	10,378	1004	9222[d]	17,100	36,700	Nowak et al. (2018)
Toronto, ON	2008	10,220	46,700	8780[e]	1905	17,146[e]	9700[e]	35,626[e]	Nowak et al. (2013b)
Philadelphia, PA	2012	2918	24,600	4625	531	4602	6900	16,127	Nowak et al. (2016c)
Chicago, IL	2007	3585	22,800	4286	806	6398	360	11,044	Nowak et al. (2010b)
Washington, DC	2004	1928	14,700	2764	379	2858	2653	8275	Nowak et al. (2006d)
Minneapolis, MN	2004	979	8100	1523	277	2242	216	3981	Nowak et al. (2006c)
Syracuse, NY	2009	1088	5300	996	101	852	1100	2948	Nowak et al. (2016a)
Scranton, PA[f]	2006	1198	3700	696	65	514	628	1838	Nowak et al. (2010a)
Morgantown, WV	2004	658	2600	489	65	489	380	1358	Nowak et al. (2012)
Casper, WY	2006	123	1100	207	34	275	−27	455	Nowak et al. (2006b)

t metric tons

[a]Carbon sequestration value is based on social cost of carbon for 2020 updated to 2018 dollar values based on the producer price index ($188/tC) (Interagency Working Group on Social Cost of Carbon 2015)

[b]Pollution removal of carbon monoxide ($1407/t), nitrogen dioxide ($9906/t), ozone ($9906/t), particulate matter less than 10 microns ($6614/t), and sulfur dioxide ($2425/t) unless otherwise noted

[c]Saving from alter building energy use due to trees. Negative values indicate increase in energy use costs

[d]Pollution removal of nitrogen dioxide, ozone, particulate matter less than 2.5 microns and sulfur dioxide with value based on local health impacts (Nowak et al. 2014)

[e]In Canadian dollars; other values are in US dollars

[f]Urban area only

	Carbon sequestration				Air pollution removal		Avoided energy use		Avoided emissions		Total
	t/year	SE	$/year	SE	t/year	$/year	$/year	SE	$/year	SE	$/year
State	$\times 10^3$	$\times 10^3$	$\times 10^6$	$\times 10^6$	$\times 10^3$	$\times 10^6$	$\times 10^6$	$\times 10^6$	$\times 10^6$	$\times 10^6$	$\times 10^6$
Alabama	915.9	147.6	118.8	19.1	16.5	82.9	49.7	3.7	27.4	2.0	278.9
Arizona	575.9	92.8	74.7	12.0	10.5	33.1	102.2	4.4	29.9	1.3	239.9
Arkansas	478.5	77.1	62.1	10.0	8.3	40.4	34.4	2.5	18.3	1.3	155.2
California	**2880.2**	464.1	**373.6**	60.2	**63.4**	**639.4**	274.1	13.2	67.1	3.2	**1354.2**
Colorado	151.0	24.3	19.6	3.2	2.2	5.1	10.3	1.1	5.2	0.5	40.2
Connecticut	766.8	123.6	99.5	16.0	15.9	94.8	117.3	6.9	22.0	1.3	333.6
Delaware	136.3	22.0	17.7	2.8	2.7	14.6	43.7	3.4	29.7	2.3	105.7
Florida	**4163.5**	670.9	**540.1**	87.0	**80.7**	**554.4**	**512.7**	30.7	**272.0**	16.3	**1879.2**
Georgia	**2832.6**	456.5	**367.5**	59.2	**62.2**	255.6	124.6	8.8	92.5	6.5	840.1
Idaho	34.3	5.5	4.4	0.7	1.5	18.8	22.0	0.8	5.9	0.2	51.1
Illinois	999.7	161.1	129.7	20.9	14.6	157.7	281.4	8.8	147.2	4.6	716.0
Indiana	566.0	91.2	73.4	11.8	13.0	88.7	156.6	9.7	113.6	7.1	432.3
Iowa	177.3	28.6	23.0	3.7	2.6	21.0	59.8	4.9	53.1	4.4	156.9
Kansas	272.7	43.9	35.4	5.7	2.8	14.5	121.3	5.1	60.7	2.6	231.8
Kentucky	410.4	66.1	53.2	8.6	10.0	55.8	113.5	9.6	85.2	7.2	307.7
Louisiana	994.9	160.3	129.1	20.8	23.9	117.7	139.9	9.3	81.1	5.4	467.7
Maine	132.4	21.3	17.2	2.8	4.3	25.1	13.5	2.9	5.6	1.2	61.4
Maryland	963.4	155.2	125.0	20.1	29.2	177.1	230.2	14.1	**188.0**	11.5	720.3
Massachusetts	1229.2	198.1	159.5	25.7	29.5	197.1	192.2	10.6	64.7	3.6	613.5
Michigan	961.5	154.9	124.7	20.1	29.5	131.5	233.0	13.4	**147.8**	8.5	637.0
Minnesota	520.6	83.9	67.5	10.9	7.7	40.0	65.0	8.9	38.8	5.3	211.3
Mississippi	496.6	80.0	64.4	10.4	12.8	66.8	39.1	1.6	22.6	1.0	192.8
Missouri	654.4	105.4	84.9	13.7	14.4	88.3	174.5	10.1	105.7	6.1	453.3
Montana	33.4	5.4	4.3	0.7	1.3	13.3	2.0	0.4	1.6	0.3	21.2
Nebraska	72.8	11.7	9.4	1.5	0.6	4.2	34.7	1.2	25.4	0.9	73.7
Nevada	58.3	9.4	7.6	1.2	2.0	8.8	21.9	0.6	5.4	0.2	43.6
New Hampshire	224.7	36.2	29.2	4.7	5.7	15.2	14.5	1.8	4.5	0.6	63.4

(continued)

Table 10.3 (continued)

	Carbon sequestration				Air pollution removal		Avoided energy use		Avoided emissions		Total
	t/year	SE	$/year	SE	t/year	$/year	$/year	SE	$/year	SE	$/year
State	×10^3	×10^3	×10^6	×10^6	×10^3	×10^6	×10^6	×10^6	×10^6	×10^6	×10^6
New Jersey	1130.2	182.1	146.6	23.6	22.0	151.4	219.8	10.0	57.0	2.6	574.8
New Mexico	104.3	16.8	13.5	2.2	3.3	6.0	20.0	1.0	10.5	0.5	50.1
New York	1371.6	221.0	177.9	28.7	**41.5**	**408.2**	**345.8**	21.0	69.5	4.2	**1001.4**
North Carolina	**2146.9**	346.0	**278.5**	44.9	**50.3**	191.8	150.3	9.5	86.3	5.5	706.9
North Dakota	11.9	1.9	1.5	0.2	0.1	0.5	2.8	0.2	2.5	0.2	7.3
Ohio	1173.7	189.1	152.3	24.5	35.0	**265.6**	**313.3**	18.3	**240.3**	14.0	**971.5**
Oklahoma	301.3	48.6	39.1	6.3	5.5	34.6	17.8	2.5	9.9	1.4	101.5
Oregon	234.9	37.8	30.5	4.9	4.3	79.1	21.4	1.7	5.1	0.4	136.1
Pennsylvania	1337.3	215.5	173.5	28.0	40.6	**441.6**	**290.2**	18.7	**164.3**	10.6	**1069.6**
Rhode Island	149.1	24.0	19.3	3.1	3.0	26.1	37.2	2.2	5.1	0.3	87.8
South Carolina	1140.4	183.8	147.9	23.8	27.6	125.3	76.4	4.2	35.9	2.0	385.6
South Dakota	28.3	4.6	3.7	0.6	0.2	1.7	6.2	0.3	5.1	0.3	16.7
Tennessee	1033.4	166.5	134.1	21.6	23.1	108.2	83.7	5.8	50.8	3.5	376.7
Texas	**2453.1**	395.3	**318.2**	51.3	33.7	185.6	**308.4**	11.2	125.5	4.6	937.8
Utah	83.3	13.4	10.8	1.7	2.6	11.4	19.6	1.2	11.7	0.7	53.5
Vermont	46.7	7.5	6.1	1.0	1.1	5.7	8.2	1.3	0.5	0.1	20.4
Virginia	1007.7	162.4	130.7	21.1	30.6	134.7	114.4	5.2	69.3	3.1	449.2
Washington	614.5	99.0	79.7	12.8	16.4	180.3	56.2	3.8	11.9	0.8	328.2
West Virginia	218.7	35.2	28.4	4.6	5.3	30.6	21.4	2.0	15.6	1.5	95.9
Wisconsin	334.1	53.8	43.3	7.0	7.4	45.3	81.5	7.8	44.8	4.3	214.9
Wyoming	12.0	1.9	1.6	0.3	0.5	2.5	1.3	0.1	1.1	0.1	6.4
US48	36,653	5325	4755	690.7	821.7	5398	5380	61.7	2743	33.6	18,274
Alaska	59.3	9.6	7.7	1.2	na	na	na	na	na	na	Na
Hawaii	270.4	43.6	35.1	5.7	na	na	na	na	na	na	Na
US50	36,983	5361	4798	695.5	na	na	na	na	na	na	Na

From Nowak and Greenfield (2018)

Bold numbers indicate that the state is within the top five highest values for that category

SE standard error, *US48* conterminous US, *US50* entire US, *na* not analyzed, *t* tons

(Robine et al. 2008). The issue of heat-related morbidity and mortality is expected to increase substantially with climate change (Gasparrini et al. 2017). Both pollution and increased temperatures impact human health, but they may also interact to produce an even greater negative impact on health (Harlan and Ruddell 2011).

Trees, through their interaction with the atmosphere, affect air quality and consequently human health, particularly when in close association with people (e.g., in cities). For centuries, it has been known that trees affect the atmospheric environment. In the 1800s, parks in cities were referred to as "Lungs of the city" due to the ability of the park vegetation to produce oxygen and remove industrial pollutants from the atmosphere (Compton 2016). In addition to this "lung" capacity of vegetation, a cooling capacity of vegetation has also long been known to affect the local environment. Historical home designs dating back over a millennia often included trees and water features to help cool the environment (Laurie 1986). Trees and forests can be used to improve air quality and reduce heat, and consequently improve human health.

To help understand how trees affect air quality, it is important to understand the different types of air pollutants. Some pollutants, both gaseous and particulate, are directly emitted into the atmosphere and include sulfur dioxide (SO_2), nitrogen oxides (NO_x), carbon monoxide (CO), particulate matter (PM) and volatile organic compounds (VOC). Other pollutants are not directly emitted; rather they are formed through chemical reactions. For example, ground-level ozone is often formed when emissions of NO_x and VOCs react in the presence of sunlight. Some particles are also formed from other directly emitted pollutants. Trees affect these air pollutants in three main ways: they (1) alter local air temperatures, microclimates, and building energy use; (2) remove air pollution; and (3) emit various chemicals.

10.4.1 Trees' Effects on Air Temperatures, Local Microclimate, and Building Energy Use

Increased air temperatures can lead to increased building energy demand in the summer, increased air pollution, and heat-related illness. Trees alter microclimates and cool air temperatures through evaporation from tree transpiration, blocking winds, and shading various surfaces. Vegetated areas can cool the surroundings by several degrees C, with higher tree and shrub cover leading to cooler air temperatures (Chang et al. 2007). Although trees usually contribute to cooler summer air temperatures, their presence can increase air temperatures in some instances (Myrup et al. 1991). For example, reduced windspeeds due to trees can increase temperatures in treeless impervious areas on sunny days as cooler air is prevented from mixing with or dispersing the warm air coming off the impervious surfaces. Reduced air temperature due to trees can improve air quality because the emission of many pollutants and/or ozone-forming chemicals is temperature dependent.

Tree transpiration and tree canopies also affect radiation absorption and heat storage, relative humidity, turbulence, surface albedo, surface roughness, and mixing-layer height (i.e., height within which wind and surface substances (e.g., pollution) are dispersed by vertical mixing processes). These changes in local meteorology can alter pollution concentrations in urban areas (Nowak et al. 2000).

Changes in wind speeds can lead to both positive and negative effects related to air pollution. On the positive side, reduced wind speeds will tend to reduce winter-time heating energy use in buildings (and associated pollutant emissions from power plants) by reducing cold air infiltration into buildings. On the negative side, reductions in wind speed can reduce the dispersion of pollutants, which will tend to increase local pollutant concentrations. In addition, with lower winds, the height of the atmosphere in which the pollutant mixes is often reduced. This reduction in the "mixing height" will tend to increase pollutant concentrations as the same amount of pollution is now mixed within a smaller volume of air.

In addition, reduced air temperatures and shading of buildings can reduce the amount of energy used to cool buildings in the summer-time. However, shading of buildings in winter can lead to increased building energy use (e.g., Heisler 1986). This altered energy use consequently leads to altered pollutant emissions from power plants. Proper tree placement near buildings is critical to achieve maximum building energy conservation benefits. Urban forests in the conterminous United States annually reduce residential building energy use by $5.4 billion per year and avoid the emission of thousands of tonnes of pollutants valued at $2.7 billion per year (Table 10.3).

Methods for estimating tree effects on building energy use are given in McPherson and Simpson (1999) and coded within the i-Tree Eco model (www.itreetools.org). Methods for estimating tree effects on air temperatures (Yang et al. 2013) are also integrated within i-Tree.

10.4.2 Removal of Air Pollutants

Trees remove gaseous air pollution primarily by uptake through leaf stomata, though some gases are removed by the plant surface. Once inside the leaf, gases diffuse into intercellular spaces and may be absorbed by water films to form acids or react with inner-leaf surfaces (Smith 1990), which can be a source of the essential plant nutrients of sulfur and nitrogen (NAPAP 1991). Trees also directly affect particulate matter in the atmosphere through the interception of particles, emission of particles (e.g., pollen), and resuspension of particles captured on the plant surface. Many of the particles that are intercepted are eventually resuspended back to the atmosphere, washed off by rain, or dropped to the ground with leaf and twig fall. Consequently, vegetation is only a temporary retention site for many atmospheric particles. The removal of gaseous pollutants is more permanent as the gases are often absorbed and transformed within the leaf interior (Smith 1990). Some pollutants under high concentrations can damage leaves (e.g., sulfur dioxide, nitrogen dioxide, ozone) (e.g., Nowak 1994; Nowak et al. 2015), particularly of pollutant-sensitive species. Given the pollution concentration in most cities, these pollutants would not be expected to cause visible leaf injury, but could in cities or areas with high pollutant concentrations.

At the species level, pollution removal of gaseous pollutants will be affected by tree transpiration rates (gas exchange rates) and amount of leaf area. Particulate matter removal rates will vary depending upon leaf surface characteristics and area. Species with dense and fine textured crowns and complex, small, and rough leaves would capture and retain more particles than open and coarse textured crowns, and simple, large, smooth leaves (Little 1977; Smith 1990). Evergreen trees provide for year-round removal of particles. A species ranking of trees in relation to pollution removal is estimated in i-Tree Species (www.itreetools.org).

Healthy trees in cities can remove significant amounts of air pollution. Areas with a high proportion of tree cover (e.g., forest stands) will remove more pollution and have the potential to have greater reductions in air pollution concentrations in and around these areas. One hectare of tree cover has a US average pollution removal of about 75 kg/year in urban areas, but this value could range up to over 200 kg per year in more polluted areas with long growing seasons (e.g., Los Angeles) (Fig. 10.1). Large healthy trees (>76 cm in stem diameter) remove approximately 60–70 times more air pollution annually than small healthy trees (<7.6 cm in stem diameter), with large trees removing about 1.4 kg per year (Nowak 1994). Pollution removal rates by vegetation differ among regions according to the amount of vegetative cover, the amount of air pollution, length of in-leaf season, precipitation, and other meteorological variables.

There are numerous studies that link air quality to human health effects, but only a limited number of studies have looked at the estimated health effects of air pollution removal by trees. In the United Kingdom, woodlands are estimated to reduce between 5 and 7 deaths and between 4 and 6 hospital admissions per year due to reduced sulfur dioxide and particulate matter less than 10 microns (PM_{10}) (Powe and Willis 2004). In London, it is estimated that the city's 25% tree cover removes 90.4 tonnes of PM_{10} pollution per year, which equates to a reduction of 2 deaths and 2 hospital stays per year (Tiwary et al. 2009). Nowak et al. (2013a) reported that the total amount of $PM_{2.5}$ removed annually by trees in 10 US cities in 2010 varied from 4.7 tonnes in Syracuse to 64.5 tonnes in Atlanta, with health values ranging from $1.1 million in Syracuse to $60.1 million in New York City.

Although the individual tree and per acre tree cover values may be relatively small, the combined effects of large numbers of trees and tree cover in aggregate can lead to significant effects. Pollution removal by trees in cities can range up to 11,100 tons per year with societal values ranging up to $89 million per year in Jacksonville, FL due to its large land area and tree cover (Nowak et al. 2006a). Trees and forests in the conterminous United States removed 22.4 million tonnes of air pollution in 2010, with human health effects valued at $8.5 billion. Most of the pollution removal occurred in rural areas, while most of the health benefits were within urban areas. In urban areas, trees removed 822,000 tonnes per year valued at $5.4 billion (Table 10.3). Nationwide, health impacts included the avoidance of more than

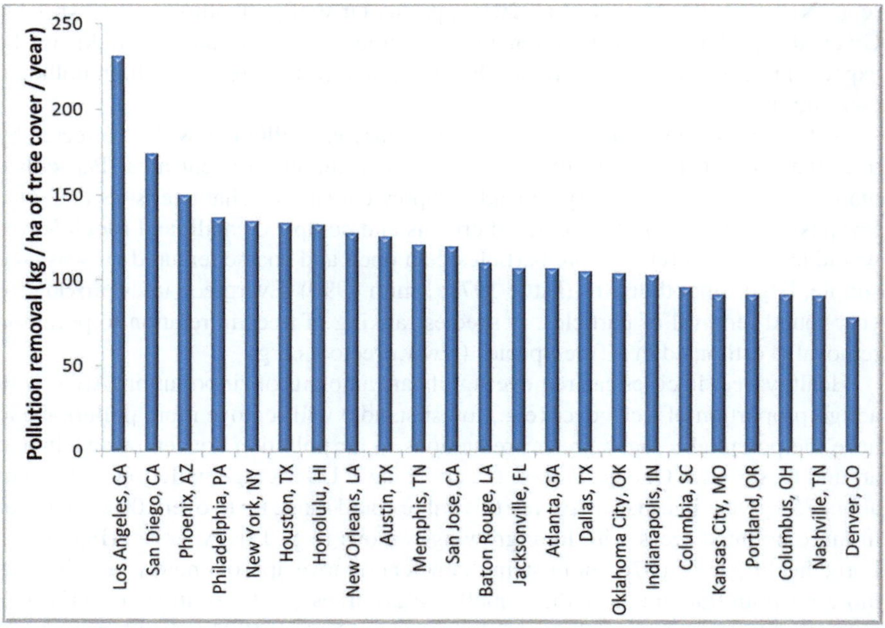

Fig. 10.1 Pollution removal values per acre of tree cover in select cities. Estimates assume a leaf area index of 6 and 10% evergreen species. Leaf area index is per unit tree cover and calculated as total leaf area (m^2) divided by tree cover (m^2). (Derived from Nowak et al. 2006a)

850 incidences of human mortality. Other substantial health benefits include the reduction of more than 670,000 incidences of acute respiratory symptoms, 430,000 incidences of asthma exacerbation and 200,000 school loss days (Nowak et al. 2014).

Though the amount of air pollution removal by trees may be substantial, the percent air quality improvement in an area will depend upon the amount of vegetation and meteorological conditions. Average air quality improvement due to pollution removal by trees in cities during daytime of the in-leaf season is less than 1%. However, in areas with 100% tree cover, hourly air pollution improvements average around 4 times greater and can reach up to 16% (Nowak et al. 2006a). From a public health perspective, it is important to consider that even though percent air quality improvement from trees may not be very large, a small percent change in air quality can have a substantial impact on human health (Cohen et al. 2017).

Methods of Estimating Pollution Removal by Trees

Hourly pollution removal by vegetation can be estimated with information regarding tree cover (m^2), leaf area index (total one-sided leaf/total projected ground area of canopy), leaf type (deciduous or evergreen), hourly meteorological data (e.g., air temperature, wind speed, cloud cover), and air pollution concentrations. Pollution removal or downward pollutant flux (F; in $g/m^2/s$) is calculated as the product of the deposition velocity (V_d; in m/s) and the pollutant concentration (C; in g/m^3):

$$F = V_d C$$

Deposition velocity is calculated as the inverse of the sum of the aerodynamic (R_a), quasi-laminar boundary layer (R_b), and canopy (R_c) resistances (Baldocchi et al. 1987).

$$V_d = 1/(R_a + R_b + R_c)$$

Hourly estimates of R_a and R_b are calculated using standard resistance formulas (Killus et al. 1984; Pederson et al. 1995; Nowak et al. 1998) and hourly weather data. Hourly canopy resistance values for O_3, SO_2, and NO_2 can be calculated based on a modified hybrid of big-leaf and multilayer canopy deposition models (Baldocchi et al. 1987; Baldocchi 1988). Canopy resistance (R_c) has three components: stomatal resistance (r_s), mesophyll resistance (r_m), and cuticular resistance (r_t), such that

$$1/R_c = 1/(r_s + r_m) + 1/r_t$$

In the i-Tree model, mesophyll resistance is set to zero s/m for SO_2 (Wesely 1989) and 10 s/m for O_3 (Hosker and Lindberg 1982). Mesophyll resistance is set to 100 s/m for NO_2 to account for the difference between transport of water and NO_2 in the leaf interior, and to bring the computed deposition velocities in the range typically exhibited for NO_2 (Lovett 1994). Base cuticular resistances are set at 8000 s/m for SO_2, 10,000 s/m for O_3, and 20,000 s/m for NO_2 to account for the typical variation in r_t exhibited among the pollutants (Lovett 1994). Deposition velocities are sensitive to leaf area index, with velocities increasing as the index increases (Hirabayashi et al. 2011).

As removal of CO and particulate matter by vegetation is not directly related to transpiration, R_c for CO is set to a constant for in-leaf season (50,000 s/m) and leaf-off season (1,000,000 s/m) based on data from Bidwell and Fraser (1972). For PM_{10}, the median deposition velocity from the literature (Lovett 1994) is 0.0128 m/s for the in-leaf season. Base particle V_d is set to 0.064 based on an LAI of 6 and a 50% resuspension rate of particles back to the atmosphere (Zinke 1967). The base V_d is adjusted according to actual LAI and in-leaf versus leaf-off season parameters. For $PM_{2.5}$, hourly deposition velocities and resuspension rates vary with wind speed and leaf area as detailed in Nowak et al. (2013a).

To limit deposition estimates to periods of dry deposition, deposition velocities in i-Tree are set to zero during periods of precipitation. The model is run at the population scale to estimate pollution removal effects. Hourly pollutant flux (g/m^2 of tree canopy coverage) among the pollutant monitor sites is multiplied by total tree-canopy coverage (m^2) to estimate total hourly pollutant removal by trees across the study area.

10.4.3 Emission of Chemicals

While trees reduce air pollution by reducing air temperatures and directly removing pollution, trees also emit various chemicals that can contribute to air pollution (Sharkey et al. 1991). Trees emit varying amounts of volatile organic compounds (e.g., isoprene, monoterpenes) (Geron et al. 1994; Guenther 2002). These compounds are natural chemicals that make up essential oils, resins, and other plant products, and may be useful in attracting pollinators or repelling predators (Kramer and Kozlowski 1979). Oxidation of volatile organic compounds is an important component of the global carbon monoxide budget (Tingey et al. 1991). VOCs emitted by trees can also contribute to the formation of ozone and particulate matter (Sharkey et al. 1991). Because VOC emissions are temperature dependent and trees generally lower air temperatures, increased tree cover can lower overall VOC emissions and, consequently, ozone levels in urban areas (e.g., Cardelino and Chameides 1990). Ozone inside leaves can also be reduced due to the reactivity with biogenic compounds (Calfapietra et al. 2009).

VOC emission rates vary by species. Nine tree genera that have the highest standardized isoprene emission rate and therefore the greatest relative effect on increasing ozone, are: beefwood (*Casuarina* spp.), *Eucalyptus* spp., sweetgum (*Liquidambar* spp.), black gum (*Nyssa* spp.), sycamore (*Platanus* spp.), poplar (*Populus* spp.), oak (*Quercus* spp.), black locust (*Robinia* spp.), and willow (*Salix* spp.). However, just because these genera have relatively high emission rates, does not mean that they lead to a net production of ozone as they also remove ozone and lower air temperatures.

Other factors to consider in addition to VOC emissions are tree maintenance and pollen emission. Because some vegetation, particularly urban vegetation, often requires relatively large inputs of energy for maintenance activities, resulting pollutant emissions from maintenance equipment need to be considered. Pollen particles from trees can lead to allergic reactions (e.g., Cariñanos et al. 2014). Examples of some of the most allergenic species are *Acer negundo* (male), *Ambrosia* spp., *Cupressus* spp., *Daucus* spp., *Holcus* spp., *Juniperus* spp. (male), *Lolium* spp., *Mangifera indica*, *Planera aquatica*, *Ricinus communis*, *Salix alba* (male), *Schinus* spp. (male), and *Zelkova* spp. (Ogren 2000).

Methods for Calculating VOC Emissions by Trees

Tree VOC emissions can be estimated using procedure from the EPA's Biogenic Emissions Inventory System (BEIS) (U.S. EPA 2017). The amount of VOC emissions depends on tree species, leaf biomass, air temperature, and other environmental factors. Species leaf biomass is multiplied by genus-specific emission factors (e.g., Nowak et al. 2002a) to produce emission levels standardized to 30 °C and photosynthetically active radiation (PAR) flux of 1000 μmol m^{-2} s^{-1}. Standardized emissions are converted to actual emissions based on light and temperature correction factors (Geron et al. 1994) and local meteorological data.

VOC emission (E) (in μgC/tree/hr at temperature T (K) and PAR flux L (μmol/m^2/s)) for isoprene, monoterpenes, and OVOC is estimated as follows:

$$E = B_E \times B \times \gamma$$

where B_E is the base genus emission rate in μgC (g leaf dry weight)/hr at 30 °C and PAR flux of 1000 μmol/m^2/s; B is species leaf dry weight biomass (g); and

$$\gamma = \left[\alpha \cdot c_{L1} L / \left(1 + \alpha^2 \cdot L^2\right)^{\frac{1}{2}}\right] \cdot \left[\exp\left[c_{T1}\left(T - T_S\right)/R \cdot T_S \cdot T\right] / \left(0.961 + \exp\left[c_{T2}\left(T - T_M\right)/R \cdot T_S \cdot T\right]\right)\right]$$

for isoprene where L is PAR flux; $\alpha = 0.0027$; $c_{L1} = 1.066$; R is the ideal gas constant (8.314 K^{-1} mol^{-1}); T(K) is leaf temperature, which is assumed to be air temperature; T_S is standard temperature (303 K); and $T_M = 314$ K, $C_{T1} = 95{,}000$ J mol^{-1}, and $C_{T2} = 230{,}000$ J mol^{-1} (Geron et al. 1994; Guenther et al. 1995; Guenther 1997).

For monoterpenes and OVOC,

$$\gamma = \exp\left[\beta\left(T - T_s\right)\right]$$

where $T_S = 303$ K and $\beta = 0.09$.

10.4.4 Overall Effects of Trees on Air Pollution

There are many factors, both positive and negative, that determine the ultimate effect of trees on pollution. While pollution removal, reduced air temperatures, and general reduction in energy use improve air quality, the emission of VOCs and changes in wind speed can offset some of the improvement.

One model simulation illustrated that a 20% loss in forest cover in the Atlanta area due to urbanization led to a 14% increase in ozone concentrations (Cardelino and Chameides 1990). Although there were fewer trees to emit volatile organic compounds, an increase in Atlanta's air temperatures, due to tree loss and the urban heat island, increased VOC emissions from trees and other sources and altered ozone chemistry such that concentrations of ozone increased. Another model simulation of California's South Coast Air Basin suggests that the air quality impacts of increased urban tree cover may be locally positive or negative with respect to ozone. However, the net basin-wide effect of increased urban vegetation was a decrease in ozone concentrations if the additional trees are low VOC emitters (Taha 1996).

Modeling the effects of increased urban tree cover on ozone concentrations from Washington, DC to central Massachusetts, revealed that urban trees generally reduce ozone concentrations in cities, but tend to slightly increase average ozone concentrations regionally. The dominant tree effects on ozone were due to pollution removal and change in air temperatures, wind fields, and mixing-layer heights (Nowak et al. 2000). Modeling of the New York City metropolitan area also revealed that increasing tree cover by 10% reduced maximum ozone levels by about 4 ppb. This reduction was about 37% of the amount needed for attainment of the ozone air quality standard, revealing that increased tree cover can have a significant impact on reducing peak ozone concentrations in this region (Luley and Bond 2002).

Though reduction in wind speeds can increase local pollution concentrations due to reduced dispersion of pollutants and lowering of mixing heights, altering of wind patterns can also have a potential positive effect. Tree canopies can potentially prevent pollution in the upper atmosphere from reaching ground-level air space. Measured differences in ozone concentration between above- and below-forest canopies in California's San Bernardino Mountains have exceeded 50 ppb (40% lower concentration below the canopy) (Bytnerowicz et al. 1999). Forest canopies can limit the mixing of upper air with ground-level air, leading to significant below-canopy air quality improvements. However, where there are numerous pollutant sources below the canopy (e.g., automobiles), the forest canopy could increase concentrations by minimizing the dispersion of the pollutants away at the ground level (Fig. 10.2). This effect could be particularly important in heavily treed areas where automobiles drive under tree canopies. At the local scale, pollution concentrations can be increased if trees: (a) trap pollutants beneath tree canopies near emission sources (e.g., along road ways) (Gromke and Ruck 2009; Wania et al. 2012; Salmond et al. 2013; Vos et al. 2013); (b) limit dispersion by reducing wind speeds; and/or (c) lower mixing heights by reducing wind speeds (Nowak et al. 2000, 2014). However, standing in the interior of stands of trees can offer cleaner air if there are no local ground sources of emissions (e.g., from automobiles) nearby. Various studies (e.g., Dasch 1987; Cavanagh et al. 2009) have illustrated reduced pollutant concentrations in the interior of forest stands compared to outside of the forest stand.

While increased tree cover will enhance pollution removal and reduce summer air temperatures, local scale forest designs need to consider the location of pollutant sources relative to the distribution of human populations to minimize pollution concentrations and maximize air temperature reduction in heavily populated areas. Forest designs also need to consider numerous other tree impacts that can affect human health and well-being (e.g., impacts on ultraviolet radiation, water quality, aesthetics, etc.).

Fig. 10.2 Design of vegetation near roadways is important to minimize potential negative effects, such as trapping of pollutants. (Image source: D. Nowak)

10.5 Software to Assess Urban Forest Effects and Values

Computer models have been developed to assess forest composition and its associated effects on environmental quality and human health. While research is still needed regarding many of the environmental services that trees provide, resource managers can utilize existing models to better understand the role of vegetation in improving human health and environmental quality, lower costs of maintenance, and increase resource stewardship as an effective means to provide substantial economic savings to society.

Structure is a key variable as it is what managers manipulate to influence forest benefits and values. Structure represents the physical attributes of the urban forest, such as abundance, size, species, health, and location of trees. Managers often choose what species to plant, where and when to plant it, and what trees are removed from the landscape. These actions directly influence structure and consequently the benefits derived from the urban forest.

Field data on urban forest structure can be obtained from either inventories or sampling of the local urban forest. For large tree populations, field data in conjunction with aerial-based assessments will likely provide the best and most cost-effective means to assess forest structure. The most important tree characteristics to measure are species, diameter, crown dimensions, and tree condition. This information is helpful to managers regarding population management and assessing risks to the forest, but is also essential for estimating forest benefits and costs. The most important tree attribute is leaf area. While not directly measured in the field, this variable can be modeled from species, crown, and condition information. Diameter measures are also essential for estimating carbon storage. Leaf and tree biomass are other important variables that can be modeled from the core tree variables. Other information that is important for estimating forest benefits is crown competition (important for tree growth estimation and carbon sequestration) and location around buildings (important of energy conservation). Numerous forest benefits can currently be modeled from these tree variables, in conjunction with other local information (e.g., weather, pollution concentrations, and population data). Once the benefits are quantified, various methods of market as well as nonmarket valuation can be applied to characterize their monetary value (e.g., Hayden 1989).

There are various models that quantify forest benefits. Some free models include InVEST (Natural Capital Project 2016), Biome-BGC (Numerical Terradynamic Simulation Group 2016), and numerous other tools to assess forest carbon (e.g., U.S. Forest Service 2016). However, few models quantify urban forests. To date, the most comprehensive model developed to quantify urban forest structure, benefits, and values is i-Tree (www.itreetools.org). This freely available suite of tools was developed by the US Forest Service through a public-private partnership. The model is based on peer-reviewed science and can be used globally, with over 750,000 users in 180 countries. i-Tree was designed to accurately assess local forest structure and its impacts on numerous benefits, costs, and values (Table 10.4). Model results have been validated against numerous field measurements (e.g., Morani et al. 2014) to provide sound estimates of urban forest benefits. The model focuses on estimating forest structure and the magnitude of services received (e.g., tons removed). It then relies on economic valuation (e.g., $/ton removed) to estimate a value of the service. Various economic estimates are used and many can be adjusted by the users if local economic values are available.

The core program is *i-Tree Eco* – this model uses sample or inventory data and local environmental data to assess and forecast forest structure, benefits, threats, and values for any tree population (Nowak et al. 2008). The program includes plot selection tools, mobile data entry applications, table and graphic reporting and exporting, and automatic report generation. Urban forest assessments have been conducted in numerous cities globally (e.g., Barcelona, Spain; Calles, Mexico; Chicago, IL, USA; Medellin, Colombia; Milan, Italy; London, England; New York, NY, USA; Perth, Australia; Porto, Portugal; Santiago, Chile; Seoul, South Korea; Strasbourg, France; Toronto, Canada; – Chaparro and Terradas 2009; Escobedo et al. 2006; Graca et al. 2017; Nowak et al. 2010b, 2013b, 2018; Rogers et al. 2015; Selmi et al. 2016). See Table 10.2 for results from US cities.

Table 10.4 Ecosystem effects of trees currently quantified and in development in i-Tree

Ecosystem effect	Attribute	Quantified	Valued
Atmosphere	Air temperature	o	o
	Avoided emissions	•	•
	Building energy use	•	•
	Carbon sequestration	•	•
	Carbon storage	•	•
	Human comfort	o	
	Pollen	•	
	Pollution removal	•	•
	Transpiration	•	
	UV radiation	•	o
	VOC emissions	•	
Community/Social	Aesthetics/property value	o	o
	Food/medicine	o	
	Health Index[a]	o	
	Forest products[b]	•	•
	Underserved areas	•	
Terrestrial	Biodiversity	o	
	Invasive plants	•	
	Nutrient cycling	•	
	Wildlife habitat	•	
Water	Avoided runoff	•	•
	Flooding	o	o
	Rainfall interception	•	
	Water quality	•	o

Many of the listed ecosystem effects are both positive and negative depending on specific conditions or perspective. For example, trees can increase or decrease energy use depending upon location; pollen can be positive in terms of food production or negative in terms of allergies depending upon species

• Attribute currently quantified or valued in i-Tree

o Attribute in development in i-Tree

[a]Developing a health index based on mapping of green viewing ("forest bathing")

[b]Estimating product potential based on forest structure (e.g., timber, wood pellets, ethanol)

Other tools in i-Tree include,

- *i-Tree Species*: This tool selects the most appropriate tree species based on desired environmental functions and geographic area.
- *i-Tree Hydro*: This tool simulates the effects of changes in tree and impervious cover on runoff, stream flow, and water quality.
- *i-Tree Canopy*: This tool allows users to easily photo-interpret Google aerial images to produce statistical estimates of land cover types. Use of historical imagery in Google Earth can also be used to aid in change analyses of land cover types.

- *i-Tree Design*: This tool links to Google Maps and allow users to quantify the current and future benefits of trees on their property.
- *MyTree*: This tool easily assesses the benefits of one to a few trees using a phone via a mobile web browser.
- *i-Tree Landscape*: This tool allows users to explore tree canopy, land cover, tree benefits, forest and health risks, and basic demographic information anywhere in the United States and prioritize areas for tree planting and protection.

Many new forest benefits and costs are currently being added to the model (Table 10.4). i-Tree is developed using a collaborative effort among numerous partners to better understand and quantify how changes in forest structure will affect numerous benefits and values, and to aid in urban forest management and planning.

10.6 Conclusion

Urban vegetation provides numerous benefits to society regarding physical, mental, and environmental health. Many benefits and costs remain to be quantified, but science and science-based tools are aiding our understanding of the myriad of vegetation benefits. By understanding these benefits and how vegetation affect these benefits, urban systems can be better engineered using plants and other natural elements and processes to help improve human and environmental health for current and future generations.

Disclaimer The use of trade names in this chapter is for the information and convenience of the reader. Such does not constitute an official endorsement or approval by the US Department of Agriculture or Forest Service of any product or service to the exclusion of others that may be suitable.

References

Albaugh, J.M., Dye, P.J., King, J.S. 2013. Eucalyptus and Water Use in South Africa. International Journal of Forestry Research, 2013: Article 852540, 11 p.

Anderson, L.M., Mulligan, B.E., Goodman, L.S. 1984. Effects of vegetation on human response to sound, Journal of Arboriculture. 10(2): 45–49.

Annavarapu, R.N., Kathi, S. 2016. Cognitive disorders in children associated with urban vehicular emissions. Environmental Pollution, 208:74–78.

Aylor, D.E., 1972, Noise reduction by vegetation and ground, J. Acoust. Soc. Am. 51(1):197–205.

Baldocchi, D. 1988. A multi-layer model for estimating sulfur dioxide deposition to a deciduous oak forest canopy. Atmospheric Environment. 22: 869–884.

Baldocchi, D.D., Hicks, B.B., Camara, P. 1987. A canopy stomatal resistance model for gaseous deposition to vegetated surfaces. Atmospheric Environment. 21: 91–101.

Basu, R. 2009. High ambient temperature and mortality: a review of epidemiologic studies from 2001 to 2008. Environmental Health, 8:40 doi:https://doi.org/10.1186/1476-069X-8-40.

Basu, R., Ostro, B.D. 2008. A Multicounty Analysis Identifying the Populations Vulnerable to Mortality Associated with High Ambient Temperature in California. American Journal of Epidemiology, 168:632–637.

Bidwell, R.G.S., Fraser, D.E. 1972. Carbon monoxide uptake and metabolism by leaves. Canadian Journal of Botany. 50: 1435–1439.

Bratkovich, S., Fernholz, K. 2010. Urban Wood Utilization and Industrial Clusters: A Twin Cities Case Study. Dovetail Partners, Inc. http://www.dovetailinc.org/report_pdfs/2010/werc63010finalreportsm.pdf (accessed March 2019)

Brauer, M. 2015. Air pollution, stroke, and anxiety: Particulate air pollution is an emerging risk factor for an increasing number of common conditions. BMJ (Online) 350.

Bytnerowicz A., Fenn M.E., Miller P.R., Arbaugh, M.J. 1999. Wet and dry pollutant deposition to the mixed conifer forest. In: Oxidant Air Pollution Impacts in the Montane Forests of Southern California: A Case Study of the San Bernardino Mountains (Eds. Miller PR & McBride JR) Springer-Verlag. New York. pp. 235–269.

Calderón-Garcidueñas, L., Engle, R., Mora-Tiscareño, A., Styner, M., Gómez-Garza, G., Zhu, H., Jewells, V., Torres-Jardón, R., Romero, L., Monroy-Acosta, M.E., Bryant, C., González-González, L.O., Medina-Cortina, H., and D'Angiulli, A. 2011. Exposure to severe urban air pollution influences cognitive outcomes, brain volume and systemic inflammation in clinically healthy children. Brain and Cognition 77(3):345–55.

Calfapietra, C., Fares, S., Loreto, F. 2009. Volatile organic compounds from Italian vegetation and their interaction with ozone. Environ. Pollut. 157:1478–1486.

Cardelino, C.A., Chameides, W.L. 1990. Natural hydrocarbons, urbanization, and urban ozone. J. Geophys. Res. 95(D9):13,971–13,979.

Cariñanos, P., Casares-Porcel, M.,Quesada-Rubio, J.M. 2014. Estimating the allergenic potential of urban green spaces: A case-study in Granada, Spain. Landscape and Urban Planning 123:134–144.

Cavanagh J.E., Zawar-Reza P., Wilson J.G. 2009. Spatial attenuation of ambient particulate matter air pollution within an urbanised native forest patch. Urban Forestry and Urban Greening 8:21–30.

Cesa, E.T., Lempicki, E.A., Knotts, J.H. 2003. Recycling Municipal Trees: A Guide for Marketing Sawlogs from Street Tree Removals in Municipalities. NA-TP-02-94. USDA Forest Service, Northeastern Area, State and Private Forestry, Morgantown, WV. 49p.

Chang, C., Li, M., Chang, S. 2007. A preliminary study on the cool-Island intensity of Taipei city parks. Landscape and Urban Planning, 80, 386–395.

Chaparro, L., Terradas, J. 2009. Ecological services of urban forest in Barcelona. Bellaterra Spain, Centre de Recerca Ecològica i Aplicacions Forestals, Universitat Autònoma de Barcelona.

Clark, K.H., Nicholas, K.A. 2013. Introducing urban food forestry: a multifunctional approach to increase food security and provide ecosystem services. Landscape Ecology, 28:1649–1669

Cohen, A.J., Brauer, M., Burnett, R., Anderson, H.R., Frostad, J., Estep, K., Balakrishnan, K., Brunekreef, B., Dandona, L., Dandona, R., Feigin, V., Freedman, G., Hubbell, B., Jobling, A., Kan, H., Knibbs, L., Liu, Y., Martin, R., Morawska, L., Pope, C.A, Shin, H., Straif, K., Shaddick, G., Thomas, M., van Dingenen, R., van Donkelaar, A., Vos, T., Murray, C.J.L., Forouzanfar, M.H. 2017. Estimates and 25-year trends of the global burden of disease attributable to ambient air pollution: an analysis of data from the Global Burden of Diseases Study 2015. The Lancet 389(10082):1907–1918

Compton, J.L. 2016. Evolution of the "parks as lungs" metaphor: is it still relevant? World Leisure Journal https://doi.org/10.1080/16078055.2016.1211171

Cook, D.I. 1978. Trees, solid barriers, and combinations: Alternatives for noise control, in: Proc. of the Natl. Urban For. Conf., (G. Hopkins, ed.), SUNY College of Environmental Science and Forestry, Syracuse, NY. pp. 330–339.

Cook, D.I., Van Haverbeke, D.F. 1971. Trees and shrubs for noise abatement, Res. Bull. 246. Nebraska Agricultural Experiment Station, Lincoln.

Costanza, R., de Groot, R., Sutton, P., van der Ploeg, S., Anderson, S.J., Kubiszewski, I., Farber, S., Turner, R.K. 2014. Changes in the global value of ecosystem services. Global Environmental Change, 26, 152–158.

Crowther, T.W., Glick, H.B., Covey, K.R., Bettigole, C., Maynard, D.S., Thomas, S.M., Smith, J.R., Hintler, G., Duguid, M.C., Amatulli, G., Tuanmu, M.N., Jetz, W., Salas, C., Stam, C., Piotto, D., Tavani, R., Green, S., Bruce, G., Williams, S.J., Wiser, S.K., Huber, M.O., Hengeveld, G.M., Nabuurs, G.J., Tikhonova, E., Borchardt, P., Li, C.F., Powrie, L.W., Fischer, M., Hemp, A., Homeier, J., Cho, P., Vibrans, A.C., Umunay, P.M., Piao, S.L., Rowe, C.W., Ashton, M.S., Crane, P.R., Bradford, M.A. 2015. Mapping tree density at a global scale. Nature, 525:201–205.

Dasch J.M. 1987. Measurement of dry deposition to surfaces in deciduous and pine canopies. Environmental Pollution 44:261–277.

de Vries S., van Dillen S.M.E., Groenewegen P.P., Spreeuwenberg P. 2013. Streetscape greenery and health: Stress, social cohesion and physical activity as mediators. Social Science & Medicine, 94(0):26–33.

Dematte, J.E., O'Mara, K., Buescher, J., Whitney, C.G., Forsythe, S., McNamee, T., Adiga, R.B., Ndukwu, I.M. 1998. Near-Fatal Heat Stroke during the 1995 Heat Wave in Chicago. Annals of Internal Medicine, 129(3): 173–181.

Egorov, A.I., Griffin, S.M., Converse, R.R., Styles, J.N., Sams, E.A., Wilson, A., Jackson, L.E., Wade, T.J. 2017. Vegetated land cover near residence is associated with reduced allostatic load and improved biomarkers of neuroendocrine, metabolic and immune functions. Environ Res., 158:508–521.

Escobedo, F.J, Nowak, D.J., Wagner, J.E., Luz de la Maza, C., Rodriguez, M. 2006. The socioeconomics and management of Santiago de Chile's public urban forest. Urban Forestry and Urban Greening. 4: 105–114.

Fann, N., Lamson, A.D., Anenberg, S.C., Wesson, K., Risley, D., Hubbell, B. 2012. Estimating the national public health burden associated with exposure to ambient PM2.5 and ozone. Risk Analysis 32:81–95.

Gasparrini, A., Guo, Y., Sera, F., Vicedo-Cabrera, A.M., Huber, V., Tong, S., M. Coelho, P. Saldiva, E. Lavigne, P. Correa, N. Ortega, H. Kan, S. Osorio, J. Kyselý, A. Urban, J. Jaakkola, N. Ryti, M. Pascal, P. Goodman, A. Zeka, P. Michelozzi, M. Scortichini, M. Hashizume, Y. Honda, M. Hurtado-Diaz, J. Cruz, X. Seposo, H. Kim, A. Tobias, C. Iñiguez, B. Forsberg, D. Åström, M. Ragettli, Y, Guo, C. Wu, A. Zanobetti, J. Schwartz, M. Bell, T. Dang, D. Van, C. Heaviside, S. Vardoulakis, S. Hajat, A. Haines, B. Armstrong. 2017. Projections of temperature-related excess mortality under climate change scenarios. Lancet Planet Health, https://doi.org/10.1016/S2542-5196(17)30156-0

Geron, C.D., Guenther, A.B., Pierce, T.E. 1994. An improved model for estimating emissions of volatile organic compounds from forests in the eastern United States. J. Geophys. Res. 99(D6):12773–12791.

Graca, M.S., Goncalves, J.F., Alves, P.J., Nowak, D.J., Hoehn, R., Ellis, A., Farinha-Marques, P, Cunha, M. 2017. Assessing mismatches in ecosystem services proficiency across the urban fabric of Porto (Portugal): the influence of structural and socioeconomic variables. Ecosystem Services. 23: 82–93.

Gromke C., B. Ruck. 2009. On the impact of trees on dispersion processes of traffic emissions in street canyons. Boundary-Layer Meteorology 131(1):19–34.

Guenther, A. 1997. Seasonal and spatial variation in natural volatile organic compound emissions. Ecological Applications. 7(1): 34–45.

Guenther A. 2002. The contribution of reactive carbon emissions from vegetation to the carbon balance of terrestrial ecosystems. Chemosphere; 49: 837–844.

Guenther, A., Hewitt, C.N., Erickson, D., Fall, R., Geron, C., Graedel, T., Harley, P., Klinger, L., Lerdau, M., McKay, W.A., Pierce, T., Scholes, B., Steinbrecher, R., Tallamraju, R., Taylor, J., Zimmerman, P. 1995. A global model of natural volatile organic compound emissions. Journal of Geophysical Research. 100 (D5): 8873–8892.

Harlan, S.L., Ruddell, D.M. 2011. Climate change and health in cities: impacts of heat and air pollution and potential co-benefits from mitigation and adaptation. Current Opinion in Environmental Sustainability, 3:126–134.

Hayden, F.G. 1989. Survey of methodologies for valuing externalities and public goods. EPA-68-01-7363, Environmental Protection Agency, Washington, DC.

Heisler, G.M. 1986. Energy savings with trees. Journal of Arboriculture, 12(5):113–125.

Heisler, G.M., Grant, R.H. 2000. Ultraviolet radiation in in urban ecosystems with consideration of effects on human health. Urban Ecosystems, vol 4, pp. 193–229.

Hirabayashi, S., Kroll, C., Nowak, D. 2011. Component-based development and sensitivity analyses of an air pollutant dry deposition model. Environmental Modeling and Software. 26:804–816.

Hosker, R.P., Lindberg, S.E. 1982. Review: atmospheric deposition and plant assimilation of gases and particles. Atmospheric Environment. 16(5): 889–910.

Ilstedt, U., Bargués Tobella, A., Bazié, H.R., Bayala, J., Verbeeten, E., Nyberg, G., Sanou, J., Benegas, L., Murdiyarso, D., Laudon, H., Sheil, D., Malmer, A. 2016. Intermediate tree cover can maximize groundwater recharge in the seasonally dry tropics. Scientific Reports: 6: Article number: 21930

Interagency Working Group on Social Cost of Carbon, United States Government. 2015. Technical Support Document: Social Cost of Carbon for Regulatory Impact Analysis under Executive Order 12866. Available online at https://www.whitehouse.gov/sites/default/files/omb/inforeg/scc-tsd-final-july-2015.pdf; last accessed Jan. 19, 2017.

Keys, P.W., Wang-Erlandsson, L., Gordon, L.J., Galaz, V., Ebbesson, J. 2017. Approaching moisture recycling governance. Global Environmental Change 45:15–23.

Killus, J.P., Meyer, J.P., Durran, D.R., Anderson, G.E., Jerskey, T.N., Reynolds, S.D., Ames, J. 1984. Continued research in mesoscale air pollution simulation modeling. Volume V: refinements in numerical analysis, transport, chemistry, and pollutant removal. Publ. EPA/600/3-84/095a. Research Triangle Park, NC: U.S. Environmental Protection Agency.

Kondo, M.C., Fluehr, J.M., McKeon, T., Branas, C.C. 2018. Urban green space and its impact on human health. International Journal of Environmental Research and Public Health, 15, 445; doi:https://doi.org/10.3390/ijerph15030445

Kramer, P.J., Kozlowski, T.T. 1979. Physiology of Woody Plants. Academic Press, New York.

Laurie M. 1986. An Introduction to Landscape Architecture. New York: Elsevier 248 p.

Lerman, S.B, Nislow, K.H., Nowak, D.J. DeStefano, S. King, D.I., Jones-Farrand, D.T. 2014. Using urban forest assessment tools to model bird habitat potential. Landscape and Urban Planning, vol 122, pp. 29–40.

Little, P. 1977. Deposition of 2.75, 5.0, and 8.5 mm particles to plant and soil surfaces. Environmental Pollution 12:293–305.

Lovett, G.M. 1994. Atmospheric deposition of nutrients and pollutants in North America: an ecological perspective. Ecological Applications. 4: 629–650.

Luley, C.J., Bond, J. 2002. A plan to integrate management of urban trees into air quality planning. Report to Northeast State Foresters Association. Davey Resource Group, Kent, OH. 73 p.

Marino, E., Caruso, M., Campagna, D., Polosa, R. 2015. Impact of air quality on lung health: myth or reality? Therapeutic Advances in Chronic Disease 6(5) 286–298.

McPherson, E.G., Simpson, J.R. 1999. Carbon Dioxide Reduction through Urban Forestry: Guidelines for Professional and Volunteer Tree Planters. PSW-GTR-171. Berkeley, CA: U.S. Department of Agriculture, Forest Service, Pacific Southwest Research Station.

Millennium Ecosystem Assessment, Ecosystems and Human Well- Being: Synthesis. 2005. Washington, DC: Island Press. 137 p.

Morales, D.J. 1980. The contribution of trees to residential property value. Journal of Arboriculture, 6: 305–308.

Morani, A., Nowak, D., Hirabayashi, S., Guidolotti, G., Medori, M., Muzzini, V., Fares, S., Scarascia Mugnozza, G., Calfapietra, C. 2014. Comparing modeled ozone deposition with field measurements in a periurban Mediterranean forest. Environmental Pollution 195: 202–209.

Myrup, L.O., McGinn, C.E., Flocchini, R.G. 1991. An analysis of microclimate variation in a suburban environment, in: Seventh Conference on Applied Climatology, American Meteorological Society, Boston, MA, pp. 172–179.
Na, H.R., Heisler, G.M., Nowak, D.J., Grant, R.H. 2014. Modeling of urban trees' effects on reducing human exposure to UV radiation in Seoul, Korea. Urban Forestry and Urban Greening, 13:785–792.
National Acid Precipitation Assessment Program (NAPAP). 1991. 1990 integrated assessment report. Washington, DC: National Acid Precipitation Assessment Program. 520 pp.
Natural Capital Project. 2016. InVEST: Integrated Valuation of ecosystem services and tradeoffs. http://www.naturalcapitalproject.org/invest/ accessed January 2016.
Nowak, D.J. 1994. Air pollution removal by Chicago's urban forest. In: McPherson, E.G, D.J. Nowak and R.A. Rowntree. Chicago's Urban Forest Ecosystem: Results of the Chicago Urban Forest Climate Project. USDA Forest Service General Technical Report NE-186. pp. 63–81.
Nowak, D.J. 2012. Contrasting natural regeneration and tree planting in 14 North American cities. Urban Forestry and Urban Greening. 11: 374– 382.
Nowak, D.J., Dwyer, J.F. 2007. Understanding the benefits and costs of urban forest ecosystems. In: Kuser, J. (ed.) Urban and Community Forestry in the Northeast. Springer, New York. p. 25–46.
Nowak, D.J., Greenfield, E.J. 2018. U.S. urban forest statistics, values and projections. Journal of Forestry, 116(2):164–177.
Nowak, D.J., Greenfield, E.J. 2020. The increase of impervious cover and decrease of tree cover within urban areas globally (2012-2017). Urban Forestry and Urban Greening. 49: 126638.
Nowak, D.J., Rowntree, R.A., McPherson, E.G., Sisinni, S.M., Kerkmann, E., Stevens, J.C. 1996. Measuring and analyzing urban tree cover. Lands. Urban Plann. 36:49–57.
Nowak, D.J., McHale, P.J., Ibarra, M., Crane, D., Stevens, J., Luley, C. 1998. Modeling the effects of urban vegetation on air pollution. In: Gryning, S.E. and N. Chaumerliac (eds.) Air Pollution Modeling and Its Application XII. Plenum Press, New York. pp. 399–407.
Nowak, D.J., Civerolo, K.L., Rao, S.T., Sistla, G., Luley, C.J., Crane, D.E. 2000. A modeling study of the impact of urban trees on ozone. Atmos. Environ. 34:1610–1613.
Nowak, D.J., Crane, D.E., Stevens, J.C., Ibarra, M. 2002a. Brooklyn's Urban Forest. USDA Forest Service Gen. Tech. Rep. NE-290. 107p.
Nowak, D.J., Stevens, J.C., Sisinni, S.M., Luley, C.J. 2002b. Effects of urban tree management and species selection on atmospheric carbon dioxide. Journal of Arboriculture. 28(3):113–122.
Nowak, D.J., Kurodo, M., Crane, D.E. 2004. Urban tree mortality rates and tree population projections in Baltimore, Maryland, USA. Urban Forestry and Urban Greening. 2(3):139–147.
Nowak, D.J., Crane, D.E., Stevens, J.C. 2006a. Air pollution removal by urban trees and shrubs in the United States. Urban Forestry and Urban Greening. 4:115–123.
Nowak, D.J., Hoehn, R., Crane, D.E., Stevens, J.C., Walton, J.T. 2006b. Assessing urban forest effects and values: Casper, WY's urban forest. USDA Forest Service, Northern Research Station Resource Bulletin NRS-4. Newtown Square, PA. 20 p.
Nowak, D.J., Hoehn, R., Crane, D.E., Stevens, J.C., Walton, J.T. 2006c. Assessing urban forest effects and values: Minneapolis' urban forest. USDA Forest Service, Northeastern Research Station Resource Bulletin NE-166. Newtown Square, PA. 20 p.
Nowak, D.J., Hoehn, R., Crane, D.E., Stevens, J.C., Walton, J.T. 2006d. Assessing urban forest effects and values: Washington D.C.'s urban forest. USDA Forest Service, Northern Research Station Resource Bulletin NRS-1. Newtown Square, PA. 24 p.
Nowak, D.J., Hoehn, R.H., Crane, D.E. 2007. Oxygen production by urban trees in the United States. Arboriculture and Urban Forestry. 33(3):220–226.
Nowak, D.J., Hoehn, R.E., Crane, D.E., Stevens, J.C., Walton, J.T., Bond, J. 2008. A ground-based method of assessing urban forest structure and ecosystem services. Arboric. Urb. For. 34(6): 347–358.
Nowak, D.J., Hoehn, R., Crane, D.E., Stevens, J.C., Cotrone, V. 2010a. Assessing urban forest effects and values: Scranton's urban forest. USDA Forest Service, Northern Research Station Resource Bulletin NRS-43. Newtown Square, PA. 23 p.

Nowak, D.J., Hoehn, R., Crane, D.E., Stevens, J.C., LeBlanc, C. 2010b. Assessing urban forest effects and values: Chicago's urban forest. USDA Forest Service, Northern Research Station Resource Bulletin NRS-37. Newtown Square, PA. 27 p.
Nowak, D.J., Hoehn, R., Crane, D.E., Weller, L., Davila, A. 2011. Assessing urban forest effects and values: Los Angeles's urban forest. USDA Forest Service, Northern Research Station Resource Bulletin NRS-47. Newtown Square, PA. 30 p.
Nowak, D.J., Hoehn, R., Crane, D.E., Cumming, J., Mohen, S., Buckelew-Cumming, A. 2012. Assessing urban forest effects and values: Morgantown's urban forest. USDA Forest Service, Northern Research Station Resource Bulletin NRS-70. Newtown Square, PA. 24 p.
Nowak, D.J., Hirabayashi, S., Bodine, A., Hoehn, R. 2013a. Modeled PM2.5 removal by trees in ten U.S. cities and associated health effects. Environmental Pollution. 178: 395–402.
Nowak, D.J., Hoehn, R.E., Bodine, A.R., Greenfield, E.L., Ellis, A., Endreny, T.E., Yang, Y., Zhou, T., Henry, R. 2013b. Assessing Forest Effects and Values: Toronto's Urban Forest USDA Forest Service, Northern Research Station Resource Bulletin NRS-79. Newtown Square, PA. 59 p.
Nowak, D.J., Hirabayashi, S., Ellis, A., Greenfield, E.J. 2014. Tree and forest effects on air quality and human health in the United States. Environmental Pollution 193:119–129.
Nowak, D.J., Jovan, S., Branquinho, C., Augusto, S., Ribeiro, M.C., Kretsch, C.E. 2015. Chapter 4: Biodiversity, air quality and human health. In: Connecting global priorities: biodiversity and human health: a state of knowledge review. World Health Organization, Geneva, Switzerland. p. 63–74.
Nowak, D.J., Hoehn, R.E., Bodine, A.R., Greenfield, E.J., O'Neil-Dunne, J. 2016a. Urban Forest Structure, Ecosystem Services and Change in Syracuse, NY. Urban Ecosystems. 19:1455–1477.
Nowak, D.J., Bodine, A.R., Hoehn, R.E., Edgar, C.B., Hartel, D.R., Lister, T.W., Brandeis, T.J. 2016b. Austin's Urban Forest, 2014. USDA Forest Service, Northern Research Station Resources Bulletin. NRS-100. Newtown Square, PA. 55 p.
Nowak, D.J., Bodine, A.R., Hoehn, R.E., Low, S.C., Roman, L.A., Henning, J.G., Stephan, E., Taggart, T., Endreny, T. 2016c. The Urban Forest of Philadelphia. Resource Bulletin NRS-106. Newtown Square, PA: U.S. Department of Agriculture, Forest Service, Northern Research Station. 80 p.
Nowak, D.J., Bodine, A.R., Hoehn, R.E., Edgar, C.B., Riley, G., Hartel, D.R., Dooley, K.J., Stanton, S.M., Hatfield, M.A., Brandeis, T.J., Lister, T.W. 2017. Houston's Urban Forest, 2015. USDA Forest Service, Southern Research Station Resources Bulletin. SRS-211. Newtown Square, PA. 91 p.
Nowak, D.J., Bodine, A.R., Hoehn, R.E., Ellis, A., Hirabayashi, S., Coville, R., Auyeung, D.S.N., Falxa Sonti, N., Hallett, R.A., Johnson, M.L., Stephan, E. 2018. The Urban Forest of New York City. Resource Bulletin NRS-RB-117. Newtown Square, PA: U.S. Department of Agriculture, Forest Service, Northern Research Station. 82 p.
Nowak, D.J., Coville, R., Endreny, T., Abdi, R., Van Stan, J. 2020. Chapter 15: Valuing Urban Tree Impacts on Precipitation Partitioning. In: Van Stan, J, Gutmann, E., Friesen, J. (eds.) Precipitation Partitioning by Vegetation: A Global Synthesis, Springer. 253–268
Nowak, D.J., Greenfield, E.J. 2020. The increase of impervious cover and decrease of tree cover within urban areas globally (2012–2017) Urban Forestry & Urban Greening. 49: 126638
Numerical Terradynamic Simulation Group. 2016. Biome-BGC. University of Montana. http://www.ntsg.umt.edu/project/biome-bgc accessed January 2016.
Ogren, T.L. 2000. Allergy-free gardening. Ten Speed Press, Berkeley, CA. 267 p.
Pederson, J.R., Massman, W.J., Mahrt, L., Delany, A., Oncley, S., den Hartog, G., Neumann, H.H., Mickle, R.E., Shaw, R.H., Paw U, K.T., Grantz, D.A., MacPherson, J.I., Desjardins, R., Schuepp, P.H., Pearson, R., Arcado, T.E. 1995. California ozone deposition experiment: methods, results, and opportunities. Atmospheric Environment. 29(21): 3115–3132.
Pope, C.A., Burnett, R.T., Thun, M.J., Calle, E.E., Krewski, D., Ito, K., Thurston, G.D. 2002. Lung cancer, cardiopulmonary mortality, and long-term exposure to fine particulate air pollution. Journal of the American Medical Association 287(9):1132–1141.
Powe, N.A., Willis, K.G. 2004. Mortality and morbidity benefits of air pollution (SO2 and PM10) absorption attributable to woodland in Britain. J. Environ. Manage. 70: 119–128.

Reethof, G., McDaniel, O.H. 1978. Acoustics and the urban forest, in: Proc. of the Natl. Urban For. Conf., (G. Hopkins, ed.), SUNY College of Environmental Science and Forestry, Syracuse, NY, pp. 321–329.

Robine, J.M., Cheung, S.L., Le Roy, S., Van Oyen, H., Griffiths, C., Michel, J.P., Herrmann, F.R. 2008. Death toll exceeded 70,000 in Europe during the summer of 2003. C. R. Biologies, 331:171–178.

Robinette, G.O. 1972. Plants/People/ and Environmental Quality, USDI National Park Service, Washington, DC.

Rogers, K., Sacre, K., Goodenough, J., Doick, K. 2015. Valuing London's urban forest: results of the London i-Tree Eco Project. London, Treeconomics.

Salisbury, F.B., Ross, C.W. 1978. Plant Physiology. Wadsworth Publishing Company, Belmont, CA. 422 pp.

Sallis, J.F., Cerin, E., Conway, T.L., Adams, M.A., Frank, L.D., Pratt, M., Salvo, D., Schipperijn, J., Smith, G., Cain, K.L., Davey, R., Kerr, J., Lai, P.C., Mitáš, J., Reis, R., Sarmiento, O.L., Schofield, G., Troelsen, J., Van Dyck, D., De Bourdeaudhuij, I., Owen, N. 2016. Physical activity in relation to urban environments in 14 cities worldwide: a cross-sectional study. Lancet, 28:387(10034):2207–17.

Salmond, J.A., Williams, D.E., Laing, G., Kingham, S., Dirks, K., Longley, I., Henshaw, G.S. 2013. The influence of vegetation on the horizontal and vertical distribution of pollutants in a street canyon. Science of the Total Environment 443: 287–298.

Saphores, J., Li, W. 2012. Estimating the value of urban green areas: A hedonic pricing analysis of the single family housing market in Los Angeles, CA. Landscape and Urban Planning, 104: 373–387.

Selmi, W., Weber, C., Rivière, E., Blond, N., Mehdi, L., Nowak, D. 2016. Air pollution removal by trees in public greenspace in Strasbourg city, France. Urban Forestry and Urban Greening 17: 192–201.

Sharkey, T.D., Holland, E.A, Mooney, H.A. eds. 1991. Trace gas emissions by plants. New York: Academic Press. 365 p.

Smith, W.H. 1990. Air pollution and forests. New York: Springer-Verlag. 618 p.

Taha, H. 1996. Modeling impacts of increased urban vegetation on ozone air quality in the South Coast Air Basin. Atmos. Environ. 30(20):3423–3430.

Tallamy, D.W., Shropshire, K.J. 2009. Ranking Lepidopteran use of native versus introduced plants. Conservation Biology, vol 23, pp. 941–947.

Thompson, R., Hanna, R., Noel, J., Piirto, D. 1999. Valuation of tree aesthetics on small urban-interface properties. Journal of Arboriculture, 25(5): 225–234.

Tingey, D.T., Turner, D.P., Weber, J.A. 1991. Factors controlling the emissions of monoterpenes and other volatile organic compounds, in: Trace Gas Emissions by Plants. (T.D. Sharkey, E.A. Holland, H.A. Mooney, eds.), Academic Press, New York, pp. 93–119.

Tiwary, A., Sinnett, D., Peachey, C., Chalabi, Z., Vardoulakis, S., Fletcher, T., Leonardi, G., Grundy, C., Azapagic, A., Hutchings, T.R. 2009. An integrated tool to assess the role of new plantings in PM10 capture and the human health benefits: A case study in London. Environmental Pollution 157: 2645–2653.

U.S. Environmental Protection Agency (US EPA). 2010. Our nation's air: status and trends through 2008. EPA-454 / R-09-002. Office of Air Quality Planning and Standards. Triangle Park, NC. 49 p.

U.S. Environmental Protection Agency (US EPA). 2017. Air Emission Modeling: Biogenic Emissions Inventory System (BEIS). https://www.epa.gov/air-emissions-modeling/biogenic-emission-inventory-system-beis [accessed December 2017]

U.S. Forest Service. 2016. Carbon: tools for carbon inventory, management, and reporting [online]. United States Department of Agriculture Forest Service. [Cited January 2016]. www.nrs.fs.fed.us/carbon/tools

Vieira, S. 2015. The health burden of pollution: the impact of prenatal exposure to air pollutants. International Journal of COPD 10:1111–1121.

Vos, P.E.J., Maiheu, B., Vankerkom, J., Janssen, S. 2013. Improving local air quality in cities: To tree or not to tree? Environmental Pollution 183: 113–122.

Wania, A., Bruse, M., Blond, N., Weber, C. 2012. Analysing the influence of different street vegetation on traffic-induced particle dispersion using microscale simulations. Journal of Environmental Management, 94: 91–101.

Wesely, M.L. 1989. Parameterization for surface resistance to gaseous dry deposition in regional-scale numerical models. Atmospheric Environment 23, 1293–1304.

World Health Organization. 2014. 7 million premature deaths annually linked to air pollution. Geneva (Switzerland): World Health Organization World Health Organization, Geneva. http://www.who.int/mediacentre/news/releases/2014/air-pollution/en/ (November, 2017).

World Health Organization. 2016. Ambient air pollution: a global assessment of exposure and burden of disease. Geneva (Switzerland): World Health Organization, 131 p.

Yang, X.E., Wu, X., Hao, H.L., He, Z.L. 2008. Mechanisms and assessment of water eutrophication. Journal of Zhejiang University. Science. B, 9(3): 197–209.

Yang Y., Endreny, T.A., Nowak, D.J. 2013. A physically-based local air temperature model. Journal of Geophysics Research-Atmospheres, 118: 1–15.

Zinke, P.J. 1967. Forest interception studies in the United States. In: Sopper, W.E.; Lull, H.W., eds. Forest hydrology. Oxford, UK: Pergamon Press: 137–161.

Chapter 11
Services from Agroecosystems and Their Quantification

Sami Khanal

11.1 Introduction

Agroecosystems are ecosystems on agricultural land that are primarily managed to optimize the production of food, fiber, and fuel. They cover nearly 40% of the total land area of the Earth (FAOSTAT 2016). These systems are both users and providers of ecosystem services (ES), which are the direct and indirect benefits that organisms receive from ecosystems. ES are critical for maintaining the conditions for life on Earth and are classified into four categories: provisioning, supporting, cultural, and regulating (MEA 2003) (Chap. 2). Some examples include food and fuel under provisioning services, flood and disease control under regulating services, recreational and spiritual under cultural services, and nutrient cycling and soil formation under supporting services.

11.1.1 Ecosystem Services Flowing to Agroecosystems

To maximize provisioning services, agroecosystems depend on a wide variety of supporting and regulating services provided by natural ecosystems, such as soil fertility, pollination, nutrient cycling, and water regulation. These supporting and regulating services vary across geographic locations depending upon weather, topography, and human management practices. For example, the Midwestern USA has highly fertile soils and relatively mild climatic conditions that support a higher production of corn and soybean. Western USA, on the other hand, has semiarid and

S. Khanal (✉)
Department of Food, Agricultural and Biological Engineering at The Ohio State University, Columbus, OH, USA
e-mail: khanal.3@osu.edu

B. R. Bakshi (ed.), *Engineering and Ecosystems*,
https://doi.org/10.1007/978-3-031-35692-6_11

arid climatic conditions that support fruit production. Some of the ES flowing to agroecosystems are discussed below:

Soil Fertility

Soils that are rich in organic matter and are well-aerated and well-drained supply essential nutrients to crops. Soil pore structure, soil aggregation, and decomposition of organic matter impact soil quality, and these are influenced by activities of bacteria, fungi, and macrofauna, such as earthworms, termites, and other invertebrates (Power 2010). Soil microorganisms produce different kinds of organic compounds that help to hold soil particles together, which help to control erosion and movement of water and nutrients. Soil fertility is largely influenced by agricultural management practices. For example, management practices such as plowing, disking, cultivating, and harvesting degrade soil structure and microbial communities, while management practices that incorporate crop residues, such as conservation tillage and plantation of cover crops, promote soil water retention and nutrient flows to crops.

Pollination

Pollination is the process of transferring pollen grains from the male parts of a flower to the female parts, which results in the fertilization of plant ovaries and the production of seeds. Pollination services are critical to agricultural production and are provided by a wide range of insect species, including bees, wasps, butterflies, beetles, and moths. Analysis of data from 200 countries indicated that 75% of crop species of global significance for food production rely primarily on insect pollination (Klein et al. 2007). Bees alone were estimated to contribute 11% of US agricultural gross domestic product in 2009, which equals to $14.6 billion per year. Of this, 20% is provided by wild pollinators that depend on suitable land for nesting and foraging (Koh et al. 2016). Despite the agricultural importance of pollinators, an abundance of multiple species is declining due to several factors, such as pesticide use, climate change, and habitat loss (Bartomeus et al. 2013).

Biological Pest Control

Biological control is the method of controlling pests (e.g., insects, mites, weeds, and plant disease) through predation, herbivory, parasitism, and other natural mechanisms by birds, spiders, flies, wasps, and microbial pathogens. This is another important ES provided to agroecosystems by natural ecosystems. This approach of controlling pest suppresses pest damage, improves yield, and reduces the need for pesticides. Economic valuation of these services has been estimated to be very large. For example, Losey and Vaughan (2006) estimated that natural pest control services save $13.6 billion per year in agricultural crops in the USA. Diversification of cropping practices helps to enhance natural enemies by providing diverse plant-derived resources such as pollen, nectar, and shelter. However, increased homogenization of agricultural production and the removal of natural habitat have limited the availability of this service.

Water Quantity and Quality

Availability of the right quantity and right quality of water at the right time is critical for agricultural production. Globally, agriculture accounts for 70% of freshwater withdrawals (FAO 2017). Natural ecosystems play a critical role in regulating water flow and its retention and infiltration. Trapping of water, nutrients, and sediments is controlled by plant cover, soil quality, and the amount of surface litter. For example, deep rooting plant species can improve the availability of both water and nutrients to other species in the ecosystem. Forest soils tend to have a higher infiltration rate than other soils, and forests tend to reduce peak flows and floods. With climate change, agriculture is expected to face increasing water shortages in the future. Increased variability in rainfall and increasing temperature are predicted to lead to a greater risk of drought and flood, as well as increased water demand. Some of the agricultural management practices such as mulching and conservation practices help improve water retention by reducing soil evaporation.

11.1.2 Ecosystem Disservices Flowing to Agroecosystems

In addition to these benefits, agroecosystems receive disservices from natural ecosystems. For example, crop pests and pathogens from natural ecosystems can result in crop damage and thus loss in revenue. Although the use of pesticide helps to reduce pest outbreak in the short term, it has been found that it can lead to overreliance on pesticides and emergence of certain species with genetic resistance to specific pesticide compounds, thereby triggering pest outbreaks in the long term (Zhang et al. 2007). Thus, chemical control can be costly and result in unintended negative outcomes for nontarget plants and public health.

The productivity of agroecosystems can also be affected by non-crop plants due to competition for resources. For example, in crop fields, competition of crops with weeds for sunlight, water, and nutrients can limit crops' access to required resources, thereby reducing crop growth. Trees can transpire water and reduce the recharge of aquifers, which can limit water availability for irrigation.

Climate is another important ES that can both positively and negatively impact the provisioning services of agroecosystems. Agricultural crops are sensitive to climate change, including changes in temperature, precipitation, and concentration of CO_2.

11.1.3 Ecosystem Services and Disservices from Agroecosystems

Agroecosystems are considered primarily as sources of provisioning services. However, depending upon their structure and management practices, they may contribute to a number of ES, such as pollination, soil retention, soil fertility, water quality and quantity, and nutrient cycling as described above, as well as disservices, such as biodiversity loss, water contamination, and greenhouse gas emissions (GHGe) (Fig. 11.1). For example, excessive use of fertilizer application and conventional tillage practices for agricultural production can result in GHGe, nutrient runoff, and sedimentation. However, the conservation agricultural practices, such as conservation tillage, cover crops, and riparian vegetation, help improve soil fertility, carbon sequestration, nutrient cycling, nutrient runoffs, and crop productivity.

The flow of services and disservices from agroecosystems to other ecosystems relies on rate and extent of intensification and expansion of agricultural practices. Oftentimes, there is a mismatch between benefits incurred to agricultural sectors and the costs that are typically borne by society at various geographic and time scales. For example, by adopting intensive farming practices, such as increased fertilizer uses and conventional tillage, farmers may increase crop production and

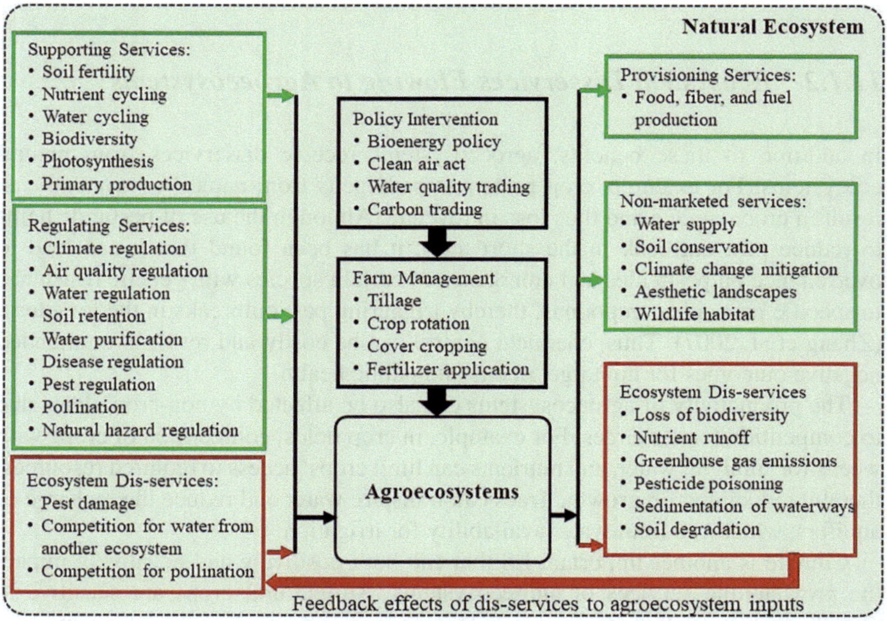

Fig. 11.1 Ecosystem services and disservices to and from agroecosystems. Red boxes and arrows indicate disservices and green boxes and arrows indicate services

reap benefits in the short term. However, those practices could lead to water and air pollution that could impact communities at local and regional scales. Some of the disservices from agroecosystems are discussed below.

Nutrient Pollution

Agricultural management practices have dramatically altered biogeochemical cycles and nutrient availability at various scales, ranging from field to watershed to regional to global. Nutrients, primarily nitrogen and phosphorus, play a critical role in agricultural production and thus are heavily applied in agroecosystems. As a result, these nutrients have entered ground and surface waters, resulting in several negative consequences for human health and the environment. For example, corn and soybean production within the Midwestern USA accounts for over 51% of the total nitrogen (N) load to the northern Gulf of Mexico (Costello et al. 2009). Impacts of increased nutrient concentration in ground and surface waters include water pollution, polluted drinking water, eutrophication, increased frequency and severity of algal blooms, hypoxia and fish kills, and loss of aquatic habitat. Other ecosystem disservices from agriculture include loss of biodiversity and pesticide residues in surface and groundwater due to applications of pesticides. Best management practices (BMPs) help minimize some of these adverse environmental effects while maintaining provisioning services. Nutrient BMPs that focus on application of right nutrients at the right rate, at the right timing, and at the right place are primary approaches to minimizing nutrients runoffs and losses from the agricultural fields. Additional BMPs, such as cover cropping, diverse crop rotations, conservation tillage, and buffer strips, reduce standing pools of nutrients that are most susceptible to loss by promoting plant uptake of nutrients.

Greenhouse Gas Emissions

The agricultural sector is the world's second largest emitter of GHGe, after the energy sector (which includes power generation and transport). In 2019, the agricultural sector contributed to 9.6% of the total US GHGe (US EPA 2021a). Additionally, within USA, they contribute to 76% of the total nitrous oxide (N_2O) emission (Costello et al. 2009). Agricultural activities contribute to emissions in several ways, including conversion of forest, pasture, and rangeland to cropland, application of nutrients to agricultural production, burning of crop residues, livestock production, and the use of machinery for farming activities. For example, deforestation or conversion of natural ecosystems to agriculture results in loss of aboveground carbon, and it is estimated that it also reduces soil carbon pool by 30–50% over 50–100 years in temperate regions and 50–75% over 20–50 years in tropics (Lal 2008). Application of nutrients can significantly increase the rate of N_2O emissions, particularly when more nitrogen is applied that can be taken up by the plants. Ruminant livestock, such as cattle, sheep, and goat, emit methane

(CH_4) as a byproduct of their digestive processes. Similarly, biological breakdown of livestock waste can release CH_4 and N_2O.

Agriculture can offset GHGe through a variety of processes, such as effective land management practices and manure management. Carbon uptake and storage in soils, i.e., carbon sequestration, can be increased using conservation measures, such as conservation tillage, no-till cultivation, and diverse crop rotations. These practices not only reduce nutrient runoffs but also maintain soil organic carbon. Integration of manure in crop production instead of synthetic nitrogen fertilizers can increase soil organic carbon and reduce GHGe. Less productive lands can be used for the production of perennial crops for bioenergy production. Compared to traditional cropping systems (i.e., corn and soybean), perennial crops offer numerous advantages including lower energy and nutrient inputs, greater soil carbon sequestration, reduced GHGe, and improved water quality (Davis et al. 2012; VanLoocke et al. 2017).

11.1.4 Ecosystem Services and Public Policy

ES are highly interdependent on one another and the relationships between them are nonlinear. They often vary spatially and temporally. Trade-offs among ES thus should be considered in terms of spatial and temporal scales. ES provided to agriculture occurs at various geographic scales, which can influence a farmer's incentive for the adoption of certain management practices for managing the ES. ES, such as soil fertility, soil water retention, and pest control, are provided at the field and farm scale, so farmers have a direct interest in managing these services. Other ES, such as air and water quality, occur at larger scales, which means that benefits are likely to accrue not only to farmers who expend their resources to support those services but also to others that do not expend effort or pay for the benefits – a free-rider problem that is commonly seen with non-excludable ES. Thus, there lack incentives to farmers to set aside the optimal amount of resources to manage these ES (Zhang et al. 2007).

To create incentives for farmers to maintain ES that accrue at larger landscapes, there exist several federal- and state-funded programs. For instance, the US Department of Agriculture (USDA) spends significant funds via programs, such as the Conservation Reserve Program (CRP) and Environmental Quality Incentive Program (EQIP), to provide financial and technical assistance to agricultural producers to plan and implement conservation practices. The 2008 Farm Bill incorporates a "Sodsaver" provision, which decreases crop insurance subsidy incentives for converting previously uncultivated land (Miao et al. 2016). To ensure reduction of GHGe associated with bioenergy production under the Renewable Fuel Standard (RFS), the Environmental Protection Agency (EPA) requires biofuels to achieve at least 20% GHGe reduction relative to conventional biofuels (corn ethanol) to qualify as a renewable fuel, 50% to qualify as advanced biofuels, and 60% to qualify as cellulosic biofuels (Schnepf and Yacobucci 2013). The RFS also contains a provision to prevent land-use change and its associated GHGe by explicitly excluding feedstocks sourced from rangeland or land converted to cropland after 2007 from qualifying for renewable credits (US EPA 2010). In the USA, there exist several voluntary emissions trading programs, such as carbon and water quality trading, to reduce air and water pollution (US EPA 2020, 2021b). Emission trading is an exchange of credits between emitters designed to reduce emissions. Under emission trading, farmers could maximize revenue through the implementation of management practices that help reduce GHGe and nutrient loads. Despite the presence of these emission trading programs across the country, the overall volume of trading however remains low, which could be attributed to a lack of technologies to quantify ES reliably, accurately, and cost effectively at the field level (*Sect.* 14.2 *for additional discussion*).

11.1.5 Functioning of Agroecosystems

The functions of ecosystem refer to the dynamic processes that control the movement of matter/materials and energy and the interactions between organisms and materials in the system. Understanding of these processes and relationships is essential in assessing efficiency, productivity, and development of agroecosystems, where function can determine the success and failure of a given crop or management practice. As agroecosystem involves human manipulation and alteration of an ecosystem for the purpose of agricultural production, it introduces several changes to the structure and functions of a natural ecosystem (Fig. 11.2). The two most fundamental processes in any ecosystem are energy and nutrient flows.

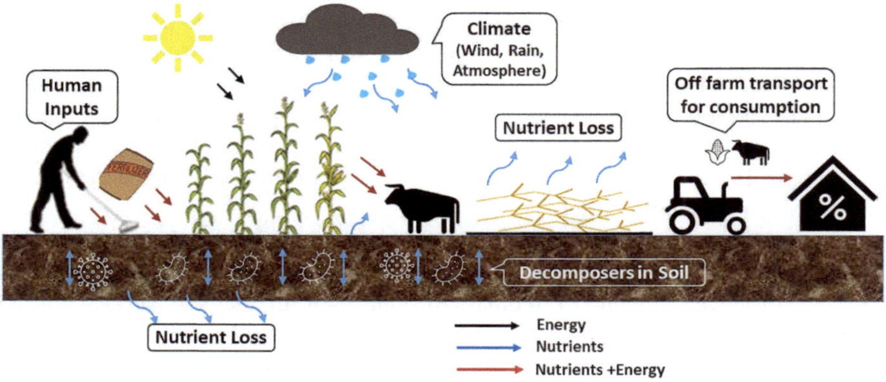

Fig. 11.2 Functional components of an agroecosystem and flow of energy and nutrients within those components

Energy Flow Solar radiation is the major source of energy in both natural ecosystems and agroecosystems. It is captured by plants and stored in the chemical bonds of plants' biomass. The total amount of energy accumulated by plants in a system in a given length of time can be determined by evaluating plant biomass. The rate at which solar energy is converted into biomass is called gross primary productivity, which is expressed in terms of kilocalories per square meter per year. When the energy plants use for photosynthesis is subtracted from gross primary productivity, it is called net primary productivity. Ecosystems vary in their ability to convert solar energy to biomass.

In agroecosystems, many energy inputs are derived from human-manufactured sources, which are unsustainable. For example, a high level of fossil fuel and labor is directed to operate machinery for planting, fertilization, and harvest. Biomass is harvested for consumption which otherwise accumulates within the system and contributes to important ecosystem processes (e.g., organic detritus returned to the soil, which serves as energy source for microorganisms that are essential for maintaining soil health) (Fig. 11.2). For sustainability of agroecosystems, the use of renewable sources of energy must be maximized to support and maintain interactions that provide ES.

Nutrient Cycling In addition to energy, organisms require nutrients to maintain and carry out living functions. Some of the important nutrients are carbon, nitrogen, oxygen, phosphorus, and water. Their availability to organisms varies depending on the form and the size of the reservoirs they are available at. Nutrients enter an ecosystem through several biogeochemical processes and are circulated within the ecosystem through several complex sets of interconnected cycles. For example, carbon easily moves between its abiotic form in the atmosphere and its biotic form in plant or animal matter as it cycles between the atmosphere as carbon dioxide and biomass as carbohydrates. Nitrogen in atmosphere however exists in less readily available form (i.e., N2) and needs to be converted to some other forms before they can be used by plants. Molecular nitrogen (N2) is converted into ammonia (NH3) through biological fixation by microorganisms, which makes nitrogen available to plants. Other nutrients such as phosphorus and sulfur are less mobile, and soil is their main abiotic reservoir. These nutrients are taken up by plant roots, stored in biomass for a certain time, and eventually returned to the soil through decomposition of biomass. In natural ecosystems, biomass productivity is closely linked to the rates at which nutrients are recycled.

In an agroecosystem, recycling of nutrients is generally minimal. A considerable amount of nutrients is lost from the system through leaching, erosion, and runoffs because of lower amount of biomass residues left in the soil surface during harvest. To replace those losses, modern agroecosystems thus rely heavily upon nutrients that are derived from petroleum-based sources. To achieve sustainability of agroecosystems, nutrient losses need to be reduced and recycling mechanisms need to be introduced. One of the recycling mechanisms includes integration of livestock and cropping systems, where crops are used to feed the livestock and nutrients are available in the form of manure through livestock.

Water Cycling Water flows in an ecosystem in various forms, such as rainfall, snowmelt, evapotranspiration from plants, and evaporation from water bodies. Plants require certain amount of water to be available to their roots in the soil for their functioning. Water helps plants to transport important nutrients from soil through its stems and leaves. In natural ecosystems, plants are adapted to certain moisture regimes that are set by climate and soil types.

In agroecosystems, whether in rainfed or in irrigated farming systems, water is one of the key factors that needs to be managed to enhance agricultural production. For example, in rainfed systems, if the soil is saturated by water, to ensure optimal water availability for crops, farmers use tile drainage systems to drain excess water. In irrigated system, if the soil is dry during the crop growing season, water is supplied to the soil through irrigation using surface water or groundwater sources. For sustainable use of water in agroecosystem, a farmer needs to be aware of various factors that influence water availability to plants, such as how water levels in the soil are influenced by weather conditions, cropping systems, and management practices and how much water is needed by crops.

11.2 Modeling and Measuring Agroecosystem Services

Quantifying agroecosystem services and disservices is a complicated task that requires a thorough understanding of the fundamental processes within the agroecosystem, including physical and biological processes. Historically, such understanding has been based on laboratory experiments and field measurements, which are often labor-intensive, expensive, and incomplete. Further, due to variability in soils, climate, land use, land cover, topography, and other site-specific attributes, ecosystem responses to external disturbances are spatially and temporally heterogeneous, making predictions of unobserved locations using existing field observations very challenging. Thus, ecosystem models have emerged as powerful alternatives to experimental science and observations when phenomena are not observable or when measurements are infeasible. Models help researchers to understand the dynamics of ecosystem processes and improve monitoring capabilities by filling spatiotemporal data gaps and predicting short- and long-term impacts of various management practices across scales.

An ecosystem model provides mathematical representations of one or several processes, such as carbon, water, and nutrient cycles that characterize the functioning of a given biological system. Although models are only an approximation of the real world and cannot reproduce all the biophysical processes occurring during crop growth, they have played important roles in the interpretation of ecosystem responses to various agricultural management practices.

Well-tested ecosystem models serve several important purposes, some of which are:

- Powerful alternatives to experimental science and observation for scientists and researchers to understand and reproduce the complex system of weather-soil-plant interactions, as well as to test scientific hypotheses when phenomena are not observable or when measurements are infeasible.
- Guide for decision-makers toward designing sustainable agricultural landscapes to produce more food, fuel, and fiber with reduced environmental impact under changing climate.
- Promising decision support tools for farmers to manage risk and opportunities through investment decisions on appropriate cultivars and practices under weather, pest, and disease hazards.

11.2.1 Types of Ecosystem Models

There exist numerous ecosystem models ranging from empirical to mechanistic, which vary from simple to complex in design, with different strengths and limitations. They usually differ in the complexity of the vegetation growth module and its interaction with the carbon, nutrient, and water cycles they represent.

Empirical Ecosystem Models These models are based on statistical relationships between environmental factors and response variables (e.g., crop yield can be one ES of interest), and they have been used for several decades now. The statistical equations produce responses in ES for different management practices. For example, Thompson (1986) used a statistical model to determine the impact of changes in climate and weather variability on corn production from 1891 to 1983 in five Corn Belt states in the USA. The study found preseason precipitation, June temperature, and temperature and rainfall in July and August to be closely related to corn yield variations. Lobell et al. (2011) used statistical models to determine the effects of weather on global corn and wheat production and concluded that temperature increase will play a large role in yield decrease under climate change. In general, empirical models are relatively easy and transparent to use. However, they are often site-specific, and their results cannot be extrapolated to other regions and time as variations in soils, landscapes, and weather information are not included in the pool of information from when the model was developed. Also, they don't capture the effects of spatial and temporal variability on ecosystem service at finer scales. Due to these reasons, they can be less flexible in handling variable management practices.

Process-Based or Mechanistic Ecosystem Models These models use mechanistic (i.e., process-oriented) equations that are developed based on substantial long-term research to represent fundamental mechanisms of plant growth, nutrient, water, and soil dynamics. These models are well represented by a large number of parameters, which are generally higher than those required by empirical models. Since these models are complex and often require a large dataset, they were developed soon after computers became available. With the recent developments in computing technologies, computational time required for running process-based models has shortened. This has also allowed incorporation of complex processes for improvement of process-based models. Process-based models can be used at local and regional scales with pros and cons, such as the following:

- Local-scale process-based model is useful for capturing fine-scale variability and dynamics of the ecosystem, but it requires significantly more site-level data inputs.
- Regional-scale process-based model covers area with similar soils and climate. This approach can be relatively simple, transparent, and low cost. However, it may not reflect the spatial and temporal variability of ES that are specific to a local site in the region.

In general, a relatively large amount of input data is required to run process-based models. Thus, the application of these models could be limited by data availability. Also, uncertainties in model inputs and process approximation make the model outcomes uncertain. For example, uncertainty in the rate and magnitude of changing climate variables at the local and regional scale is often compounded by uncertainty in the responses of ecological processes to these changes (Cuddington et al. 2013).

Functional or Hybrid Ecosystem Models These models use simplified approaches to simulate complex ecosystem processes; thus, they are of intermediate complexity. They mix both empirical and mechanistic approaches to represent the ecosystem processes they want to emphasize. For example, many hybrid ecosystem models use daily solar radiation as the amount of energy available for photosynthesis. The energy intercepted by plants is approximated using the information about leaf area index, which in turn helps approximate biomass production based on a simple concept of radiation use efficiency – the rate at which biomass is produced per unit of radiation intercepted (Basso et al. 2013).

11.2.2 A Review of Ecosystem Models for Quantifying Agroecosystem Services

Concerns regarding food security and environmental degradation under changing climate have favored the development and applications of various ecosystem models to explore and understand the agricultural practices that enhance agricultural production while lowering adverse environmental impacts. Literature offers development and application of several process models that operate either at local (i.e., field) or regional/watershed or global scale, some of which are focused on crop yield prediction, while others are focused on simulation of GHG fluxes and water quality and quantity.

Some of the crop simulation models that operate at a field scale include Crop Environment Resource Synthesis (CERES), Decision Support System for Agrotechnology Transfer (DSSAT), Agricultural Land Management Alternative with Numerical Assessment Criteria (ALMANAC), and Environmental Policy

Integrated Climate Model (EPIC). Although some of these models share similar crop and soil modules, they are designed for a specific purpose. For instance, CERES model was developed to simulate crop growth in response to climate, soil, genotypes, and management practices across locations throughout the world. As such, there are several forms of CERES model for several crop species, such as maize, rice, wheat, and soybean (Hodges et al. 1987; Otter and Ritchie 1985; Yao et al. 2007). These models were initially developed as stand-alone applications to simulate grain yield, but were later integrated as modules within other model such as DSSAT model. DSSAT comprises crop simulation models for over 42 crops along with database management program for soil, weather, crop management, and experimental data to facilitate the effective use of models (Jones et al. 1998). ALMANAC model includes subroutines and functions from the EPIC model to simulate hydrology, soils and crop growth (Kiniry et al. 1992).

Some other crop simulation models that operate at a global scale include the Joint UK Land Environment Simulator (JULES)-Crop, Lund-Potsdam-Jena managed Land (LPJmL), and Biome BioGeochemical Cycles (Biome-BGC). JULES-Crop is a large-area process-based model that simulates the fluxes of carbon, water, energy and momentum between the land surface and the atmosphere. It can be run on stand-alone mode or as a component of a coupled Earth system model (Williams et al. 2017). LPJmL is a dynamic global vegetation model (DGVM), which simulates the vegetation composition and distribution as well as stocks and land-atmosphere exchange flows of carbon and water, for both natural and agricultural ecosystems (Schaphoff et al. 2018). Similarly, Biome-BGC estimates fluxes and storage of energy, water, carbon, and nitrogen for the vegetation and soil components of terrestrial ecosystems and has been applied successfully to a variety of forest types, urban landscapes, and agricultural fields at the regional, continental, and global scales (Di Vittorio et al. 2010).

Similarly, several process-based models have been developed to quantify GHG fluxes in agriculture systems at regional and field scales. For instance, CENTURY/Daily Century Model (DAYCENT), Denitrification-Decomposition (DNDC), and EPIC/Agricultural Policy/Environmental Extender (APEX) are well-parameterized models for quantifying agricultural GHGe and have been used extensively in the USA and other countries. CENTURY is a biogeochemical model that simulates the long-term dynamics of carbon, nitrogen, phosphorus, and sulfur for different plant-soil systems at a monthly time step. DAYCENT is a daily version of CENTURY model which can be simulated at both site (Del Grosso et al. 2005, 2009) and regional scale (Lee et al. 2012) to quantify soil carbon storage, GHGe (i.e., N_2O, CO_2, CH_4). The EPIC model predicts effects of management decisions on soil loss, water quality, and crop yields for areas with homogeneous soils and management practices. It has algorithms to simulate water quality, nitrogen and carbon cycling, climate change, and the effects of atmospheric CO_2. APEX is the watershed version of EPIC model and has components for routing water, sediments, nutrients, and pesticides across complex landscape and channel systems to the watershed outlet as well as groundwater and reservoir (Texas A&M Agrilife Research 2019). DNDC is another biogeochemical model that can be used for predicting crop growth, soil temperature and moisture regimes, soil carbon dynamics, nitrogen leaching, and emission of trace gases. It can be applied at various scales, ranging from site-specific applications to county and regional scales (Olander et al. 2011).

Models such as Soil Water Assessment Tool (SWAT), Hydrologic Simulation Program – Fortran (HSPF), and SPAtially Referenced Regressions On Watershed attributes (SPARROW) are some of the most widely used water quality models that operate at a watershed scale. SWAT model (Arnold and Allen 1999) is used to simulate the quality and quantity of surface and groundwater and predict the environmental impact of land-use management practices and climate change. In SWAT, a watershed is divided into multiple sub-watersheds, which are then further subdivided into hydrologic response units (HRUs) that consist of homogeneous land use, management, and soil properties. Water discharges, sediment yield, and water quality parameters from each HRU in a sub-watershed are summed, and the resulting loads are routed through channels, ponds, and reservoirs to the watershed outlet. A wide spectrum of crop rotations can be simulated in the model, which facilitates the analyses of alternative cropping systems on water quality. HSPF is another watershed hydrology and water quality model that allows integrated simulation of land and soil contaminant runoff processes on pervious and impervious land surfaces and in streams and well-mixed impoundments (US EPA 2019). SPARROW is a hybrid model which is also designed to predict long-term changes in water characteristics, such as concentrations and amounts of selected constituents that are delivered to downstream receiving waters (Schwarz et al. 2006).

11.2.3 Selection of an Appropriate Agroecosystem Model

There exist several agroecosystem models that range from simple to complex, and thus cautions should be taken during model selection, especially regarding whether the model complexity/simplicity is appropriate to a particular use case scenario, and whether the configuration under which the model was tested can be representative of alternative environmental condition. Oftentimes, field-scale models are applied on a larger scale without a proper parameterization or ignoring the conditions for which the model was formulated. Also, the models were applied to estimate the impacts of climate change on crop productivity even if they are not tested before. Further, calibration and validation of the model outputs are constrained to limited site-specific estimates. Hence, there can be high uncertainties in model outputs related to model parameterization and calibration, and model developers need to document these details in model description and promotion, and users need to be aware of this (Di Paola et al. 2016).

The first criterion when choosing a model should be the main purpose of applying it, followed by the selection of the best compromise between accuracy and ease of use. To better understand the degree of ease/complexity of using a model and its workflow, it is important to know information such as applicability domain of the model, number of required model parameters, model uncertainties, and calibration and validation requirements. Calibration is a process of optimizing the model performance by comparing observed and simulated data, and validation is the process of confirming that the model estimates adequately represent observed physical phenomena. It is critical to validate the model if it is used to provide predictions under certain environmental, production, and management conditions. Uncertainty in the model outputs depends on the uncertainty in the model inputs, and sensitivity analysis helps examine different sources of uncertainty in the model outputs and test the robustness of the results.

Although such information is critical in model selection, they are not explicitly documented in the models' guide, and thus, oftentimes understanding a model and correctly applying it becomes very difficult. To address this concern, Di Paola et al. (2016), using a literature survey, identified almost 70 ecosystem models that differed in complexity and areas of interests (regions, scales, crops) and reported their main characteristics to enable both scientific and nonscientific users of diverse interests and levels of expertise make the best choice among models and reduce errors while they use models for specific purposes. Table 11.1 reports some of the ecosystem models that have been used for agriculture-based studies and attempts to provide a quick guideline for selection of a suitable model. This list is not exhaustive; however, it represents a diversity of ecosystem models currently being used in agriculture-based studies.

In Table 11.1, the first column provides the name of the model in the form of their acronyms. The field "crop growth model type" specifies whether the modeling approach is mechanistic, empirical, or hybrid. The field "sub-module" specifies types of biogeochemical cycles (e.g., water, nitrogen, phosphorous) introduced in the model. Typically, the presence of greater number of biogeochemical cycles in the model indicates greater complexity of the model. For example, some models consider only the water-plant interaction. Although these models have proven to be effective in studies of arid environments or water-limiting regions, they cannot be useful in scenarios where other factors such as soil fertility, GHGe, and crop rotations need to be considered. Spatial scale and temporal step of the model and the crop for which the model was primarily developed are also documented. Few models focus on a single crop, while the majority considers a broad set of crops. Some models (e.g., LPJmL, JULES-Crop) focus on crop functional types (CFTs, according to the crop physiology and vegetation structure) rather than a single crop or a crop family. Models dealing with CFTs are typically used at regional to global scales to speed up simulations on crop production, yield, and biogeochemical cycles when land-use and climate change projections need to be coupled. Large-scale simulations however do not necessarily imply the use of CFTs when focus needs to be limited to specific crops for which dedicated models are available (Di Paola et al. 2016).

There is already a large list of ecosystem models, and this is expected to grow in the future. Most of these models were developed at least two decades ago. Some models while have no subsequent developments from their original version and currently seem unused, other models have been modified or developed over time.

Table 11.1 A list of agroecosystem models and major features

Name	Crop growth model type	Sub-module[a]	Scale of applica-tion	Time step	Crop considered
CERES (maize, rice, wheat)	Hybrid	W, N	Field	Daily	Rice, maize, wheat
CROPOPRGRO	Hybrid	W, N	Field	Daily	Soybean
DSSAT*	Mechanistic (incorporates CROPGRO, CERES, and other modules)	W, N, C	Field	Daily	Over 42 crops
ALMANAC	Hybrid	W, N	Field	Daily	Switchgrass (biofuel), maize, sorghum, sunflower
DNDC	Mechanistic	W, N, C	Field to regional	Daily	Corn, wheat, soybean, beans, rice, sorghum, vegetables, and fruits
DAYCENT	Mechanistic	W, N, C	Field to regional	Daily	Major crops
EPIC	Mechanistic	W, N, P	Field	Daily	A wide range of crop rotations and vegetative systems, tillage systems, and other management strategies
APEX	Mechanistic (from EPIC)	W, N, P	Watershed	Daily	A wide range of crop rotations and vegetative systems, tillage systems, and other management strategies
SWAT	Mechanistic (incorporates features of models such as CREAMS, GLEAMS, and EPIC)	W, N, P	Watershed	Daily	A wide range of crop rotations and vegetative systems, tillage systems, and other management strategies
SPARROW	Hybrid	W, N	Regional to global	Annual	A wide range of crops rotations and management strategies
SUCROS	Empirical	W, N	Field	Daily	Cotton
HSPF	Mechanistic	W, N, P	Watershed	Daily	A wide range of crops rotations and management strategies
JULES-crop	Mechanistic (from SUCROS)	W	Regional to global	Daily	12 CFTs
LPJmL	Mechanistic	W, C	Regional to global	Daily	13 CFTs
Biome-BGC	Hybrid	W, N, C	Regional to global	Daily	General biomes

The asterisk (*) indicates that the tool is an integrated framework
*CFT – Crop functional types
[a]Biogeochemical sub-models: W water, N nitrogen, C soil organic carbon, P phosphorus (Modified from Di Paola et al. (2016))

Thus, progress needs to be made in updating models to reflect new research and findings in crop physiology, agronomy, and sol science. Future focus should be on joining and sharing expertise and skills among the numerous model developers for structuring a few but more robust and versatile models (Di Paola et al. 2016).

11.2.4 Remote Sensing Applications in Agriculture

Improving the physical description of physiological processes in an ecosystem model often requires the measurement or estimation of several biophysical (e.g., LAI, canopy height) and weather (e.g., precipitation, land surface temperature, radiation) parameters that vary in time and space (Fatichi et al. 2016; Pappas et al. 2016). Field experiments to measure some of these biophysical and weather variables at higher temporal and spatial resolutions are however time-, cost-, and labor-intensive.

One of the alternatives to collect these data is the use of remote sensing technologies, which obtain information about objects or area from a distance without being in contact with the object. Remote sensing is primarily based on measurement of reflected and emitted electromagnetic radiations (EM) from the targeted objects using sensors mounted on platforms such as satellite, aircrafts (manned and unmanned), and ground vehicles. There are two types of sensors: passive and active. Passive sensor does not have its own source of radiation and measures reflected radiations from a natural origin, such as sun. Active sensor has a built-in source of radiation and measures signals transmitted by the sensor that were reflected, refracted, or scattered by the objects.

EM comprises a spectrum of wavelengths, ranging from short-wave gamma rays to long-wave radio frequencies (Fig. 11.3). Reflection and absorption of EM wavelengths by objects depend on their unique physical and chemical properties. Visible (440–690 nm) and near-infrared (NIR) (760–900 nm) regions of the EM spectrum are most commonly used for agricultural monitoring. Figure 11.3

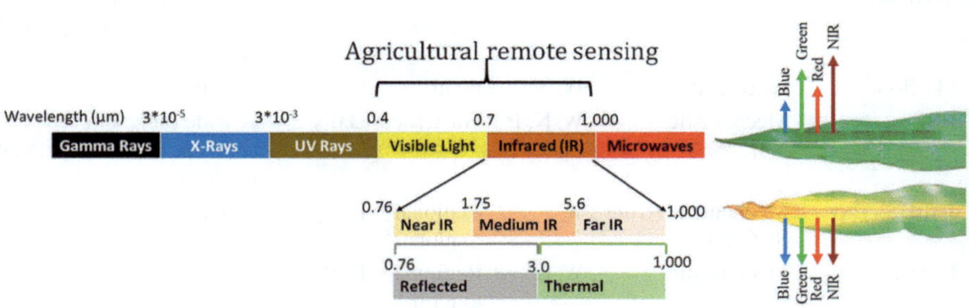

Fig. 11.3 Electromagnetic radiation (left) and spectral signatures (in the visible and NIR region) of healthy (top) versus unhealthy corn leaves (bottom) (right)

illustrates how healthy corn leaves absorb more blue (450–520 nm) and red (630–680 nm) and less green (520–600 nm) light. This results in a higher reflectance in the green band and therefore plants appear green to the human eye. Plants absorb less NIR compared to visible light. Generally, healthy plants show high values of reflectance in the NIR and low values in the visible spectrum. When crops senesce or are subjected to stress (e.g., pests, nutrients, and water), chlorophyll content decreases, causing decrease in green and increase in red and blue reflectance.

Recognizing the potential benefits of remote sensing technologies in agricultural and environmental monitoring, several satellite missions have been launched since the early 1970s (Table 11.2). Landsat 1, originally named "Earth Resources Technology Satellite 1," was the first Earth-observing satellite launched in 1972 to study and monitor planet's landmasses. Soon after Landsat 1, several satellites of varying characteristics were launched. Spatial resolution, also known as pixel size, determines the size of the smallest identifiable features in an image. As spatial resolution improves, the pixel size decreases, and homogeneity of feature within that pixel increases. Frequency of measurements, also known as temporal resolution, is important for assessing temporal changes in crop and soil characteristics. The earliest applications of remote sensing in agriculture were primarily focused on classification of agricultural landscapes. For example, imagery from Landsat 1 was initially used by Bauer and Cipra (1973) to classify agricultural landscapes in the Midwestern USA into maize or soybean fields with an accuracy of 83%.

Application of remote sensing in agricultural monitoring typically involves the use of various mathematical combinations of spectral bands, mainly red, green, and infrared, known as spectral indices or vegetation indices (VIs). VIs are designed to find functional relationships between vegetation characteristics and remote sensing observations. One of the most commonly used vegetation index is the normalized difference vegetation index (NDVI), which is calculated using reflectance ratios in the near-infrared and red portion of the spectrum. It has been used to detect crop nutrient deficiencies, crop yield prediction, patterns in insect and weed infestations, and crop diseases (Figs. 11.4 and 11.5). There are several VIs, such as Green NDVI (Gitelson et al. 1996), Modified Chlorophyll Absorption Ratio Index (Daughtry et al. 2000), Soil-Adjusted Vegetation Index (Huete 1988), and Transformed Chlorophyll Absorption Reflectance Index (Haboudane et al. 2002), that have been used to estimate crop characteristics such as chlorophyll content, leaf area index, soil reflectance, and biomass fraction.

Application of remote sensing technologies in agriculture is typically influenced by the type of platform (i.e., satellite, aerial, and ground) used to carry sensor, sensor characteristics (e.g., spectral range and width of spectral bands), and the frequency of data collection. High-resolution, high-return frequency satellites, such as IKONOS, QuickBird, SPOT-5, RapidEye, GeoEye, and WorldView, have spatial resolutions ranging from 0.3 to 6.5 meter and return frequencies ranging from 1.1 days to 5.5 days. Data collected from these platforms allow proper detection of spatial and temporal patterns in soil and crop health (Fig. 11.6). Some of the satellites offer additional spectral bands for observation, which can be used to detect patterns that could not be identified using other spectral bands. For example, red

Table 11.2 Satellite remote sensing platforms and their spectral and spatial resolution and return frequency

Satellite	Year	Spectral bands	Frequency (d)	Spatial resolution
Landsat 1	1972	G, R, two IR	18	56 * 79 m
AVHRR	1978	R, NIR, two TIR	1	1090 m
Landsat 5 TM	1984	B, G, R, two NIR, MIR, TIR	16	30 m
SPOT 1	1986	G, R, NIR	2–6	20 m
IRS 1A	1988	B, G, R, NIR	22	72 m
ERS-1	1991	Ku-band altimeter, IR	35	20 m
JERS-1	1992	L-band radar	44	18 m
RADARSAT	1995	C-band radar	1–6	30 m
IKONOS	1999	Panchromatic, B, G, R, NIR	3	1–4 m
Landsat 7 ETM	1999	Panchromatic, B, G, R, NIR, two SWIR, thermal	16	15–60 m
ASTER	1999	Three NIR, six SWIR, five TIR	16	15–90 m
Terra EOS ASTER	2000	G, R, NIR, six MIR, 5 TIR	16	15–90 m
EO-1 Hyperion	2000	400–2500 nm, 10 nm bandwidth	16	30 m
QuickBird	2001	Panchromatic, B, G, R, NIR	1–4	0.61–2.4 m
EOS MODIS	2002	36 bands in VIS-IR	1–2	250–1000 m
SPOT-5	2002	Panchromatic, R, NIR, SWIR	2–3	2.5–20 m
WorldView-1	2007	Panchromatic	1.7	0.46 m
RapidEye	2008	B, G, R, red edge, NIR	5.5	6.5 m
GeoEye-1	2008	Panchromatic, B, G, R, two NIR	2–8	1.6 m
WorldView-2	2009	Panchromatic, B, G, R, red edge, two NIR, coastal, yellow	1.1–3.7	0.46–2.4 m
Pleiades-1A	2011	Panchromatic, G, R, NIR	1	0.5–2 m
Pleiades-1B	2012	Panchromatic, G, R, NIR	1	0.5–2 m
Landsat 8	2013	Coastal aerosol, panchromatic, B, G, R, NIR, two SWIR, two TIR	16	15–100 m
WorldView-3	2014	Panchromatic, three aerosol, three water, B, G, R, red edge, two NIR, coastal, yellow, eight SWIR	<1	0.31–30 m
Sentinel-1A	2014	C-band SAR	6	5 m–20 km
Sentinel-2A	2015	Coastal aerosol, B, G, R, three red edge, NIR, three SWIR	5	10–60 m
Sentinel-1B	2016	C- band SAR	6	5 m–20 km
PlanetScope	2016	R, G, B, NIR, RedEdge	1	3 m

B refers to blue, G to green, R to red, B to blue, IR to infrared, NIR to near infrared, MIR to mid-infrared, TIR to thermal infrared waveband, and SAR to synthetic aperture radar (Modified from Mulla (2013))

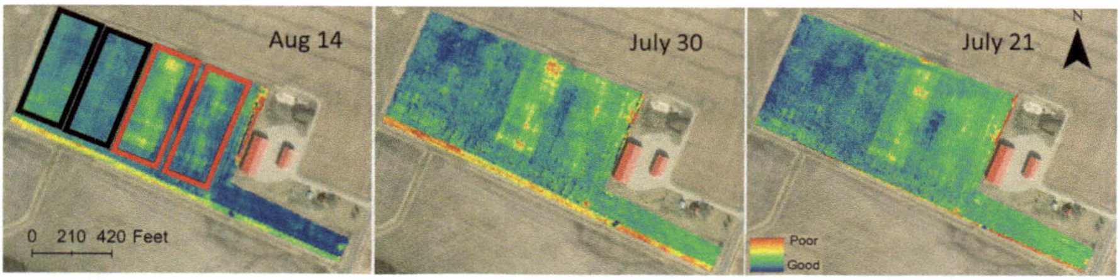

Fig. 11.4 Corn health monitoring during the growing season of 2015 using images collected via manned aircraft. *Note*: Warm and cooler colors show poor and good crop health conditions, respectively. Red and black rectangular boxes in the upper left imagery indicate sections of a field treated with lower (11–78 kg N/ha) and higher N (235–302 kg N/ha) fertilizer rates

edge band is highly sensitive to the chlorophyll status of growing crops. Currently, interest is growing in the use of low-altitude unmanned aerial vehicles (UAVs) as a remote sensing platform for precision agriculture. Imagery collected with UAVs provide opportunity to view individual plants and leaves and can be used to assess crop LAI, biomass, plant height, nutrient and water stress, weed and pest infestation, and yield and grain protein content (Khanal et al. 2017b).

Due to the technological advancements in the acquisition systems, such as launches of new satellites, UAV and related regulations, robot platforms, and Internet of Things (IoT), as well as data storage and computation (e.g., Google Earth Engine, Amazon Cloud Computing) and advanced algorithms for deep learning techniques for data processing, a majority of remote sensing technologies have become accessible and affordable for the agricultural community (Khanal et al. 2020). High-resolution images have been used in various agricultural applications, such as weed detection, crop counting, crop growth identification, and crop stress identification.

Currently, there is a lot of interest in integrating remote sensing data with ecosystem models at various geographic scales. Although some satellite remote sensing data have already been used for updating the state of global terrestrial models, e.g., for hydrological (Girotto et al. 2016) and carbon cycle (Scholze et al. 2017), the potential of integrating remote sensing data and ecosystem models for near real-time monitoring at field and local scales is yet to be fully exploited.

11.3 Case Study: Modeling the Water Quality Effects of Land-Use Management Practices in the Muskingum River Basin (MRB) at Ohio

Introduction In this case study, the hydrologic model – SWAT – was used to quantify the water quality effect of land-use practices in the Muskingum River Basin (MRB), which is the largest wholly contained watershed in the state of Ohio

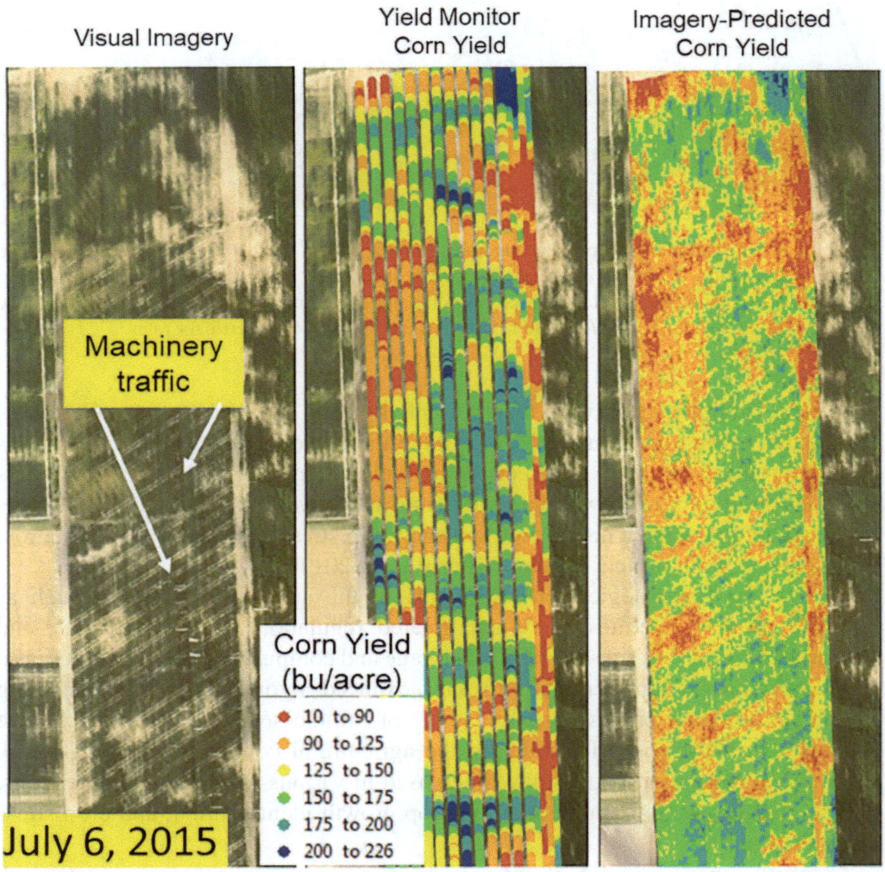

Fig. 11.5 Importance of a high-resolution data in understanding within-field yield variability in crop health due to machinery traffic. Visual image of a field in the left shows patterns due to machinery traffic. Image in the middle shows corn yield data collected using a yield monitoring system mounted on a combine at a time of harvest in late October. Image in the right shows a high-resolution corn yield map, which was prepared based on a statistical relationship between visual images, topographical variables (elevation and slope), and corn yield. Negative yield effects due to machinery traffic can easily be seen in a high-resolution image (right) compared to the image in the middle

(Fig. 11.7). Studies (Environment America 2006; Public News Service 2014) have shown that this region is the fourth most polluted watershed in the USA. Both point source pollution from industrial facilities, such as wastewater treatment plants and landfills, and nonpoint source pollution (NPSP) from croplands are the major contributors to poor water quality in the region. Poor water quality in this region is particularly concerning because this watershed drains into the Ohio River, which in turn drains into the Mississippi River and then to the Gulf of Mexico. Currently, the

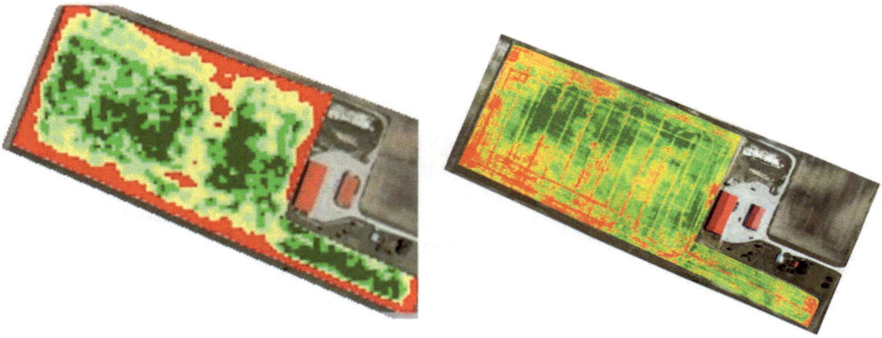

Fig. 11.6 Normalized difference vegetation index maps at two different spatial resolutions. The low spatial resolution vegetative map (left) has a pixel size of 10 meters, whereas the high-resolution map (b) has a pixel size of 0.25 meters. These data were collected midsummer from a canopied corn crop; note the difference in detail between resolutions. (Source: Khanal et al. (2017a))

dead zone in the northern Gulf of Mexico is the second largest human-caused coastal hypoxic area in the world (Turner and Rabalais 2019). Hypoxia, or low oxygen, is primarily a problem for many ES, such as fish habitat, recreational opportunities, water clarity, and public health. Nutrient (i.e., nitrogen and phosphorus) loadings from agricultural fields in the Mississippi River Basin have been identified as the primary driver for this.

Currently, agriculture covers approximately 40% of the land use in the MRB, with 22% being used to grow row crops and 18% for pasture hay. Corn, soybeans, and winter wheat are the major row crops (99% of the total row crops) planted in the watershed. Further, livestock (mainly dairy, beef cow, hog, and poultry) production is significant in the basin (USDA NASS 2014). Many agricultural conservation practices, including riparian buffers, buffer strips, reduce fertilizer application, conservation tillage, and change in fertilizer application timing, have been recommended to lower NPSP (EPA 2017). However, for cost-effective implementation of these conservation practices, identification of areas where water quality problems are prevalent is important. Water quality monitoring through field experiments is usually costly and labor-intensive and requires a number of years of monitoring for proper accounting of variability due to changing weather. The use of properly calibrated and validated watershed models can overcome the limitations associated with field studies by acting as a test bed for alternative climate and management scenarios. The model findings can be used to identify and prioritize sub-watersheds for cost-effective implementation of management practices to manage water quality and quantity concerns.

Approach Using SWAT (version 2012), the MRB was divided into 83 subbasins. The subbasins' area ranged from 11 to 501 km^2 with an average area of 251 km^2. A 30 m digital elevation model (DEM) and a stream network from the National

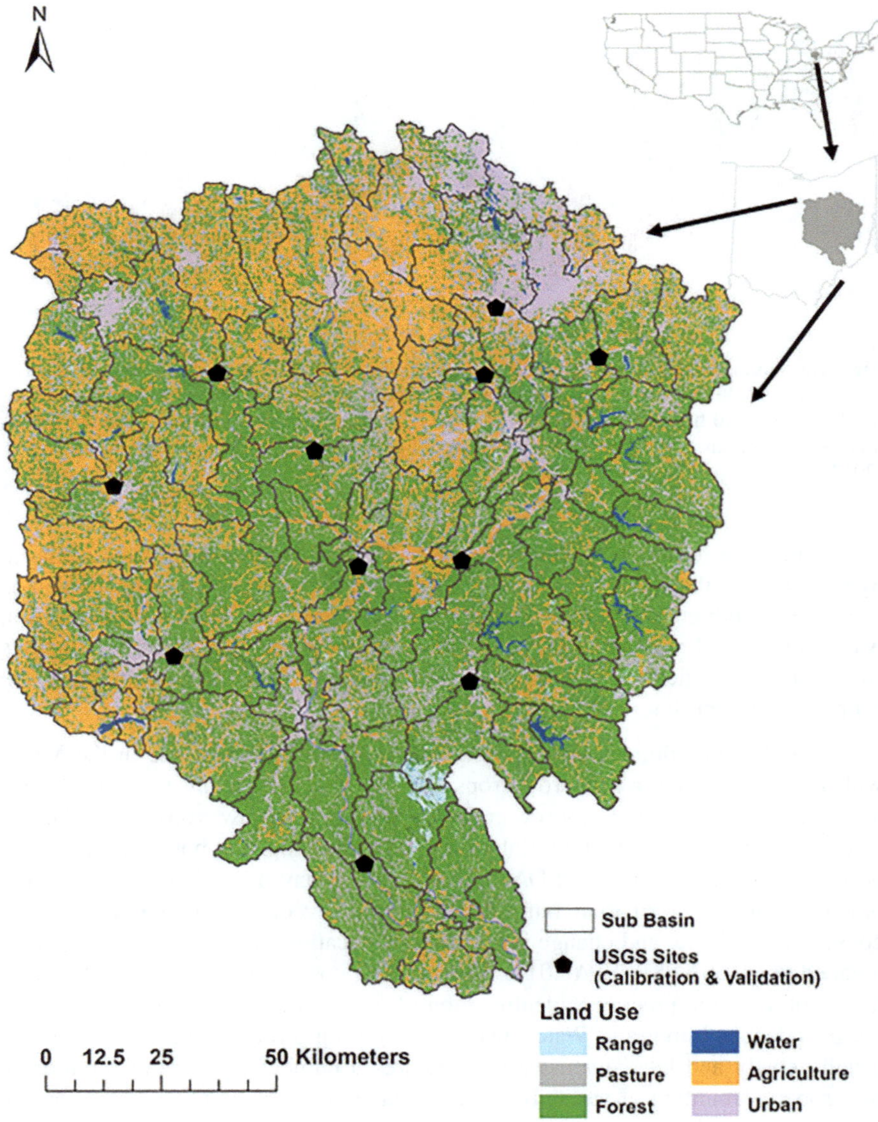

Fig. 11.7 Muskingum River Basin, Ohio

Hydrography Dataset were used to delineate watershed boundary and compute flow direction. Hydrologic response units (HRUs), which are areas consisting of homogeneous land-use, management, and soil properties, were generated by overlapping the 2006 National Land Cover Dataset, the Soil Survey Geographic

– SSURGO database, and the DEM-based slope classes. This resulted in a total of 1867 HRUs, with 431 HRUs representing cropland. SWAT model was set up using current agricultural practices, including major crop rotations, fertilizer application practices, tillage practices, and tile drainage structures. Details are provided in Khanal et al. (2018).

The model was run from 1990 to 2010, where period 1990 to 1994 served as a spin-up to minimize uncertain conditions (e.g., soil moisture and groundwater) in the model output. Period 1995 to 2004 was used for calibration, and 2005 to 2010 for validation of the model. The model was calibrated and validated for monthly water discharges using data collected at ten sites, total suspended sediment (TSS) using data from one site, and total nitrogen (TN) and total phosphorus (TP) using data from four sites. These sites were maintained by the US Geological Survey (USGS), the National Center for Water Quality Research at Heidelberg College, and the USEPA. The model performance was ensured to be good to be used for simulations of alternative management practices scenarios (Khanal et al. 2018).

To simulate the effectiveness of conservation practices on sediment and nutrient load reductions, five conservation practices were considered: (1) reduced use of commercial fertilizer, (2) reduced application of manure fertilizer, (3) implementation of conservation tillage in areas where conventional tillage practices (i.e., chisel plow and tandem-disc plow) are used, (4) the use of 6 m vegetative filter strips, and (5) terraces. These conservation practices were implemented only in the croplands of the MRB. Additional simulations were run to understand the extent to which the adoption of these conservation practices reduces nutrient loads at a watershed outlet. For this, eight scenarios considering 10% and 50% reduction in the current use of commercial and manure fertilizer for (1) all the cropland in the watershed and (2) only to croplands of a few subbasins where TN, TP, and TSS are estimated to be higher were simulated and analyzed.

Results Under the current management practices, the sediment and nutrient loads were found to be higher in the central part of the basin where agriculture and animal production are predominant. In regions with smaller footprint of agricultural lands, a higher load of sediments and nutrients was associated with steep topography and intensive grazing activities. On average, cropland contributed the most to the sediment and nutrient runoffs, followed by pasture and urban land (Table 11.3). Grazing activities in the pastureland were found to contribute to a higher fraction of sediment and nutrient loads.

Critical source areas (CSAs) at the subbasin level were identified based on SWAT model outputs. For this purpose, the sediment and nutrient yields from each subbasin were ranked in descending order based on loads per unit area. That means, the subbasin with the highest yield was ranked first and the lowest yield was ranked last. Starting from the highest ranking to the lowest, subbasins that collectively contributed 50% of the sediment, TP, or TN were considered to be CSAs. The selection of threshold of 50% in this study is truly hypothetical. Typically, in real-world scenario, actual threshold is based on the total cost required for implementing

Table 11.3 Average percent contribution to MRB's total sediment (TSS) and nutrient (TN and TP) loads by land-use types

	Contribution to total basin's loads (%)		
Land-use type	TSS	TN	TP
Cropland	70	63	71
Pasture (all)	18	15	15
Pasture (grazing)[a]	53	32	64
Urban	6	20	10

[a] Average percentage load of TSS, TN, and TP from pasture-lands

conservation practices. Under limited budget, the use of lower threshold helps to target small geographic areas that need to be prioritized.

At 50% threshold, 27%, 28%, and 19% areas were classified as CSAs for TSS, TN, and TP, respectively. Some of the subbasins that were identified as sediment CSAs were also identified as TP and TN CSAs (Fig. 11.8), suggesting that the implementation of conservation practices in those regions could result in substantial improvement in water quality.

Implementation of conservation practices, particularly filter strips and terraces, in croplands of the CSAs for both sediments and nutrients resulted in significant reduction in sediment and nutrient loads at a subbasin level. The use of 6 m vegetative filter strips alone reduced TSS, TN, and TP loads in the range of 4–44%, 4–26%, and 5–53%, respectively. Similarly, placement of terrace structures reduced TSS, TN, and TP loads in the range of 4–45%, 5–28%, and 7–45%, respectively. Reduction in commercial fertilizer use by 10% of the current level in cropland of CSAs demonstrated a very little effect in the reduction of nutrient loads. However, similar reduction in current manure application rates resulted in a higher reduction in TN and TP loads.

In MRB, upstream reduction in sediment and nutrient loads after the implementation of conservation practices however did not directly translate to equivalent reductions at downstream of the watershed. This is because, oftentimes, the substantial portion of sediment and nutrient loads in the upstream watershed may get trapped by different structures including land cover, soil, and reservoirs, and thus, only a certain portion of these loads reach downstream. For example, with the implementation of terrace, TSS, TN, and TP were reduced by only 0.9%, 3.2%, and 3.9%, respectively, at the outlet of study region, relative to current management scenario. Similarly, the use of filter strips resulted in reduction of TSS, TN, and TP by 0.9%, 2.1%, and 2.9%, respectively, relative to current management practices.

There was a higher reduction in TN and TP loads at the MRB outlet with the reduction in manure application rates than with the commercial fertilizer application. Reduction in fertilizer use in the croplands of CSAs (identified based on both sediment and nutrients) lowered TN and TP loads in the range of 0.1–3.6% and 0.1–2.0%, respectively. When fertilizer use in all the croplands of the basin was reduced, TN and TP loads were reduced in the range of 0.5–10.9% and 0.5–5%, respectively (Fig. 11.9).

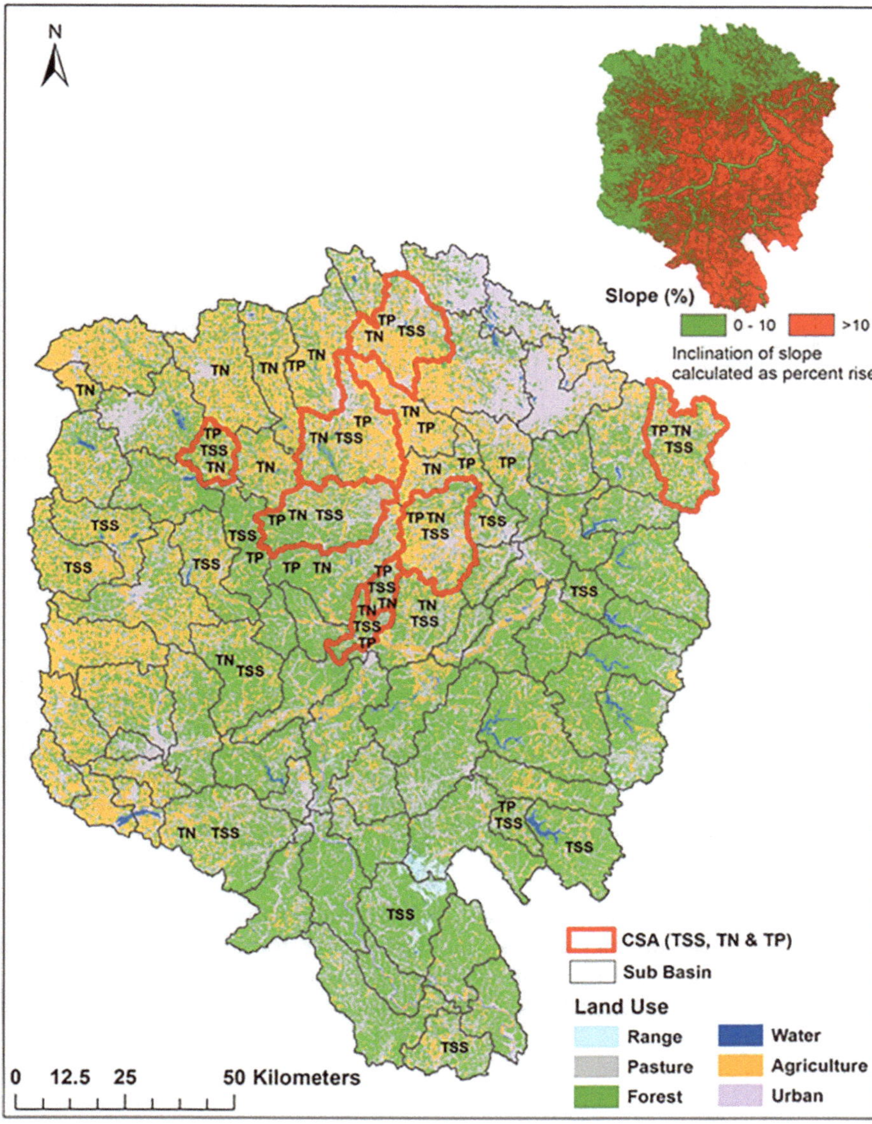

Fig. 11.8 Critical source areas (CSAs) based on total suspended sediments (TSS), total nitrogen (TN), and total phosphorus (TP) classified separately and together at 50% threshold

As expected, adoption of all five conservation practices together reduced TN and TP loads significantly. When these conservation practices were adopted only in the CSAs, TN and TP loads at the basin outlet reduced by 6% and 5% compared to the current practices. But when these conservation practices were extended to all

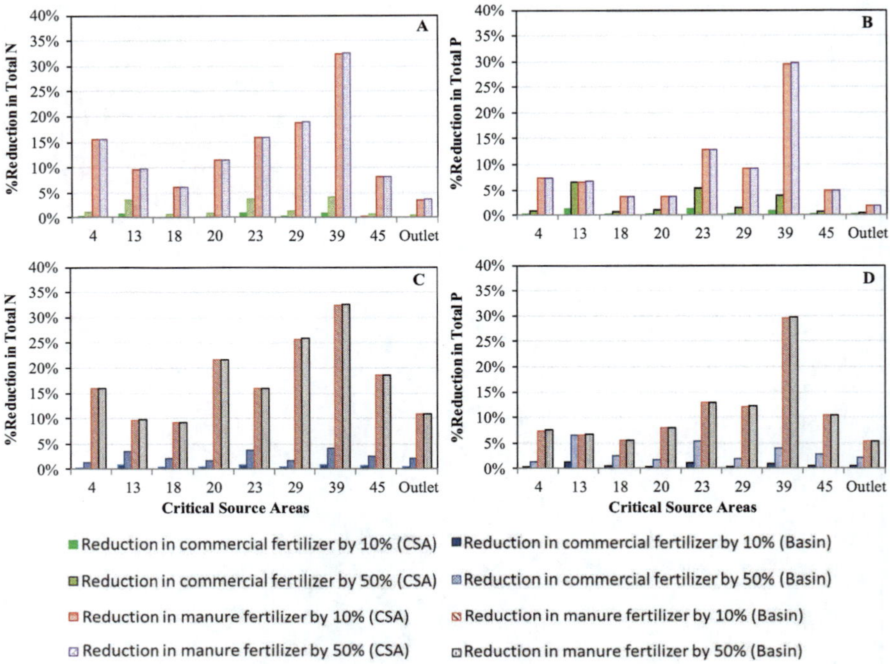

Fig. 11.9 Percent reduction in nitrogen (A, C) and phosphorus loads (B, D) in critical source areas (CSAs) and at the outlet of the MRB under 10% and 50% reduction in current commercial and manure fertilizer application rates* *Average commercial fertilizer application rates considered in the CSAs were 170 (±10) kg N ha^{-1} and 67 (±2) kg P ha^{-1} for corn, 76 (±8) kg N ha^{-1} and 53 (±2) kg P ha^{-1} for wheat, and 16.8 kg N ha^{-1} and 48 (±2) kg P ha^{-1} for soybean. Average manure application rates were 165 (±123) kg N ha^{-1} and 122 (±73) kg P ha^{-1}

croplands in the MRB basin, TN and TP at the outlet reduced by 26% and 22%, respectively. These findings suggest the need to look for additional conservation practices, as well as expansion of conservation practices to areas such as pasture and urban to lower nutrient loads more than 26% relative to the current nutrient loads.

11.4 Summary

This chapter provided an overview of various ecosystem services and disservices to and from agroecosystems and approaches such as ecosystem modeling and remote sensing to model and measure them. Recent advances in technology, data collection, data analytics, and modeling routines have led to improved ecosystem models, and these improved models could be used as reliable and cost-effective alternatives to gain some perspectives on the types and extent of agricultural conservation practices required to enhance ES. Since ecosystem models are increasingly used in scientific research, farm management, and policy decision, it is critical to pay attention to model limitations and limits of applicability, as well as to enhance the robustness of model performance by reducing sources of uncertainties, including model structure and inputs. An improved understanding of the limitations of the ecosystem model save time often devoted to iterations of tuning steps to force a wrong model to get the right answer.

References

Arnold, J.G., Allen, P.M., 1999. Automated methods for estimating baseflow and ground water recharge from streamflow records. American Water Resources Association. 2(35), 411-424.

Bartomeus, I., Ascher, J.S., Gibbs, J., Danforth, B.N., Wagner, D.L., Hedtke, S.M., Winfree, R., 2013. Historical changes in northeastern US bee pollinators related to shared ecological traits. Proc. Natl. Acad. Sci. 110, 4656–4660. https://doi.org/10.1073/PNAS.1218503110

Basso, B., Cammarano, D., Carfagna, E., 2013. Review of Crop Yield Forecasting Methods and Early Warning Systems. In Proceedings of the first meeting of the scientific advisory committee of the global strategy to improve agricultural and rural statistics, FAO Headquarters, Rome, Italy, vol. 241.

Bauer, M.E., Cipra, J.E., 1973. Identification of Agricultural Crops by Computer Processing of ERTS MSS Data. In NASA. Goddard Space Flight Center Symposium on Significant Results obtained from the ERTS-1, vol. 1.

Costello, C., Griffin, W.M., Landis, A.E., Matthews, H.S., 2009. Impact of biofuel crop production on the formation of hypoxia in the Gulf of Mexico. Environ. Sci. Technol. 43, 7985–7991.

Cuddington, K., Fortin, M.J., Gerber, L.R., Hastings, A., Liebhold, A., O'connor, M., Ray, C., 2013. Process-based models are required to manage ecological systems in a changing world. Ecosphere 4, 1–12. https://doi.org/10.1890/ES12-00178.1

Daughtry, C.S.., Walthall, C.., Kim, M.., de Colstoun, E.B., McMurtrey, J.., 2000. Estimating Corn Leaf Chlorophyll Concentration from Leaf and Canopy Reflectance. Remote Sens. Environ. 74, 229–239. https://doi.org/10.1016/S0034-4257(00)00113-9

Davis, S.C., Parton, W.J., Grosso, S.J. Del, Keough, C., Marx, E., Adler, P.R., DeLucia, E.H., 2012. Impact of second-generation biofuel agriculture on greenhouse-gas emissions in the corn-growing regions of the US. Front. Ecol. Environ. 10, 69–74. https://doi.org/10.1890/110003

Del Grosso, S.J., Mosier, A.R., Parton, W.J., Ojima, D.S., 2005. DAYCENT model analysis of past and contemporary soil N 2 O and net greenhouse gas flux for major crops in the USA. Soil Tillage Res. 83, 9–24.

Del Grosso, S.J., Ojima, D.S., Parton, W.J., Stehfest, E., Heistemann, M., DeAngelo, B., Rose, S., 2009. Global scale DAYCENT model analysis of greenhouse gas emissions and mitigation strategies for cropped soils. Glob. Planet. Change 67, 44–50.

Di Paola, A., Valentini, R., Santini, M., 2016. An overview of available crop growth and yield models for studies and assessments in agriculture. J. Sci. Food Agric. 96, 709–714. https://doi.org/10.1002/jsfa.7359

Di Vittorio, A. V., Anderson, R.S., White, J.D., Miller, N.L., Running, S.W., 2010. Development and optimization of an Agro-BGC ecosystem model for C4 perennial grasses. Ecol. Modell. 221, 2038–2053. https://doi.org/10.1016/J.ECOLMODEL.2010.05.013

Environment America. 2006. Wasting Our Waterways: Toxic Industrial Pollution and Restoring the Promise of the Clean Water Act. Boston, MA: Environment Amer-

ica Research & Policy Center. http://environmentamericacenter.org/sites/environment/files/reports/US_wastingwaterways_scrn061814_0.pdf.
EPA, 2017. Mississippi River/Gulf of Mexico Watershed Nutrient Task Force: 2017 Report to Congress. https://www.epa.gov/sites/default/files/2017-11/documents/hypoxia_task_force_report_to_congress_2017_final.pdf
FAO, 2017. Water for Sustainable Food and Agriculture A report produced for the G20 Presidency of Germany. Rome. http://www.fao.org/3/i7959e/i7959e.pdf
FAOSTAT, 2016. FAOSTAT - Land Use [WWW Document]. URL http://www.fao.org/faostat/en/#data/RL/visualize (accessed 8.26.21)
Fatichi, S., Pappas, C., Ivanov, V.Y., 2016. Modeling plant-water interactions: an ecohydrological overview from the cell to the global scale. Wiley Interdiscip. Rev. Water 3, 327–368. https://doi.org/10.1002/wat2.1125
Gitelson, A.A., Kaufman, Y., Merzlyak, M.N., 1996. Use of a Green Channel in Remote Sensing of Global Vegetation from EOS-MODIS 298, 289–298.
Girotto, M., De Lannoy, G. J., Reichle, R. H., Rodell, M., 2016. Assimilation of gridded terrestrial water storage observations from GRACE into a land surface model. Water Resources Research, 52(5), 4164–4183.
Haboudane, D., Miller, J.R., Tremblay, N., Zarco-Tejada, P.J., Dextraze, L., 2002. Integrated narrow-band vegetation indices for prediction of crop chlorophyll content for application to precision agriculture. Remote Sens. Environ. 81, 416–426.
Hodges, T., Botner, D., Sakamoto, C., Hays Haug, J., 1987. Using the CERES-Maize model to estimate production for the U.S. Cornbelt. Agric. For. Meteorol. 40, 293–303. https://doi.org/10.1016/0168-1923(87)90043-8
Huete, A.R., 1988. A soil-adjusted vegetation index (SAVI). Remote Sens. Environ. 25, 295–309.
Jones, J.W., Tsuji, G.Y., Hoogenboom, G., Hunt, L.A., Thornton, P.K., Wilkens, P.W., Imamura, D.T., Bowen, W.T., Singh, U., 1998. Decision support system for agrotechnology transfer: DSSAT v3, in: Tsuji G.Y., Hoogenboom G., T.P.K. (Eds.), Understanding Options for Agricultural Production. Systems Approaches for Sustainable Agricultural Development. Springer, Dordrecht, pp. 157–177. https://doi.org/10.1007/978-94-017-3624-4_8
Khanal, S., Fulton, J., Hawkins, E., Port, K., Klopfenstein, A., 2017a. Remote Sensing in Precision Agriculture. Ohio State University Extension. FABE 554.1. https://ohioline.osu.edu/factsheet/fabe-5541
Khanal, S., Fulton, J., Shearer, S., 2017b. An overview of current and potential applications of thermal remote sensing in precision agriculture. Comput. Electron. Agric. 139, 22–32.
Khanal, S., KC, K., Fulton, J.P., Shearer, S., Ozkan, E., 2020. Remote Sensing in Agriculture—Accomplishments, Limitations, and Opportunities. Remote Sens. 12. https://doi.org/10.3390/rs12223783
Khanal, S., Lal, R., Kharel, G., Fulton, J., 2018. Identification and classification of critical soil and water conservation areas in the muskingum river basin in ohio. J. soil water Conserv 73(2), 213-226. https://doi.org/10.2489/jswc.73.2.213
Kiniry, J.R., Williams, J.R., Gassman, R.W., Debaeke, R., 1992. A general, process-oriented model for two competing plant species. ASAE 35.
Klein, A.-M., Vaissière, B.E., Cane, J.H., Steffan-Dewenter, I., Cunningham, S.A., Kremen, C., Tscharntke, T., 2007. Importance of pollinators in changing landscapes for world crops. Proc. R. Soc. B Biol. Sci. 274, 303–313. https://doi.org/10.1098/rspb.2006.3721
Koh, I., Lonsdorf, E. V, Williams, N.M., Brittain, C., Isaacs, R., Gibbs, J., Ricketts, T.H., 2016. Modeling the status, trends, and impacts of wild bee abundance in the United States. Proc. Natl. Acad. Sci. U. S. A. 113, 140–5. https://doi.org/10.1073/pnas.1517685113
Lal, R., 2008. Soil carbon stocks under present and future climate with specific reference to European ecoregions. Nutr. Cycl. Agroecosystems 81, 113–127. https://doi.org/10.1007/s10705-007-9147-x
Lee, J., Pedroso, G., Linquist, B.A., Putnam, D., 2012. Simulating switchgrass biomass production across ecoregions using the DAYCENT model 521–533. https://doi.org/10.1111/j.1757-1707.2011.01140.x

Lobell, D.B., Schlenker, W., Costa-Roberts, J., 2011. Climate Trends and Global Crop Production Since 1980. Science 333 (6042), 616–620. https://doi.org/10.1126/SCIENCE.1204531

Losey, J.E., Vaughan, M., 2006. The Economic Value of Ecological Services Provided by Insects. Bioscience 56 (4), 311–323.

MEA, 2003. Ecosystems and Human Well-being Millennium Ecosystem Assessment Ecosystems and Human Well-being, vol 5. p563. https://www.millenniumassessment.org/documents/document.356.aspx.pdf

Miao, R., Hennessy, D.A., Feng, H., 2016. The Effects of Crop Insurance Subsidies and Sodsaver on Land-Use Change, Journal of Agricultural and Resource Economics. 41 (2), 247–265.

Mulla, D.J., 2013. Twenty five years of remote sensing in precision agriculture: Key advances and remaining knowledge gaps. Biosyst. Eng. 114, 358–371.

Olander, L.P., Haugen-Kozyra, K., Grosso, S. Del, Izaurralde, C., Malin, D., Paustian, K., Salas, W., 2011. Using Biogeochemical Process Models to Quantify Greenhouse Gas Mitigation from Agricultural Management Projects.Technical working group on agricultural greenhouse gases (T-AGG) Supplemental Report. https://nicholasinstitute.duke.edu/sites/default/files/publications/using-biogeochemical-process-paper.pdf

Otter, S., Ritchie, J.T., 1985. Validation of the Ceres-Wheat Model in Diverse Environments, in: Wheat Growth and Modelling. Springer US, Boston, MA, pp. 307–310. https://doi.org/10.1007/978-1-4899-3665-3_28

Pappas, C., Fatichi, S., Burlando, P., 2016. Modeling terrestrial carbon and water dynamics across climatic gradients: does plant trait diversity matter? New Phytol. 209, 137–151. https://doi.org/10.1111/nph.13590

Power, A.G., 2010. Ecosystem services and agriculture: tradeoffs and synergies. Philos. Trans. R. Soc. B Biol. Sci. 365, 2959–2971. https://doi.org/10.1098/rstb.2010.0143

Public News Service. 2014. Ohio Waterways Ranked 10th in the Nation for Toxic Chemicals. http://www.publicnewsservice.org/2014-06-24/water/ohiowaterways-ranked-10th-in-the-nation-for-toxicchemicals/a40095-1

Schaphoff, S., Forkel, M., Müller, C., Knauer, J., von Bloh, W., Gerten, D., Jägermeyr, J., Lucht, W., Rammig, A., Thonicke, K., Waha, K., 2018. LPJmL4 – a dynamic global vegetation model with managed land – Part 2: Model evaluation. Geosci. Model Dev. 11, 1377–1403. https://doi.org/10.5194/gmd-11-1377-2018

Schnepf, R., Yacobucci, B.D., 2013. Renewable Fuel Standard (RFS): overview and issues, CRS Report for Congress. https://sgp.fas.org/crs/misc/R40155.pdf

Scholze, M., Buchwitz, M., Dorigo, W., Guanter, L., Quegan, S., 2017. Reviews and syntheses: Systematic Earth observations for use in terrestrial carbon cycle data assimilation systems. Biogeosciences 14, 3401–3429. https://doi.org/10.5194/bg-14-3401-2017

Schwarz, G.E., Hoos, A.B., Alexander, R.B., Smith, R.A., 2006. The SPARROW surface water-quality model: theory, application and user documentation. US Geol. Surv. Tech. methods report, B. 6, 248.

Texas A&M Agrilife Research, 2019. EPIC & APEX Models [WWW Document].URL https://epicapex.tamu.edu/ (accessed 8.1.21)

Thompson, L.M., 1986. Climatic Change, Weather Variability, and Corn Production. Agron. J. 78, 649–653. https://doi.org/10.1126/science.188.4188.535

Turner, R.E., Rabalais, N.N., 2019. The Gulf of Mexico. World Seas an Environ. Eval. 445–464. https://doi.org/10.1016/B978-0-12-805068-2.00022-X

USDA NASS (National Agricultural Statistics Service). 2014. Quick Stats. https://quickstats.nass.usda.gov/

US EPA, 2021a. Inventory of U.S. Greenhouse Gas Emissions and Sinks 1990 - 2019. https://www.epa.gov/sites/default/files/2021-04/documents/us-ghg-inventory-2021-main-text.pdf?VersionId=wEy8wQuGrWS8Ef_hSLXHy1kYwKs4.ZaU (accessed 8.1.21)

US EPA, 2021b. How Do Emissions Trading Programs Work? [WWW Document]. URL https://www.epa.gov/emissions-trading-resources/how-do-emissions-trading-programs-work (accessed 8.1.21).

US EPA, 2020. Water Quality Trading [WWW Document]. URL https://www.epa.gov/npdes/water-quality-trading (accessed 8.1.21).
US EPA, 2019. Hydrological Simulation Program - FORTRAN (HSPF).[WWW Document].URL https://www.epa.gov/ceam/hydrological-simulation-program-fortran-hspf (accessed 8.26.21)
US EPA, 2010. Renewable Fuel Standard (RFS2) program (40 CFR 80 Subpart M). Code of Federal Regulations Title 40: Protection of Environment Part 80—Regulation of Fuels And Fuel Additives.
VanLoocke, A., Twine, T.E., Kucharik, C.J., Bernacchi, C.J., 2017. Assessing the potential to decrease the Gulf of Mexico hypoxic zone with Midwest US perennial cellulosic feedstock production. GCB Bioenergy 9, 858–875. https://doi.org/10.1111/gcbb.12385
Williams, K., Gornall, J., Harper, A., Wiltshire, A., Hemming, D., Quaife, T., Arkebauer, T., Scoby, D., 2017. Evaluation of JULES-crop performance against site observations of irrigated maize from Mead, Nebraska. Geosci. Model Dev 10, 1291–1320. https://doi.org/10.5194/gmd-10-1291-2017
Yao, F., Xu, Y., Lin, E., Yokozawa, M., Zhang, J., 2007. Assessing the impacts of climate change on rice yields in the main rice areas of China. Clim. Change 80, 395–409. https://doi.org/10.1007/s10584-006-9122-6
Zhang, W., Ricketts, T.H., Kremen, C., Carney, K., 2007. Ecosystem services and dis-services to agriculture. Ecol. Econ. 64, 253–260. https://doi.org/10.1016/J.ECOLECON.2007.02.024

Part IV
Including Nature in Engineering

Chapter 12
Seeking Synergies Between Technological and Ecological Systems: Challenges and Framework

Bhavik R. Bakshi

12.1 Motivation

The previous chapters have described approaches for quantifying the direct and indirect dependence of human activities on goods and services from nature, and approaches for determining the capacity of ecosystems to supply these goods and services. Seeking synergies between human and natural systems requires such information to develop ways by which human activities can benefit from nature's services while ensuring that nature's carrying capacity is respected. Such information is also essential for allowing engineering to participate in the transformation to a nature-positive world. The next several chapters will describe various efforts that seek synergies between human and natural systems. They will convey the many benefits of such synergies, which include the following:

- *Encouraging nature-positive decisions:* Accounting for ecosystems by keeping nature in the system boundary can result in solutions that respect nature's capacity by being aware of when and by how much this capacity is exceeded. This can encourage conventional engineering efforts for improving technological efficiency along with efforts toward protection and restoration of ecosystems.
- *Discovering innovative solutions:* Often, techno-ecologically synergistic solutions can provide goods and services to support human activities in a manner that can be economically feasible, environmentally superior, and socially more desirable than what is possible with conventional or techno-centric approaches.

B. R. Bakshi (✉)
School for Engineering of Matter, Transport and Energy, School of Sustainability, School of Complex Adaptive Systems, Arizona State University, Tempe, AZ, USA

Lowrie Department of Chemical and Biomolecular Engineering, The Ohio State University, Columbus, OH, USA
e-mail: bhavik.bakshi@asu.edu

B. R. Bakshi (ed.), *Engineering and Ecosystems*,
https://doi.org/10.1007/978-3-031-35692-6_12

Realizing such benefits requires overcoming many challenges and barriers, which are described in this chapter. For example, ecosystems tend to require less resources, rely on renewable energy, and can be more resilient to perturbations. However, ecosystems tend to exhibit intermittency and are often more difficult to control than human-designed systems. Approaches for overcoming some of the challenges are described in subsequent chapters.

The framework of techno-ecological synergy (TES) [1] is presented as a way to bridge the gaps between techno-centric and eco-centric methods to result in a novel approach for developing innovative designs that enable harmony between human and natural systems. This framework provides the foundation for assessing and designing coupled natural and human systems while explicitly accounting for their dependence and impact on each other. It provides science-based metrics for determining absolute environmental sustainability by comparing the demand and supply of ecosystem services across multiple spatial scales for a single region, activity, or life cycle. It forms the basis for designing networks of coupled human and natural systems based on managing the trade-offs while benefiting from the synergies. Through the resulting designs, human systems benefit from nature's contributions while natural systems receive respect and protection for their role and may be restored. The net outcome can be nature-positive decisions.

12.2 Challenges in Developing Synergistic Natural-Human Systems

12.2.1 Engineering Attitude Toward Nature

As described in Chap. 1, over the last two centuries, engineering, like most other disciplines, has developed while taking nature for granted and aiming to control its behavior. This has opened a wide chasm between how engineered systems are developed and expected to operate and how ecosystems function and provide goods and services. For example, technology is usually expected to behave in a consistent and predictable manner. When we put a light switch in the on position, we expect to get light instantaneously. We expect it to stay on and provide light with constant luminosity until we put the switch in the off position. In contrast, ecosystems tend to be intermittent and less predictable. For example, the availability of natural light depends on the season, whether it is day or night, and the amount of cloud cover. Controlling this is difficult if not impossible, and its predictability is also quite limited. In general, engineering systems follow the paradigm of "imposed-design" while ecosystems are "self-designed." Also, technological systems are usually designed and operated to exhibit homeostasis, that is, to stay at a desired set point like the bulb steadily providing light. In contrast, ecosystems, at scales larger than an organism, exhibit homeorhesis or intermittent and pulsing behavior like the fluctuations in riverine water levels due to floods, droughts, and other seasonal events. Furthermore, technology is usually developed without accounting for their dependence on ecosystems and without considering nature's constraints. For example, activities that combust fossil fuels such as power plants, manufacturing facilities, automobiles, and airplanes are not designed or operated while accounting for the fact that human activities have exceeded nature's capacity to sequester carbon. Similarly, traditional urban and industrial development rarely account for the total amount of renewable water available in the watershed and the effect of land-use change on large-scale water circulation through the water cycle. Instead, they rely on engineering to move water from other watersheds and even over mountains if necessary.

The goal of seeking synergies with nature may impose constraints on engineering design that may be difficult to satisfy and may require a change in the engineering mindset. For example, a manufacturing process that aims to respect the capacity of local vegetation to mitigate its air emissions may need to modify its operation so that its emissions adapt to intermittency in the pollution uptake capacity of trees. Similarly, synergistic designs for meeting water requirements will need to work with the homeorhetic nature of river flow and designs for maintaining indoor comfort will have to adapt to fluctuations in outdoor temperature and opportunities for using natural light.

In addition to such technical challenges, there are also educational challenges since engineers will need knowledge about ecosystems and their goods and services, including the importance of maintaining their basic structure and holistic properties. For example, a typical response from engineering regarding the role of trees to sequester carbon is to improve their efficiency of capturing carbon by innovations such as genetically modified trees and their monoculture plantation. Such an approach would be narrow, reductionist and unlikely to contribute to the goal of nature-positive engineering. It is also unlikely to be sustainable because of their susceptibility to pests, need for high external inputs like fertilizers and water, and little to no support for biodiversity. Thus, engineering will have to get over its traditional approach of maximizing reductionist efficiency with the attitude of wanting to dominate nature and engineer it to meet a narrow set of goals. In addition, engineers will need to become more ecologically literate and even learn the basics of ecosystem ecology in their academic programs. Studies convey the low ecological literacy of engineers [2]. Practitioners and researchers in other disciplines related to the proposed synergies such as ecology and environmental science will need to work with engineers to ensure that ecosystem characteristics such as biodiversity and resilience are maintained.

12.2.2 Learning from Nature

Ideas about working with nature and accounting for the role of ecosystems in engineering are not new. Many approaches have been developed in this direction, as shown in Fig. 12.1. Over the last few decades, with the introduction and enforcement

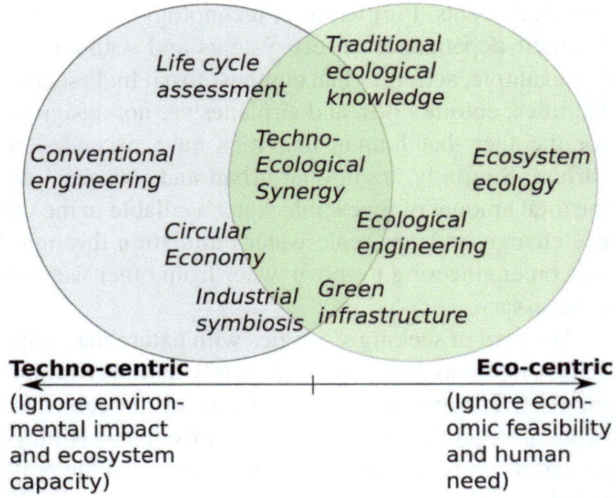

Fig. 12.1 Methods from engineering and ecology and their relevance to seeking techno-ecological synergies [3]

of environmental regulations, industrial processes focused on minimizing the generation of wastes and pollution. Such efforts were initially restricted to an individual process, but in the last two decades pollution from processes in the entire life cycle is being considered. These approaches are depicted in Fig. 12.2a. Thus, from the purely techno-centric direction, conventional engineering has made efforts toward environmental sustainability by expanding its system boundary beyond the product or process of interest. Methods like life cycle assessment (LCA) include activities in the supply and demand chain or value chain of a product or process and account for the environmental impact over this larger system. The goal of LCA is to reduce the chance of burden shifting to other parts of the life cycle and encourage reduction of the total or life cycle environmental impact. Details about LCA are provided in Chaps. 3 and 20.

Efforts toward developing a circular economy aim to keep products and materials such as plastics and metals in the economy and prevent them from entering the environment. The goal is to transform the mostly linear economy of "take-make-use-dispose" to a circular one by including tasks such as reuse, refurbishment, and recycling. This is depicted by the gray loop in Fig. 12.2b. Industrial symbiosis or byproduct synergy focuses on keeping waste in industrial networks by finding solutions where waste from one industry is used as a resource in another. This is depicted by the red loop in Fig. 12.2b. These efforts of circular economy and industrial symbiosis are inspired by nature's ability to sustain itself by intense cycling of materials and by using waste from one activity as a resource in another. Even though inspired by ecosystems, these efforts still take nature for granted and ignore its role and need not respect nature's limits or its intermittent character. Therefore, it is unlikely that they will encourage the development of nature-positive solutions.

From the direction of ecology, approaches to connect with human systems and engineering include efforts such as use of traditional ecological knowledge to

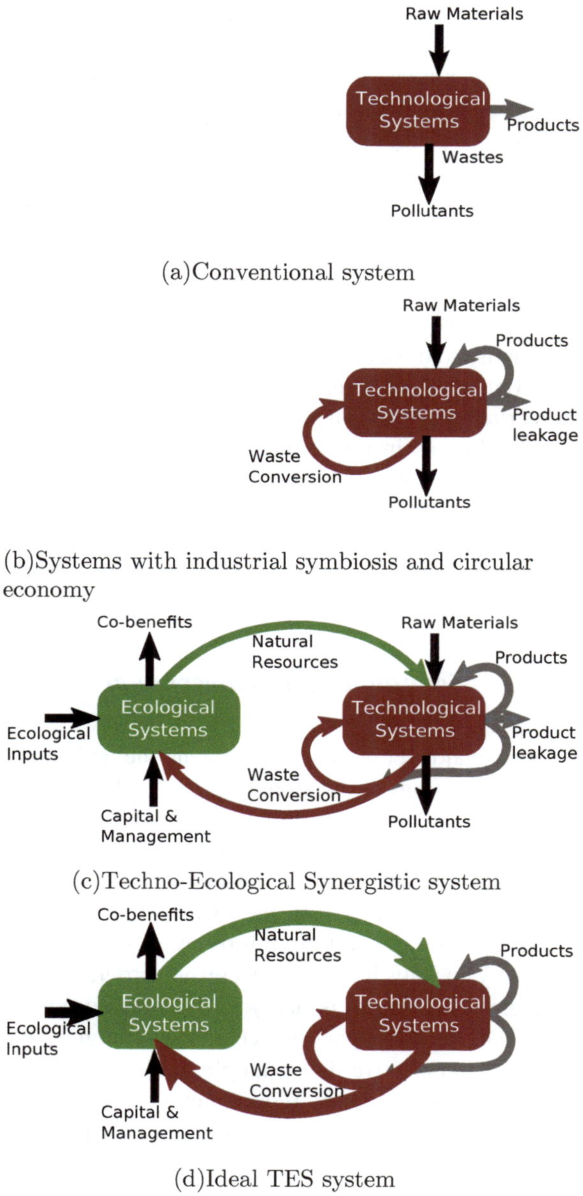

Fig. 12.2 (**a**) Conventional engineering design focuses on linear systems; (**b**) industrial symbiosis and circular economy aim to circulate economic goods in human-designed systems; (**c**) techno-ecologically synergistic systems account for the role of ecosystems in mitigating emissions and supplying goods and services to human-designed systems; (**d**) eventually, TES designs may result in fully circular and synergistic systems that meet human needs while respecting nature's capacity

meet human needs, ecological engineering, and inclusion of green infrastructure in engineered systems. Traditional ecological knowledge predates modern engineering by many centuries. It refers to knowledge systems for managing natural resources that were developed and handed down over generations, usually by indigenous cultures. Such systems are usually tailored to a specific location based on a deep understanding of local conditions and ecosystems that are often part of the community's religious or spiritual practices. Ecological engineering develops ecosystems to meet human needs while maintaining their structure and function [4]. A popular example of ecological engineering is the use of wetlands to treat wastewater. Such efforts are also ancient but are being rediscovered and updated for modern societies. For example, modern treatment wetlands may be designed to treat household wastewater as was done traditionally but also industrial wastewater that contains novel chemical entities such as heavy metals, insecticides, and other industrial chemicals. Green infrastructure refers to the use of ecosystems or nature-based solutions in urban planning and design. Examples include the use of rain gardens, bioswales, and wetlands to manage water runoff from buildings and parking lots and planting of trees and constructing walls with climbing plants to reduce heating and cooling requirements. These efforts are usually not fully integrated with technological systems and often represent "end of pipe" solutions.

12.2.3 Economics and Governance

Another major set of challenges facing the inclusion of nature in engineering decisions comes from the economic system and government policies. The conventional free market economy as practiced today in most parts of the world focuses on maximizing monetary benefits to consumers and producers. In an ideal free market, the market equilibrium maximizes the benefit to both. However, ecosystems are usually outside the market due to which goods and services from nature have no monetary value. Therefore, efforts by a corporation to seek synergies with nature are likely to only incur a cost but not provide any direct revenue. Thus, seeking synergies with nature is likely to decrease corporate profit, which means there is no reason for a company to seek such synergies, at least not from a conventional monetary perspective. For example, if a facility that emits air pollutants and meets environmental regulatory requirements, plants trees to mitigate emissions beyond regulatory limits, the cost of planting trees will be borne by the company while the benefit will be to society. In addition, regulatory agencies often do not allow the use of nature-based solutions to meet environmental regulations. Also, regulations are rarely based on accounting for nature's capacity but are based on estimated impact and political considerations. There is a need for approaches to verify and monitor such solutions before they can be used with confidence in their ability to deliver the benefits. Insight is also needed about the corporate benefits of protecting and restoring nature such as the effect on reputation, license to operate, and other intangible contributions.

12.3 Assessing and Designing Techno-Ecological Synergies

Assessment and design of synergistic techno-ecological systems need to quantify how human activities depend on nature's contributions and the extent to which nature is capable of supporting these activities. For this, we rely on determining the *demand* for ecosystem services imposed by human activities on nature and the *supply* of goods and services that ecosystems are capable of providing for human use, that is, after ecological needs are met. The demand of goods such as ores, fossil fuels, and water is quantified by the amount used by human activities. The demand for services is quantified by the emissions into the environment. For example, the demand for the carbon sequestration ecosystem service is quantified by the quantity of carbon dioxide emitted and the demand for insect pollination is quantified by the insects needed to yield fruits of typical quality and quantity. Table 12.1 lists the demand and supply of various ecosystem goods and services.

This table also lists the largest spatial scale that is relevant to each ecosystem service. This region is called the serviceshed. It is the region from which the selected ecosystem service can be available to an activity in that region. For example, the serviceshed for carbon sequestration is global since an emitted molecule of CO_2 can be taken up anywhere in the world. As shown in Table 12.1, the serviceshed for air quality regulation is regional, for water provisioning is the watershed, and for pollination is the region covered by pollinators. Approaches relevant to quantifying the demand of ecosystem services are described in Chaps. 3–7 of this book and for quantifying the supply are in Chaps. 8–11.

As discussed in Sect. 12.2.2 and depicted in Figs. 12.2a and b, the conventional linear approach and more recent approaches based on industrial symbiosis and circular economy ignore the role of ecosystems and their capacity. The framework of techno-ecological synergy (TES) explicitly accounts for the role of ecosystems in sustaining technological and other human systems. As shown in Fig. 12.2c, a TES system considers the role of ecosystems in mitigating emissions, enabling circularity of products, and providing resources to the technological system. As shown in this figure, in addition to meeting the needs of the technological system, ecosystems can also provide many cobenefits to society. They rely on inputs from nature and may

Table 12.1 Demand and supply for some ecosystem services

Ecosystem service, k	Demand, $D_{i,j,k}$	Supply, $S_{i,j,k}$	Largest scale, J
Carbon sequestration	CO_2 emission	Ecological sequestration capacity	Global
Water provisioning	Water withdrawal	Renewable water from rivers, aquifers, etc.	Watershed
Air quality regulation	Air pollutants	Cleaning capacity of trees, wind, etc.	Regional
Pollination	Flowers needing pollination	Availability of pollinators	Local

also need economic inputs such as capital and land. Eventually, it is expected that the system in Fig. 12.2c will evolve to Fig. 12.2d where the TES system is highly circular and fully synergistic with minimum input and emissions.

12.3.1 Assessment

Assessing the extent of synergies between technological and ecological systems relies on comparing the demand and supply for specific ecosystem services. For the k-th ecosystem service, the sustainability metric may be defined as,

$$V_k = \frac{S_k - D_k}{D_k}$$

where S_k is the supply of that ecosystem service and D_k is its demand. Sustainability is indicated by $V_k \geq 0$. Note that V_k can have values between -1 and $+\infty$. This metric may be determined at a specific spatial scale, with the serviceshed being the largest scale of interest for that ecosystem service. The metric at the serviceshed scale is represented as V_k^*. For example, V_k for a house and yard may be determined by its water consumption as the demand and precipitation as the supply from nature. Absolute environmental sustainability for the k-th ecosystem service requires $V_k^* \geq 0$. Depending on the values of V_k and V_k^*, four situations are possible as depicted in Fig. 12.3. In the first quadrant, V_k and V_k^* are positive, implying that the system is sustainable locally and in the serviceshed. This is the best situation from the perspective of environmental sustainability. Systems in the second quadrant are locally unsustainable but sustainable in the serviceshed. Such systems are islands of unsustainability in a serviceshed that is sustainable. In the third quadrant neither local nor absolute sustainability are achieved, which is the worst situation. In the fourth quadrant we have an island of sustainability in an unsustainable serviceshed. Application of these metrics can help in determining the degree of ecological overshoot of a given activity, as demonstrated in more detail in Chap. 20.

Metrics such as V_k that compare the demand and supply of ecosystem services indicate absolute environmental sustainability since their comparison is with nature's carrying capacity. Most of the commonly used metrics for environmental sustainability assessment are based on resource use or environmental impact. Examples include global warming potential which indicates the impact on climate change and ecotoxicity that indicates ecological impact due to toxicity of pollutants. Such metrics do not compare the impact with nature's carrying capacity and are best for comparing alternative activities. Thus, they measure relative sustainability.

Relative sustainability metrics are best for improving existing products and processes by reducing their environmental impact. Such metrics encourage doing "less bad." However, since relative sustainability metrics do not account for nature's carrying capacity and the degree of overshoot, they tend to be ignorant of the extent

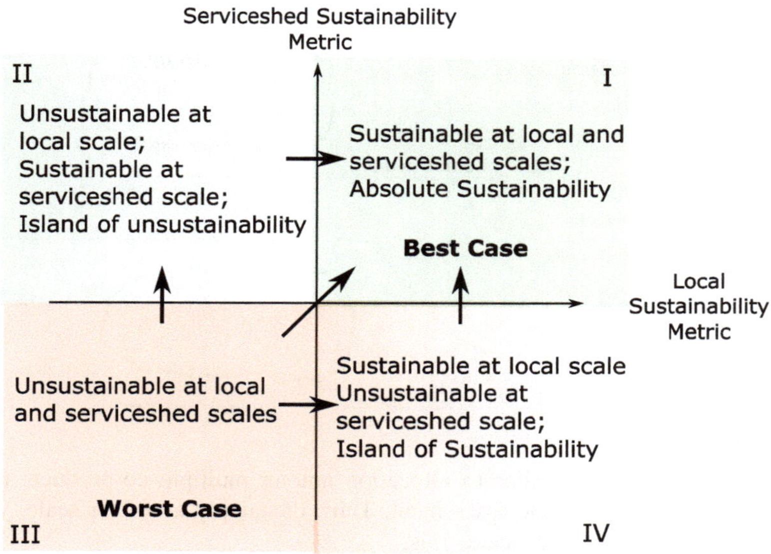

Fig. 12.3 Insight from local (V_k) and serviceshed scale (V_k^*) metrics. (Reproduced with permission from [8])

to which environmental impact needs to be reduced to operate within nature's constraints. Thus, doing less bad may not be good enough for environmental sustainability. In contrast, absolute sustainability metrics can encourage doing "less bad" by reducing impact, but they could also encourage doing "more good" by protecting and restoring ecosystems to maintain or enhance the supply of ecosystem services. Thus, absolute sustainability metrics can encourage the establishment of techno-ecological synergies at multiple spatial scales. Practical examples to demonstrate this characteristic are in Part IV of this book.

The supply of ecosystem services in a region is used by multiple activities in the region. For example, the capacity of trees to sequester carbon dioxide and regulate air quality is used by all activities that emit pollutants such as carbon dioxide, particulate matter, and oxides of sulfur and nitrogen. Assessment of individual activities requires allocation or partitioning of the supply of regional ecosystem services to each activity. A variety of approaches are available for allocation of ecosystem services as summarized here and illustrated in Fig. 12.4. The simplest approach is to distribute the supply of the ecosystem service to its users in proportion to a quantity such as population, land area, demand for the ecosystem service, gross domestic product, or reciprocal of GDP. The allocation or partitioning coefficient is calculated as the ratio of the quantity selected for allocation in the i-th activity divided by the total quantity in the region. Thus,

$$P_{i,j,k} = \frac{x_{i,j,k}}{\sum_i x_{i,j,k}} \tag{12.1}$$

Fig. 12.4 Partitioning of publicly owned ecosystem services to users in the region [9]

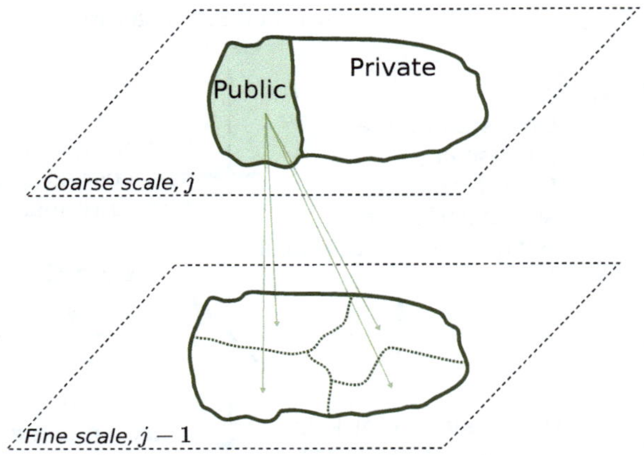

This approach is very similar to allocation among multiple co-products that is commonly used in life cycle assessment. Thus, the supply at a finer scale, $j-1$, by allocation from a coarser scale, j, is,

$$S_{i,j-1,k} = P_{i,j,k} S_{i,j,k} \tag{12.2}$$

As an example, the finer scale could be users in a region and coarser scale would be the region. Two approaches that are commonly used for allocation are described below. Planetary boundaries (PB) have been quantified [5, 6] to determine the "safe operating space" for humanity. This space may also be interpreted as the supply of ecosystem services available for human use. Many efforts [7] for quantifying absolute environmental sustainability metrics rely on directly downscaling the planetary boundary by Eq. 12.2 where $S_{i,j,k}$ is the safe operating space. Another approach [8] combines downscaling with the locally available supply of ecosystem services. Consider a farmer who uses conservation tillage practices and is able to sequester carbon in the soil of his farm. Downscaling by Eq. 12.2 distributes the available supply of ecosystem service equally among all users according to the selected criterion. It does not give any credit to the farmer for the use of conservation tillage on his farm. Therefore, another way of determining the supply is as follows, [9]

$$S_{i,j-1,k} = S^{loc}_{i,j-1,k} + P_{i,j,k} S^{pub}_{i,j,k} \tag{12.3}$$

Here, $S^{loc}_{i,j-1,k}$ is the local ecosystem service supply available from the specific farm, while $S^{pub}_{i,j,k}$ is the supply of that service from public land at the coarser scale, j. Thus, this approach distinguishes between services from private and public land. Services from private land are considered to belong to the landowner and are not allocated to anyone else. Services from public land such as government-owned land including state and national parks are considered to belong to everyone and may

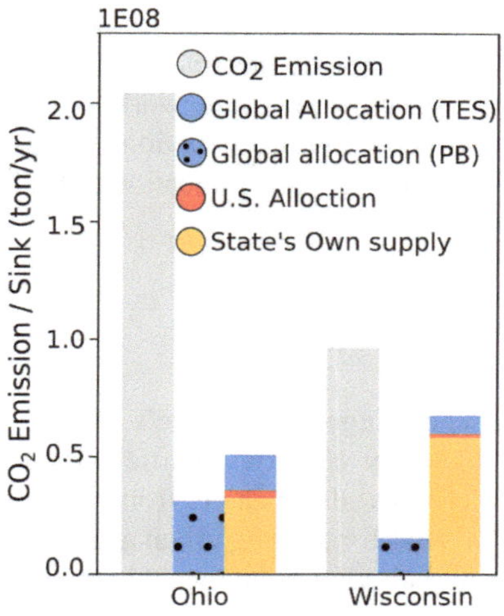

Fig. 12.5 Demand and supply of carbon sequestration ecosystem service for Ohio and Wisconsin [9]. Gray bar is the demand in each state. Blue bar with dots is the supply by downscaling planetary boundary (Eq. 12.2). Multicolored bar considers local supply in each state and downscaled supply from federally owned land in the U.S. and global oceans (Eq. 12.3)

be downscaled or distributed among all users. Thus, this approach distinguishes between public and private land and assumes that ecosystem services from private land belong to the landowner. In contrast, the direct downscaling approach based on Eq. 12.2 considers all ecosystem services, whether on private or public land, to belong to everyone. Such an approach may disincentivize protection and restoration of ecosystems since such activities will not directly benefit the landowner but will be distributed between the landowner and everyone across the serviceshed.

Application of this assessment approach to the carbon sequestration ecosystem service in the U.S. states of Ohio and Wisconsin is depicted in Fig. 12.5. In this figure, the gray bar represents emissions from the state. The blue bar with black dots is the global carbon sequestration capacity or safe operating space allocated to these states in proportion to their population. The V_k metric for states using this approach of downscaling planetary boundaries (PB) is −0.85 for Ohio and −0.84 for Wisconsin. The multicolored bars represent the carbon sequestration capacity calculated by the TES approach [9]. The yellow bar is the local sequestration capacity in forests in the state. The other bars are the allocated capacity from publicly owned land in the U.S. and international waters in the world. Values of the V_k metric using this TES approach are −0.75 for Ohio and −0.31 for Wisconsin. The key difference between the metrics from the two approaches is that the TES-based approach gives each state full credit for its local sequestration capacity whereas the PB-based approach distributes the state's capacity across the entire world. Since Ohio and Wisconsin have a larger sequestration capacity than the world average, V_k based on TES is larger than the value based on direct downscaling of planetary boundaries. In addition, the larger sequestration capacity in Wisconsin is also reflected in its larger value than for Ohio. Thus, PB-based approaches like those in [7] may disincentivize local action for protection and restoration of ecosystems since the benefits of such efforts are distributed, while the cost is going to be to the landowner or stakeholder.

Among methods for sustainability assessment such as carbon footprint and life cycle analysis, the traditional system boundary does not account for the role of ecosystems. The framework for techno-ecological synergy can be used to overcome this shortcoming, as is done by the approach of TES-LCA [8]. Details about this approach and an application to biofuels are provided in Chap. 20.

12.3.2 Design

Benefitting from nature's ability to provide various services has been popular for many centuries, particularly in traditional societies [10]. Some engineering activities also benefit from ecosystems. For example, wetlands are increasingly popular for treating urban and industrial wastewater [11] due to their environmental and economic benefits [12]. More recently, cities and manufacturing sites have considered the role of trees in improving air quality and various projects have been initiated in that direction [13]. Such inclusion of ecosystems in industrial and urban settings is described in Chaps. 13 and 14.

Most existing designs that include ecosystems use nature as an end-of-the-pipe solution. The use of wetlands to treat water and of trees to clean air is considered after the polluted water and air leave the process that generates them. Thus, the process and ecosystem are designed separately in a disintegrated manner. Often, the process is designed first, and the ecosystem is added later to address the impact of the pollutant stream. An integrated design would consider the process and ecosystem together and design them simultaneously such that the process adapts to the ecosystem's ability to treat pollution and the ecosystem is designed to meet the process needs. Such an integrated design is likely to perform the same or better than any end-of-pipe design.

One way of developing integrated designs of manufacturing processes and ecosystems is to consider ecosystems to be analogous to technological equipment or unit operations and then apply conventional engineering design methods. Thus, for a single-family house and yard, ecosystems in the yard could be considered just as technologies in the house are considered during the design phase. Then the ability of trees around the house to reduce cooling needs in the summer could result in the installation of a smaller capacity air-conditioning system than what would be installed without accounting for the effect of trees. Similarly, for a chemical process, ecosystems such as forests, soil, and wetlands that provide services to the process could be "designed" together with conventional process equipment. Then, if a wetland is available for treating polluted water in a manner that is less expensive than a technological approach, the process may rely less on purification by using expensive equipment such as membranes or distillation columns. Such an integrated solution could have a smaller overall cost to the company and may also reduce the cost incurred by society due to the pollutants. Such solutions can also encourage industry and society to protect and restore ecosystems and contribute to developing a nature-positive world. Several examples of such benefits are described in Part IV of this book.

Optimization is a popular and powerful and popular approach for the design of conventional and sustainable engineering solutions. It determines optimal values of design variables that maximize or minimize specified objectives while meeting constraints. An important characteristic of sustainable design is that multiple objectives often need to be considered to account for environmental, social, and ecological aspects. For TES design, in addition to accounting for the behavior of technological systems, it is also necessary to consider the behavior of ecosystems. The typical formulation of a multiobjective TES design problem is as follows:

$$
\begin{array}{lll}
\max & P(x_t, x_e); & \text{maximize profit} \\
\min & \phi(x_t, x_e) & \text{minimize impact} \\
\text{s.t.} & f_t(x_t, x_e) = 0 & \text{technology models} \\
 & f_e(x_t, x_e) = 0 & \text{ecosystem models}
\end{array}
$$

This formulation focuses on the economic objective of maximizing profit and the environmental objective of minimizing environmental impact. It is subject to constraints that capture available technological and ecological alternatives. Thus, for the house and yard design problem, the constraints represent models of the house, effect of trees, etc. Decision variables in this problem include characteristics of technological and ecological systems such as orientation of the house, type of appliances selected, type of insulation and light fixtures, location and species of trees, ways of using the available land, etc. Other variables such as heating and cooling set point may also be included. They depend on human preferences and behavior. The result shown in Fig. 12.6 is for the design that minimizes the carbon footprint of the overall system [14].

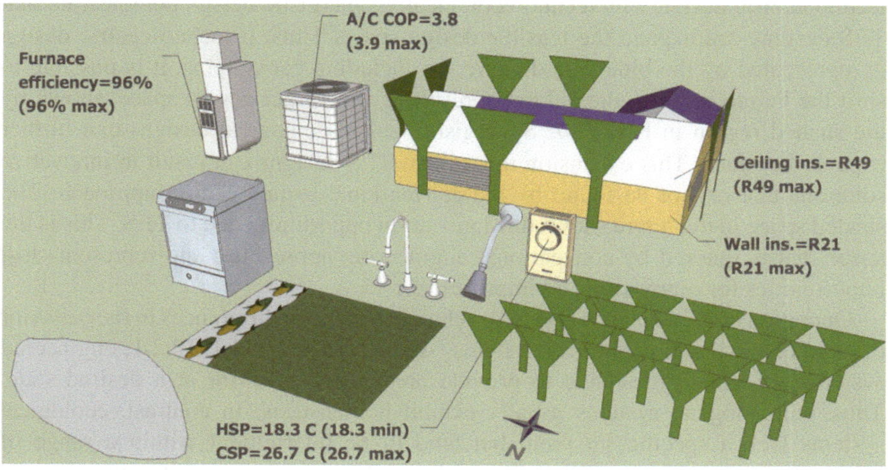

Fig. 12.6 Technological, ecological, and behavioral options for synergistic design of a house and yard. This design is for minimizing the carbon footprint [14]

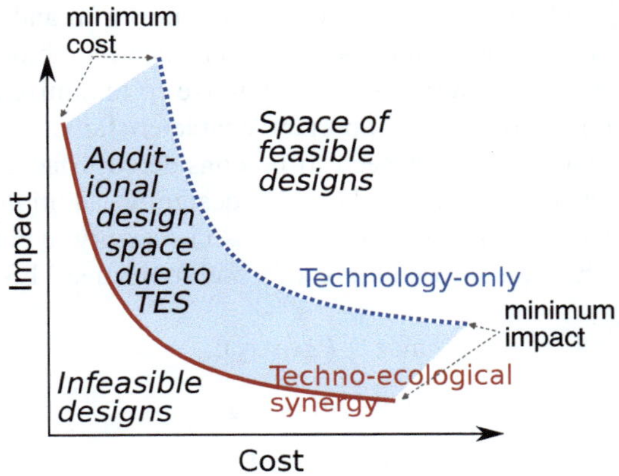

Fig. 12.7 Pareto curves for technology-only and techno-ecological synergy designs. TES designs expand the design space to include solutions that are not attainable by conventional techno-centric approach

Multiobjective optimization provides insight into the trade-off between multiple objectives. For the dashed-blue "technology-only" curve in Fig. 12.7, the two objectives are minimization of cost and minimization of impact. The right extreme point for each curve is the optimum for minimum impact and the left extreme of the same curve is for minimum cost. The curve between these two extremes represents the trade-off between the two objectives. Designs that lie on this curve have cost and impact values between the extreme values, and no design can be found that does better for both objectives. To the right and above the curve is the space of feasible designs. For values of the objectives in this space, it is possible to find feasible designs. However, the other side of the curve is not attainable: no design can be found that has values of objectives represented by any point in the unattainable region.

Several studies like those in Chaps. 16, 17, 19 that compare conventional or techno-centric design with techno-ecologically synergistic design demonstrate that TES designs can expand the feasible design space. Thus, if techno-centric design is represented by the blue dotted curve, by including ecosystems, it is possible to shift the Pareto curve to the red curve resulting in a larger design space, shown by the shaded region in Fig. 12.7. Synergistic techno-ecological designs can further expand the space. This expansion means that TES designs can result in innovative solutions that cannot be found by the conventional techno-centric approach. The shaded space in the figure shows this innovation opportunity due to TES. This is the space that is ignored by conventional engineering approaches and represents lost opportunities for innovation and improvement.

Operation of TES designs also poses challenges due to differences in the behavior of technological and ecological systems. As discussed in Sect. 12.2.1, engineered systems are usually designed to provide predictable behavior at a desired state. Thus, technological systems usually exhibit homeostasis. In contrast, ecological systems lack a specific set point but tend to be intermittent within a range of values. Such behavior is called homeorhesis. Due to nature's intermittency, the services provided by ecosystems to human-designed systems may change over time and across seasons. For example, the capacity of trees to take up pollutants increases as the tree grows. Further, deciduous trees lose this capacity in the Fall and Winter seasons since such trees shed their leaves. The designed TES system will need to be operated while taking nature's intermittency into account. This can be accomplished by relying more on technological pollution control or reducing the production rate according to the extent to which ecosystem services are available. In such an approach, rather than implicitly assuming that nature can take care of industrial pollutants, TES systems will be operated so that technology adapts to nature's intermittency. Some initial work in this direction and its potential benefits are described in Chap. 19.

Given the inherent resilience of most natural ecosystems, it is often expected that TES systems will be more resilient than conventional techno-centric systems. Thus, wetlands used for treating water are likely to be more resilient to flooding and power outages than conventional technological solutions. Similarly, vegetation is likely to be more resilient to perturbations such as storms than scrubbers and filters. However, ecosystems may be less resilient to other perturbations such as forest fires and pest outbreaks. In addition, a trade-off exists between efficiency and resilience: efficient systems may be economically more attractive but less capable of recovering from perturbations. Such considerations need to be included in the design of TES systems, as discussed in Chap. 21.

12.4 Conclusions

Ecosystems provide many goods and services that are essential for technological and other human activities. In this chapter, we focus on the challenges and potential benefits of techno-ecologically synergistic systems. The two key benefits are that TES systems are likely to be aware of the role of nature and ecological constraints. This can result in greater incentive to protect and restore ecosystems. Secondly, since nature can provide many services at little monetary cost, TES systems can be economically and environmentally superior to conventional techno-centric systems. For society to benefit from synergies between human and natural systems, it is essential for engineering to shift its paradigm toward accounting for nature and respecting its limits. This is in direct contrast to the paradigm followed in modern engineering for about three centuries. There is also a need to learn from nature and emulate its ways. This faces the challenge of ecological illiteracy among engineers. In addition, various economic and governance challenges also exist for wide adoption of TES designs.

Methods for assessing the extent of techno-ecological synergy compare the demand and supply of specific ecosystem services in a specified location. Such TES metrics may be calculated at a local and serviceshed scale. At a local scale, the supply of ecosystem services may be quantified by means of ecological models or data. In addition to this local supply, the supply from larger scales, that is, publicly owned, may also be allocated to the local system. Absolute environmental sustainability requires the demand to be within the supply of and is calculated at the serviceshed scale. Such an approach may be used for a single process or a life cycle.

Design of TES systems is likely to be most beneficial when the designed system integrates the characteristics of technological and ecological systems. Current use of ecosystems in industrial, urban, and other systems tends to be as end-of-pipe solutions. Integrated TES designs may be developed by solving multiobjective optimization problems. Existing case studies indicate that TES designs are likely to expand the design space to enable innovative and win-win designs. Furthermore, to address the intermittency of ecosystems, technological systems could be operated so that they adapt to this intermittency. Case studies in subsequent chapters provide details about the approach for TES analysis and design, their attractive characteristics, and research challenges.

References

1. Bhavik R. Bakshi, Guy Ziv, and Michael D. Lepech. Techno-ecological synergy: A framework for sustainable engineering. *Environmental Science & Technology*, 49(3):1752–1760, 2015.
2. Shih-Yun Kuo and Nancy L. Jackson. Influence of an environmental studies course on attitudes of undergraduates at an engineering university. *The Journal of Environmental Education*, 45(2):91–104, 2014.
3. B. R. Bakshi. Toward sustainable chemical engineering: The role of process systems engineering. *Annual Review of Chemical and Biomolecular Engineering*, 10:265–288, 2019.
4. W. J. Mitsch and S. E. Jørgensen. *Ecological Engineering and Ecosystem Restoration*. John Wiley & Sons, Inc., 2004.
5. J. Rockstrom, W. Steffen, K. Noone1, A. Persson, F. S. Chapin, E. F. Lambin, T. M. Lenton, M. Scheffer, C. Folke, H. J. Schellnhuber, B. Nykvist, C. A. de Wit, T. Hughes, S. van der Leeuw, H. Rodhe, S. Sörlin, P. K. Snyder, R. Costanza, U. Svedin, M. Falkenmark, L. Karlberg, R. W. Corell, V. J. Fabry, J. Hansen, B. Walker, D. Liverman, K. Richardson, P. Crutzen, and J. A. Foley. A safe operating space for humanity. *Nature*, 461(7263):472–475, 2009.
6. Will Steffen, Katherine Richardson, Johan Rockström, Sarah E. Cornell, Ingo Fetzer, Elena M. Bennett, Reinette Biggs, Stephen R. Carpenter, Wim de Vries, Cynthia A. de Wit, Carl Folke, Dieter Gerten, Jens Heinke, Georgina M. Mace, Linn M. Persson, Veerabhadran Ramanathan, Belinda Reyers, and Sverker Sörlin. Planetary boundaries: Guiding human development on a changing planet. *Science*, 347(6223), 2015.
7. Ryberg, Morten W., et al. Downscaling the planetary boundaries in absolute environmental sustainability assessments-A review. *Journal of Cleaner Production*, 276: 123287, 2020.
8. Xinyu Liu and Bhavik R. Bakshi. Ecosystem services in life cycle assessment while encouraging techno-ecological synergies. *Journal of Industrial Ecology*, 23(2):347–360, 2019.
9. Ying Xue and Bhavik R. Bakshi. Metrics for a nature-positive world: A multiscale approach for absolute environmental sustainability assessment. *Science of the Total Environment*, 2022. accepted.
10. Fikret Berkes. *Sacred Ecology*. Routledge, third edition, 2012.
11. R. H. Kadlec and S. D. Wallace. *Treatment Wetlands*. CRC Press, second edition, 2009.
12. Johnathan L. DiMuro, France M. Guertin, Rich K. Helling, Jessica L. Perkins, and Scanlon Romer. A financial and environmental analysis of constructed wetlands for industrial wastewater treatment. *Journal of Industrial Ecology*, 18(5):631–640, 2014.
13. Katherine J. Willis and Gillian Petrokofsky. The natural capital of city trees. *Science*, 356(6336):374–376, 2017.
14. R. A. Urban and B. R. Bakshi. Techno-ecological synergy as a path toward sustainability of a North American residential system. *Environmental Science & Technology*, 47(4):1985–1993, 2013.

Chapter 13
Making the Business Case for Nature-Based Solutions

Martha Rogers, France Guertin, Eric Lonsdorf, Chris Nootenboom, Lianna McFarlane-Connelly, and Todd Guidry

13.1 Introduction

Historically, for-profit companies have justified capital investments based on maximizing profits. An investment's success would largely be determined through return on investment and net present value calculations. When it comes to the environment, the main consideration would often be how to minimize negative impacts, but the full range of environmental impacts and services were not central to the investment decision. Tools like social and environmental impact assessments and biodiversity action plans can help assess the level of environmental impact but do not identify opportunities to leverage or improve the benefits of environmental services in a project's design. To truly incorporate nature and nature-based decisions into planning and designs, however, companies will need to shift their focus to one that values nature's contributions and that prioritizes projects that benefit nature.

M. Rogers · L. McFarlane-Connelly
The Nature Conservancy, Arlington, VA, USA

F. Guertin (✉)
The Dow Chemical Company, Midland, MI, USA
e-mail: FMGuertin@dow.com

E. Lonsdorf
Environmental Sciences, Emory University, Atlanta, GA, USA

C. Nootenboom
Institute on the Environment at the University of Minnesota, Natural Capital Insights, St Paul, MN, USA

T. Guidry
Global Technology Center Civil Engineering Department, The Dow Chemical Company, Chennai, India

B. R. Bakshi (ed.), *Engineering and Ecosystems*,
https://doi.org/10.1007/978-3-031-35692-6_13

Over the last decade, companies have moved toward a better understanding of the full suite of benefits that environmental assets provide to their operations and business. Often called nature-based solutions (NBS), these solutions utilize natural and modified ecosystems in a way that effectively addresses business challenges and benefits human well-being and biodiversity (IUCN 2021). A company can also make decisions, such as a facility's activities, products, or services, that impact the environment in both positive and negative ways.

In 2011, The Dow Chemical Company (Dow) and The Nature Conservancy (TNC) announced the beginning of a groundbreaking collaboration aimed, in part, at demonstrating the value of NBS to business (Business Wire 2011). Their research together has shown that floodplain restoration can be a cost-effective means of addressing industrial water scarcity issues along the Brazos River in Texas (Reddy et al. 2015) and that reforestation can be a cost-effective means of reducing high ambient ozone concentrations in areas of the United States not in attainment with National Ambient Air Quality Standards (Kroeger et al. 2014). In addition, Dow has demonstrated on-the-ground savings exceeding $282 million from a tertiary treatment wetland in Seadrift, Texas, when compared to a conventional wastewater treatment facility (sequential batch reactor) delivering the same treatment capability (DiMuro et al. 2014).

However, moving from a singular case study or project to incorporating nature into basic business and engineering principles is not a simple transfer of traditional conservation science and values. Many other companies have noted that NBS, especially when used in combination with grey infrastructure, are often overlooked even though they are an essential element in a company's portfolio of solutions (TNC et al. 2013). To drive adoption of NBS in business settings, early research on the value of NBS must be adapted and translated to complement ongoing or existing organizational processes of strategy and decision-making.

Often, operationalizing NBS within a company can take years (TNC 2019). To fully operationalize NBS, companies need to understand when NBS can deliver value, how to quantify NBS' financial and nonfinancial benefits, and how to engineer innovative NBS. Fully bringing NBS into a company's operations delivers benefits to multiple groups; the company wins because NBS can offer financial savings relative to traditional grey projects, either through cost savings or additional revenue generation, and the broader community wins because NBS often provide accompanying ecosystem service benefits. In the case of the Seadrift wetlands example, the design parameter was total suspended solids (TSS) as specified in the Texas Commission on Environmental Quality's water quality permit. To date, the wetlands have met the TSS while also improving air quality and local habitat creation; they have also supported a diverse species of birds in search of freshwater habitat not captured within the stated $282 million savings.

In this chapter, we review the processes that TNC and Dow developed to help scale NBS across the company. While the process was designed for Dow's 2025 Valuing Nature Goal, which aims to deliver $1B in net present value through projects that are good for business and better for the environment, it is easily transferable across companies (Dow 2021). In the end, projects that count toward this goal will include both NBS and man-made (i.e., grey) solutions that are shown to be better for nature without significant trade-offs, such as projects that use resources in more productive ways.

Specifically, this chapter focuses on the development of what we have termed a "nature scorecard" – a tool used to complement financially driven decision-making by providing easily comparable nature metrics for a project. In the following sections, we review Dow's overall process for operationalizing the Nature Goal and driving further adoption of NBS within the organization which includes the development and implementation of the nature scorecard.

13.2 A Methodology for Valuing Nature

When Dow issued its Valuing Nature Goal in 2015, it did not yet have a process for driving uptake of the goal across the organization. In rolling out the goal, there were three main challenges faced:

1. Quickly and easily identifying potential Nature Goal projects.
2. Measuring and comparing the ecosystem impacts, positive or negative, of potential Nature Goal project options.
3. Understanding the importance of these ecosystem impacts to Dow and the local environment.

To address these challenges, Dow and TNC developed a three-tiered project valuation methodology (Guertin et al. 2019):

1. An initial screen to identify potential opportunities at a very early stage of a project.
2. A subsequent analysis to identify and quantify environmental impacts of the proposed project. This analysis may use various tools and methodologies as long as the results are able to effectively evaluate the project's overall impact on ecosystem services with and without the use of NBS alternatives.
3. A final step that considers and compares the financial and natural capital returns associated with the various project alternatives.

For each step in this valuation methodology, the idea was to develop tools that are easy to use and scientifically credible. Developing an approach to value nature can be daunting as natural systems are both highly complex and dynamic. As proposed by Walker and Salt (2010) in their book entitled *Resilience Practice*, valuation requires an adaptive and dynamic approach with a focus on what is important. A key concept in resilience thinking is requisite simplicity while being rigorous.

Dow also developed the term "engineered natural technologies" to reference NBS internally with the goal of making NBS communication more relatable to project managers, engineers, and business leaders across Dow. Engineered natural technologies is defined as "engineered systems that use or mimic natural processes able to deliver the same design functionality as a man-made solution while affording benefits to people and the planet."

In the initial screen, project engineers are asked a few simple yes or no questions about their project's potential impacts on nature. The screen introduces the concept of "ecosystem services" to project staff, using the definition from the Millennium Ecosystem Assessment (2005) summarized as the benefits people get from ecosystems. For the purposes of Dow's goal, TNC and Dow also constructed an internal working definition of nature:

> Earth's collective inhabitants and nonliving environments interacting as functional ecosystems and providing services such as clean water, clean air, and healthy soil.

With this working definition established, the Collaboration built the screening tool to reflect the four key elements of nature highlighted in the definition: (i) an ecosystem's overall functioning, (ii) clean water, (iii) clean air, and (iv) healthy soil. The screening tool includes one question related to each of these elements and is designed to identify projects at Dow that have a significant impact on at least one of the four elements. In addition, the screening tool promotes the consideration of NBS in Dow projects. An example of the screen is included in the Appendix corresponding to the Dow Riverside Old Ash Pond Closure project, which is presented in the case study section.

The second step of the valuation methodology involves the quantification of the ecosystem service impacts of the project using such methods as a life cycle assessment or the Ecosystem Services Identification and Inventory Tool (ESII Tool). Life cycle assessment is a very mature methodology for assessing the environmental aspects associated with a product over its life cycle. It is used to analyze the overall environmental impact of a product or process associated with the various stages of its production.

The ESII Tool is an ecosystem service modeling tool developed by Dow, EcoMetrix Solutions Group (ESII 2021), and TNC designed to help decision-makers understand ecological consequences of proposed changes to natural areas. The outputs focus on measuring ecosystem service performance for 12 distinct ecosystem services: water provisioning, nitrogen removal in air, particulate removal in air, air temperature regulation, carbon uptake, erosion regulation, nitrogen removal in water, water filtration, water temperature regulation, water quantity control, noise attenuation, and visual screening. With the ESII Tool, users can evaluate the ecosystem service performance of various project alternatives and use the tool to enhance the design of a particular project to increase the overall ecosystem service performance.

In the third step of the valuation methodology, Dow and TNC aimed to create a scorecard that could be used to clearly translate a proposed project's impact on nature into a distinct metric, or set of metrics, that quantify their value. This metric should complement the financial metrics that are traditionally used to evaluate proposed projects (e.g., net present value and return on investment) and should be easily comparable across proposed projects globally while capturing the environmental impact locally. The finished scorecard must be able to answer two basic questions:

1. How might a Dow project impact the local environment?
2. How vulnerable is the local environment to the identified impacts?

The output produced by the scorecard must effectively communicate findings to non-conservation professionals including engineers, site managers, and upper-level executives.

While both the screen and the ESII Tool have been published externally and discussed in detail, the nature scorecard has only been preliminarily deployed internally (Guertin et al. 2019; Dow and TNC 2017). In the following section, we discuss in detail the methods used to develop the scorecard.

13.3 Nature Scorecard

To develop a scorecard that summarizes the impact on multiple environmental metrics, we use multi-objective decision analysis (MODA), a structured approach that uses a formal, mathematical method of making trade-offs when objectives potentially conflict (Kirkwood 1997; Merrick et al. 2004) based on the concept of value-focused thinking. *Value-focused thinking* leverages a stakeholder-engaged process to elicit the values and objectives behind decisions with a goal of fully illuminating the competing interests, trade-offs, and potential synergies between interest groups (Keeney 1996). First, decision-makers and stakeholders list their various values and objectives as they relate to the decision; these values and objectives are categorized either as *means* (as in, the means to achieve the ends) or *fundamental.* Any fundamental objective should be explicit and decomposable into measurable attributes (Keeney and Gregory 2005). Decision-makers organize these objectives into structures that reflect group preferences, such as an objective hierarchy (Keeney 1992). Once this framework is in place, decision-makers can assess the performance of various actions in terms of their impact on the attributes that compose the objectives within the hierarchy. This is particularly useful in decisions with complex objective structures where a formalized assessment of multiple criteria enables decision-makers to fully evaluate the consequences of action (Huang et al. 2011).

We slightly modify Merrick and Garcia's (2004) five-step process for developing and applying the MODA:

1. Defining a hierarchy of objectives that describe how the multiple metrics contribute to the overall environmental score (termed the "Nature Score") (Fig. 13.1 box a).
2. Using global data to define weights for the objectives to reflect their relative value to the decision-maker, i.e., the project manager (Fig. 13.1 box b), referred to as *environmental diagnostics*).
3. Choosing site-level measurable attributes for the achievement of the objectives (Fig. 13.1 box c).

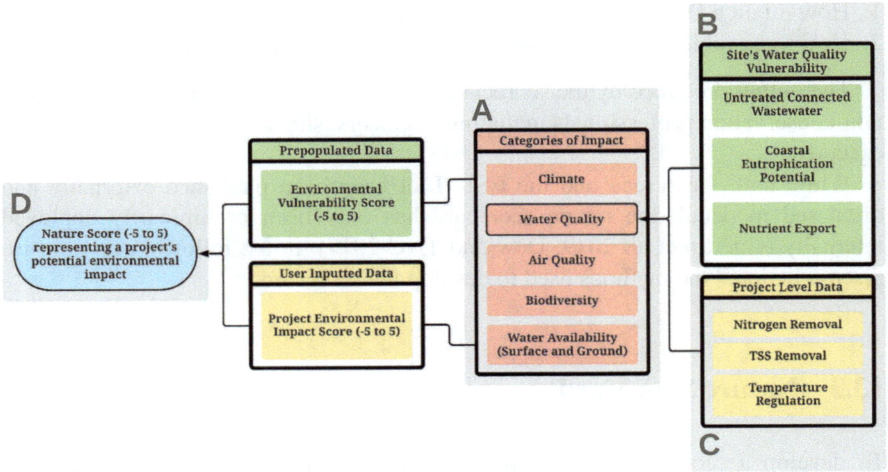

Fig. 13.1 The conceptual framework of the scorecard, showing how global environmental data (in green) (b) combines with user-inputted projections on project impact (in yellow) (c) for each vulnerability objective (in red) (a) to form an overall nature score for a proposed project (in blue) (d)

4. Transforming the measurable attributes into common units of value (Fig. 13.1 box c).
5. Integrating the evaluation and weights to determine the nature score (Fig. 13.1 box d).

The result of this process integrates multiple goals into a single overall score for evaluating each management option. Following Walker and Salt's guidance to keep things simple, we use the Simple Multi-Attribute Rating Technique (SMART), a product of multi-attribute utility theory to provide this overall score as well as scores for each component in the scorecard. SMART requires decision-makers to assign weights to each node in an objective hierarchy based on relative importance. These weights allow for an accordion-like compression of a hierarchy – a broad category such as "water quality," which is composed of multiple attributes such as "untreated connected wastewater" and "nitrogen export," receives a single score by combining its component attributes using their relative weights. This process is otherwise known as additive value function (Kirkwood 1997; Merrick et al. 2005) such that the nature score for any Dow site, N, is simply: $N = \sum G_i * S_i$, where G_i is the global importance weight of attribute i and S_i is the site-specific impact on attribute i. Here, we lay out each step of the MODA process (steps 1–4) and then describe how the results are integrated using SMART to provide the overall score (step 5).

13.3.1 Step 1: Hierarchy Defining Components of Environmental Vulnerability in a Global Context

The scorecard applies value-focused thinking to approach two decision contexts: (1) where to prioritize environmentally beneficial actions across a global portfolio of land holdings and manufacturing sites and (2) which actions to take at a local level given the globally informed priorities. Dow's broad goal for the scorecard is to enhance ecosystem services across the company's value chain while understanding potential trade-offs and negative impacts at a global and site scale.

With the scorecard, Dow defines the primary aspects of its potential environmental impact using five broad environmental categories of impact: (i) water quality, (ii) water availability, (iii) greenhouse gas emissions, (iv) air quality, and (v) biodiversity (Fig. 13.2). Within each category of impact, the area of each component in Fig. 13.2 corresponds to its relative weight, with all component weights summing to one.[1] For example, the six components that comprise biodiversity are equally weighted, while the three components comprising water quality are unequally weighted. We refer to this data extraction and combination process as environmental diagnostics. The relative weights shown in Fig. 13.2 were all derived from the scientific literature and are detailed in subsequent steps.

13.3.2 Step 2: Input Data and Objective Weights

To calculate the relative importance of each category of impact, the scorecard relies on multiple global datasets and models. Below we list these data inputs, grouped by category, and explain the relative weights used to combine them (see Fig. 13.3). Note that, as the scorecard aims to be transparent and flexible to user inputs, these weights are subject to change by an individual decision-maker, such as when an individual project manager has better local data on a specific metric (e.g., baseline water stress).

Each site receives a raw value for every data metric that is then normalized to a − 5 to 5 metric score. We defined a score of 5 as the best environmental outcome for each metric, such that sites with low environmental vulnerability receive scores approaching 5, while sites with high vulnerability receive scores approaching −5. The scorecard calculates a site's percentile ranking relative to all Dow sites to normalize metric values into metric scores: a score of 5 corresponds to the 100th

[1] An exception was made to the water availability metric, which has two subcomponents (surface water and groundwater) that each have their own components. Which subcomponent is analyzed for a given site depends on that site's water intake methods (e.g., groundwater or surface water sources) – sites that intake from both surface and groundwater sources take an average of both subcomponents.

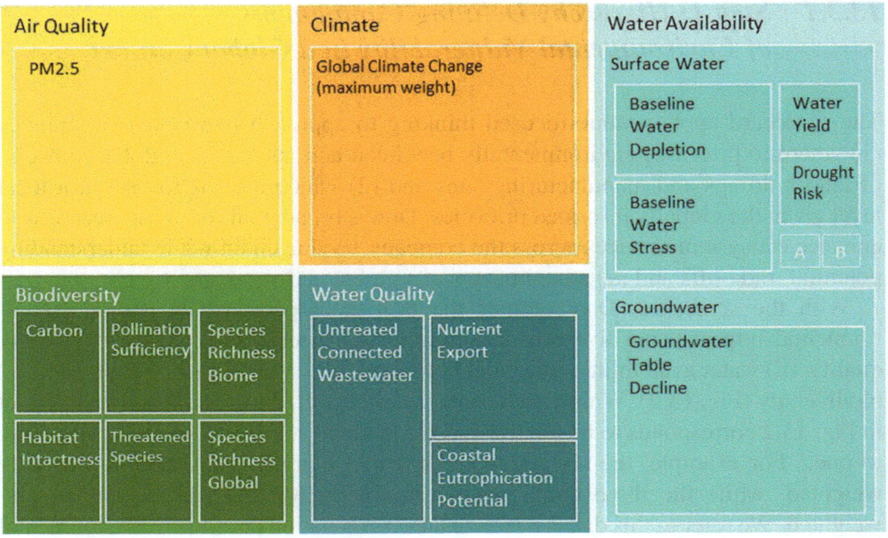

Fig. 13.2 Treemap diagram illustrating the mathematical underpinnings of the scorecard's environmental diagnostics. Components of the global weights for each Dow site's categories of impact are displayed with areas equal to their relative weights within an overall impact score. Components not listed include (A) inter-annual variability and (B) seasonal variability

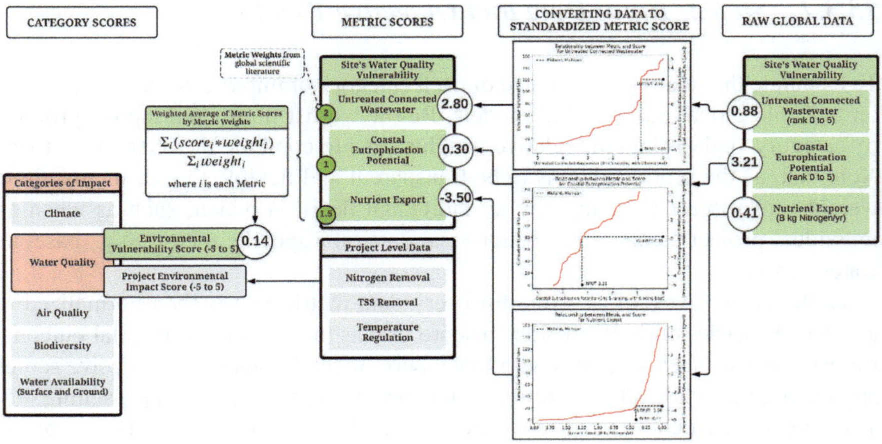

Fig. 13.3 Conceptual flowchart showing the conversion of raw global environmental data into metric scores for each data metric (−5 to 5, with 5 representing the best case) and the combination of these metric scores into an environmental vulnerability score (−5 to 5, with 5 representing the best case) using weights derived from the scientific literature. This figure is populated with data from the Dow site in Midland, Michigan. While the figure showcases water quality as an example, the same process is repeated for each category

percentile, i.e., the most environmentally beneficial value for that metric across all sites – and a score of −5 corresponds to the 0th percentile (Fig. 13.3).

Environmental Vulnerability Data Sources

Water Availability

We split water availability into two subcategories to reflect both surface and groundwater availability. With this split, we are also able to create scores that are specific to each site's reported water source, surface and/or groundwater. For surface water availability, we relied on two global environmental data sources: the World Resources Institute's Aqueduct 3.0 (Hofste et al. 2019) and a global version of the Natural Capital Project's InVEST Water Yield model (Johnson et al. 2020). Aqueduct 3.0 provides numerous metrics pertaining to water availability and water quality reported globally at the sub-basin level, alongside several suggested weighting schemes to aggregate individual metrics into composite scores. We used the *baseline water stress*, *baseline water depletion*, *inter-annual variability*, *seasonal variability*, and *drought risk* metrics from Aqueduct, selecting their default weighting scheme (4, 4, 0.5, 0.5, and 2, respectively) for eventual combination.

The InVEST Water Yield model calculates the cubic meters of water expected to runoff each pixel of landmass throughout a single year, accounting for groundwater recharge and vegetative uptake. A site's raw water yield score is the average runoff in a 300 m buffer around each site. Sites with high runoff were given low water availability weights. We gave the water yield metric score a weight of 2.2 (the average weight of the existing Aqueduct weighting scheme) and combined both Aqueduct and InVEST metrics together using a weighted average.

For groundwater availability, we used the Aqueduct *groundwater table decline* metric.

Water Quality

To estimate water quality, we supplemented Aqueduct 3.0 data with nutrient export modeled using the InVEST Nutrient Delivery Ratio as calculated globally for the Intergovernmental Science-Policy Platform on Biodiversity and Ecosystem Services (IPBES) (Chaplin-Kramer et al. 2019). We used Aqueduct's *untreated connected wastewater* and *coastal eutrophication potential* metrics with their corresponding default weights (2 and 1, respectively).

The InVEST Nutrient Delivery Ratio model uses data on precipitation, nutrient deposition rates, slope, and vegetative uptake to calculate expected amounts of nutrients (in this case, kilograms of nitrogen) flowing into watersheds from the landscape (Natural Capital Project 2023). The global version of the model was run at 1 decimal degree resolution (Chaplin-Kramer et al. 2019); we extracted the average nitrogen export in the nine pixels surrounding each site. We gave the nutrient export component score a weight of 1.5 (the average weight of the existing Aqueduct weighting scheme) and combined both Aqueduct and InVEST metrics together using a weighted average.

Climate

As greenhouse gas emissions contribute equally to the deterioration of a global environmental commons, the environmental vulnerability is estimated to be the same globally.

Air Quality

We used globally gridded atmospheric particulate matter (PM2.5) data as a proxy for air quality on the local scale (Hammer et al. 2020). PM2.5 is defined as any airborne particulate matter with diameters of less than 2.5 micrometers and is a common measure of air quality, as PM2.5 exposure has been linked to human health outcomes (Feng et al. 2016). For each of the nine pixels surrounding a site, we extracted the PM2.5 load and then averaged them together to reach a single value.

Biodiversity

For biodiversity estimates, we relied on data from TNC's web-based biodiversity assessment tool – Biodiversity and Ecosystem Service Trends and Conditions Assessment Tool (BESTCAT) (Oakleaf et al. 2014). From BESTCAT, we used *habitat intactness*, *threatened species*, *species richness (global)*, and *species richness (biome)*.

We complemented the BESTCAT data with two variables from the InVEST model: Pollinator Sufficiency and Carbon Sequestration. The InVEST Pollinator Sufficiency model predicts the relative pollinator abundance near agricultural development and has been used in assessing the global economic impacts of pollinator decline (Chaplin-Kramer et al. 2019). The InVEST Carbon Sequestration model has been run globally to assess global vegetative and soil carbon stocks (Johnson et al. 2021); it functions as a proxy for vegetative density in combination with data on species abundance and richness from BESTCAT. For each of these data sources, we calculated the average score for the nine-pixel buffer surrounding each site. As there was no guidance on appropriate weights to use when combining these data, we evenly weighted each metric.

13.3.3 Steps 3 and 4: Evaluation Measures and Standardization

The scorecard evaluates potential actions on a given site based on those actions' impacts on each category and the environmental vulnerability of that category. The scorecard output provides guidance for two decision contexts: which actions to take at a local level and where to prioritize action across Dow's global sites.

The scorecard provides a list of typical impacts a project may have on the environment; examples of these include water temperature regulation, water recycle/capture efforts, greenhouse gas and NO_X emission reduction schemes, and land preservation and restoration. When a project manager selects an impact to evaluate using the scorecard, they input the units used to evaluate that action (e.g., kg of total nitrogen removal or hectares of land preservation) alongside the current and proposed impacts of the action (e.g., current vs future rates of total nitrogen removal or land preservation). The scorecard then calculates the percent difference of impact between the current and proposed scenario.

For the nature scorecard, it was important to maintain flexibility around the scorecard's ability to perform an environmental diagnostic that is both localized and able to translate the improvement into a global perspective. This flexibility is achieved by having the project managers and others at Dow focused on the Nature Goal review the project's local performance while providing a global perspective of how that site "ranks" as far as the current level of emissions or dependency on resources. This iterative and collaborative process ensures the end score is consistent and aligned to the global strategy while being tailored to local site needs.

For example, project managers indicate in the scorecard whether and to what degree that percentage difference benefits the local environment by selecting one of a series of impact levels from "large improvement" to "large decline." To assist project managers in selecting an impact level, the scorecard includes information on how a Dow site's yearly emissions (e.g., greenhouse gases, NO_X, priority compounds, waste) and intakes (e.g., surface water, groundwater) rank across all Dow sites. Project managers can also use the environmental site ranking to position the projects in a global perspective within business decisions. The scorecard then transforms the reported impact of that action into a common score between -5 and 5 (with 5 equating to "large improvement" with a high percentage change in projected measurable outcomes from the action).

13.3.4 Step 5: Integration

Once the project manager has inputted all the proposed impacts for evaluation, the scorecard calculates an overall project environmental impact score for each category by averaging the individual scores of each action in the category. Then, to obtain the comprehensive Nature Score, the scorecard uses SMART guidelines to combine all

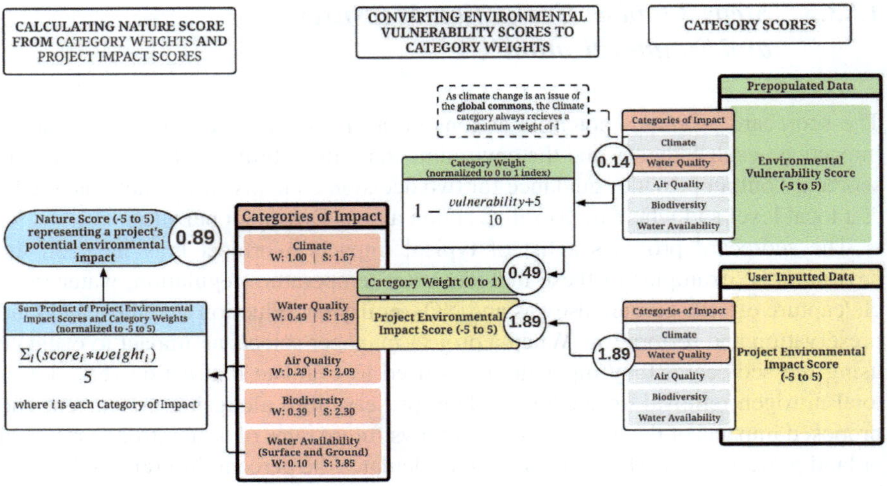

Fig. 13.4 Conceptual flowchart showing the conversion of category scores (environmental vulnerability and project impact) for each category of impact into a single nature score representing the overall impact a proposed project will have on the environment, with example numbers from the Midland, Michigan project detailed below. Category scores for environmental vulnerability (−5 to 5, with 5 representing the best case) are combined into a category weight (0 to 1, where 1 indicates sites with critical environmental vulnerability) for each category of impact. While the figure showcases water quality as an example, the same process is repeated for each category. The scorecard calculates the nature score by taking the weighted average of project impact across each category using the category weights

project environmental impact scores (one from each category of impact) using each category's environmental vulnerability score as a relative importance weight (Fig. 13.4).

13.3.5 Data Visualization

The methods presented above allow the scorecard to synthesize complex environmental data to evaluate innumerable possible projects aimed at improving Dow's environmental impact. An additional benefit of this approach is the capacity to visualize the current environmental vulnerabilities Dow has across all sites. This benefit addresses the second of our two decision contexts, that of where to prioritize action across Dow's global sites. Below we present histograms depicting how each of the 156 Dow sites, of which 104 are manufacturing sites, included in the scorecard measures in terms of environmental vulnerability for each category of impact (Fig. 13.5).

The scorecard provides a single number (the nature score) so that decision-makers can compare projects' environmental impacts with varying local environ-

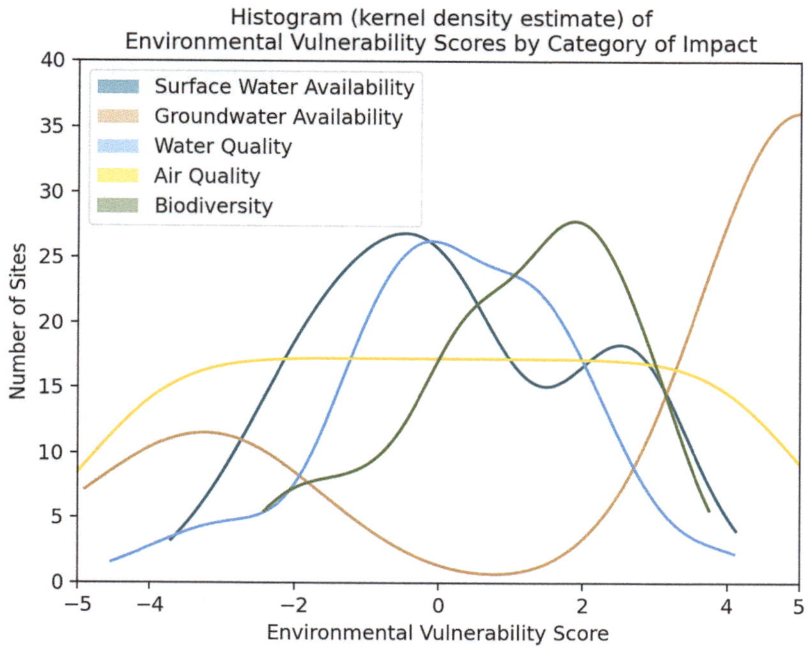

Fig. 13.5 Histogram (smoothed via kernel density estimate) of Dow sites by scores for environmental vulnerability (where −5 indicates a highly vulnerable site) across all categories of impact

ments and credibly show that their proposed project is better for the environment than alternatives. In the next section, we present a case study highlighting one of the primary ways in which the scorecard can be used to enhance decision-making processes around NBS.

13.4 Case Study: Dow Riverside from Old Ash Pond to Wetland Restoration

A project manager needing to evaluate two alternative project designs can use the environmental diagnostics process embedded in the nature scorecard to help inform trade-offs for a project. Without the scorecard, a project manager may be able to understand the physical differences between two alternative designs but would not be able to contextualize these physical differences from the site's environmental perspective. Additionally, the project manager may not be aware of how this site is ranked within Dow's global portfolio for a specific environmental category. For example, how critical is water quantity in that specific location? Should the proposed design then focus on capturing additional rainwater in a wetland?

PROJECT PERFORMANCE (SELECTED CRITERIA)

Component	Units	Alternative A	Alternative B	Percent Difference	Impact on Nature	Score (-5 to 5)
Nitrogen Removal	pp	0.27	0.4	48%	Large Improvement	3.70
TSS Removal	pp	0.52	0.46	-12%	Small Decline	-1.11
Temperature Regulation	pp	0.14	0.22	57%	Improvement	3.08
					Project Performance on improving Water Quality	1.89

Fig. 13.6 Project impact score in the water quality category. The project manager inputs the data for each relevant category of impact

ENVIRONMENTAL DIAGNOSTICS: MIDLAND, MI

CATEGORIES OF IMPACT	Environmental Vulnerability*	Project's Environmental Impact
Water Quality	0.14	1.89
Water Availability	4.03	3.85
Climate	—	1.67
Air Quality	2.10	2.09
Biodiversity	1.08	2.30

*Scores range from -5 to 5, with -5 being the most vulnerable location in DOW's portfolio and 5

NATURE SCORE

	Adjusted for project size
0.89	0.89

Calculated by taking the weighted sum of the Project's Environmental Impact, with Environmental Vulnerability (normalized to a 0-1 index) as weights.

Fig. 13.7 Excavate and Restore Wetlands project in Midland, Michigan. The figure shows scores across categories for environmental vulnerability and project's environmental impact. The project impact scores reflect the benefit of a wetland restoration project instead of the traditional cap-and-treat solution. The project environmental impact score for water quality is 1.89, which is the averaged value seen in Fig. 13.6. The final nature score was 0.89

In this case study, a Dow project team in Midland, Michigan, needed to address groundwater and soil contamination from an old coal fired power plant pond containing ash located along the Tittabawassee River in Midland, Michigan. Using the ESII Tool, the team was able to capture the various levels of ecosystem service performance across the two different closure alternatives (additional details of this project can be found in Guertin et al. (2019)). The first closure alternative involved a traditional cap and treat. The second involved excavating the concern materials within the pond and restoring the pond and overall site into a functioning wetland and riparian buffer.

Using the scorecard, we can better understand the value of the project for the specific characteristics of Midland, Michigan. Figure 13.6 shows the values generated for the projected restored wetland (Alternative B) compared to the traditional cap and treat (Alternative A) along the three impacted components in the water quality category.

This process is done for each project impact across the five impact categories (Fig. 13.7). As we can see in the full scorecard for this project (included in the Appendix), the restored wetland alternative delivers a higher level of nitrogen removal and a higher level of temperature regulation relative to the traditional cap and treat. It also delivers a lower level of TSS removal, although the difference is small.

Using this information, a decision-maker has a better understanding of how a project aligns with the concerns at the local site. With the wetland restoration project scoring process, a project manager can better understand exactly *how* much better the proposed alternative is compared to a traditional cap-and-treat solution. The project's impact scores result from comparing the cap-and-treat and wetland solutions; a positive number indicates that the wetland will perform better. We can also see that the project has the greatest positive impact (3.85) in water availability. The site will offer relief to the local flooding that occurs downstream from the site during heavy rains impacting the city of Midland.

In the final scoring, we can see that the wetland solution is projected to perform better in all the categories of impact (project's potential in mitigating impact in the figure), which leads to its high weight for nature final score. The final project impact score for the proposed restored wetland across all five categories was 0.89. This positive score (scaled from −5 to +5) indicates that the proposed project offers significant environmental benefits and could easily be compared to the final project impact score of other proposed projects across Dow's portfolio.

The final scorecard output can be used by site-level and global decision-makers to better understand the environmental profile of their portfolio and how those preferences can inform capital allocation. Armed with this tool, project managers can better understand environmental impacts of investments and operationalize NBS where applicable.

13.5 Lessons Learned and Next Steps

Driving the adoption of NBS in business settings is a complex problem that requires all parties to approach each individual piece of the puzzle thoughtfully. For example, one needs to understand the local environment enough to know when NBS can be effectively incorporated into a project design. Or, one needs to know how to accurately account for the full suite of benefits of NBS as they are often very different than the benefits created through traditional grey infrastructure. The TNC-Dow collaboration team conceived of the nature scorecard to address one particular piece of the puzzle: how do we arm project managers and other decision-makers with the information needed to advocate for a particular project based on its environmental impact?

In designing the nature scorecard, the project team came away with three principles to help ensure the success of the finished project.

First, the nature scorecard must be able to accurately contextualize the local environmental impacts of any proposed project. All locations are not created equal and the biomes and environmental stressors will vary across a company's asset base. In order for the scorecard to be effective and informative, it needs to accurately reflect these differences.

Second, the nature scorecard must be sensible and defensible. The uptake and shelf life of the nature scorecard will be a direct function of the level of confidence that the scorecard users have in its output. Any information included in the scorecard must be backed by science yet understandable to the scorecard's users.

Third, the nature scorecard must be transparent and useable. Not only do scorecard users need to have confidence in the output of the scorecard, but they also need to be able to easily access this information. Project engineers are already overloaded with project design components to consider and financial and performance models to run. If one is to ask them to take on the burden of an additional project evaluation metric, the proposed tool needs to be easily understood and accessed.

In implementing these principles, the TNC-Dow collaboration realized how important diverse partnerships are to this work (Davis et al. 2021). In partnering with the Natural Capital Project, we were able to tap into state-of-the-art global environmental data and models that would have been otherwise out of our areas of expertise. By integrating global data, project managers at Dow could assess a project's environmental impact in new and informative ways. In addition, incorporating the expertise of the scientific community helped to ensure that the scorecard accurately reflected the environmental conditions of a given site.

Moving forward, we recognize that any one model and setup will have its limitations. Thus, we view the scorecard and its underlying framework as an iterative process that will be adapted as the science evolves. Sustained multidisciplinary partnerships and approaches will be key to the continued enhancement of the scorecard.

As we move into the next phase of work, we are shifting our focus to the third principle noted above. The scorecard is in the process of being internally deployed across Dow; with this deployment, we will be able to gather critical feedback on its usability and how the final score compares across a large set of diverse projects. The project team will then take this feedback and use it to improve the design of the scorecard from a usability standpoint. Our ultimate goal is to make all of this work publicly available for other companies to use and learn from.

To operationalize NBS, companies need to be able to understand the environmental impacts of investments and other opportunities. The scorecard provides one means of informing that decision process. The final scorecard output can be used by site-level and global decision-makers to better understand the environmental profile of their portfolio and how those preferences can inform capital allocation. As more companies use these methods and tools, greater investment in conservation should follow because such investment makes good business sense.

A.1 Appendix

1. Nature Screen for wetlands project in Midland, Michigan

2025 Sustainability Goal: Valuing Nature

We will apply a business decision process that values nature. We will deliver business value and natural capital value through projects that are good for business and good for ecosystems.

If you need immediate assistance, please contact the nature team at

*Required fields.

Submitter Information

*Submitter Name:	Guertin, France (FM) Lookup »
*Submitter User ID:	
*Submitter Email Address:	

Project Information

*Project Name:	Dow Riverside Old Ash Pond Closure
*Project Site:	Midland, Michigan
*Geographic Area:	North America
*Project Manager Name:	Betsy Witt Lookup »
*Project Manager Email:	

*Project Type:
Select one or more.

- [x] Active Project
- [ ] Project Idea / Concept
- [x] Engineered Natural Technology
- [ ] Process Improvement
- [x] Land Management
- [ ] Complete (projects completed after May 2015 may count toward the goal)

Project Objective:
Optional

Define and evaluate the cost/benefit of a nature based alternative to the current traditional cap and treat option for the Dow Riverside old ash pond closure.

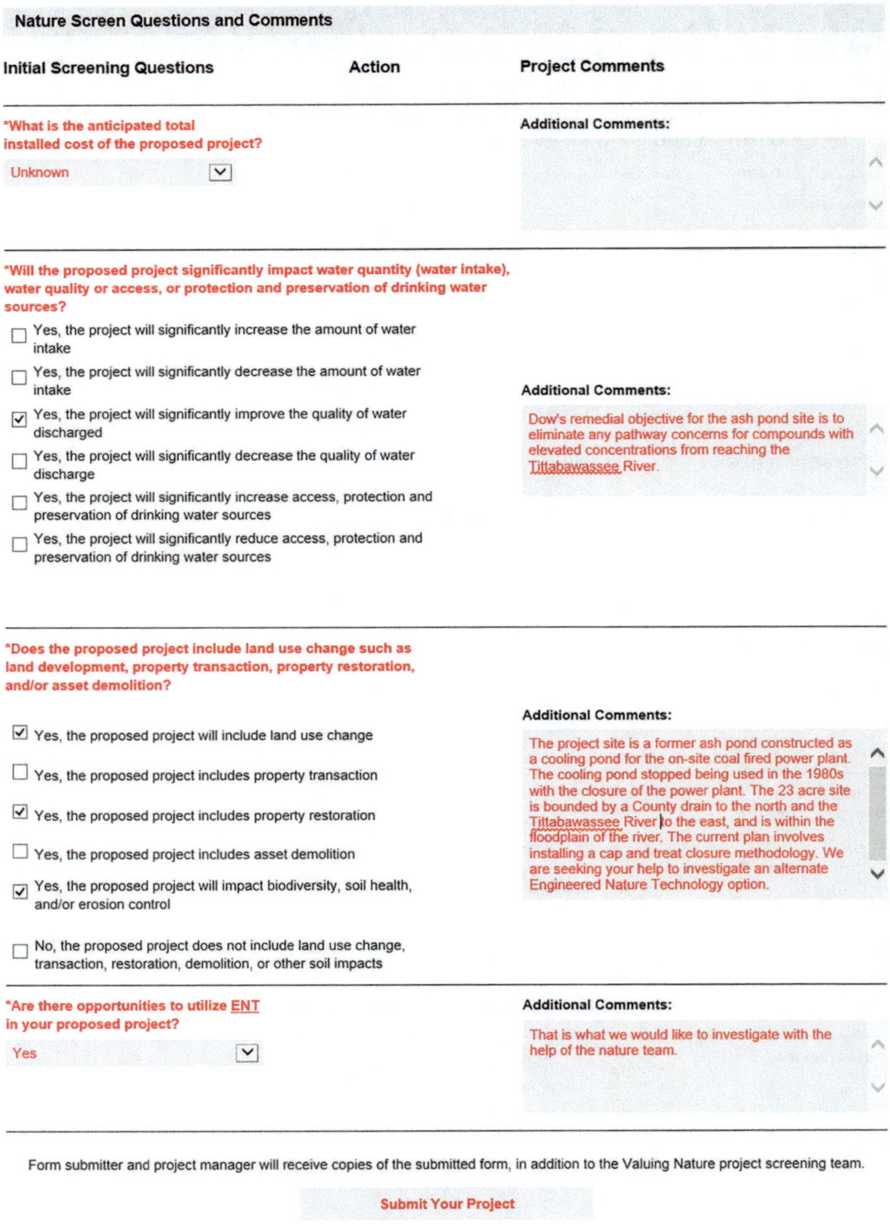

Nature Screen Questions and Comments

Initial Screening Questions	Action	Project Comments
*What is the anticipated total installed cost of the proposed project?	Unknown	Additional Comments:
*Will the proposed project significantly impact water quantity (water intake), water quality or access, or protection and preservation of drinking water sources?	☐ Yes, the project will significantly increase the amount of water intake ☐ Yes, the project will significantly decrease the amount of water intake ☑ Yes, the project will significantly improve the quality of water discharged ☐ Yes, the project will significantly decrease the quality of water discharge ☐ Yes, the project will significantly increase access, protection and preservation of drinking water sources ☐ Yes, the project will significantly reduce access, protection and preservation of drinking water sources	Additional Comments: Dow's remedial objective for the ash pond site is to eliminate any pathway concerns for compounds with elevated concentrations from reaching the Tittabawassee River.
*Does the proposed project include land use change such as land development, property transaction, property restoration, and/or asset demolition?	☑ Yes, the proposed project will include land use change ☐ Yes, the proposed project includes property transaction ☑ Yes, the proposed project includes property restoration ☐ Yes, the proposed project includes asset demolition ☑ Yes, the proposed project will impact biodiversity, soil health, and/or erosion control ☐ No, the proposed project does not include land use change, transaction, restoration, demolition, or other soil impacts	Additional Comments: The project site is a former ash pond constructed as a cooling pond for the on-site coal fired power plant. The cooling pond stopped being used in the 1980s with the closure of the power plant. The 23 acre site is bounded by a County drain to the north and the Tittabawassee River to the east, and is within the floodplain of the river. The current plan involves installing a cap and treat closure methodology. We are seeking your help to investigate an alternate Engineered Nature Technology option.
*Are there opportunities to utilize ENT in your proposed project?	Yes	Additional Comments: That is what we would like to investigate with the help of the nature team.

Form submitter and project manager will receive copies of the submitted form, in addition to the Valuing Nature project screening team.

Submit Your Project

2. Scorecard for wetlands restoration project in Midland, Michigan

Nature Scorecard

Inputs Tab

Responsible: <insert name>

Date: Tuesday, August 31, 2021

ASSET INFORMATION	
Name	MIDLAND, MI
Country	United States
Ecoregion	Southern Great Lakes forests
Ecoregion Status	None
Water Intake Type	PURCHASED

IMPACTED AREA BY PROJECT		[TOTAL]
Land (acres)	23	23
Water Bodies (acres)		

PROJECT INFORMATION	
Project Name:	Old ash pond closure
Project Comparison Type:	Alternative A vs. Alternative B
Alternative A Project Description:	Cap and treat traditional closure methodology
Alternative B Project Description:	Wetlands restoration
Project Type:	
Data Source:	ESII Tool
Other Ancillary Benefits:	connection to city of Midland's parks and trails
Alternative A Project NPV:	
Alternative B Project NPV:	

IMPACT ON NATURE

Only input change from relevant factors

PROJECT IMPACTS

Water Quality

Does your project impact...

- ☑ Nitrogen Removal
- ☑ TSS Removal
- ☑ Temperature Regulation
- ☐ Other <insert name>
- ☐ Other <insert name>
- ☐ Other <insert name>

	Env Vulnerability	Site Category Rank
Water Quality in Midland, MI	0.14	76 of 156

PROJECT PERFORMANCE (SELECTED CRITERIA)

Component	Units	Alternative A	Alternative B	Percent Difference	Impact on Nature	Score (-5 to 5)
Nitrogen Removal	PP	0.27	0.4	48%	Large Improvement	3.70
TSS Removal	PP	0.52	0.46	-12%	Small Decline	-1.11
Temperature Regulation	PP	0.14	0.22	57%	Improvement	3.08
Project Performance on improving Water Quality						1.89

Water Availability

Does your project impact...

- ☐ Water Intake (freshwater or groundwater)
- ☐ Water Recycle/Capture
- ☐ Water Provisioning
- ☑ Water Quantity Control
- ☐ Other <insert name>
- ☐ Other <insert name>
- ☐ Other <insert name>

		Env Vulnerability	Site Category Rank
Water Availability in Midland, MI	Surface	3.06	150 of 156
	Ground	5.00	149 of 156

Component	Units	0	0	0	Impact on Nature	
Water Quantity Control	PP	0.39	0.6	54%	Large Improvement	3.85
Project Performance on improving Water Availability						3.85

Climate

Does your project impact...

- ☐ Scope 1 Emissions
- ☐ Scope 2 Emissions
- ☐ Scope 3 Emissions
- ☑ Carbon Uptake
- ☑ Other <insert name>
- ☐ Other <insert name>
- ☐ Other <insert name>

	Env Vulnerability	Site Category Rank
Climate in Midland, MI	–	66 of 156

Component	Units	0	0	0	Impact on Nature	
Carbon Uptake	PP	0.1	0.14	40%	Small Improvement	1.83
Climate Air temp. regulation	PP	0.33	0.42	27%	Small Improvement	1.51
Project Performance on improving Climate						1.67

Air Quality

Does your project impact...

- ☐ NOX Emissoins
- ☐ VOC Emissions
- ☐ Priority Compounds Emissions
- ☑ Nitrogen Removal
- ☑ Particulate Removal
- ☐ Air Temperature
- ☐ Other <insert name>
- ☐ Other <insert name>
- ☐ Other <insert name>

	Env Vulnerability	Site Category Rank
Air Quality in Midland, MI	2.10	70 of 156

Component	Units	0	0	0	Impact on Nature	
Nitrogen Removal	PP	0.12	0.31	158%	Improvement	4.15
Particulate Removal	PP	0.63	0.64	2%	No Change	0.04
Project Performance on improving Air Quality						2.09

Biodiversity

Does your project impact...

- ☐ Preserved Land Area
- ☑ Restored Land Area
- ☐ Created Land Area
- ☐ Preserved Water Body/River Coastline Area
- ☑ Restored Water Body/River Coastline Area
- ☑ Created Water Body/River Coastline Area
- ☐ Raw Material Consumption
- ☐ Solid Waste Generation
- ☐ Recycling
- ☐ Organic Waste Composting
- ☑ Landfill Disposal
- ☑ Erosion Regulation
- ☑ Erosion Regulation - Mass Wasting
- ☑ Aesthetics - Noise
- ☑ Aesthetics - Visual
- ☐ Other <insert name>
- ☐ Other <insert name>
- ☐ Other <insert name>

	Env Vulnerability	Site Category Rank
Biodiversity in Midland, MI	1.08	76 of 156

Component	Units	0	0	0	Impact on Nature	
Restored Land Area	acres	0	23		Large Improvement	
Restored Water Body/River	miles	0	1		Improvement	
Created Water Body/River Coastline					Improvement	
Landfill Disposal					Decline	
Erosion Regulation	PP	0.69	0.68	-1%	No Change	0.04
Erosion Regulation - Mass Wasting	PP	0.09	0.09	0%	No Change	0.00
Aesthetics - Noise	PP	0.15	0.43	187%	Improvement	4.15
Aesthetics - Visual	PP	0.19	0.41	116%	Large Improvement	5.00
Project Performance on improving Biodiversity						2.30

* The ESII tool was used to model the ecological performance of each alternative. The various component units are expressed in percent performance (PP) of functions and services. It is a measure of how well an area performs relative to optimal conditions.

References

Business Wire. 2011. "Dow and The Nature Conservancy Announce Collaboration to Value Nature." Press Release, Midland, MI (January 24). Retrieved from https://www.businesswire.com/news/home/20110124006239/en/Dow-Nature-Conservancy-Announce-Collaboration-Nature.

Chaplin-Kramer, Rebecca, Richard P. Sharp, Charlotte Weil, Elena M. Bennett, Unai Pascual, Katie K. Arkema, Kate A. Brauman, et al. 2019. "Global Modeling of Nature's Contributions to People." *Science* 366 (6462): 255–58. https://doi.org/10.1126/science.aaw3372.

Davis, Jarod, France Guertin, Todd Guidry, Martha Rogers, Zen Saunders, Michael Uhl. 2021. "Lessons Learned from a Corporate Manufacturer on Driving the Adoption of Nature-Based Solutions." *Integrated Environmental Assessment and Management*. https://doi.org/10.1002/ieam.4442.

DiMuro, Johnathan, Guertin, France, Helling, Rich, Perkins, Jessica and Romer, Scanlon. 2014. "A Financial and Environmental Analysis of Constructed Wetlands for Industrial Wastewater Treatment." *Journal of Industrial Ecology* 18: 631–640. https://doi.org/10.1111/jiec.12129.

The Dow Chemical Company. 2021. "Valuing Nature." Retrieved from https://corporate.dow.com/en-us/science-and-sustainability/2025-goals/nature.html.

The Dow Chemical Company and The Nature Conservancy. 2017. "Working Together to Value Nature: 2017 Summary Report." Retrieved from https://www.nature.org/content/dam/tnc/nature/en/documents/Dow-TNC%202017%20Collaboration%20Report.pdf.

(ESII) Ecosystem Services Inventory and Identification Tool. 2021. "ESII: The tool that helps you make better decisions. Better for your organization and better for nature." Retrieved from https://www.esiitool.com/.

Feng, S., Gao, D., Liao, F., Zhou, F. and Wang, X., 2016. "The health effects of ambient PM2. 5 and potential mechanisms." *Ecotoxicology and environmental safety*, 128, pp.67–74.

Guertin, France, Kevin Halsey, Thomas Polzin, Martha Rogers, and Betsy Witt. 2019. "From Ash Pond to Riverside Wetlands: Making the Business Case for Engineered Natural Technologies" Science of the Total Environment 651 (1):419–426.

Hammer, Melanie S., Aaron van Donkelaar, Chi Li, Alexei Lyapustin, Andrew M. Sayer, N. Christina Hsu, Robert C. Levy, et al. 2020. "Global Estimates and Long-Term Trends of Fine Particulate Matter Concentrations (1998–2018)." *Environmental Science & Technology* 54 (13): 7879–90. https://doi.org/10.1021/acs.est.0c01764.

Hofste, Rutger, Samantha Kuzma, Sara Walker, Edwin Sutanudjaja, Marc Bierkens, Marijn Kuijper, Marta Faneca Sanchez, et al. 2019. "Aqueduct 3.0: Updated Decision-Relevant Global Water Risk Indicators." *World Resources Institute*. https://doi.org/10.46830/writn.18.00146.

Huang, Ivy B., Jeffrey Keisler, and Igor Linkov. 2011. "Multi-Criteria Decision Analysis in Environmental Sciences: Ten Years of Applications and Trends." *Science of The Total Environment* 409 (19): 3578–94. https://doi.org/10.1016/j.scitotenv.2011.06.022.

IUCN (International Union for Conservation of Nature). 2021. "Nature-Based Solutions for People and Planet." Retrieved from https://www.iucn.org/theme/nature-based-solutions.

Johnson, J. A., Ruta, G., Baldos, U., Hertel, T., Liu, J., Nootenboom, C., Polasky, S., & Roxburgh, T. (2020). *Global futures: Modelling the global economic impacts of environmental change to support policy-making*. Woking, UK: WWF, GTAP and the Natural Capital Project.

Johnson, Justin Andrew; Ruta, Giovanni; Baldos, Uris; Cervigni, Raffaello; Chonabayashi, Shun; Corong, Erwin; Gavryliuk, Olga; Gerber, James; Hertel, Thomas; Nootenboom, Christopher; Polasky, Stephen. 2021. The Economic Case for Nature: A Global Earth-Economy Model to Assess Development Policy Pathways. World Bank, Washington, DC. © World Bank. https://openknowledge.worldbank.org/handle/10986/35882 License: CC BY 3.0 IGO.

Keeney, Ralph L. 1992. *Value-Focused Thinking: A Path to Creative Decisionmaking*. Cambridge, Mass.: Harvard Univ. Press.

———. 1996. "Value-Focused Thinking: Identifying Decision Opportunities and Creating Alternatives." *European Journal of Operational Research* 92 (3): 537–49. https://doi.org/10.1016/0377-2217(96)00004-5.

Keeney, Ralph L., and Robin S. Gregory. 2005. "Selecting Attributes to Measure the Achievement of Objectives." *Operations Research* 53 (1): 1–11. https://doi.org/10.1287/opre.1040.0158.

Kirkwood, Craig W. 1997. *Strategic decision making: Multiobjective Decision Making with Spreadsheets.* Wadsworth Publishing Company, US.

Kroeger, Timm, Francisco J. Escobedo, José L. Hernandez, Sebastián Varela, Sonia Delphin, Jonathan RB Fisher, and Janice Waldron. "Reforestation as a Novel Abatement and Compliance Measure for Ground-Level Ozone." *Proceedings of the National Academy of Sciences* 111, no. 40 (2014): E4204-E4213.

Merrick, Jason R. W., and Margot W. Garcia. 2004. "Using Value-Focused Thinking to Improve Watersheds." *Journal of the American Planning Association* 70 (3): 313–27. https://doi.org/10.1080/01944360408976381.

Millennium Ecosystem Assessment. 2005. *Ecosystems and Human Well-Being: Synthesis*. Island Press: Washington, DC.

Natural Capital Project, 2023. InVEST 3.13.0. Stanford University, University of Minnesota, Chinese Academy of Sciences, The Nature Conservancy, World Wildlife Fund, Stockholm Resilience Centre and the Royal Swedish Academy of Sciences. *https://naturalcapitalproject.stanford.edu/software/invest*

Oakleaf, James R., Christina M. Kennedy, Timothy Boucher, and Joseph Kiesecker. 2014. "Tailoring Global Data to Guide Corporate Investments in Biodiversity, Environmental Assessments, and Sustainability." *Sustainability* 5(10): 4444–4460.

Reddy, Sheila MW, Robert I. McDonald, Alexander S. Maas, Anthony Rogers, Evan H. Girvetz, Jeffrey North, Jennifer Molnar, Tim Finley, Genā Leathers, and Johnathan L. DiMuro. "Finding Solutions to Water Scarcity: Incorporating Ecosystem Service Values into Business Planning at The Dow Chemical Company's Freeport, TX Facility." *Ecosystem Services* 12 (2015): 94–107.

TNC (The Nature Conservancy), Dow, Swiss Re, Shell and Unilever. 2013. "The Case for Green Infrastructure." Joint-Industry White Paper. Retrieved from https://www.nature.org/content/dam/tnc/nature/en/documents/the-case-for-green-infrastructure.pdf.

TNC (The Nature Conservancy). 2019. "Strategies for Operationalizing Nature-Based Solutions in the Private Sector." The Nature Conservancy Business Council Natural Infrastructure Working Group White Paper. Retrieved from https://www.nature.org/content/dam/tnc/nature/en/documents/NBSWhitePaper.pdf.

Walker, Brian and David Salt. 2010. *Resilience Practice*. Island Press: Washington, DC.

Chapter 14
Greenprinting: Urban Planning for Ecosystem Services

Robert I. McDonald and Misty Edgecomb

14.1 The Urban Century

We are living in what could truly be called the urban century, the period of fastest urban growth in human history. By 2050, there are expected to be more than two billion additional urban dwellers in cities globally (UNPD 2018). Getting our cities right – creating people-centric, green, sustainable, resilient, and livable cities – may be the most important action we can take to ensure the survival of our civilization. However, to get to that more verdant urban world, we need a plan. This chapter will focus on the potential for nature to provide ecosystem services in cities, presenting techniques for better planning ("greenprinting") that can help cities better manage urban biodiversity and human health.

14.2 Key Environmental Problems Facing Cities

Even in this urban century, at their period of fastest growth, cities face profound challenges. Many of these challenges concern the environment, broadly construed, and for some of these, there is a natural infrastructure solution, a way ecosystem services can help alleviate the problem.

In this subsection, we list some of the major environmental challenges facing cities that have a natural infrastructure solution. Each of these environmental

R. I. McDonald (✉)
The Nature Conservancy in Europe, Berlin, Germany
e-mail: mcdonald@tnc.org

M. Edgecomb
Las Vegas, VA, USA

B. R. Bakshi (ed.), *Engineering and Ecosystems*,
https://doi.org/10.1007/978-3-031-35692-6_14

challenges gives rise to slightly different urban planning processes that can be entry points for the central topic of this chapter: the incorporation of ecosystem service information about the value of natural infrastructure in the planning process. Put another way, these environmental challenges are what municipal governments are hoping that natural infrastructure can, in part, solve.

14.2.1 Health

Cities globally have several major health concerns. Many of these do not have a strong nexus for natural infrastructure and nature-based solutions. For instance, cities struggling with the spread of coronavirus (Covid-19) or problems with opioid addiction arguably do not need to spend meaningful time considering links to nature-based solutions when planning their response (although see Spotswood et al. 2021 for a discussion of Covid-19 and urban nature). However, there are a few significant urban health problems that do have such a nexus with nature-based solutions.

Air pollution is one such health problem. Outdoor particulate matter (PM) is the ambient pollutant with the greatest health toll, killing globally more than three million people per year, in both rural and urban areas (Lim et al. 2012). Other major air pollutants include ground-level ozone, carbon monoxide, sulfur dioxide, and nitrogen dioxide. The good news is that governments have made significant progress on reducing air pollution over time. In the United States, for instance, particulate matter emissions (PM_{10}) fell by 57% between 1980 and 2016 (EPA 2016).

Another health problem is obesity, a growing problem in both rural and urban areas. Worldwide, more than 600 million people are obese, double what it was in 1980 (WHO 2017). Obesity rates rise when there is an increased intake of calorie-rich foods and less physical activity (Swinburn et al. 2011). Both phenomena are more common in cities, where greater incomes allow more calorie intakes and a larger fraction of jobs are sedentary. Obesity in turn leads to many other negative health outcomes for urban residents, including diabetes and heart disease.

Cities also struggle from what has been called an urban psychological penalty (McDonald et al. 2018b). Some aspects of mental health have been shown to be worse in urban areas than in rural areas (Gruebner et al. 2017). For instance, urban life has been associated with greater stress than rural life, as well as with changes to brain function (Lederbogen et al. 2011). The causes of the urban psychological penalty is a complex, varying by disease and by the urban cultural context (Pedersen and Mortensen 2001). However, on average the greater population densities found in cities are associated with higher incidences of specific mental health diseases. For example, one study of more than four million adults in Sweden (2004) found a significant increase in rates of psychosis and depression in higher densities in cities than in lower densities in rural areas. Moreover, rates of mental illness are rising over time in societies around the globe (Marcus et al. 2012), including in many cities. For all these reasons, many cities are struggling to find ways to maintain the mental health of their citizens.

14.2.2 *Climate Change*

Climate change is another substantial challenge that many cities are facing. Climate change is already profoundly affecting society in myriad ways, and whole books have been written on climate adaptation planning (Davoudi et al. 2009). However, two climate change-related challenges are perhaps most often discussed in the context of nature-based solutions to climate change, often called ecosystem-based adaptation (EBA).

First, heat waves are already a significant public health threat, killing an estimated 12,000 people annually worldwide in a typical year (McMichael et al. 2004). One heat wave in Europe in 2003 estimated to have killed more than 70,000 people (Robine et al. 2008). Higher air temperatures cause heat stroke and exhaustion and exacerbate existing cardiovascular, pulmonary, and renal diseases (McMichael et al. 2004; McDonald et al. 2016). Climate change is projected to increase average summertime air temperatures, as well as the frequency and severity of heat waves, thus potentially leading to large increases in mortality (Hales et al. 2014). Cities are increasingly focused on strategies that can minimize the impact of high air temperature on their residents' health (McDonald et al. 2016, 2019).

Second, climate change is causing a rise in sea levels, as ice on land melts and flows into the seas, and as the sea's water warms and expands in volume. Estimates are that 10% of people live globally in the low-elevation coastal zone that may be at threat from sea-level rise (McGranahan et al. 2007). Additionally, climate change may increase the intensity or frequency of coastal storms, which could further increase the risks in the low-elevation coastal zone. Cities and communities along coasts are beginning to plan for ways to mitigate the risks of coastal hazards and adapt to the coming rise in sea levels.

14.2.3 *Urban Stormwater Management*

Another major challenge facing cities is appropriate management of urban stormwater. In just the United States, more than 700 cities have combined sewer systems that could overflow when it rains (EPA 2014). No comprehensive global figure exists (McDonald 2015), but out of the tens of thousands of cities that exist globally, the majority likely has combined sewer systems. Even cities that use separate sanitary sewer systems are still plagued by concerns about the water quality of stormwater that is being dumped in rivers and streams. Stormwater mitigation has become one of the central urban challenges of the twenty-first century. Replacing combined sewer systems with separate sanitary systems is too expensive an option for most cities. Many cities have tried therefore to find ways to reduce the amount of stormwater entering a system. If they can do that, then they can reduce combined sewer overflow (CSO) events. Slowing the flow of water also reduces the rapid flush of contaminants and sediment after a big rain event, improving stormwater quality.

14.2.4 Biodiversity Protection

Urban growth has been a major cause of natural habitat loss historically (McDonald et al. 2018a). Urban growth was responsible for the loss of 190,000 km^2 of natural habitat between 1992 and 2000, around 16% of all the natural habitat lost over this period. Biomes with large amounts of natural habitat lost due to urban growth include temperate forests, deserts and xeric shrublands, and tropical moist forests. In the future, this trend will continue, especially in tropical moist forests. Urban growth, if not properly planned, could threaten 290,000 km^2 of natural habitat by 2030. However, preventing habitat conversion and increasing land protection are key goals of many municipal governments, which are trying to maintain biodiversity even in the face of urban growth.

14.3 Urban Ecosystem Services

14.3.1 Key Ecosystem Services in Cities

This book describes the many ecosystem services that are important to human well-being. Only a subset of these are commonly planned for in urban planning processes (Table 14.1). Below, we discuss each of these urban-relevant ecosystem services briefly, grouping them in the categories made customary by the Millennium Ecosystem Assessment (2003). See McDonald (2015) for a book-length treatment of these ecosystem services, or Keeler et al. (2019) for a more recent review of the state of knowledge of these urban ecosystem services.

One category of ecosystem services is provisioning services, the products people obtain from ecosystems such as food, fuel, or fiber. Agricultural crop production, livestock production, and aquaculture are clear examples of provisioning services. Many of these provisioning services occur in more rural areas, and their products are transported into urban areas. However, there is burgeoning interest in urban agriculture in many cities, and urban agriculture is sometimes one element of urban conservation planning.

Providing sufficient quantity of water is one of the most important provisioning services for cities. Water utilities supply water to municipal residents, who need water for drinking, sanitation, cleaning, and water lawns. Water is also crucial for energy production, particularly the cooling of thermoelectric plants. Cities depend

Table 14.1 Ecosystem services of greatest relevance to cities, as well as the spatial scale at which they operate

Ecosystem service	Spatial scale
Provisioning services	
Agriculture (crops, livestock, aquaculture, etc.)	Regional to global
Water (quantity)	100's km – upstream source watershed
Cultural services	
Aesthetic benefits	10's km – area of daily travel by urbanites
Recreation and tourism	10's km – area of daily travel by urbanites
Physical health	10's km – area of daily travel by urbanites
Mental health	10's km – area of daily travel by urbanites
Spiritual value and sense of place	Varies – often local, but can be up to global
Biodiversity	Varies – global for existence value, local for direct interaction
Regulating services	
Drinking water protection (water quality)	100's km – upstream source watershed
Stormwater mitigation	100's m – downstream stormwater system
Mitigating flood risk	100's km – downstream flood-prone areas
Coastal protection	10's km – coastal zone
Air purification (particulates, ozone)	100's km – regional airshed
Shade and heat wave mitigation	<100 m – varies with solar angle

Table adapted from McDonald (2015)

on the natural water cycle to provide sufficient water to their water intake points, and if there isn't sufficient water, there are serious economic consequences.

A second category of ecosystem services is cultural services, the nonmaterial benefits people obtain from ecosystems. For instance, the aesthetic benefits of natural areas can be very important to urban dwellers. These aesthetic considerations have been often quite important during urban planning processes. Recreation opportunities for urban residents are another important benefit of urban natural areas, and such recreation has important health benefits, including reductions in obesity and increases in mental health.

Another category is regulating services, the benefits people obtain from the regulation of ecosystem function. For instance, natural floodplains play an important risk-reduction role by allowing floodwaters to spread out, lessening peak flows and reducing flooding risk downstream. Similarly, some natural coastal habitats like wetland, oyster reefs, mangroves, and coral reefs may mitigate the risk of flooding to cities during storms. The natural world plays an important climate regulation role, affecting surface temperature, evapotranspiration, wind flow, and other climate variables. Finally, natural habitat may help reduce air pollution and keep air quality within acceptable limits.

14.3.2 Urban Ecosystem Services and Market Failure

If ecosystem services are so important to urban residents, why are so many ecosystem services not adequately provided by the free market to urban residents? The dysfunctional provision for most ecosystem services an example of market failure, which occurs systematically for certain types of goods and services (McDonald 2020).

An economic good can be either rival or non-rival. A rival good is one where one person's use of the good prevents others for using it. Non-rival goods, by contrast, are not used up, and one person's use does not prevent another's use. Economic goods can also be classified as either excludable or non-excludable. An excludable good is one where it is possible to control who has access to the good. Non-excludable goods, on the other hand, are ones where it is not feasible to control access to the good (Kolstad 2000).

Private goods, like food, timber, and fiber, are rival and excludable: I cannot bring home produce from the store unless I pay (excludable) and my purchase of the produce prevents others from purchasing it (rival). However, most of the ecosystem services in Table 14.1 are *public goods*, defined as non-rival and non-excludable. Urbanites' enjoyment of the aesthetic beauty of a row of street trees does not prevent others from enjoying it (non-rival), and public streets are open to all (non-excludable). Another important category of goods is *common goods*, those that are rival but non-excludable. Recreation in some crowded urban parks might be an example, since anyone can use the park (non-excludable) but the amount of space available is finite (rival).

Private goods are well-provided to cities. Companies have financial incentives to bring private goods to cities, and the amount of private goods supplied is brought in line with the quantity of the good demanded via changes in price. In contrast, public goods and common goods are generally underprovided to those in cities. As they are non-excludable, no company could make money off the provision of these goods, since users who have not paid would just use the good for free.

Most ecosystem services that cities depend on are public or common goods (Table 14.1), and therefore their provision by natural habitat will not be adequately maintained by free market actors. For instance, urban parks for recreation, a common good, will tend not to be provided by private land developers. For ecosystem services that are public or common goods, there is no functioning market, so society has to find other ways to ensure those needs are met.

If the private market has little incentive to consider many ecosystem services in their decision-making, then governments or other social organizations are justified in stepping in to ensure provision, either directly through policy or indirectly by giving firms incentive to consider ecosystem services in their decisions. The solution to market failure is collective action to promote the public good. Urban planning and zoning is one of the key places where ecosystem service provision can be ensured. Thus, the economic justification of urban conservation planning for ecosystem services is collective planning to ensure the provision of public or common goods (McDonald 2020).

14.3.3 The Spatial Scale of Ecosystem Services Varies

Natural infrastructure need to be within a certain distance from people to provide an ecosystem service (McDonald 2009). One common mistake in urban planning is to focus on mapping ecosystem area (e.g., urban tree cover) and then treat such zones as simple overlays in planning decisions. This approach misses a very important spatial dimension of ecosystem services, which is the importance of proximity between natural habitat and beneficiary. This zone of provision is sometimes called the "serviceshed" (Tallis and Wolny 2011), after the familiar concept of a watershed.

The serviceshed of different ecosystem services varies widely (Table 14.1), which affects where urban planners need to protect or restore natural infrastructure. Some services are very local, operating over the scale of meters, like the shade from the street trees outside. Others, like the provision of parks for day-to-day recreation, operate over the scale of tens of kilometers. Water provision operates within watersheds, which can vary from small to quite big, and has a unique element of upstream/downstream directionality: Actions upstream affect water downstream, and actions downstream do not affect points upstream. Similarly, air quality in a region's "airshed" depends on regional wind patterns, which define an upwind/downwind directionality. As these examples illustrate, the transportability of an ecosystem service is not a simple function of Euclidean distance, but is determined by the physics of the ecosystem service in question, which controls how useful a particular patch of natural infrastructure is for a particular person's well-being.

Cities characteristically have a dense core, and then lower population density as one moves from the core into suburban areas. Most people live and work in cities, and by definition ecosystem services benefit people, so cities are centers of ecosystem service demand (McDonald 2009). Natural infrastructure may provide greater ecosystem services when it is closer to the dense urban core than if it was in a remote rural area. However, protecting or creating natural infrastructure in the center city has high opportunity costs, since the land could be used for many other purposes.

A mathematical theory of how to evaluate this trade-off, based off of bid rent theory, is described in McDonald (2009). For the purposes of this chapter, however, it is enough to say that an urban planning process has to consider what ecosystem services the municipal government or private organization wants to provide, to whom, and what it would cost to work in different locations. All of these considerations would affect where the institution should protect or create natural infrastructure.

14.4 Greenprinting: Urban Planning for Ecosystem Services

Urban greenprinting refers to planning how natural habitats or natural features (e.g., street trees, parks, open space, constructed wetlands) can be protected, restored, or created to maximally protect biodiversity and enhance human well-being (Fig. 14.1). Sometimes working landscapes such as croplands and ranchlands are also a conservation target of greenprints. Urban greenprinting is not fundamentally different from the "ordinary" conservation planning that conservationists have done for a long time, although the focus is on ecosystem services rather than biodiversity, and there are unique challenges to planning in urban areas (McDonald 2015).

Fig. 14.1 An example output from a greenprint. This example is from an assessment of the Bay Area Greenprint (California, USA). The Nature Conservancy, Greenbelt Alliance, GreenInfo Network, American Farmland Trust, and Bay Area Open Space Council developed a toolkit that supports improved conservation-focused policy for local and regional planning in the Bay Area. One major planning goal in the Bay Area is to accommodate the expected two million additional residents by 2030 while losing no natural habitat and expanding human access to nature to enhance human well-being and health. This image shows a portion of the website that allows users to define the relative importance of different goals and then shows the Bay Area landscape prioritized for conservation accordingly. More info can be found online at www.bayareagreenprint.org

The term *greenprint* was popularized in the United States by the Conservation Fund in the 1990s. There are alternative terms in use, such as *urban natural resource planning*, *eco-urban assessments*, or *urban conservation planning*. The most appropriate depends on the language and region of the greenprint.

Greenprinting seeks to do two things:

- *Bring biodiversity and ecosystem service information into spatial planning*: By incorporating information on key natural features for biodiversity and ecosystem services into plans that affect how cities develop and expand, conservationists hope to shape how cities grow. The goal here is to bring knowledge to bear on key decisions so that natural resources and working lands are protected, restored, and valued.
- *Silo busting*: Key stakeholders in cities are often in separate silos. By bringing groups together to craft a joint spatial vision (a greenprint), conservationists can overcome the lack of coordination metro areas suffer from.

14.5 A Framework for Greenprinting

Once a city has decided to conduct a greenprint, how should it evaluate the possibilities for natural infrastructure to satisfy the needs of its citizens? The following framework (Table 14.2) has been used by cities as they create greenprints. The steps in this framework are derived from the rational planning model commonly used in urban planning, as well as the first two steps of the Manual for Cities published by The Economics of Ecosystems and Biodiversity (TEEB) program, which presents a related framework that cities can use to assess the value of ecosystem services to their residents. See McDonald (2015) for a book-length presentation of this format, with case studies from multiple cities.

A greenprint is only as good as the people who create it. Having a broad set of stakeholders involved throughout a greenprint is essential to ensuring that the plan will best provide ecosystem services that meet the needs of all. Each city has a unique political, socioeconomic, and ecological context, and cities will have to modify these stages to fit into their particular circumstances.

Table 14.2 A framework for greenprint planning

Phase	Goal
1	Define the problem or policy issue
2	Take inventory: what ecosystem services matter in your city?
3	What natural infrastructure provides these services?
4	Identify options for action
5	Assess options and implement

14.5.1 Define the Problem or Policy Issue

First, those leading the planning process need to have a dialogue with key stakeholders about the problem or policy issue that ecosystem services can address. A city that evaluates natural infrastructure in the context of climate change adaptation planning will define the problem one way: What actions should the city take to increase our resilience to climate change? In contrast, a government agency in charge of managing coastal hazards might define the problem in a more focused way: What actions should be taken to reduce the risk of coastal flooding damages? Getting clarity on the key problem or policy issue to be addressed is essential and will shape the actions taken at every other stage in the framework.

14.5.2 Take Inventory: What Ecosystem Services Matter in Your City?

Good greenprints begin broad, considering the full suite of ecosystem services and determining which services matter to their stakeholders for their focal problem or policy issue. Even when the focal problem seems to point toward one ecosystem service as of paramount importance, a full consideration of the other ecosystem services that might be important in a city will be crucial, at a minimum, for identifying important co-benefits that should be included as part of the planning process. The goal of this phase is to quickly get from a large list of potentially important ecosystem services to a short list of which ecosystem services really matter and will be further evaluated in the planning process.

14.5.3 What Natural Infrastructure Provides These Services?

The next step is figuring out which patches of natural infrastructure, of various types (e.g., street trees, parks, wetland, floodplains, remnant natural habitats), provide one or more of the important ecosystem services currently. Select thoughtfully the types of natural infrastructure to map, based upon the key ecosystem services of interest. This map of important natural infrastructure is the baseline, status quo case today. If possible, it is helpful to have quantitative estimates of the ecosystem service benefits provided, either in physical units (e.g., tons of sediment not eroded due to vegetation) or in monetary units (e.g., USD). Not all habitat patches are equally important and having some way to at least rank their importance is important for later steps in this framework. Having quantitative estimates of benefits provided by each patch allows for a transparent, defensible way to choose which habitat patches to try to protect or to restore. If advanced modeling efforts are not possible to rank patches, then sometimes expert opinion can help provide a semiquantitative ranking of patches.

14.5.4 Identify Options for Actions

The next step is to figure out what actions the city could take to maintain or enhance ecosystem service provision. Many greenprinting processes begin by defining the threats that may reduce or destroy the effectiveness of current natural infrastructure. For situations in which the restoration of degraded natural infrastructure or the creation of novel patches of natural infrastructure is a possibility, the planning process must consider where spatially restoration or creation would be most appropriate. This step is key because the whole point of natural infrastructure planning is to increase ecosystem service provision relative to a baseline, status quo scenario of no action (i.e., if a city took no action, what would ecosystem service provision be). If there is little threat to an existing natural area, then efforts to protect that critical natural area have little impact on levels of future ecosystem service provision. Conversely, if a piece of critical habitat is very likely to be lost under the baseline, status quo scenario, then conservation action significantly increases future ecosystem service provision above the baseline. The effectiveness of a restoration action can similarly be evaluated against what the ecosystem service provision would be without the restoration action, under the status quo scenario.

Then, cities can more easily identify the opportunities or strategies that mitigate the threats to critical natural systems. Land protection is one common strategy, but there are many others. Incentives to provide natural habitat and ecosystem services on private land, for instance, could be another cost-effective strategy to mitigate threats or even restore some habitat. For situations where restoration or creation of new natural infrastructure is a possibility, specific opportunities need to be defined. The outcome of this stage is a finite, well-defined set of proposed natural infrastructure options that seem worthy of further evaluation.

14.5.5 Assess Options and Implement

Now cities must evaluate the various potential options and pick the best one. Sometimes this is done using formal cost-benefit analyses. In order to evaluate the return on investment of a strategy, an analysis has to integrate information on the economic value of the ecosystem services provided, the threat to those services under the baseline scenario, and the costs of implementing the strategy. Sometimes, however, an opportunity or strategy just makes the most sense to urban leaders and is selected without a formal cost-benefit strategy. After selecting the best opportunity or opportunities, it is usually released publicly as a document, the "greenprint" for what the city plans to so. Key stakeholder can then move to

Fig. 14.2 A view on the park along the Yarra River, in Melbourne, Australia. The metro area recently completed the Living Melbourne report, an example of a greenprint. (Photo Credit: pasukaru76 (public domain))

implement the document. This often takes leadership by key municipal officials, since many successful strategies to protect ecosystem services require working across multiple departments in a city and asking staff to do new jobs that they may be hesitant to do.

14.6 Living Melbourne: A Case Study of Greenprinting

Melbourne is often ranked among the world's most livable cities – a community of parks and gardens, with a plethora of brightly colored native bird species flitting through the treetops. But that reputation has attracted hundreds of thousands of additional people to Melbourne. The city's growth rates more closely match those of the developing world, with the metro region's five million people expected to exceed eight million by 2050. As the city grows, it risks losing the very natural assets that make it desirable.

In response, leaders from more than 30 municipalities within the Greater Melbourne metro area came together, under the Resilient Melbourne initiative funded by 100 Resilient Cities, to develop a plan for how nature will be an integral part of their future community (Fig. 14.2).

The Nature Conservancy, Living Melbourne, and other partners started by establishing a vision for urban forests in Melbourne: Why do urban forests benefit people and nature? Where do they exist today? What threats do they face? What solutions could help protect and restore the forest canopy? Through countless community workshops and stakeholder interviews, they settled on a common vision: "Our thriving communities are resilient, connected through nature." Realizing such a future will provide three significant benefits to Melburnians: healthy people, abundant nature, and natural infrastructure, according to the plan. To support that vision, authors identified six specific actions that will be required to realize their ideal future through 2050:

- Protect and restore species habitat and improve connectivity.
- Set targets and track progress.
- Scale up greening in the private realm.
- Collaborate across sectors and regions.
- Build a toolkit of resources to underpin implementation.
- Fund the protection and enhancement of the urban forest.

The Living Melbourne report represents one of the most ambitious greenprints that has ever been attempted globally, in terms of geography and in the broad diversity of municipal leaders involved. It is an inclusive model for both data analysis and collaborative planning that could be implemented in many cities of various sizes and across cultures and governmental structures.

A Melbourne that reflects the outcomes delineated in this plan will be able to manage its growing population, balancing urban density with the need for natural assets across the landscape. Wildlife – birds in particular – will have corridors to migrate as climate change shifts their ranges, offering some degree of protection for rare or threatened species. And natural assets, including parks, street trees, and trees on private property, will help the city adapt to a hotter, wetter climate regime, providing shade and helping the city manage stormwater and the flooding and pollution that it can cause. Public health and recreation benefits will also flow from this work, making Melbourne a desirable city to call home.

Specific metrics that will be used to track success against these goals include:

- Reduced habitat fragmentation.
- Increased habitat connectivity and created corridors.
- Improved soil moisture, water quality, and flood management through water-sensitive urban design.
- Cooler urban landscapes.
- Increased percentage of public and private land that has canopy cover.

The Living Melbourne plan was developed with support from 100 Resilient Cities, and the $1 billion cost of implementing the plan will be distributed across the dozens of partner organizations. The plan also includes examples of collaboration, financing, and policy mechanisms from across the world to guide cost-effective strategies for implementation of specific regional targets by 2030, 2040, and 2050.

References

Davoudi, S., J. Crawford, and A. Mehmood. 2009. Planning for climate change: strategies for mitigation and adaptation for spatial planners. Earthscan.

EPA. 2014. Combined Sewer Overflows (CSO). Environmental Protection Agency, Washington, DC.

EPA. 2016. Air Quality- National Summary. Environmental Protection Agency., Online at https://www.epa.gov/air-trends/air-quality-national-summary.

Gruebner, O., M. A. Rapp, M. Adli, U. Kluge, S. Galea, and A. Heinz. 2017. Cities and Mental Health. Deutsches Ärzteblatt International **114**:121.

Hales, S., S. Kovats, S. Lloyd, and D. Campbell-Lendrum. 2014. Quantitative Risk Assessment of the effects of Climate Change on Selected Causes of Death, 2030s and 2050s. World Health Organization, Geneva.

Keeler, B. L., P. Hamel, T. McPhearson, M. H. Hamann, M. L. Donahue, K. A. Meza Prado, K. K. Arkema, G. N. Bratman, K. A. Brauman, J. C. Finlay, A. D. Guerry, S. E. Hobbie, J. A. Johnson, G. K. MacDonald, R. I. McDonald, N. Neverisky, and S. A. Wood. 2019. Social-ecological and technological factors moderate the value of urban nature. Nature Sustainability **2**:29–38.

Kolstad, C. D. 2000. Environmental Economics. Oxford University Press, New York.

Lederbogen, F., P. Kirsch, L. Haddad, F. Streit, H. Tost, P. Schuch, S. Wust, J. C. Pruessner, M. Rietschel, M. Deuschle, and A. Meyer-Lindenberg. 2011. City living and urban upbringing affect neural social stress processing in humans. Nature **474**:498–501.

Lim, S. S., T. Vos, A. D. Flaxman, G. Danaei, K. Shibuya, H. Adair-Rohani, M. A. AlMazroa, M. Amann, H. R. Anderson, K. G. Andrews, M. Aryee, C. Atkinson, L. J. Bacchus, A. N. Bahalim, K. Balakrishnan, J. Balmes, S. Barker-Collo, A. Baxter, M. L. Bell, J. D. Blore, F. Blyth, C. Bonner, G. Borges, R. Bourne, M. Boussinesq, M. Brauer, P. Brooks, N. G. Bruce, B. Brunekreef, C. Bryan-Hancock, C. Bucello, R. Buchbinder, F. Bull, R. T. Burnett, T. E. Byers, B. Calabria, J. Carapetis, E. Carnahan, Z. Chafe, F. Charlson, H. Chen, J. S. Chen, A. T.-A. Cheng, J. C. Child, A. Cohen, K. E. Colson, B. C. Cowie, S. Darby, S. Darling, A. Davis, L. Degenhardt, F. Dentener, D. C. Des Jarlais, K. Devries, M. Dherani, E. L. Ding, E. R. Dorsey, T. Driscoll, K. Edmond, S. E. Ali, R. E. Engell, P. J. Erwin, S. Fahimi, G. Falder, F. Farzadfar, A. Ferrari, M. M. Finucane, S. Flaxman, F. G. R. Fowkes, G. Freedman, M. K. Freeman, E. Gakidou, S. Ghosh, E. Giovannucci, G. Gmel, K. Graham, R. Grainger, B. Grant, D. Gunnell, H. R. Gutierrez, W. Hall, H. W. Hoek, A. Hogan, H. D. Hosgood, III, D. Hoy, H. Hu, B. J. Hubbell, S. J. Hutchings, S. E. Ibeanusi, G. L. Jacklyn, R. Jasrasaria, J. B. Jonas, H. Kan, J. A. Kanis, N. Kassebaum, N. Kawakami, Y.-H. Khang, S. Khatibzadeh, J.-P. Khoo, C. Kok, F. Laden, R. Lalloo, Q. Lan, T. Lathlean, J. L. Leasher, J. Leigh, Y. Li, J. K. Lin, S. E. Lipshultz, S. London, R. Lozano, Y. Lu, J. Mak, R. Malekzadeh, L. Mallinger, W. Marcenes, L. March, R. Marks, R. Martin, P. McGale, J. McGrath, S. Mehta, Z. A. Memish, G. A. Mensah, T. R. Merriman, R. Micha, C. Michaud, V. Mishra, K. M. Hanafiah, A. A. Mokdad, L. Morawska, D. Mozaffarian, T. Murphy, M. Naghavi, B. Neal, P. K. Nelson, J. M. Nolla, R. Norman, C. Olives, S. B. Omer, J. Orchard, R. Osborne, B. Ostro, A. Page, K. D. Pandey, C. D. H. Parry, E. Passmore, J. Patra, N. Pearce, P. M. Pelizzari, M. Petzold, M. R. Phillips, D. Pope, C. A. Pope, III, J. Powles, M. Rao, H. Razavi, E. A. Rehfuess, J. T. Rehm, B. Ritz, F. P. Rivara, T. Roberts, C. Robinson, J. A. Rodriguez-Portales, I. Romieu, R. Room, L. C. Rosenfeld, A. Roy, L. Rushton, J. A. Salomon, U. Sampson, L. Sanchez-Riera, E. Sanman, A. Sapkota, S. Seedat, P. Shi, K. Shield, R. Shivakoti, G. M. Singh, D. A. Sleet, E. Smith, K. R. Smith, N. J. C. Stapelberg, K. Steenland, H. Stöckl, L. J. Stovner, K. Straif, L. Straney, G. D. Thurston, J. H. Tran, R. Van Dingenen, A. van Donkelaar, J. L. Veerman, L. Vijayakumar, R. Weintraub, M. M. Weissman, R. A. White, H. Whiteford, S. T. Wiersma, J. D. Wilkinson, H. C. Williams, W. Williams, N. Wilson, A. D. Woolf, P. Yip, J. M. Zielinski, A. D. Lopez, C. J. L. Murray, and M. Ezzati. 2012. A comparative risk assessment of burden of disease and injury attributable to 67 risk factors and risk factor clusters in 21 regions, 1990–2013: a systematic analysis for the Global Burden of Disease Study 2010. The Lancet **380**:2224–2260.

Marcus, M., M. T. Yasamy, M. v. van Ommeren, D. Chisholm, and S. Saxena. 2012. Depression: A global public health concern.

McDonald, R., M. L. Colbert, M. Hamann, R. Simkin, B. Walsh, F. Ascensao, M. Barton, K. Crossman, M. Edgecomb, T. Elmqvist, A. Gonzalez, B. Guneralp, D. Haase, O. Hillel, K. Huang, D. Maddox, A. V. Mansur, H. Pereira, J. R. Pierce, R. Weller, K. Seto, M. Tan, and C. Ziter. 2018a. Nature in the Urban Century: A global assessment of where and how to conserve nature for biodiversity and human wellbeing. The Nature Conservancy., Washington, DC.

McDonald, R. I. 2009. Ecosystem service demand and supply along the urban-to-rural gradient. Journal of Conservation Planning **5**:1–14.

McDonald, R. I. 2015. Conservation for cities: How to plan & build natural infrastructure. Island Press, Washington, DC.

McDonald, R. I. 2020. Putting biodiversity and ecosystem services into urban planning and conservation. *in* K. Seto and W. D. Solecki, editors. Handbook on urbanization and global environmental change. Taylor & Francis, Okon, UK.

McDonald, R. I., T. Beatley, and T. Elmqvist. 2018b. The green soul of the concrete jungle: The urban century, the urban psychological penalty, and the role of nature. Sustainable Earth **1**:3.

McDonald, R. I., T. Kroeger, T. Boucher, L. Wang, and R. Salem. 2016. Planting Healthy Air: A global analysis of the role of urban trees in addressing particulate matter pollution and extreme heat. The Nature Conservancy, Arlington, VA.

McDonald, R. I., T. Kroeger, P. Zhang, and P. Hamel. 2019. The Value of US Urban Tree Cover for Reducing Heat-Related Health Impacts and Electricity Consumption. Ecosystems:1–14.

McGranahan, G., D. Balk, and B. Anderson. 2007. The rising tide: assessing the risks of climate change and human settlements in low elevation coastal zones. Environment and Urbanization **19**:17–37.

McMichael, A., D. Campbell-Lendrum, S. Kovats, S. Edwards, P. Wilkinson, T. Wilson, R. Nicholls, S. Hales, F. Tanser, D. Le Sueur, M. Schlesinger, and N. Andronova. 2004. Global Climate Change. *in* M. Ezzati, A. D. Lopez, A. Rodgers, and C. Murray, editors. Comparative Quantification of Health Risks: Global and regional burden of disease attributable to selected major risk factors. World Health Organization, Geneva.

MEA. 2003. Ecosystems and Human Well-Being: A Framework for Assessment. Island Press, Washington, D.C.

Pedersen, C. B., and P. B. Mortensen. 2001. Evidence of a dose-response relationship between urbanicity during upbringing and schizophrenia risk. Archives of general psychiatry **58**:1039–1046.

Robine, J., S. Cheung, S. Le Roy, H. Van Oyen, C. Griffiths, J. Michel, and F. Herrmann. 2008. Death toll exceeded 70,000 in Europe during the summer of 2003. Comptes Rendus Biologies **331**:171–178.

Spotswood, E. N., M. Benjamin, L. Stoneburner, M. M. Wheeler, E. E. Beller, D. Balk, T. McPhearson, M. Kuo, and R. I. McDonald. 2021. Nature inequity and higher COVID-19 case rates in less-green neighbourhoods in the United States. Nature Sustainability **4**:1092–1098.

Swinburn, B. A., G. Sacks, K. D. Hall, K. McPherson, D. T. Finegood, M. L. Moodie, and S. L. Gortmaker. 2011. The global obesity pandemic: shaped by global drivers and local environments. The Lancet **378**:804–814.

Tallis, H., and S. Wolny. 2011. Including people in the mitigation hierarchy: Mapping ecosystem service winners and losers in Colombia.*in* Ecological Society of America. Ecological Society of America, Austin, TX.

UNPD. 2018. World Urbanization Prospects: The 2018 Revision. United Nations Population Division, New York.

WHO. 2017. Obesity and overweight factsheet. World Health Organization, Available online at http://www.who.int/mediacentre/factsheets/fs311/en/.

Chapter 15
Wetlaculture: Solving Harmful Algal Blooms with a Sustainable Wetland/Agricultural Landscape

William J. Mitsch, Bingbing Jiang, Samuel K. Miller, Kyle D. Boutin, Li Zhang, Andrew Wilson, and Bhavik R. Bakshi

15.1 Introduction

Harmful algal blooms have increased dramatically in the world in the past few decades, mostly due to over-enrichment of freshwater and coastal waters by nutrients, primarily nitrogen and phosphorus. Recent worldwide increases in coastal and freshwater nutrient enrichment, referred to terms such as red tide, cyanobacteria or blue-green algae blooms, hypoxia, or simply cultural eutrophication, have been

W. J. Mitsch (✉)
School of Environment and Natural Resources, The Ohio State University, Columbus, OH, USA

The Water School, Florida Gulf Coast University, Naples, FL, USA
e-mail: wmitsch@fgcu.edu; wmitsch.1@gmail.com

B. Jiang
Marine and Environmental Science at Zhejiang Ocean University, Zhoushan, Zhejiang, China

S. K. Miller
Mendoza College of Business at the University of Notre Dame, Notre Dame, IN, USA

K. D. Boutin
North Dakota State University, Fargo, ND, USA

L. Zhang
Research Programs in The Water School Administration at Florida Gulf Coast University in Naples, Naples, FL, USA

A. Wilson
Florida Department of Environmental Protection's South District Office, Fort Myers, FL, USA

B. R. Bakshi
School for Engineering of Matter, Transport and Energy, School of Sustainability, School of Complex Adaptive Systems, Arizona State University, Tempe, AZ, USA

Lowrie Department of Chemical and Biomolecular Engineering, The Ohio State University, Columbus, OH, USA

B. R. Bakshi (ed.), *Engineering and Ecosystems*,
https://doi.org/10.1007/978-3-031-35692-6_15

described as being due to dramatic increases in fertilizer use in the last 50–70 years coupled with landscape saturation of these fertilizers and loss of wetlands as buffer systems (Mitsch and Gosselink 2015; Mitsch 2017b,c). Climate-driven changes of increased precipitation and subsequent discharges from nutrient-saturated landscapes, higher sustained water temperatures, and even higher frequency of tropical and subtropical cyclones that may cause upwelling of marine sediments have been suggested as reasons why algal blooms are increasing (Paerl and Huisman 2008; Michalak 2016; Aumann et al. 2018; Richardson et al. 2019; Havens et al. 2019). It has been suggested that the use of urea-based fertilizers increased worldwide more than 100-fold in the last four decades of the twentieth century (Glibert et al. 2006; Glibert 2017; Glibert and Burkholder 2018). Freshwater algal blooms by cyanobacteria and other freshwater algal species around the world have been attributed to phosphorus fertilization in lakes around the world, including severely eutrophic Taihu in China, western Lake Erie and Lake Okeechobee in North America, and the Baltic Sea in Europe, all of which are fed by significant nutrient fluxes from agricultural and urban runoff (Duan et al. 2009; Stumpf et al. 2012; Michalak et al. 2013; Paerl et al. 2014, 2019; Mitsch 2017b; Scavia et al. 2017, 2018; Havens et al. 2019). Wurtsbaugh et al. (2019) summarized that "annual costs of eutrophication have indicated $1 billion losses for European coastal waters and $2.4 billion for lakes and streams in the United States." These are undoubtedly low estimates as the publications quoted for these data that are now almost decade old.

15.1.1 Wetlands and Water Quality

Polluted wastewater and runoff treatment by wetlands have been considered one of the key examples of the field of ecological engineering in the western world since its principles were first introduced 30 years ago (Mitsch and Jørgensen 1989, 2004; Mitsch 2017c). Ecological engineering has been reinvented recently under new terms such as "nature-based solutions" (Schaubroeck 2017), but the concept remains the same: a partnership between humanity (e.g., treating polluted waters) and an ecosystem (e.g., wetlands).

Wetlands, both natural and created, have been shown to be sinks for a great number of chemicals (Mitsch and Gosselink 2015; Kadlec 2020). Researchers in the USA in the 1970s investigated the role of natural wetlands, particularly in regions where they are found in abundance, to treat wastewater and thus recycle clean water back to groundwater and surface water (see some of the earliest studies in cypress swamps in Florida by Odum et al. (1977) and peatlands in Michigan by Kadlec and Kadlec (1979)). Earlier than this in Europe, German scientists investigated the use of constructed basins with macrophytes (*höhere Pflanzen*) for purification of wastewater (Seidel, 1964, 1966). The two different approaches, one initially utilizing natural wetlands and the other using artificial systems, have converged into the general field of *treatment wetlands*. The field now encompasses the construction and/or use of wetlands for a myriad of water quality applications including municipal wastewater, small-scale rural wastewater, acid mine drainage, landfill leachate, and nonpoint source pollution from both urban and agricultural runoff. While water quality improvement by treatment wetlands is the primary goal of treatment wetlands, these also provide habitat for a wide diversity of plants and animals and can support many of the other wetland ecosystem services.

15.1.2 Stormwater Treatment Wetlands

One of the most important applications of wetland treatment systems is the use of wetlands for treating stormwater and runoff from agricultural fields and urban landscapes. Research projects illustrating the effects and functioning of these types of wetlands in agricultural watersheds have been summarized in large data assessments (Land et al. 2013, 2016) and conducted decades ago in many locations such as southeastern Australia (Raisin and Mitchell 1995; Raisin et al. 1997), Europe (Leonardson et al. 1994; Jacks et al. 1994; Arheimer and Wittgren 1994; Comin et al. 1997), and North America (Kadlec and Hey 1994; Phipps and Crumpton 1994; Mitsch et al. 1995; Kovacic et al. 2000; Wang and Mitsch 2000; Larson et al. 2000; Hoagland et al. 2001; Fink and Mitsch 2004).

15.1.3 Experimental and Large-Scale Stormwater Treatment Wetlands

Several wetland sites have received the equivalent of nonpoint source pollution but under somewhat more controlled or experimental hydrologic conditions (e.g., river overflow to riparian wetland) over several years of study. Boney Marsh, a constructed wetland located along the Kissimmee River in southern Florida, was investigated for nutrient retention of river water for more than a decade (Moustafa 1999; Moustafa et al. 1996), and it was found to be a consistent sink of nitrogen and phosphorus but at relatively low levels. In central Ohio, Mitsch et al. (1998, 2008, 2012, 2014) and Fink and Mitsch (2007) described nutrient retention for almost two decades on created floodplain wetlands located on the Olentangy River floodplain on the campus of The Ohio State University. These wetlands were consistent sinks of phosphorus and nitrogen with early trends of decreasing P and N retention and more recent trends of increased or stable nutrient retention as the created wetlands matured. The Ohio State wetland program then moved to Florida where more recent studies of nutrient retention from urban stormwater wetlands, similar to the Ohio site except for its subtropical location, showed some reduction in the nutrient retention over a decade (Griffiths and Mitsch 2017; Nesbit and Mitsch 2018; Mitsch et al. 2019).

Probably the largest assemblage of treatment wetlands anywhere in the world are 23,000 ha of created wetlands, locally called stormwater treatment areas (STAs), that have been created for phosphorus control from upstream agricultural areas before the effluent flows into the low-nutrient Florida Everglades. From their start, these wetlands reduced phosphorus loads significantly and, after several years of operation by 80% or more, lowered the average phosphorus concentrations in the outflows consistently below 20 ppb (Juston and DeBusk 2006, 2011; Pietro 2012; Dierberg and DeBusk 2008; Paudel et al. 2010; Paudel and Jawitz 2012; Entry and Gottlieb 2014; Mitsch et al. 2015, 2018; Juston and Kadlec 2019). Submerged aquatic vegetation wetlands and emergent vegetation wetlands restored from historic wetlands rather than agriculture and with loading rates at or below 2 g P m^{-2} yr^{-1} have resulted in outflow P concentrations consistently between 10 and 20 ppb and mass removal efficiencies consistently above 85%. A multi-year mesocosm study that investigated the effect that different plant communities on reducing phosphorus concentrations in the outflow of the STAs achieved 10 ppb in some of the plant communities. This may be possible when the inflow is the effluent coming from the STAs and the hydraulic loading rates are lower than those in the current STAs (Mitsch et al. 2015).

15.2 What Is Wetlaculture?

Wetlaculture is a landscape-scale integration of wetlands designed for the retention of nutrients (phosphorus and nitrogen) from polluted agricultural and urban runoff with systematic recycling of those nutrients to agriculture, horticulture, and/or forestry. Our long-term research plan involves the development of interlinking physical, mathematical, and business models to optimize design parameters in different climates, soils, landscapes, and waterscapes. The wetlaculture term comes from wetlands + agriculture. The nutrient, energy, and water fluxes in a conventional agricultural-urban landscape are compared to agriculture with wetlaculture systems added in Fig. 15.1. In conventional agriculture (Fig. 15.1a), manufactured fertilizers are added to agricultural fields that produce food for the human economy. Both runoff from agricultural fields and treated wastewater from urban environments discharge nutrients to lakes, river, and estuaries directly with little to no recycling of the nutrients. In this system, there is eventual oversaturation of both land and water with nutrients, sometimes referred to as legacy nutrients. These nutrients accumulate year after year in farm fields and in sediments of lakes, rivers, and estuaries.

Wetlaculture (Fig. 15.1b) utilizes wetlands to reduce some of the nutrient fluxes from agriculture and cities that otherwise would go directly to lakes, rivers, and estuaries. When designed properly, treatment wetlands are a key way to reduce nutrients flowing to downstream aquatic ecosystems (pathway 1 in Fig. 15.1b).

The second aspect of wetlaculture is what distinguishes it from a linear combination of agriculture and treatment wetlands. There is overwhelming evidence that wetlands can retain nitrogen and phosphorus, some for many years (Mitsch and

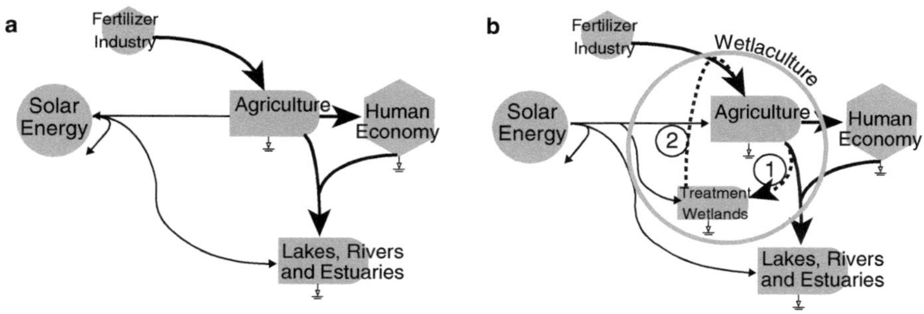

Fig. 15.1 Diagram of (**a**) current landscape-urban system that leads to excessive nutrients reaching our lakes, rivers, and estuaries and (**b**) treatment wetlands, integrated with agriculture and cities (wetlaculture landscape) that can contribute to both (1) water quality improvement by wetlands and (2) recycling of fertilizer leading to reduction of new fertilizer applications

Gosselink, 2015; Kadlec 2020), with perhaps 2 or 3 years needed for the wetland to become a sink if the wetland is constructed on high-nutrient agricultural fields (Mitsch et al. 1998, 2015).

The unique feature of a wetlaculture landscape is that a treatment wetland, after some number of years, could be "flipped" to an agricultural field, with the idea that the food or fiber production crop would grow well without adding any additional fertilizers on the nutrients that the wetland has accumulated over those "x" number of years (pathway 2 in Fig. 15.1b). Then after "y" years, the agricultural field would be "flipped" back to being a wetland. Both physical and mathematical models especially will help us begin to understand what those "x" and "y" years are for different climates, soils, and nutrient loading rates.

15.2.1 Mesocosm Models

We designed and constructed three mesocosm compounds, two in Ohio and one in Florida, where nutrient eutrophication of lakes, river, and/or estuaries and coastal waters are major issues. We are testing the wetlaculture concept at these three sites in two distinct climates (Fig. 15.2): subtropical climate of south Florida and temperate climate of central and northwestern Ohio. Preliminary descriptions of these three wetlaculture mesocosm experiments (Table 15.1) are by Mitsch (2017a, b, c, d, e, 2018), Jiang and Mitsch (2020), and Balster (2018).

Hydrologic Experiment

The same wetlaculture mesocosm experiment is being run for multiple years at each of the three locations shown in Fig. 15.2 and described in Table 15.1 and by

Fig. 15.2 Locations of three current wetlaculture experimental sites in eastern USA

Table 15.1 Three independent estimates of the global loss of wetlands in the world

1. The Economics of Ecosystems and Biodiversity (TEEB) study (Russi et al. 2013) suggested that the world lost half of its wetlands in the twentieth century alone
2. Davidson (2014), in a meta-analysis, determined that the world lost 53.5 percent of its wetlands "long-term" (i.e., multi-century), with higher loss rates in inland vs. coastal wetlands
3. 87% of world's wetlands have been lost globally in the last 300 years according to an Assessment Report of the Intergovernmental Science-Policy Platform on Biodiversity and Ecosystem Services (Sushma 2018)

Jiang and Mitsch (2020). Photos of the three experimental sites are shown in Fig. 15.3. Twenty-eight 380-L (100-gallon) Rubbermaid tubs (1 m^2 in surface area) have been installed in the ground for a 2 × 2 × 7 experiment (Fig. 15.4) with plumbing designed to deliver the desired hydraulic loading rates (HLR) to each of the 28 tubs. The principal initial experiment at each mesocosm compound will be to compare high and low loading rates to these wetland mesocosms.

Fig. 15.3 Completed mesocosm compounds at (**a**) Buckeye Lake site in in Ohio River Basin in central OH; (**b**) Black Swamp/Defiance site in Lake Erie Basin, northwest OH; and (**c**) Freedom Park site at urban runoff treatment wetlands in Greater Everglades region, Naples, FL

Mesocosm water levels are maintained at two different levels: saturated soil level and 10 cm standing water to introduce a second hydrologic variable. In each site, half of the mesocosms (14) are fed weekly with a high hydraulic loading rate (H) of 30 cm/week, and the other 14 are fed with a low hydraulic loading rate (L) of

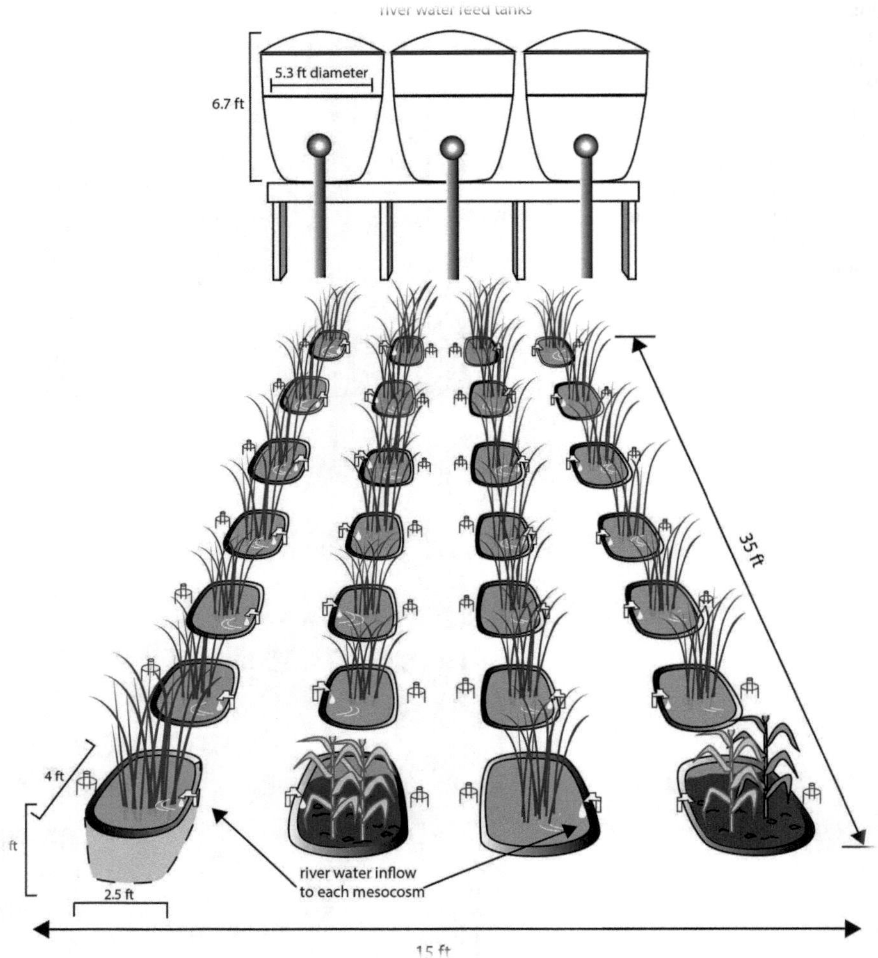

Fig. 15.4 Sketch of experimental mesocosm compound designed with 28 mesocosms for $7 \times 2 \times 2$ hydrologic experiments

10 cm/week. Two different water levels are maintained for each of the two loading rates: deep standing water (D) of 8 to 10 cm and shallow water (S) with moist soil conditions. This results in a $2 \times 2 \times 7$ experiment with four different hydrologic control treatments each condition replicated seven times. Filtered and unfiltered water samples (250 mL) are taken in acid-washed bottles from the selected inflows representing the two HLR inflows and from all 28 mesocosm outflows on a biweekly basis. Soil samples will be taken every other year (surface to 20 cm) and analyzed for nutrients (Table 15.2).

Table 15.2 Summary of three mesocosm wetlaculture experiments in eastern USA

Site name	Freedom Park, Naples, Florida	Buckeye Lake, Ohio	Defiance County (near Lake Erie), Ohio
Status	Constructed in 2018	Constructed in 2016–2017	Constructed in 2017–2018
Location	Freedom Park, Naples, FL, a 20-ha treatment wetland complex 20 km upstream of the Gulf of Mexico	Village of Buckeye Lake, Central Ohio, adjacent to eutrophic lake	A private farm in the region of the former Great Black Swamp, Defiance, Ohio, and upstream of Maumee River and western basin of Lake Erie
Climate	Subtropical	Temperate, continental	Temperate, continental
Soil unit name	Blanton fine sand	Algiers silt loam	Hoytville clay loam
Hydric soil?	No	Yes	Yes
Wetland species planted initially in all mesocosms	*Cladium jamaicense* (sawgrass)	*Schoenoplectus tabernaemontani* (bulrush)	*Schoenoplectus tabernaemontani* (bulrush)
Agricultural crop to be cycled into mesocosms	Corn	Corn	Corn
Water source	Urban runoff from drainage ditch from Naples, FL	High nutrient river (South Fork of Licking River) discharging from Buckeye Lake	Drainage ditch with agricultural runoff in former Great Black Swamp
Citations	Mitsch (2018), Wilson et al. (2020), Hartzler et al. (2021), Wilson (2021)	Mitsch (2017d), Jiang (2020), Jiang and Mitsch (2020)	Mitsch (2017b), Jiang (2020), Jiang et al. (2021)

Water Quality Improvement

The two Ohio wetlands were significant sinks for P and N within months of their start (Tables 15.3 and 15.4). The newest mesocosm compound in Florida (started in late summer 2018) was a major sink for nitrate-nitrogen but was a source for phosphorus in the early years of the study (Tables 15.3 and 15.4). We investigated the cause for this phosphorus sink and found that it was probably because the lawn soils that we used in this park for the mesocosms had been fertilized for a decade by an irrigation system which used treated wastewater from local water treatment plants in Naples, Florida. More recent results show that after about 2 years of operation while receiving stormwater runoff as inflow waters, the Florida mesocosms became sinks for phosphorus (Fig. 15.5). Though baseline soil conditions play an important role in how quickly newly created treatment wetlands become nutrient sinks, the

Table 15.3 Concentration (average ± standard error (number of samples)) of TP, SRP, TN, and nitrate+nitrite during hydroperiod in three wetlaculture mesocosm sites

Site	Years		TP (ppb)	SRP (ppb)	TN (mg N L^{-1})	Nitrate+Nitrite (mg N L^{-1})
Buckeye Lake, OH	3	Inflow	0.325 ± 0.018 (51)	0.219 ± 0.018 (69)	3.426 ± 0.154 (57)	2.483 ± 0.12 (66)
Defiance, OH	2	0.154 ± 0.011 (52)	0.029 ± 0.004 (60)	5.284 ± 0.177 (64)	3.7 ± 0.178 (64)	
Freedom Park, FL	1	0.138 ± 0.021 (17)	0.044 ± 0.004 (18)	1.443 ± 0.04 (15)	0.324 ± 0.039 (18)	
Buckeye Lake, OH	3	Outflow	0.18 ± 0.008 (452)	0.07 ± 0.003 (624)	1.78 ± 0.046 (506)	0.753 ± 0.041 (591)
Defiance, OH	2	0.04 ± 0.002 (334)	0.008 ± 0.0004 (390)	3.178 ± 0.106 (426)	2.061 ± 0.102 (428)	
Freedom Park, FL	1	0.178 ± 0.008 (164)	0.069 ± 0.003 (167)	1.496 ± 0.047 (140)	0.02 ± 0.002 (168)	

NSW no standing water with saturated soil, *SW* standing water

Table 15.4 Mass retention (average ± standard error (number of samples)) of total phosphorus and total nitrogen at three wetlaculture mesocosm sites in 2019

Site	Nutrient species	Hydroperiod (week)	2019 Mass retention (g m^{-2})			
			HLR_H		HLR_L	
			SW	NSW	SW	NSW
BuckeyeLake, OH		23	0.674±0.128(7)	1.077±0.107(7)	0.24±0.093(7)	0.48±0.038(7)
Defiance, OH	Total P	18	0.679±0.035(7)	0.614±0.103(7)	0.255±0.005(7)	0.248±0.005(7)
Freedom Park, FL		30	-0.312±0.299(7)	-0.202±0.114(7)	-0.325±0.044(7)	-0.046±0.07(7)
BuckeyeLake, OH		23	12.623±0.908(7)	11.644±0.607(7)	3.748±0.348(7)	3.414±0.613(7)
Defiance, OH	Total N	18	13.103±1.381(7)	9.817±1.812(7)	5.485±0.176(7)	4.162±0.197(7)
Freedom Park, FL		30	-0.616±1.128(7)	0.864±0.745(7)	-1.172±0.71(7)	0.452±0.182(7)

NSW = no standing water with saturated soil; *SW* = standing water, HLR_H = hydraulic loading rate of 30 cm $week^{-1}$; HLR_L = hydraulic loading rate of 30 cm $week^{-1}$. Yellow flagging indicates net export of P or N.

results of this study demonstrate that even unfavorable soil conditions can result in effective nutrient retention in wetlands, given enough time.

Nutrient Recycling Experiment

Once the wetland mesocosm soils begin to increase in nutrients (P and N) as indicated by the inflow-outflow nutrient budgets and or by significant increases in soil nutrients, we will convert 4–8 mesocosms in that year to simulate agricultural fields by turning off the inflowing water, allowing the soil to drain, and introducing agricultural crops (see two drained mesocosms in the first row in Fig. 15.4; see Table 15.1 for possible crops).

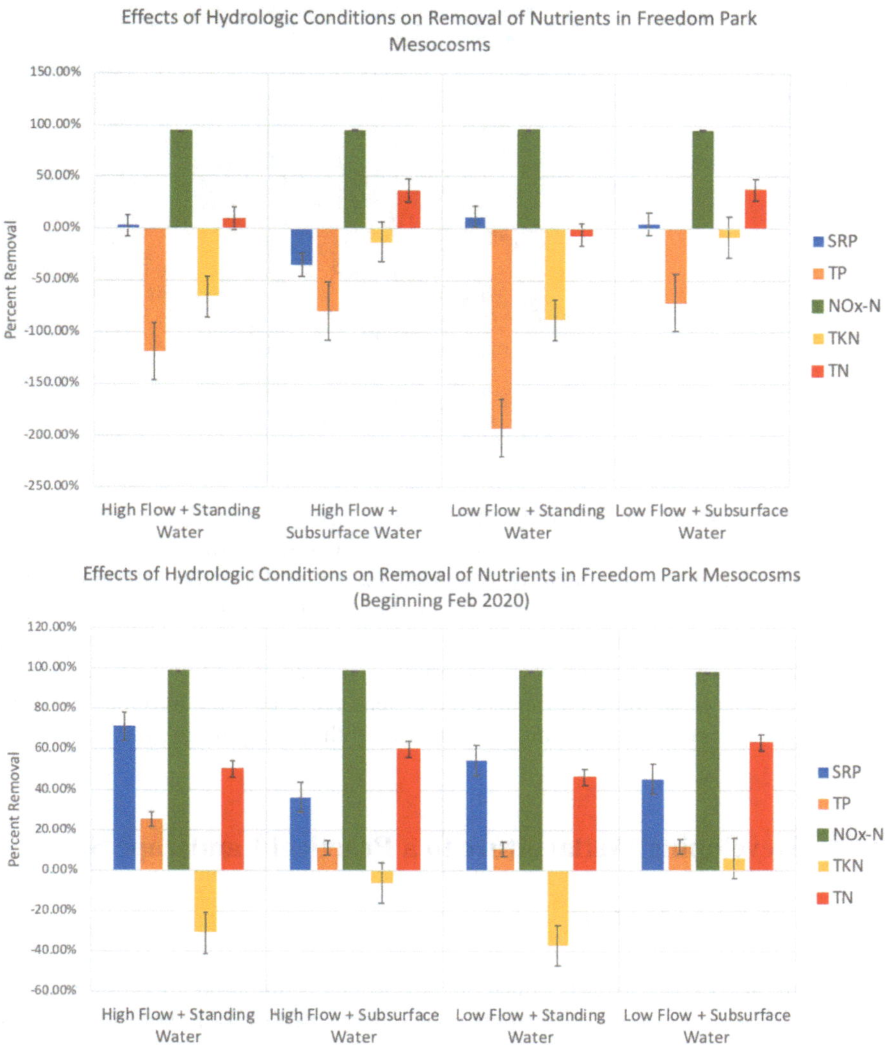

Fig. 15.5 Average removal efficiency ± standard error for each nutrient species for four separate hydrologic conditions for Freedom Park mesocosms for (**a**) full 2.5-year study and (**b**) final year Feb 2020 to Feb 2021. The bottom chart excludes data before February 2020 to illustrate the significant improvement in nutrient retention in the last year of the study

In the summer of 2020, the first agricultural trial of Wetlaculture took place at the Buckeye Lake Mesocosm Site in central Ohio (Boutin et al. 2021). Eight of the wetland mesocosms were drained and planted with field corn, a crop selected due to its dominance in the region. Grain yields were compared to those achieved by a local farm employing the conventional methods of fertilizer and pesticide applications.

Fig. 15.6 First corn crop grown in wetlaculture experiments. Site was Buckeye Lake Ohio in 2020

Despite flooding, herbivory, competition from weeds, and comparatively limited room for root growth in the mesocosms, the wetlaculture corn achieved yields of 58 ± 9.5 bu./ac. While significantly lower than the 160 ± 15.3 bu./ac attained by the conventional farm, this wetlaculture corn, as shown in Fig. 15.6, did not receive any chemical inputs but was instead fed by nutrient pollution captured in the mesocosm soil during the wetland portion of the experiment. Instead of continuing downstream to fuel algae blooms, these nutrients were instead recycled back into the economy in the form of corn, demonstrating the potential of the wetlaculture system.

15.3 Developing Wetlaculture to a Practical Landscape Scale

15.3.1 Spatial Modeling

Our team has extensive experience in developing simulation models used for investigating ecological theories and management strategies for Lake Erie wetlands (Mitsch and Reeder 1991), created riverine wetlands in the Midwest (Wang and Mitsch 2000), and Florida STA treatment wetlands (Marois and Mitsch 2015).

But initially we chose to use spatial (GIS) modeling to explore for optimum conditions for nutrient retention in the eutrophic western basin of Lake Erie. This site is now plagued by harmful algal blooms annually due to nutrient discharges primarily from this basin, and water quality was impacted so significantly with toxic cyanobacteria affecting hundreds of thousands of residents. Agricultural runoff from the Western Lake Erie Basin is the main nutrient source into Lake Erie. Restoring at least 10% of the historic Black Swamp and developing wetlaculture have been proposed as a potential landscape solution for the landscape problem in the Lake Erie basin. Thanks to the wide availability of geological data and desktop

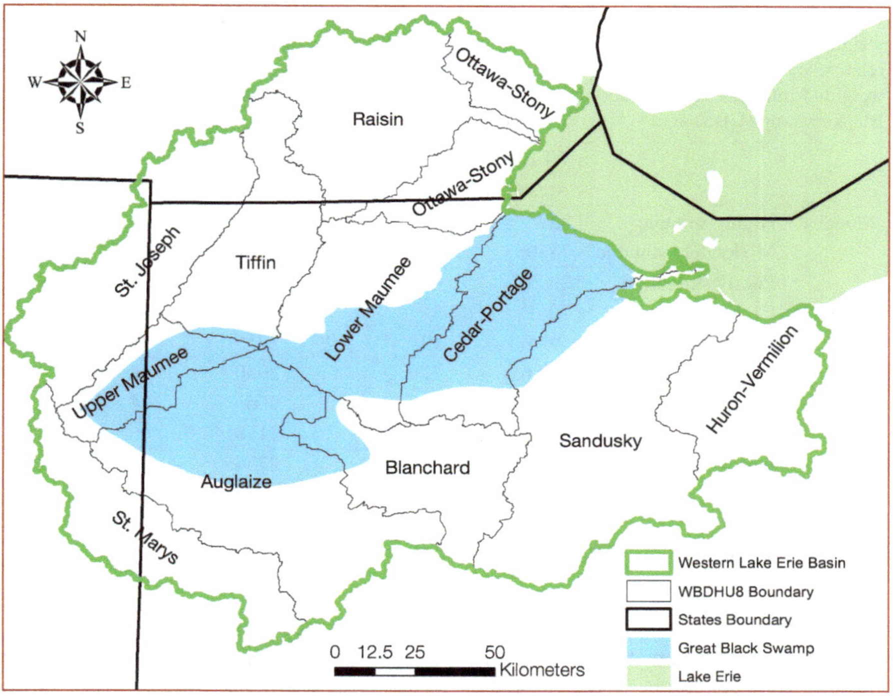

Fig. 15.7 Study area of the spatial models: the Western Lake Erie Basin (WLEB), based on the Watershed Boundary Dataset Hydrologic Unit 4-digit (WBDHU8), that consists of 13 WBDHU8 sub-watershed areas and the Great Black Swamp

GIS applications, GIS modeling can be a very useful tool in developing spatial models for building a decision support system. Evaluation of potential wetlaculture sites can provide meaningful information for developing long-term sustainable agricultural systems. This landscape investigation is focused on finding the most suitable wetland/wetlaculture restoration sites within the former 6700 km^2 Great Black Swamp in the western basin of Lake Erie, the shallowest of the Laurentian Great Lakes in North America. A potential indicator GIS model was developed, with various data layers of hydrology, soils, and topography combined, to identify and classify potential wetland areas in the now-drained Great Black Swamp region that could mitigate nutrient inflows to Lake Erie (Fig. 15.7). The models were developed from all the reclassified score layers with a different weight index combination (Table 15.5). Overall, the estimated area of highly suitable potential wetland restoration areas in the Western Lake Erie Basin and in the Great Black Swamp area is approximately 1000 km^2 (3%) and 800 km^2 (12%), respectively, much larger than the 400 km^2 of wetlands that have been suggested as necessary to control the algal blooms in Lake Erie.

Table 15.5 The estimated area of suitability potential wetlaculture by three models with different weight index: model 1(M1) has the equal weighted influences of all criteria, model 2 (M2) has relatively higher weighted influence of Compound Topographic Index (CTI) and relatively lower weighted influence of CTI model 3 (M3) in the Western Lake Erie Basin (WLEB) and the Great Black Swamp (GBS) area

		WLEB		GBS	
		Area (km^2)	Percentage (%)	Area (km^2)	Percentage (%)
Model 1	Highly suitable	852	3	726	11
	Moderately suitable	7840	25	3311	49
	Poorly suitable	20,205	66	2381	35
Model 2	Highly suitable	1061	3	713	11
	Moderately suitable	5975	19	2755	41
	Poorly suitable	21,861	71	2950	44
Model 3	Highly suitable	1248	4	966	14
	Moderately suitable	8011	26	3118	46
	Poorly suitable	19,638	64	2334	35

15.3.2 Pilot-Scale Projects in Wetlaculture

Simultaneous to this modeling effort, we will investigate collaborations with farmers or resource managers to use recently created wetlands or agricultural fields for unreplicated yet much larger pilot-scale projects to see if the flipping from farms to wetlands and wetlands to farms is a viable approach at the pilot project scale. A prototype of the Everglades STAs, a 1544-ha treatment wetland complex called the Everglades Nutrient Removal (ENR) project, was first designed and tested before the STAs were created (Reddy et al. 2006). Over its first 6-year operating schedule (1994–1999), this wetland complex decreased total phosphorus and total nitrogen by 79 and 26 percent, respectively (Gu et al. 2006), with an average outflow concentration of 21 parts per billion (ppb ~ μg-P/L) over that period. As a result of the success of the ENR project, six full-scale stormwater treatment areas (STAs) treating agricultural runoff from the EAA south of Lake Okeechobee have since been created. Some of these systems, described above, have now been in operation for over 25 years.

15.3.3 Establishing a Viable Business Model

In the Fall of 2018, a team of researchers from the University of Notre Dame, Florida Gulf Coast University, and the Ohio State University collaborated to develop, from scratch, a preliminary business model that was presented formally at a lake and reservoir conference in early November 2018 (Miller and Mitsch 2018). Subsequent wetlaculture workshops were held in Huron Ohio, adjacent to Lake Erie, in August 2019 (https://www.fgcu.edu/thewaterschool/centers/ewrp/workshop_ohio_

Table 15.6 Revenue flows for payment for ecosystem services scheme

	Sequestration (lb/acre/year)	Price range	Price per lb	Annual P.E.S. per acre
Nitrogen	359	\$0.51–5.09	\$1.30	466.70
Phosphorous	17.8	\$0.33–12.00	\$10.00	178.00
Carbon[a]	1391	\$20–40/ton	\$0.02	24.34
				669.04
	Value @ 2:1 trading ratio			334.52
CRP	70.00 / acre			70.00
Total Baseline P.E.S. Value per Acre				404.52

[a]Additional, net of corn farming seq. of 304 lb./ac/yr

2019) and in Naples, Florida, adjacent to the Florida Everglades in February 2020 (https://www.fgcu.edu/thewaterschool/centers/ewrp/symposiumonwetlands2020).

A conceptual business model was envisioned that can leverage emerging financial mechanisms to create and capture economic value from the nutrient sequestration services of the restored wetlaculture wetland (Fig. 15.8). The three-stage process could work as follows:

1. Financial investors provide investment capital in the form of *Environmental Impact Bond* (EIB) funding. This capital funds the up-front restoration of wetlands. The EIBs earn a market rate of return and are paid off at the end of the term (say 10 years).
2. Landowners (farmers) are incentivized to ensure successful restoration and stewardship of the restored wetlands through a process known as *Payment for Success* (PFS). These stakeholders are paid based on the actual nutrient capture performance of their wetlands.
3. *Payment for Ecosystem Services* (PES) revenue streams provide the funding to pay the farmers for the "productive" use of their land and to pay the interest and principal on the EIBs. Table 15.6 summarizes our early estimates of revenue flows for payment for ecosystem services related to the retention of nitrogen, phosphorus, and carbon by the wetlands.
4. These PES markets (both voluntary and compliance-based) are gaining traction, although their presence is still uncertain. Barriers to adoption include unpredictable market volatility and mistrust caused by lack of standards. Bipartisan legislation that has recently been proposed at the federal level to help remove these barriers provides a signal that there is public support for establishing market mechanisms to address this problem of externalized costs as illustrated by the Growing Climate Solutions Act developed in 2020 by the US Congress.

A summarized list of assumptions underlying the structure of this financial model is as follows:

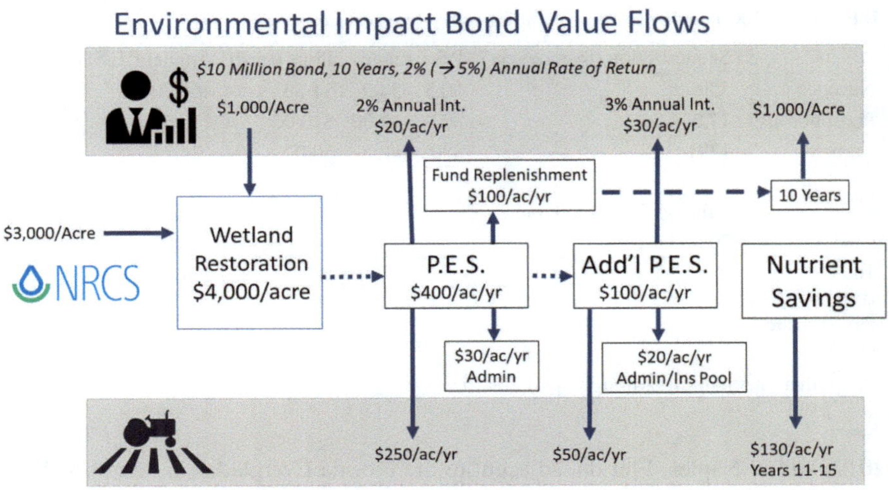

Fig. 15.8 A schematic of a business model for wetlaculture that includes economic benefits to the farmer or land owner and to investors for environmental impact bonds (from Miller and Mitsch 2018)

- Wetlaculture sites will serve 10-year cycles as wetlands – before being flipped back to agricultural use.
- EIB funding will be for 10-year term bonds, earning rates comparable to those paid by municipal bonds (~2% annual). PFS incentives can boost these returns to perhaps 5% per annum.
- The average wetlaculture site size will be ~200 ha (500 acres).
- Net restoration costs will be ~ $1000 per acre, after NCRS Wetland Reserve Program (WRP) reimbursement 75% of restoration costs.
- Average profit from farming (corn in Ohio) is ~ $340 per acre per year. Less productive land (with estimated profits of $240 per acre per year) will likely be prioritized for wetlaculture projects. This is the benchmark return farmers will seek.
- PES revenue flows are computed as shown in Table 15.6, and assume "stacking" of credits is permitted:
- A trading ratio of 2:1 is utilized to account for risk of possible ecosystem failure. With the performance incentives in the model, this trading ratio is expected to improve to 1.5:1.
- USDA Conservation Program funding of $70 per acre applies.

An updated schematic of our business model is shown in Fig. 15.8.

One final business model consideration is the funding mechanism for the PES payments. There are two potential sources of this essential funding:

1. Polluter Pays. Under this model, all farmers are held economically responsible for the nutrient runoff issue. A modest fee (say 0.5%) is assessed on all

agricultural outputs, and this funding is then available to pay the PES needs of the program.
2. Beneficiary Pays is an option whereby the downstream stakeholders see a benefit from avoided costs from harmful algae blooms and may take a favorable stance toward this approach to eliminating the problem. In this model, a modest fee of say 0.1% (which could likely be passed through to end customers) is assessed to resort and hospitality businesses in impacted regions which would in turn be used to cover the PES payment costs.

Our business model suggests an approach where farmers could make profits comparable to crop income and investment funding from return-seeking investors could provide the necessary up-front capital

15.4 Broader Impacts

Figure 15.9 illustrates a spatial pattern of wetlands and agricultural fields that could result when both wetlands and agricultural fields provide income for both farmers and investors in our business model summarized in Fig. 15.8. The research described here of integrating wetland nutrient removal with recycling of those nutrients to agriculture provides a middle ground that allows for ecosystem services of wetlands to be utilized in conjunction with sustainable agriculture, providing cleaner water in our lakes, reservoirs, estuaries, and rivers, but also economic incentives for farmers. The net result could be increased wetland creation, restoration, and conservation, more sustainable food production, and lower energy costs for water quality improvement.

The three mesocosm projects in Florida and Ohio have been very much in the public eye through their early years of this half-decade study. With only a few growing seasons of research completed, all three sites have been extensively written up the *Columbus Dispatch*, the *Toledo Blade* in Ohio, and the *Naples Daily News* in southwestern Florida. Our idea of Great Black Swamp restoration to save Lake Erie has become a lively topic in social media and was featured in an UNDARK blog story called "Learning to Love the Great Black Swamp" (Levy 2017) and described in a subsequent book (Levy 2019).

Restored and created wetlands can remove significant amounts of nitrogen and phosphorus from agricultural and stormwater runoff and sequester large quantities of carbon from the atmosphere. Wetlaculture could provide a sustainable business approach to lessen our fertilizer excesses. Farmers and investors could make a profit either by farming or creating wetlands, and downstream waters will be cleaner too. Wetlaculture has the potential of being a win-win-win proposition for a new way of managing agricultural landscapes.

Fig. 15.9 Sketch of wetlaculture landscape of agriculture fields interspersed with wetlands

References

Arheimer, B., and H. B. Wittgren. 1994. Modelling the effects of wetlands on regional nitrogen transport. *Ambio* 23: 378–386.

Aumann, H. H., Behrangi, A., & Wang, Y. 2018. Increased frequency of extreme tropical deep convection: AIRS observations and climate model predictions. *Geophysical Research Letters* 45: 13,530–13,537.

Balster, L. 2018. Black Swamp savior: How bringing back conquered wetlands could help solve harmful algal blooms. *Environmental Monitor*, Fondriest Environmental, Fairborn, OH

Boutin, K.D., W. J. Mitsch, E. Everham, B. Bakshi, and L. Zhang. 2021. An evaluation of corn production within a Wetlaculture^{TM} system at Buckeye Lake, Ohio. *Ecological Engineering* 171, 106366.

Comin, F. A., J. A. Romero, V. Astorga, and C. Garcia. 1997. Nitrogen removal and cycling in restored wetlands used as filters of nutrients for agricultural runoff. *Water Science and Technology* 35: 255–261.

Davidson, N.C. 2014 How much wetland has the world lost? Long-term and recent trends in global wetland area. *Marine and Freshwater Research* 65: 934–941.

Dierberg, F. E., and DeBusk, T. A. 2008. Particulate phosphorus transformations in south Florida stormwater treatment areas used for Everglades protection. *Ecological Engineering* 34, 100–115.

Duan, H.R., X. Ma, F. Xu, F. Kong, S. Zhang, W. Kong, J. Hao, and L. Shang. 2009. Two decade reconstruction of algal blooms in China's Lake Taihu. *Environmental Science & Technology* 43: 3524–3528.

Entry, J. A. and A. Gottlieb. 2014. The impact of stormwater treatment areas and agricultural best management practices on water quality in the Everglades Protection Area. *Environmental Monitoring and Assessment* 186: 1023–1037.

Fink, D.F. and W.J. Mitsch. 2004. Seasonal and storm event nutrient removal by a created wetland in an agricultural watershed. *Ecological Engineering* 23: 313–325.

Fink, D.F. and W.J. Mitsch. 2007. Hydrology and biogeochemistry in a created river diversion oxbow wetland. *Ecological Engineering* 30: 93–102.

Glibert, P.M., J. Harrison, C. Heil, and S. Seitzinger. 2006. Escalating worldwide use of urea – a global change contributing to coastal eutrophication. *Biogeochemistry* 77: 441–463.

Glibert, P.M. 2017. Eutrophication, harmful algae and biodiversity — Challenging paradigms in a world of complex nutrient changes. Mar Pollut Bull 124: 591–606.

Glibert, P.M. and J. M. Burkholder. 2018. Causes of Harmful Algal Blooms. Pages 1 – 38 In: *Harmful Algal Blooms: A Compendium Desk Reference*, First Edition. S. E. Shumway, J. M. Burkholder, and S. L. Morton, eds., John Wiley & Sons Ltd., Chichester, UK.

Griffiths, L.N. and W.J. Mitsch, 2017. Reduction of nutrients from urban stormwater runoff by storm-pulsed and seasonally pulsed created wetlands in the subtropics. *Ecological Engineering* 108: 414–424.

Gu, B., M. J. Chimney, J. Newman, and M. K. Nungesser. 2006. Limnological characteristics of a subtropical constructed wetland in south Florida (USA). *Ecological Engineering* 27: 345–360.

Hartzler, H., M. Ruppert, A. Wilson, and W. J. Mitsch. 2021. Plans for comparing wetlaculture™ wetland mesocosms rotated to horticulture in temperate and tropical regions in 2021. 13th International Symposium of Biogeochemistry of Wetlands, zoom meeting, March 21–25, 2021.

Havens, K.E., Ji, G.H., Beaver, J.R., Fulton, R.S., Teacher, C.E., 2019. Dynamics of cyanobacteria blooms are linked to the hydrology of shallow Florida lakes and provide insight into possible impacts of climate change. *Hydrobiologia* 82943–59.

Hoagland, C. R., L. E. Gentry, M. B. David, and D. A. Kovacic. 2001. Plant nutrient uptake and biomass accumulation in a constructed wetland. *Journal of Freshwater Ecology* 16: 527–540.

Jacks, G., A. Joelsson, and S. Fleischer. 1994. Nitrogen retention in forested wetlands. *Ambio* 23: 358–362.

Jiang, B.B. 2020. Exploring the Potential of Nutrient Retention and Recycling with Wetlaculture Systems in Ohio with Physical and Landscape Models. Ph.D. Dissertation, University of South Florida, Tampa, FL.

Jiang, B.B. and W.J. Mitsch. 2020. Influence of hydrologic conditions on nutrient retention and soil and plant development in a former central Ohio swamp: A wetlaculture mesocosm experiment. *Ecological Engineering* 157: 105969

Jiang, B.B., W.J. Mitsch and C. Lenhart. 2021. Estimating the importance of hydrologic conditions on nutrient retention and plant richness in a wetlaculture mesocosm experiment in a former Lake Erie basin swamp. *Water* 2021, 13, 2509.

Juston, J.M., and DeBusk, T. A. 2006. Phosphorus mass load and outflow concentration relationships in stormwater treatment areas for Everglades restoration. *Ecological Engineering* 26, 206–223.

Juston, J.M., and DeBusk, T.A. 2011. Evidence and implications of the background phosphorus concentration of submerged aquatic vegetation wetlands in Stormwater Treatment Areas for Everglades restoration. *Water Resources Research* 41,430–446.

Juston, J.M., and R.H. Kadlec, 2019. Data-driven modeling of phosphorus (P) dynamics in low-P stormwater wetlands. *Environmental Modelling and Software* 118: 226–240.

Kadlec, R.H. 2020. Treatment Marshes for Runoff and Polishing. CRC Press, Taylor & Francis Group, 6000, Boca Raton, FL

Kadlec, R.H. and J.A. Kadlec. 1979. Wetlands and water quality. Pages 436–456 In: *Wetlands Functions and Values: The State of Our Understanding*. P.E. Greeson, J.R. Clark, and J.E. Clark, eds. American Water Resources Association, Minneapolis, MN.

Kadlec, R. H., and D. L. Hey. 1994. Constructed wetlands for river water quality improvement. *Water Science and Technology* 29: 159–168.

Kovacic, D. A., M. B. David, L. E. Gentry, K.M. Starks, and R. A. Cooke. 2000. Effectiveness of constructed wetlands in reducing nitrogen and phosphorus export from agricultural tile drainage. *Journal of Environmental Quality* 29: 1262–1274.

Land, M., W. Granéli. A. Grimvall, C. C. Hoffmann, W. J. Mitsch, K. S. Tonderski, and J.T.A. Verhoeven. 2016. How effective are created or restored freshwater wetlands for nitrogen and phosphorus removal? A systematic review. *Environmental Evidence* 5,9 doi https://doi.org/10.1186/s13750-016-0060-0

Land, M., W. Granéli, A. Grimvall, C. C. Hoffmann, W.J. Mitsch, K. Tonderski, and J.T.A. Verhoeven. 2013. How effective are created or restored freshwater wetlands for nitrogen and phosphorus removal? A systematic review protocol. *Environmental Evidence* 2:16 doi: https://doi.org/10.1186/2047-2382-2-16

Larson, A. C., L. E. Gentry, M. B. David, R. A. Cooke, and D. A. Kovacic. 2000. The role of seepage in constructed wetlands receiving tile drainage. *Ecological Engineering* 15: 91–104.

Leonardson, L., L. Bengtsson, T. Davidsson, T. Persson, and U. Emanuelsson. 1994. Nitrogen retention in artificially flooded meadows. *Ambio* 23: 332–341.

Levy, S. 2017. Learning to Love the Great Black Swamp. UNDARK. https://undark.org/2017/03/31/great-black-swamp-ohio-toledo/

Levy, S. 2019. *The Marsh Builders: The Fight for Clean Water, Wetlands, and Wildlife*. Oxford University Press, New York, NY.

Marois, D.E. and W.J. Mitsch. 2015. Modeling phosphorus retention at low concentrations in Florida Everglades mesocosms. *Ecological Modelling* 319:42–62.

Michalak, A. M. 2016. Study role of climate change in extreme threats to water quality. *Nature* 535: 349–350.

Michalak, A.M., E.J. Anderson, D. Beletsky, S. Boland, N.S. Bosch, T.B. Bridgeman, J.D. Chaffin, K. Cho, R. Confesor, I. Daloğlu, J.V. DePinto, M.A. Evans, G.L. Fahnenstiel, L. He, J.C. Ho, L. Jenkins, T.H. Johengen, K.C. Kuo, E. LaPorte, X. Liu, M.R. McWilliams, M.R. Moore, D.J. Posselt, R.P. Richards, D. Scavia, A.L. Steiner, E. Verhamme, D.M. Wright, and M.A. Zagorski. 2013. Record-setting algal bloom in Lake Erie caused by agricultural and meteorological trends consistent with expected future conditions. *Proceedings of the National Academy of Sciences USA* 110: 6448–6452.

Miller, S. and W.J. Mitsch. 2018. The Business of Wetlaculture – Seeking Viable Business Models for Landscape-Scale Nutrient Capture. Abstracts, 38th Annual Symposium of North American Lake Management Society (NALMS), Cincinnati, OH

Mitsch, W.J. 2017a. Wetlaculture. YouTube video; Published January 21, 2017 https://www.youtube.com/watch?v=U7xOm8RBEdU

Mitsch, W.J. 2017b. Solving Lake Erie's harmful algal blooms by restoring the Great Black Swamp in Ohio. *Ecological Engineering* 108: 406–413.

Mitsch, W.J. 2017c. Ecologically engineered wetlands as solutions for nutrient pollution in lakes, wetlands, and coastlines" International Workshop on Eutrophication: Synthesis of Knowledge, CNRS Headquarter, Paris, France https://eutrophication.sciencesconf.org/data/pages/Mitsch_Bill_Eutrophication_Workshop.pdf

Mitsch, W.J. 2017d. Solving harmful algal blooms: Progress in 2016-2017 at a wetlaculture experiment at Buckeye Lake, Ohio. Ohio Wetland Association *Wetland Trumpeter Newsletter*, September 2017. pp. 8–9, 11.

Mitsch, W. J., 2017e. Solving Lake Erie's harmful algal blooms by restoring the Great Black Swamp in Ohio. *Ecological Engineering, 108*, 406–413.

Mitsch, W.J. 2018. Experimental wetlaculture (wetlands + agriculture) mesocosm compound established in Naples, Florida, to restore wetlands, solve harmful algal blooms, and develop sustainable agriculture. *Wetland Science and Practice* 35 (2): 79–80.

Mitsch, W.J. and S.E. Jørgensen (eds.) 1989. *Ecological Engineering: An Introduction to Ecotechnology.* John Wiley & Sons, New York. 472 pp.

Mitsch, W.J. and B.C. Reeder. 1991. Modelling nutrient retention of a freshwater coastal wetland: estimating the roles of primary productivity, sedimentation, resuspension and hydrology. *Ecological Modelling* 54:151–187.

Mitsch, W.J., J.K. Cronk, X. Wu, R.W. Nairn and D.L. Hey. 1995. Phosphorus retention in constructed freshwater riparian marshes. *Ecological Applications* 5:830–845.

Mitsch, W.J., X. Wu, R.W. Nairn, P.E. Weihe, N. Wang, R. Deal, C.E. Boucher. 1998. Creating and restoring wetlands: A whole-ecosystem experiment in self-design. *BioScience* 48: 1019–1030.

Mitsch, W.J. and S.E. Jørgensen. 2004. *Ecological Engineering and Ecosystem Restoration.* John Wiley & Sons, Inc., New York. 411 pp.

Mitsch, W.J., L. Zhang, D.F. Fink, M.E. Hernandez, A.E. Altor, C.L. Tuttle and A.M. Nahlik. 2008. Ecological engineering of floodplains. *Ecohydrology & Hydrobiology* 8: 139–147.

Mitsch, W.J, L. Zhang, K. C. Stefanik, A. M. Nahlik, C. J. Anderson, B. Bernal, M. Hernandez, and K. Song. 2012. Creating wetlands: Primary succession, water quality changes, and self-design over 15 years. *BioScience* 62: 237–250.

Mitsch, W.J., L. Zhang, E. Waletzko, and B. Bernal. 2014. Validation of the ecosystem services of created wetlands: Two decades of plant succession, nutrient retention, and carbon sequestration in experimental riverine marshes. *Ecological Engineering* 72: 11–24.

Mitsch, W.J. and J.G. Gosselink. 2015. *Wetlands, 5th ed.* John Wiley & Sons, Inc., Hoboken, NJ. 744 pp.

Mitsch, W.J., L. Zhang, D. Marois, and K. Song. 2015. Protecting the Florida Everglades wetlands with wetlands: Can stormwater phosphorus be reduced to oligotrophic conditions? *Ecological Engineering* 80: 8–19.

Mitsch, W.J., Marois, D., L. Zhang, and K. Song. 2018. Protecting water quality of the Florida Everglades: Sustainably achieving low phosphorus concentrations with wetlands. *Ecological Engineering* 116: 174–177.

Mitsch, W.J. L. Zhang, L. Griffiths, T. Nesbit, and L. Li. 2019. *Freedom Park Stormwater Wetlands: Towards a Long-Term Monitoring and Management Plan.* Final Report to Stormwater Management Division, Collier County, Naples FL

Moustafa, M. Z. 1999. Nutrient retention dynamics of the Everglades nutrient removal project. *Wetlands* 19: 689–704.

Moustafa, M. Z., M. J. Chimney, T. D. Fontaine, G. Shih, and S. Davis. 1996. The response of a freshwater wetland to long-term "low level" nutrient loads—marsh efficiency. *Ecological Engineering* 7: 15–33.

Nesbit, T. A. and W. J. Mitsch. 2018. Hurricane and seasonal effects on hydrology and water quality of a subtropical urban stormwater wetland. *Ecological Engineering* 120: 134–145.

Odum, H.T., K.C. Ewel, W.J. Mitsch and J.W. Ordway. 1977. Recycling treated sewage through cypress wetlands. In, F.M. D'Itri (ed.), *Wastewater Renovation and Reuse.* New York: Marcel Dekker Press. pp. 35–67.

Paudel, R., J.-K. Min, and J. W. Jawitz. 2010. Management scenario evaluation for a large treatment wetland using a spatio-temporal phosphorus transport and cycling model. *Ecological Engineering* 36: 1627–1638.

Paudel, R., and J.W. Jawitz. 2012. Does increasedmodel complexity improve descriptionof phosphorus dynamics in a large treatment wetland? Ecological Engineering 42: 283–294.

Paerl, H. W., & Huisman, J. 2008. Blooms like it hot. *Science* 320(5872): 57–58.

Paerl H., Xu, H., Hall N., Zhu G., Qin B., Wu, Y., et al., 2014. Controlling cyanobacterial blooms in hypertrophic Lake Taihu, China: will nitrogen reductions cause replacement of non-N_2 fixing by N_2 fixing taxa? *PloS One* 9 (11),1–13.

Paerl, H.W., N.S. Hall, K.E. Havens, H. Xu, G. Zhu, B. Qin, M.J. McCarthy, S.E. Newell, J.T. Scott, and T.G. Otten. 2019. Mitigating eutrophication and toxic cyanobacterial blooms in large lakes: The evolution of a dual nutrient (N and P) reduction paradigm. *Hydrobiologia* xx: 1–17 https://doi.org/10.1007/s10750-019-04087-y

Phipps, R. G., and W. G. Crumpton. 1994. Factors affecting nitrogen loss in experimental wetlands with different hydrologic loads. *Ecological Engineering* 3: 399–408.

Pietro, K. 2012. Synopsis of the Everglades stormwater treatment areas, water year 1996–2012. Technical Publication ASB-WQTT-12–001, South Florida Water Management District, West Palm Beach, FL. 56 pp.

Raisin, G. W., and D. S. Mitchell. 1995. The use of wetlands for the control of non-point source pollution. *Water Science and Technology* 32: 177–186.

Raisin, G. W., D. S. Mitchell, and R. L. Croome. 1997. The effectiveness of a small constructed wetland in ameliorating diffuse nutrient loadings from an Australian rural catchment. *Ecological Engineering* 9: 19–35.

Reddy, K. R., R. H. Kadlec, M. J. Chimney, and W. J. Mitsch, eds. 2006. The Everglades Nutrient Removal Project. Special issue of *Ecological Engineering* 27: 265–379.

Richardson J., Feuchtmayr H., Miller C., Hunter P.D., Maberly S.C., Carvalho L. 2019. Response of cyanobacteria and phytoplankton abundance to warming, extreme rainfall events and nutrient enrichment. *Global Change Biol.* 25: 3365–3380.

Russi D., P. ten Brink, A. Farmer, T. Badura, D. Coates, J. Förster, R. Kumar, and N. Davidson. 2013. *The Economics of Ecosystems and Biodiversity for Water and Wetlands*. IEEP, Ramsar Secretariat, Gland, London and Brussels.

Scavia, D., M. Kalcic, R. L. Muenich, J. Reed, N. Aloysius, I. Bertani, C. Boles, R. Confesor, J. DePinto, M. Gildow, J. Martin, T. Redder, S. Sowa, Y.-C. Wang, and H. Yen. 2017. Multiple models guide strategies for agricultural nutrient reductions. *Front. Ecol. Environ.* 15, 126–132.

Schaubroeck T. 2017. Nature-based solutions: sustainable? *Nature* 543, 315 doi:https://doi.org/10.1038/543315c

Seidel, K. 1964. Abbau von Bacterium Coli durch höhere Wasserpflanzen. *Naturwissenschaften* 51: 395.

Seidel, K. 1966. Reinigung von Gewässern durch höhere Pflanzen. *Naturwissenschaften* 53: 289–297.

Stumpf, R.P., T.T. Wynne, D.B. Baker, and G.L. Fahnenstiel. 2012. Interannual variability of cyanobacterial blooms in Lake Erie. *PLoS ONE*, 7: e42444.

Sushma, M. 2018. World lost 87 per cent wetlands in 300 years. Down-to-earth report on IPBES Report on Global Assessment on Biodiversity and Ecosystem Services https://www.downtoearth.org.in/news/wildlife-biodiversity/world-lost-87-per-cent-wetlands-in-300-years-60033

Wang, N. and W.J. Mitsch. 2000. A detailed ecosystem model of phosphorus dynamics in created riparian wetlands. *Ecological Modelling* 126: 101–130.

Wilson, A. 2021. Examining the effects of hydrology and reclaimed water application on nutrient retention in Wetlaculture mesocosms in southwest Florida. Ph.D. dissertation, University of South Florida, Tampa, FL.

Wilson, A., L. Zhang, and W.J. Mitsch. 2020. Examining hydrologic variables and external factors on nutrient retention in wetland mesocosms in southwest Florida. Abstracts, University of Florida Symposium on Sustainable Water Resources, Gainesville, FL

Wurtsbaugh, W.A., Paerl, H.W., and Dodds, W.K. 2019 Nutrients, eutrophication and harmful algal blooms along the freshwater to marine continuum. *WIREs Water*. 2019;6:e1373. https://doi.org/10.1002/wat2.1373

Chapter 16
Design of Agroecological Landscapes

Rebecca J. Hanes, Varsha Gopalakrishnan, and Bhavik R. Bakshi

16.1 Introduction

Demand for agricultural products and arable farmland is increasing, driven by the need to provide food for a growing population and by an ongoing shift from fossil fuels to biomass-derived biofuels. These increasing demands are currently being met by converting nonagricultural land to farmland and by using intensive farming practices to increase production from existing farmland, but both strategies cause ecological degradation at multiple scales through fertilizer- and pesticide-laden runoff, increased water demand, and greenhouse gas emissions. These strategies are therefore not ecologically sustainable in the long term.

This chapter presents a method based on techno-ecological synergy (TES) for making decisions around effective and sustainable land use for food and energy production [1]. Such decisions must be made with regard to three conflicting objectives: agricultural productivity (including both the quantities and types of crops produced), economic sustainability, and ecological sustainability. Activities upstream and downstream of land use, including the production of inputs to the land use activities, the conversion of energy feedstocks into fuels and other energy carriers, and the production of inputs to the conversion processes, must also be considered. These upstream and downstream activities contribute to the ecological

R. J. Hanes · V. Gopalakrishnan
William G. Lowrie Department of Chemical and Biomolecular Engineering, The Ohio State University, Columbus, OH, USA

B. R. Bakshi (✉)
School for Engineering of Matter, Transport and Energy, School of Sustainability, School of Complex Adaptive Systems, Arizona State University, Tempe, AZ, USA

Lowrie Department of Chemical and Biomolecular Engineering, The Ohio State University, Columbus, OH, USA
e-mail: bhavik.bakshi@asu.edu

B. R. Bakshi (ed.), *Engineering and Ecosystems*,
https://doi.org/10.1007/978-3-031-35692-6_16

sustainability of the overall food and energy production system, and the conversion processes impact the energy productivity and the economic sustainability of the overall system as well.

Effective agroecological landscapes should balance economic and ecological sustainability with the production of food crops, energy feedstocks, and/or electricity. Energy crop productivity should be combined with process-level information on fuel yield to determine if sufficient energy carriers can be produced. Economic sustainability should be quantified by an economic indicator such as net present value (NPV) or annual profits for both land use and the conversion of energy feedstocks to fuels and energy carriers should be. Finally, the overall system should be ecologically sustainable, meaning there is a balance between ecosystem service demand—the resources required from nature and the pollutants released into nature—and ecosystem service supply—the ability of nature to supply resources and absorb pollutants. For a system to achieve ecological sustainability, ecosystem service supply must equal or exceed demand for all relevant ecosystem services at the largest scale at which each service is relevant. For some services such as climate regulation, the largest relevant scale is global, while other services such as air quality regulation, water supply, and pollination are relevant at local and regional scales. Because agriculture depends on a variety of ecosystem services, multiple services must be considered with the objective of achieving ecological sustainability at the appropriate scale for each relevant ecosystem service.

The central concept of the TES-based method presented in this chapter is that productive, ecologically sustainable, and economically feasible landscapes must incorporate a variety of land use types and in particular must combine agricultural land uses with ecological land uses to achieve ecological sustainability. Conservation agricultural practices involving less soil tilling and/or lower amounts of fertilizer and other chemicals applied to crops are becoming more common in the USA and have been shown to improve the ecological sustainability of farming; however, even conservation agriculture requires more ecosystem services than can be supplied by the farmland alone. Land use options must be chosen with respect to the local meteorological and soil conditions, and for energy production should include both energy feedstock production and technological options, such as wind farms, solar panels, and so on. These technological options can in some cases reduce the acreage required to produce energy, freeing up area that can be used for food production or for engineered ecosystems. Moreover, land use decisions must not be made by considering only the landscape itself but must take both upstream and downstream activities, particularly the conversion of energy feedstocks into energy carriers, into account, to avoid shifting ecological impacts and to ensure that the system as a whole is economically and ecologically sustainable.

This work applies the ϵ-constraint method for multi-objective optimization to a mixed-integer linear program (MILP) previously developed in [2] and [3]. The MILP models food- and energy-producing land use activities in central Ohio, along with a set of biomass conversion processes that produce energy carriers. The life cycles of each land use activity and conversion process are modeled and included in the MILP using the process-to-planet (P2P) modeling framework

[4]. The land use options include conventional and conservation farming practices for several common cropping systems, the installation of wind turbines and solar panels, and ecological land use options consisting of reforestation with a variety of Ohio native trees and an engineered wetland. The ϵ-constraint method for multi-objective optimization is used to quantify trade-offs between food production, energy production, and system economics under an "unsustainable" scenario and a "high sustainability" scenario. Under the unsustainable scenario, ecosystem service demand is allowed to exceed supply by any amount for each of three ecosystem services considered. Under the high sustainability scenario, ecosystem service supply is constrained to be greater than ecosystem service demand for two out of three services. Initial results indicate that including ecological land use options allows sustainability in land use to be achieved with relatively minor sacrifices in the anthropocentric objectives and that trade-offs between the three anthropocentric objectives are not worsened by improving the sustainability.

16.2 Method

The agroecological landscape design in this work is done using optimization to select combinations of land use options from a predetermined set, or superstructure, of agricultural, ecological, and energy-producing land use activities shown in Fig. 16.1. Land use options were chosen based on known feasibility in central Ohio, the region under consideration. Along with land use options, several options for downstream conversion processes that produce energy carriers from biomass feedstock are included, and the system as a whole is modeled as a MILP. The MILP was first developed in [2] and was extended to include food production and an engineered wetland in [3]. This section gives an overview of the data used to build the MILP, the optimization formulation, and the application of the ϵ-constraint method. Additional details on the background data used to develop the MILP are available in [2] and in [3].

As discussed in the Introduction, an agroecological landscape should be productive, economically sustainable, and ecologically sustainable. Food and energy productivity and economic sustainability are the three objective functions considered in this work; ecological sustainability is addressed through net ecosystem service supply constraints discussed later in the section. All three objective functions are calculated based on the agroecological landscape itself (decision variables $\underline{\mathbf{s}}$ and $\underline{\mathbf{t}}$, both of which quantify the amount of land use options implemented) and on the biomass conversion processes (decision variables $\mathbf{s}$, which quantify the scale of biomass conversion process implemented). Food and energy productivity is quantified with food calories produced and gasoline gallon equivalents (GGEs) produced, respectively, and the economic sustainability is quantified with the system's net present value (NPV). NPV and productivity of the life cycle of these activities are not considered, but ecosystem service demand created within the life

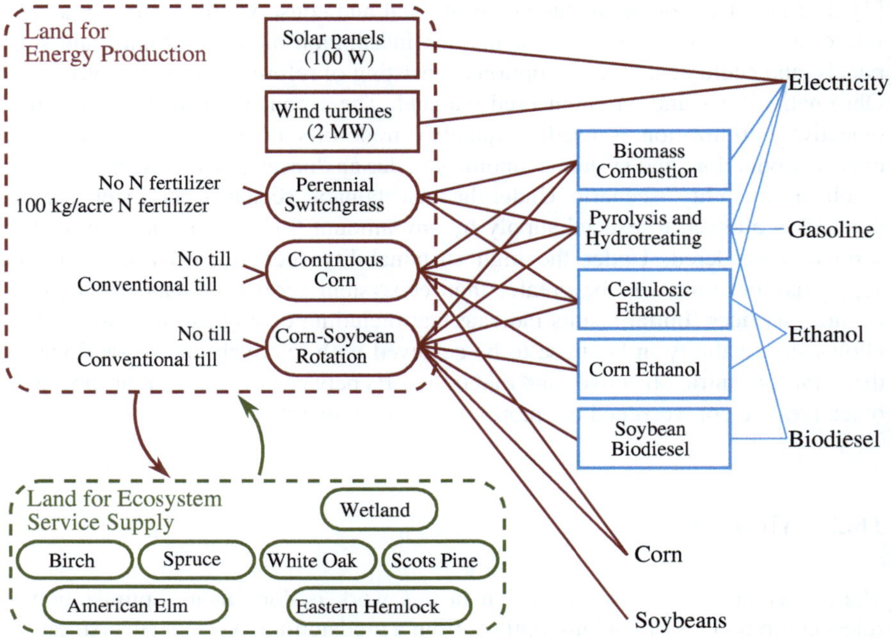

Fig. 16.1 The system modeled by the MILP consists of an agroecological landscape with land use options shown in red and green, plus conversion processes that produce energy carriers from biomass feedstocks, shown in blue. The P2P modeling framework is used to include the life cycle of land use and biomass conversion activities (not shown). (Reproduced with permission from [3])

cycle is included. Decision variables used throughout the following equations are listed, with definitions, in Table 16.7. The energy objective function is:

$$Z_{\mathbf{E}} = \mathbf{E} \cdot \begin{bmatrix} \underline{\mathbf{s}} \\ \underline{\mathbf{t}} \\ \underline{\mathbf{s}} \end{bmatrix} \text{ GGE,} \tag{16.1}$$

the food production objective function is

$$Z_{\mathbf{F}} = \mathbf{F} \cdot \begin{bmatrix} \underline{\mathbf{s}} \\ \underline{\mathbf{t}} \\ \underline{\mathbf{s}} \end{bmatrix} \text{ kcal,} \tag{16.2}$$

and the economics objective function is

$$Z_{\mathbf{N}} = \mathbf{N} \cdot \begin{bmatrix} \underline{\mathbf{s}} \\ \underline{\mathbf{t}} \\ \underline{\mathbf{s}} \end{bmatrix} \text{ USD.} \tag{16.3}$$

Table 16.1 Land used (acres), food produced (food cal), energy produced (GGE), and net present value (USD) for each component model. (Reproduced with permission from [3])

Component	Land Used (acre) $\mathbf{\Lambda}$	Food (food cal) $\mathbf{F}$	Energy (GGE) $\mathbf{E}$	NPV (USD) $\mathbf{N}$	
Cont. Corn, No Till	1	4.98×10^7	0	5.87×10^2	Per acre
Cont. Corn, Conv. Till	1	4.82×10^7	0	6.67×10^2	
Corn-Soybean, No Till	1	6.24×10^7	0	3.74×10^2	
Corn-Soybean, Conv. Till	1	5.78×10^7	0	2.67×10^2	
Switchgrass, No N Fert.	1	0	0	-4.88×10^2	
Switchgrass, With N Fert.	1	0	0	-5.31×10^2	
Wind turbine	0.1633	0	4.27×10^6	-3.01×10^5	Per unit
Solar panel	1.73×10^{-4}	0	7.9×10^1	-7.23×10^2	
White Oak	1.66×10^{-3}	0	0	-3.65×10^{-1}	Per tree
Scots Pine	8.15×10^{-4}	0	0	-1.79×10^{-1}	
American Elm	1.12×10^{-3}	0	0	-2.47×10^{-1}	
Spruce	1.12×10^{-3}	0	0	-2.47×10^{-1}	
Birch	7.91×10^{-4}	0	0	-1.74×10^{-1}	
Eastern Hemlock	1.12×10^{-3}	0	0	-2.47×10^{-1}	
Corn Ethanol	–	-4.98×10^{11}	8.02×10^7	-6.88×10^6	Per plant
Soybean Biodiesel	–	-3.99×10^{11}	5.94×10^6	2.61×10^6	
Switchgrass ethanol*	–	0	8.11×10^8	-9.17×10^6	
Stover ethanol*	–	0	1.18×10^9	-9.21×10^6	
Switchgrass pyrolysis*	–	0	5.91×10^9	-3.17×10^6	
Stover pyrolysis*	–	0	5.54×10^9	-5.96×10^6	
Switchgrass combustion	–	0	1.43×10^{10}	-1.03×10^7	
Stover combustion	–	0	1.97×10^{10}	-1.03×10^7	

* Includes energy and revenue from byproduct electricity

Components of **E**, **F**, and **N** are given in Table 16.1.

Constraints in the MILP consist of balance equations, an upper limit on the amount of land used, the three ϵ-constraints imposed on the objective functions, and a set of constraints that define the sustainability scenarios. Equation (16.4) is the set of balance equations, based on the P2P modeling framework:

$$\begin{bmatrix} \bar{\mathbf{I}} - \bar{\mathbf{A}}^* & -\underline{\mathbf{X}}_u & -\mathbf{X}_u^E \\ \underline{\mathbf{0}} & \underline{\mathbf{X}} & -\mathbf{X}_u^V \\ \mathbf{0} & \mathbf{0} & \mathbf{X} \end{bmatrix} \begin{bmatrix} \bar{\mathbf{s}} \\ \underline{\mathbf{s}} \\ \mathbf{s} \end{bmatrix} \begin{matrix} \geq \\ \geq \\ \geq \end{matrix} \begin{bmatrix} \bar{\mathbf{0}} \\ \underline{\mathbf{0}} \\ \mathbf{0} \end{bmatrix} \tag{16.4}$$

In the P2P framework, component models at local, regional, and national scales are integrated to allow for small-scale decision-making and optimization while accounting for environmental impacts at a national scale [4]. Equation (16.4) contains the agricultural and energy-producing land use component models $\underline{\mathbf{X}}$, the biomass conversion process models $\mathbf{X}$, and the national-scale model $\bar{\mathbf{I}} - \overline{\mathbf{A}}^*$, a 328-sector input–output model of the 2002 U.S. economy [5, 6]. Equation (16.4) involves the land use decision variables $\underline{\mathbf{s}}$ and the biomass conversion decision variables $\mathbf{s}$, as well as the national-scale decision variables $\bar{\mathbf{s}}$ that quantify the amount of economic activity within each sector. Unlike $\underline{\mathbf{X}}$ and $\mathbf{X}$, both of which contain material exchanges in physical units, $\bar{\mathbf{I}} - \overline{\mathbf{A}}^*$ contains exchanges between U.S. sectors in monetary units. The remaining matrices in Eq. (16.4) quantify the exchanges between component models at different scales. $\underline{\mathbf{X}}_u$ contains inputs to the land use activities and $\mathbf{X}_u^E$ contains inputs to the biomass conversion processes that originate within the economy and are quantified in monetary units. Similarly, $\mathbf{X}_u^V$ contains inputs to the biomass conversion processes that originate within the land use activities—namely, biomass—and are quantified in physical units. All component models in Eq. (16.4), except for the national-scale model, are derived from data sources given in Table 16.2, and further details on the derivation of each model may be found in [2].

The agroecological landscape is constrained to be at most 10,000 acres:

$$\mathbf{\Lambda} \cdot \left[\underline{\mathbf{s}}\ \underline{\mathbf{t}}\right]^T + \mathbf{\Omega}_i \cdot \underline{\mathbf{s}} \leq 10{,}000 \text{ acres} \tag{16.5}$$

All land being considered in this work is assumed to begin as central Ohio farmland. Equation (16.5) includes land used for agriculture and direct electricity production ($\underline{\mathbf{s}}$), land used for reforestation ($\underline{\mathbf{t}}$), and land used to establish a wetland ($\mathbf{\Omega}_i\underline{\mathbf{s}}$). Wetland acreage is quantified using multipliers, $\mathbf{\Omega}_i$, that define the acreage of wetland required to supply water quality regulation to one acre of land use. The subscript i refers to the sustainability scenario; the wetland is excluded from the unsustainable scenario in which no ecosystem service supply is required to be produced. Values of $\mathbf{\Omega}_i$ under the two sustainability scenarios are given in Table 16.3.

Table 16.2 Component model data sources

Component model	Data sources
Cropping systems (all)	[7, 8, 9, 10, 11, 12, 13, 14]
Wind turbine	[15, 16, 17, 18]
Solar panel	[19, 15]
Reforestation	[20, 21, 22, 23]
Wetland	[24, 25]
Corn ethanol	[26, 27]
Soybean biodiesel	[28, 29, 30, 31, 32]
Stover ethanol	[33, 26, 27]
Switchgrass ethanol	
Stover pyrolysis	[34, 35, 36]
Switchgrass pyrolysis	
Stover combustion	[32, 37]
Switchgrass combustion	

Table 16.3 Calculated wetland acreage per acre of land use option under three sustainability constraints

Land use option	$\boldsymbol{\Omega}_{\text{unsust.}}$	$\boldsymbol{\Omega}_{\text{high sust.}}$	Units
Cont. corn, conv. till	0	0.504	Acre/acre
Cont. corn, no till	0	0.734	
Corn-soybean, conv. till	0	0.489	
Corn-soybean, no till	0	0.590	
Switchgrass, no N fertilizer	0	0.051	
Switchgrass, with N fertilizer	0	0.052	
Wind turbine	0	0	
Solar panel	0	0	

Trade-offs between the three objectives are analyzed using multi-objective optimization via the ϵ-constraint method, in which individual objectives are optimized while the remaining objectives are constrained. The ϵ-constraint values were specified as fractions of each objective's optimal (maximum) value using the parameters ϵ_N, ϵ_E, and ϵ_F for the economic, energy production, and food production constraints, respectively. These parameters were assigned the following values:

$$\epsilon_E, \epsilon_F, \epsilon_N \in \{0.0, 0.17, 0.34, 0.51, 0.68, 0.85\} \tag{16.6}$$

The MILP was solved under all combinations of ϵ parameters (216 combinations total), although not all combinations yielded feasible designs. The energy production ϵ-constraint is:

$$\mathbf{E} \cdot \begin{bmatrix} \underline{\mathbf{s}} \\ \underline{\mathbf{t}} \\ \mathbf{s} \end{bmatrix} \geq \epsilon_E E_{\max} \text{ GGE} \tag{16.7}$$

The food production ϵ-constraint is:

$$\mathbf{F} \cdot \begin{bmatrix} \underline{\mathbf{s}} \\ \underline{\mathbf{t}} \\ \mathbf{s} \end{bmatrix} \geq \epsilon_F F_{\max} \text{ kcal} \tag{16.8}$$

And the NPV ϵ-constraint is:

$$\mathbf{N} \cdot \begin{bmatrix} \underline{\mathbf{s}} \\ \underline{\mathbf{t}} \\ \mathbf{s} \end{bmatrix} \geq \epsilon_N N_{\max} \text{ USD} \tag{16.9}$$

The values of $E_{\max}$, $F_{\max}$, and $N_{\max}$ are given in Table 16.4.

Table 16.4 Maximum values of energy production, food production, and NPV used in the ϵ-constraints

	Maximum value	Unit
Energy production	9.22×10^7	GGE
Food production	6.24×10^{11}	kcal
NPV	6.49×10^6	USD

Ecological sustainability constraints are imposed using the net ecosystem service supplies of climate regulation and of air quality regulation. For both services, net supply is defined as

$$\Sigma = S - D \tag{16.10}$$

in which S is the absolute supply of ecosystem service and D is the demand for the same ecosystem service. Net supply is also the numerator of an ecosystem service's sustainability index V [1]:

$$V = \frac{S - D}{D} \tag{16.11}$$

A net supply or sustainability index that is greater than or equal to zero indicates ecological sustainability for that particular ecosystem, defined as a supply that meets or exceeds demand. Net supply is used in the sustainability constraint rather than the sustainability index to preserve the linearity of the model; however, ecosystem service results are shown as sustainability indexes in the following section.

Two sustainability scenarios, $i \in \{\text{unsust., high sust.}\}$, are defined with constraints on the net ecosystem service supplies of climate regulation and of air quality regulation:

$$\mathbf{\Sigma}_C \cdot \begin{bmatrix} \bar{\mathbf{S}} \\ \underline{\mathbf{S}} \\ \underline{\mathbf{t}} \\ \mathbf{s} \end{bmatrix} \geq C_i \text{ kg CO}_2\text{-eq} \tag{16.12}$$

$$\mathbf{\Sigma}_A \cdot \begin{bmatrix} \underline{\mathbf{S}} \\ \underline{\mathbf{t}} \\ \mathbf{s} \end{bmatrix} \geq A_i \text{ kg NO}_2 \tag{16.13}$$

Under the unsustainable scenario, both net service supply values are constrained to be greater than or equal to their lowest possible value; thus, the net supply values are effectively unconstrained and ecosystem service demand may exceed supply by any amount. Under the high sustainability scenario, both climate regulation and air quality regulation supplies are constrained to be greater than zero, resulting in excess ecosystem service supply. Values of C_i and A_i under the unsustainable and high sustainability scenarios were calculated as 10% of the highest obtainable

Table 16.5 Net supply constraint values under the sustainability scenarios

Constraint	Value	Units
$C_{\text{unsust.}}$	-7.1×10^8	kg CO_2-eq
$C_{\text{high sust.}}$	9.99×10^9	
$A_{\text{unsust.}}$	-3.64×10^5	kg NO_2
$A_{\text{high sust.}}$	2.46×10^6	

Table 16.6 Net supplies of climate regulation and air quality regulation for local- and regional-scale component models. Positive values indicate ecological sustainability for that component and ecosystem service, while negative values indicate unsustainability. (Reproduced with permission from [3])

Component model	Climate regulation $\mathbf{\Sigma}_C$ (kg CO_2-eq)	Air quality regulation $\mathbf{\Sigma}_A$ (kg NO_2)	Units
Cont. Corn, No Till	-6.57×10^4	-2.28×10^1	Per acre
Cont. Corn, Conv. Till	-7.05×10^4	-3.72×10^1	
Corn-Soybean, No Till	-6.10×10^4	-2.26×10^1	
Corn-Soybean, Conv. Till	-5.70×10^4	-3.36×10^1	
Switchgrass, No N Fert.	-5.35×10^4	-1.10×10^1	
Switchgrass, With N Fert.	-5.75×10^4	-1.28×10^1	
Wind turbine	0	0	Per unit
Solar panel	0	0	
White Oak	1.61×10^4	2.96	Per tree
Scots Pine	8.65×10^3	2.94	
American Elm	1.23×10^4	2.96	
Spruce	1.31×10^4	2.97	
Birch	1.19×10^4	2.93	
Eastern Hemlock	5.24×10^3	3.00	
Corn ethanol	-4.30×10^7	-2.63×10^4	Per plant
Soybean biodiesel	-9.52×10^5	-2.26×10^4	
Switchgrass ethanol	-5.00×10^6	-3.15×10^5	
Stover ethanol	-5.32×10^6	-3.28×10^5	
Switchgrass pyrolysis	-1.18×10^8	-1.09×10^5	
Stover pyrolysis	-1.26×10^8	-1.16×10^5	
Switchgrass combustion	-1.38×10^7	-7.87×10^5	
Stover combustion	-1.50×10^7	-8.53×10^5	

net supply values for each service and are given in Table 16.5. A value of 10% was chosen to illustrate the effects of producing excess ecosystem service supply while not constraining the system to become infeasible. Elements of $\mathbf{\Sigma}_C$ and $\mathbf{\Sigma}_A$ for the land use options and biomass conversion processes are given in Table 16.6; elements of $\mathbf{\Sigma}_C$ for the national-scale component model are derived from [38] and are available from the authors on request.

Two constraints are imposed on the elements of $\underline{\mathbf{s}}$ that correspond to wind turbines and solar panels:

$$\underline{s}^{I}_{\mathrm{WT}} \leq 500 \text{ turbines}$$
$$\underline{s}^{I}_{\mathrm{SP}} \leq 57{,}812{,}156 \text{ solar panels} \tag{16.14}$$

These constraints ensure a safe operating distance between turbines and that the area covered by solar panels does not exceed the allowable land area.

Finally, all decision variables (Table 16.7) are constrained to be greater than or equal to zero, and the maximum number of each biomass conversion process is constrained to be one:

$$\begin{aligned} \mathbf{s} &\geq \mathbf{0} \\ \mathbf{s} &\leq \mathbf{1} \\ \underline{\mathbf{s}} &\geq \mathbf{0} \\ \overline{\mathbf{s}} &\geq \mathbf{0} \end{aligned} \tag{16.15}$$

The MILP was implemented in Python and solved using lp_solve [39]. Solutions were attempted under each combination of objective function, ϵ-constraint values, and sustainability scenario, although not all these combinations were feasible. A total of 164 optimal solutions were found and are discussed in the following section.

16.3 Results and Discussion

Figures 16.2, 16.3, and 16.4 summarize the multi-objective results obtained using the ϵ-constraint method. The point color in each figure indicates the sustainability index for one of the ecosystem services: Fig. 16.2 shows the climate regulation index, Fig. 16.3 the air quality regulation index, and Fig. 16.4 the water quality regulation index (Table 16.7). Numbers on each figure correspond to specific optimal designs shown in more detail in Figs. 16.5, 16.6, 16.7, and 16.8. Objectives, scenarios, and ϵ-constraint values for the numbered designs are listed in Table 16.8

The key finding of Figs. 16.2, 16.3, and 16.4 is that while increasing the ecological sustainability of the system restricted the design space, designs with moderately high productivity and nonzero net present values could still be found under all objectives and many combinations of ϵ-constraints. This can be seen by comparing the area covered by the circular points (unsustainable scenario) to the area covered by the triangular points (high sustainability scenario), and by comparing the energy and food productivity levels and the point size, which indicates NPV, between the two scenarios. Note that the scale of both the x- and y-axes is quite large, and productivity reductions that appear to be 50% or greater are in fact much less.

While all three ecosystem services vary under the two sustainability scenarios, climate regulation is able to achieve a much higher sustainability index (Eq. (16.11)) than either air quality regulation or water quality regulation. This is due to the

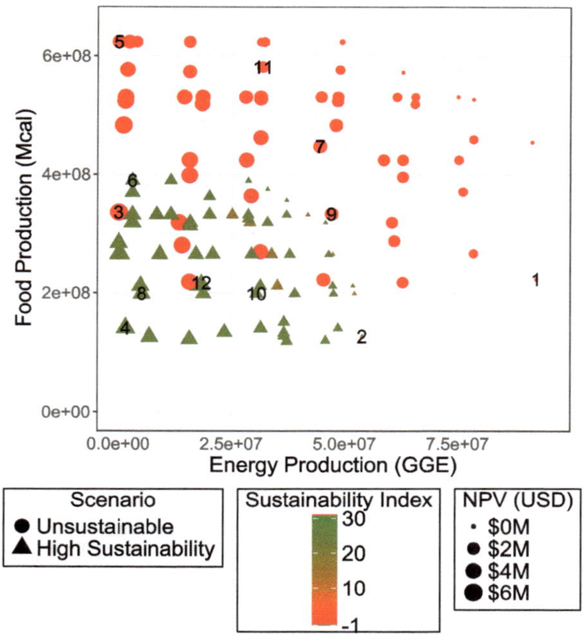

Fig. 16.2 Distribution of food production, energy production, NPV, and climate regulation sustainability index for all optimal designs found. The numbered points correspond to optimal designs shown in Figs. 16.5– 16.8

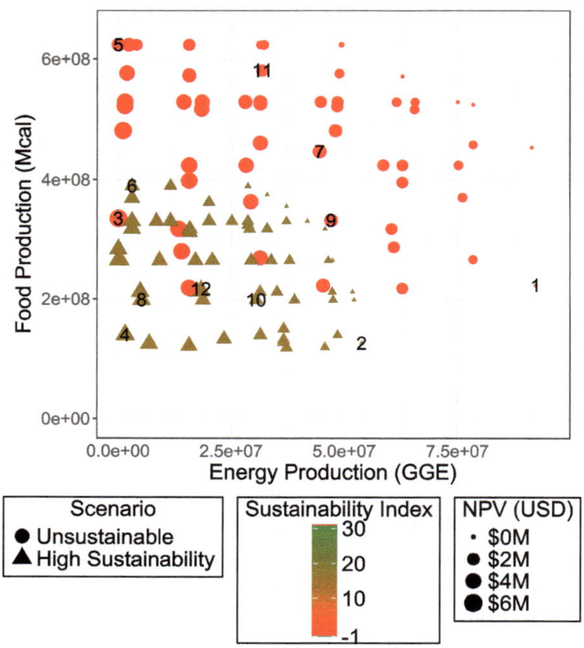

Fig. 16.3 Distribution of food production, energy production, NPV, and air quality regulation sustainability index for all optimal designs found. The numbered points correspond to optimal designs shown in Figs. 16.5– 16.8

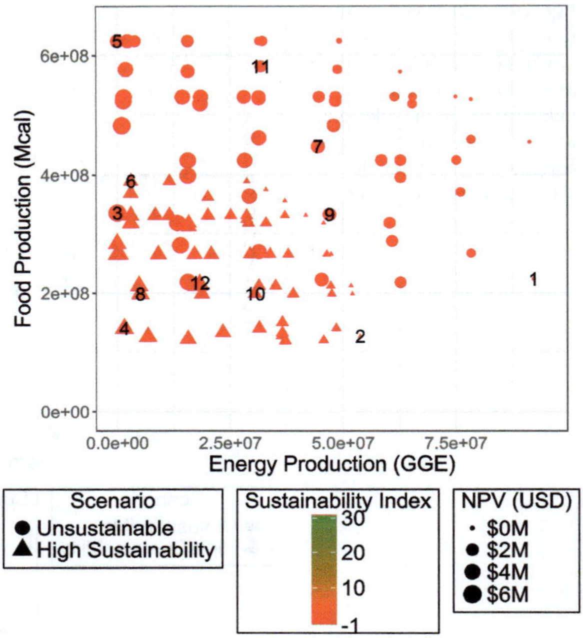

Fig. 16.4 Distribution of food production, energy production, NPV, and water quality regulation sustainability index for all optimal designs found. The numbered points correspond to optimal designs shown in Figs. 16.5– 16.8

quantities of climate regulation supply provided by the reforested land (purple bar in Figs. 16.5 and 16.7), combined with the much smaller amount provided by the agricultural land. These results are particularly promising because climate regulation is a global service and was the only one of the three services to have demand quantified at the national scale using the P2P model. That climate regulation is still able to reach large positive sustainability index and net service supply values despite the additional sources of demand is a promising result for further larger-scale design studies. Air quality regulation fared less well due to the lower quantity of supply provided by the reforested land and due to none of the agricultural activities providing any of this service. Water quality regulation was the poorest performing ecosystem service, primarily due to the way water quality regulation service was supplied. An engineered wetland is capable of improving the water quality by removing nitrates from the water, but it is not possible for any wetland to remove 100% or more of the nitrates, which would correspond to a net supply value of zero or greater. As a result, the net water quality regulation supply values approached zero but did not exceed it under the high sustainability scenario, and no excess water quality regulation service was produced.

Figures 16.5 and 16.6 provide more details about the three single-objective designs, obtained by maximizing each objective function individually with all ϵ-constraint parameters set to zero. The land use options under the high sustainability scenario are virtually identical across the three objectives, with only the number of wind turbines (not visible due to low acreage) varying significantly. The energy

Table 16.7 Decision variable definitions. All variables are continuous except those marked with a superscript I

Term	Definition
$\overline{\mathbf{s}}$	Millions of dollars of economic activity in each economic sector
$\underline{s}_{\mathrm{CCNT}}$	Acres of continuous corn, no tillage
$\underline{s}_{\mathrm{CCCT}}$	Acres of continuous corn, conventional tillage
$\underline{s}_{\mathrm{CSNT}}$	Acres of corn-soybean, no tillage
$\underline{s}_{\mathrm{CSCT}}$	Acres of corn-soybean, conventional tillage
$\underline{s}_{\mathrm{SNF}}$	Acres of switchgrass, no N fertilizer
$\underline{s}_{\mathrm{SWF}}$	Acres of switchgrass with N fertilizer
$\underline{s}^{I}_{\mathrm{WT}}$	Number of wind turbines
$\underline{s}^{I}_{\mathrm{SP}}$	Number of solar panels
s_{CornEth}	Corn ethanol plant scaling factor
s_{SoyBiod}	Soybean biodiesel plant scaling factor
s_{SwiEth}	Switchgrass ethanol plant scaling factor
s_{StoEth}	Stover ethanol plant scaling factor
s_{SwiPyr}	Switchgrass pyrolysis plant scaling factor
s_{StoPyr}	Stover pyrolysis plant scaling factor
s_{SwiCom}	Switchgrass combustion plant scaling factor
s_{StoCom}	Stover combustion plant scaling factor
$\underline{t}^{I}_{\mathrm{WO}}$	Number of white oak trees
$\underline{t}^{I}_{\mathrm{ScP}}$	Number of Scots pine trees
$\underline{t}^{I}_{\mathrm{AE}}$	Number of American elm trees
$\underline{t}^{I}_{\mathrm{Sp}}$	Number of spruce trees
$\underline{t}^{I}_{\mathrm{Bi}}$	Number of birch trees
$\underline{t}^{I}_{\mathrm{EH}}$	Number of Eastern hemlock trees

and food production quantities and NPVs, shown in Fig. 16.6, are much more variable across objectives. Energy production is very low for both the food and NPV objectives, which indicates a strong trade-off between energy and the other two objectives. NPV was likewise essentially zero when energy was maximized, although moderate NPV values were achieved when food was maximized. A similar but much weaker trade-off was observed for food production, which remained moderately high under all objectives. This is likely due to the economic benefit of producing food. Unlike energy production, which requires a large initial capital investment followed by regular operating expenses, food production has no costs beyond the agricultural activities. Producing food is therefore a way to raise the NPV of the system as a whole and offset the large expenditures involved in energy production.

The sustainability scenario had only a moderate impact on the single-objective designs. The land use options remained as the no-till corn-soybean rotation combined with wind turbines, although just over a third of the total land area was used to establish either a forest or wetland ecosystem. Food and energy production and NPV also decreased under the high sustainability scenario compared to the low

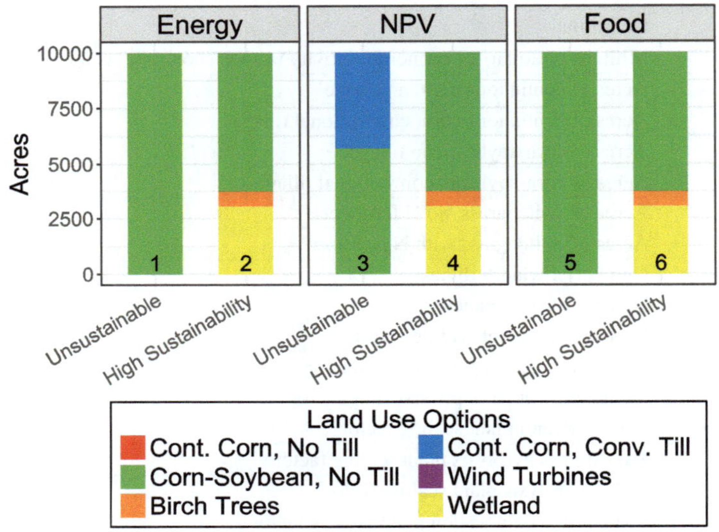

Fig. 16.5 Optimal land use that maximizes each objective individually

sustainability scenario, although the reductions were generally less than half of the maximum value achieved under the unsustainable scenario.

Six more designs that represent compromises between the three objectives are shown in Figs. 16.7 and 16.8. As was true for the single-objective designs, the designs under the high sustainability scenario are virtually identical across objectives. Once again, the same land use options are selected under all objectives, with only the proportions varying. This set of designs does use a different mix of land use options compared to the single-objective designs. Continuous corn under the no tillage practice appears in all three of the high sustainability designs, while it did not appear in any of the single-objective designs. Continuous corn under no tillage is the agricultural land use option with the highest NPV, and so its presence is necessary to meet the ϵ-constraint on the economic objective function.

The strong trade-off between energy production and NPV that was found in the single-objective designs is also visible in the compromise designs shown in Fig. 16.8. Under the maximum NPV design with ϵ-constraints on the energy and food objectives (unsustainable scenario), a slightly higher energy production was achieved than when energy was maximized under constraints on the NPV and food objectives. Conversely, the maximum energy compromise designs had slightly higher NPVs than the maximum NPV compromise designs. The designs were, however, very similar, with slightly more soybean biodiesel being produced under the maximum NPV design and a different proportion of the continuous corn and corn-soybean cropping systems.

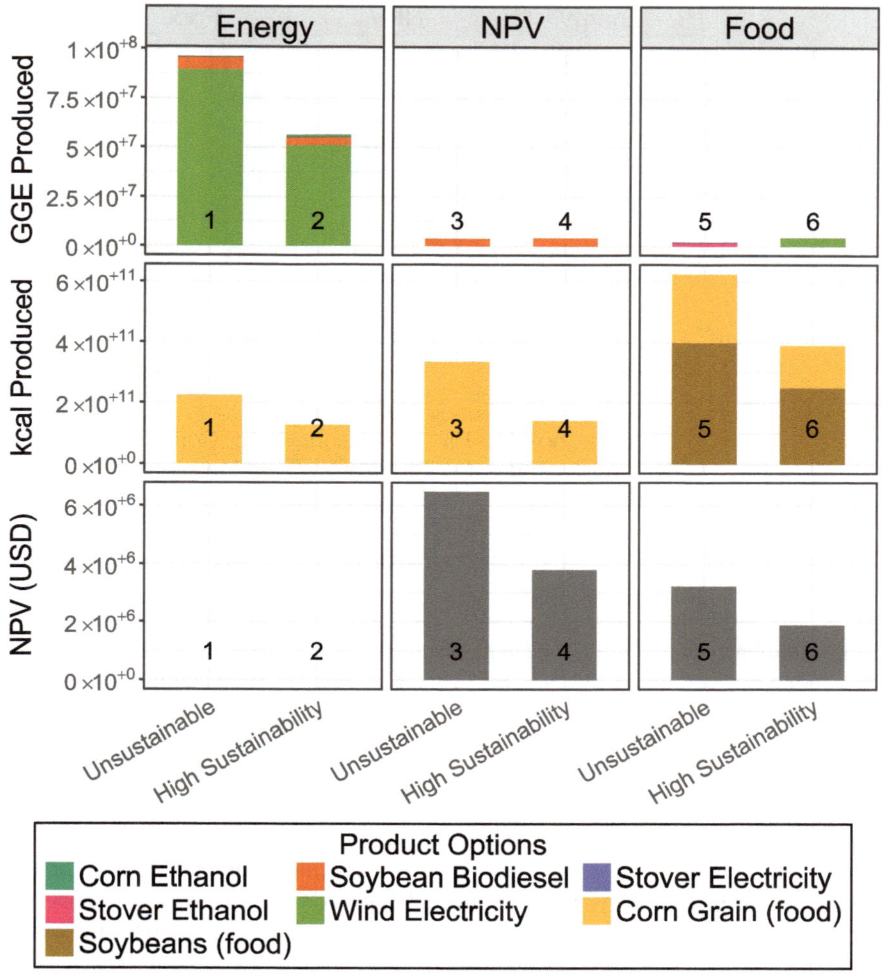

Fig. 16.6 Optimal mix of energy carriers (top) and food products (middle) that maximizes each objective individually, along with the NPV (bottom) of each design

16.4 Conclusions

This work used multi-objective optimization via the ϵ-constraint method to design an agroecological landscape for food production, energy production, and NPV. Land use activities within the landscape, including agricultural activities, energy production activities, and the establishment of two types of ecosystem, were modeled along with a set of biomass conversion processes to produce energy carriers and the life cycles of each land use and biomass conversion activity. The

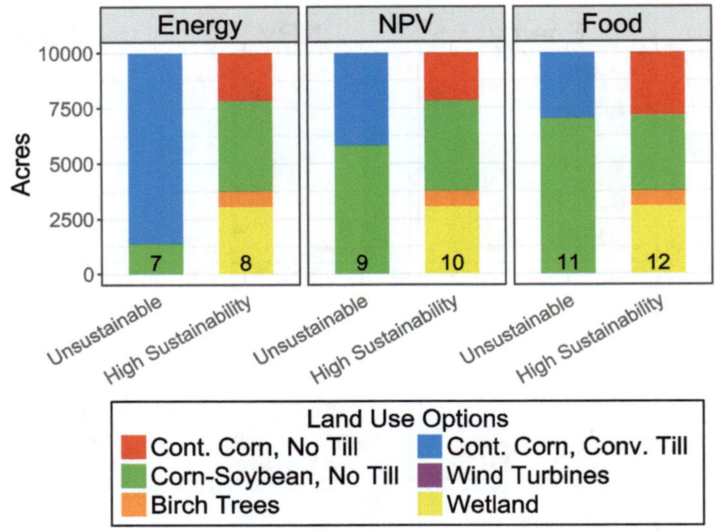

Fig. 16.7 Optimal land use that maximizes each objective under ϵ-constraints on the other two objectives

results indicate that agroecological landscapes can be optimized for productivity and economic sustainability and still achieve ecological sustainability in multiple ecosystem services. Within the system modeled, a strong trade-off existed between energy production and NPV, with weaker trade-offs between food production and the other objectives, and between the anthropocentric objectives and the system's ecological sustainability. Ongoing development efforts that seek to reduce the capital required for biomass conversion processes and increase such process' efficiencies will mitigate this trade-off to some extent. Another way to achieve systems with high energy productivity and high NPV is to monetize ecosystem services that can be provided by the agroecological landscape, and treat the excess services as another product along with food and energy. The additional revenue source will improve the economic sustainability of the system modeled and will additionally provide an incentive for adopting less intensive conservation farming practices.

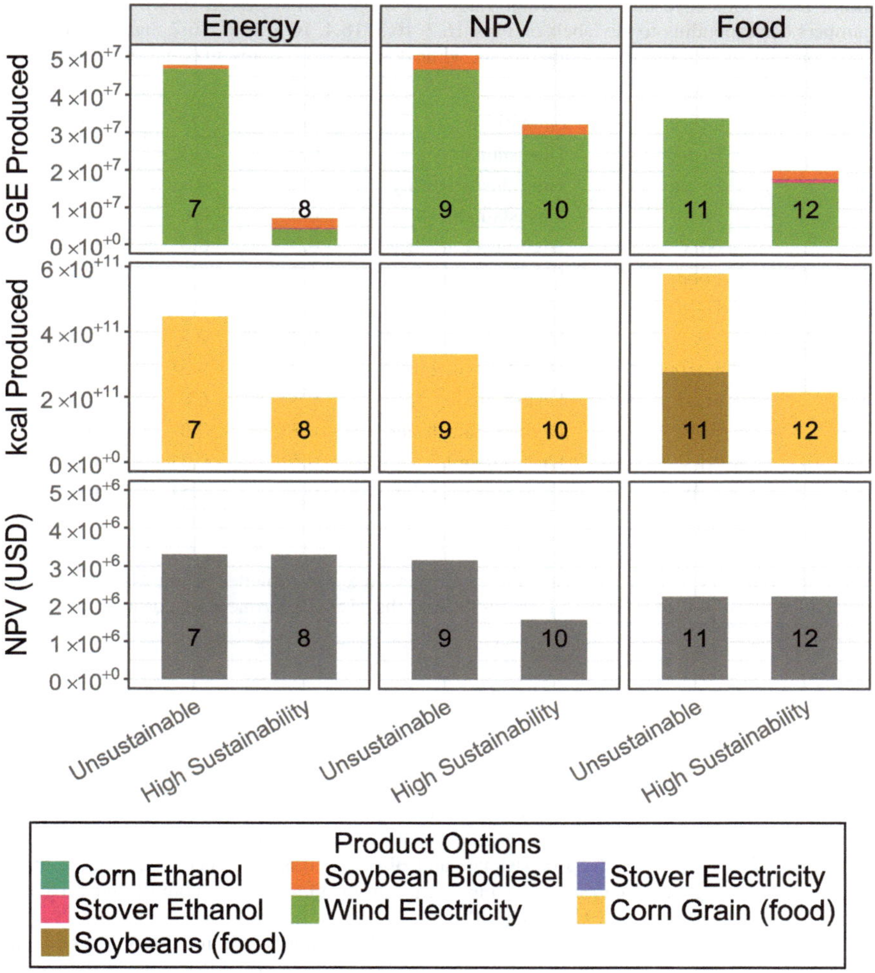

Fig. 16.8 Optimal mix of energy carriers (top) and food products (middle) that maximizes each objective under ϵ-constraints on the other two objectives, along with the NPV (bottom) of each design

Table 16.8 Objective and ϵ-constraint values for each optimal design shown in detail, with numbers corresponding to the labels on Figs. 16.2, 16.3, 16.4, 16.5, 16.6, 16.7, and 16.8. Designs 1–6 represent single-objective optimizations, while designs 7–12 represent compromises between the three objectives

Number	Objective	Scenario	ϵ_E	ϵ_F	ϵ_N
1	Energy	Unsustainable	0.0	0.0	0.0
2	Energy	High Sustainability	0.0	0.0	0.0
3	NPV	Unsustainable	0.0	0.0	0.0
4	NPV	High Sustainability	0.0	0.0	0.0
5	Food	Unsustainable	0.0	0.0	0.0
6	Food	High Sustainability	0.0	0.0	0.0
7	Energy	Unsustainable	0.0	0.51	0.51
8	Energy	High Sustainability	0.0	0.51	0.51
9	Food	Unsustainable	0.34	0.0	0.34
10	Food	High Sustainability	0.34	0.0	0.34
11	NPV	Unsustainable	0.51	0.51	0.0
12	NPV	High Sustainability	0.51	0.51	0.0

Acknowledgments Partial financial support for this work was provided by the National Science Foundation (CBET-1336872, CBET-1404956) and the U.S. Department of Agriculture (BRDI-2012-38202-19288).

References

1. Bhavik R Bakshi, Guy Ziv, and Michael D Lepech. Techno-ecological synergy: A framework for sustainable engineering. *Environmental science & technology*, 49(3):1752–1760, 2015.
2. Rebecca J Hanes, Varsha Gopalakrishnan, and Bhavik R Bakshi. Synergies and trade-offs in renewable energy landscapes: Balancing energy production with economics and ecosystem services. *Applied Energy*, 199:25–44, 2017.
3. R. J. Hanes, V. Gopalakrishnan, and B. R. Bakshi. Including nature in the food-energy-water nexus can improve sustainability across multiple ecosystem services. *Resources, Conservation and Recycling*, 137:214–228, 2018.
4. Rebecca J Hanes and Bhavik R Bakshi. Process to planet: A multiscale modeling framework toward sustainable engineering. *AIChE Journal*, 61(10):3332–3352, 2015.
5. Ronald E. Miller and Peter D. Blair. *Input-output analysis*. Englewood Cliffs, N.J. : Prentice-Hall, 1985.
6. U.S. Bureau of Economic Analysis. Interactive tables: Input-output. Accessed July 25, 2014.
7. TJ Gerik, WL Harman, JR Williams, L Francis, J Greiner, M Magre, A Meinardus, and E Steglich. Researcher's guide for WinEPIC, version 3.0. *Blackland Research and Extension Center, Temple, TX*, 2006.
8. Texas A&M University Blackland Research Center. WinEPIC 3.0, 2006.
9. NOAA NCDC. Climate data online [http://cdo.ncdc.noaa.gov]. *CDO/cdo*, 2004.
10. Stephen Wilcox. National solar radiation database 1991–2005 update: User's manual. Technical report, National Renewable Energy Laboratory (NREL), Golden, CO., 2007.
11. USDA. Arms farm financial and crop production practices. Technical report, U.S. Department of Agriculture, 2015.

http://www.ers.usda.gov/data-products/arms-farm-financial-and-crop-production-practices/documentation.aspx.

12. Roque Lemus, E Charles Brummer, Kenneth J Moore, Neil E Molstad, C Lee Burras, and Michael F Barker. Biomass yield and quality of 20 switchgrass populations in southern Iowa, USA. *Biomass and Bioenergy*, 23(6):433–442, 2002.
13. J Douglas, J Lemunyon, R Wynia, and P Salon. Planting and managing switchgrass as a biomass energy crop. USDA-NCRS. Technical report, Technical note, 2009.
14. Soil Survey Staff. Web soil survey, 2015. http://websoilsurvey.nrcs.usda.gov.
15. OpenEI. Transparent cost database. Open Energy Information, 2016. Accessed July 9, 2016: http://en.openei.org/wiki/Transparent_Cost_Database.
16. Caroline Draxl, Andrew Clifton, Bri-Mathias Hodge, and Jim McCaa. The wind integration national dataset (wind) toolkit. *Applied Energy*, 151:355–366, 2015.
17. Erich Hau and Horst von Renouard. *Wind turbines: Fundamentals, Technologies, Application, Economics*, chapter Wind turbine economics, pages 845–870. Springer, 2013.
18. Wind Powering America. How to calculate wind power output, January 2010. Accessed July 8, 2016: http://www.windpowerengineering.com/construction/calculate-wind-power-output/.
19. National Renewable Energy Laboratory. Pvwatts calculator. http://pvwatts.nrel.gov/. Accessed April 2015.
20. USDA Forest Service. iTree design. http://www.itreetools.org/, 2016.
21. David R Larsen and Ian R Scott. Developments to the sylvan stand structure model to describe wood-quality changes in southern bottom-land hardwood forests because of forest management. *Forest Growth and Timber Quality: Crown Models and Simulation Methods for Sustainable Forest Management*, page 125, 2009.
22. ODNR. Common Ohio trees. Ohio Department of Natural Resources: Division of Forestry, 2016. Accessed March 2015: http://forestry.ohiodnr.gov/trees.
23. iTree: Tools for assessing and managing community forests, 2013. http://www.itreetools.org/. Accessed October 2013.
24. RH Kadlec. Deterministic and stochastic aspects of constructed wetland performance and design. *Water Science and Technology*, 1997.
25. Diederik PL Rousseau, Peter A Vanrolleghem, and Niels De Pauw. Model-based design of horizontal subsurface flow constructed treatment wetlands: a review. *Water research*, 38(6):1484–1493, 2004.
26. Andrew McAloon, Frank Taylor, Winnie Yee, Kelly Ibsen, and Robert Wooley. Determining the cost of producing ethanol from corn starch and lignocellulosic feedstocks. Technical Report NREL/TP-580-28893, National Renewable Energy Laboratory, Golden, CO, 2000.
27. A Aden, M Ruth, K Ibsen, J Jechura, K Neeves, J Sheehan, and B Montague Wallace. Lignocellulosic biomass to ethanol process design and economics utilizing co-current dilute acid prehydrolysis and enzymatic hydrolysis for corn stover. Technical Report NREL/TP-510-32438, National Renewable Energy Laboratory, Golden, CO, 2002.
28. Michael J Haas, Andrew J McAloon, Winnie C Yee, and Thomas A Foglia. A process model to estimate biodiesel production costs. *Bioresource technology*, 97(4):671–678, 2006.
29. EC Baker and DA Sullivan. Development of a pilot-plant process for the extraction of soy flakes with aqueous isopropyl alcohol. *Journal of the American Oil Chemists' Society*, 60(7):1271–1277, 1983.
30. Chemstations, Inc. CHEMCAD Version 6.5, 2013. http://www.chemstations.com/. Accessed September 2013.
31. John Sheehan, Vince Camobreco, James Duffield, Michael Graboski, and Housein Shapouri. Life cycle inventory of biodiesel and petroleum diesel for use in an urban bus. final report. Technical report, National Renewable Energy Lab., Golden, CO (US), 1998.
32. Argonne National Laboratory. Argonne GREET Model, 2015. 2015 Version. Downloaded October 2015.
33. David D Hsu, Daniel Inman, Garvin A Heath, Edward J Wolfrum, Margaret K Mann, and Andy Aden. Life cycle environmental impacts of selected us ethanol production and use pathways in 2022. *Environmental science & technology*, 44(13):5289–5297, 2010.

34. Diego Iribarren, Jens F Peters, and Javier Dufour. Life cycle assessment of transportation fuels from biomass pyrolysis. *Fuel*, 97:812–821, 2012.
35. Nicole Tröger, Daniel Richter, and Ralph Stahl. Effect of feedstock composition on product yields and energy recovery rates of fast pyrolysis products from different straw types. *Journal of Analytical and Applied Pyrolysis*, 100:158–165, 2013.
36. M. Ringer, V. Putsche, and Scahill J. Large-scale pyrolysis oil production: A technology assessment and economic analysis. Technical Report TP-510-37779, National Renewable Energy Laboratory, November 2006.
37. R. L. Bain, W.A. Amos, M. Downing, and R.L. Perlack. Biopower technical assessment: State of the industry and technology. Technical Report TP-510-33123, National Renewable Energy Laboratory, March 2003.
38. Shweta Singh and Bhavik R Bakshi. Accounting for emissions and sinks from the biogeochemical cycle of carbon in the us economic input-output model. *Journal of Industrial Ecology*, 18(6):818–828, 2014.
39. Peter Notebaert Michel Berkelaar, Kjell Eikland. lpsolve. GNU LGPL (Lesster General Public Licence), May 2005. Open source (Mixed-Integer) Linear Programming system. Multi-platform, pure ANSI C/POSIX source code, Lex/Yacc based parsing. Available at www.lpsolve.sourceforge.net/5.5/lpsolve.sourceforge.net/5.5/.

Chapter 17
Sustainable Supply Chains by Integrating Life Cycle Modeling and Techno-ecological Synergy with Application to Mitigation of Harmful Algal Blooms

Tapajyoti Ghosh and Bhavik R. Bakshi

17.1 Introduction

Most industrial concerns, manufacturing conglomerates, and service providers today include sustainability information in their business reports. Many of them publish reports specifically addressing the sustainability challenges and solutions. Over the past two decades, there has been a significant shift from satisfying only the economic bottom line to including the triple bottom line of economic, environmental, and social goals. Inclusion of overall sustainability is a major challenge for the industries. While reduction of environmental impacts is easier at the plant site for an industry, it has been found that a large fraction of impact could occur even before the production phase. Upstream emissions, as observed from hundreds of life cycle assessment (LCA) studies, clearly show the impact of emissions in various stages of the life cycle of an industrial process or product. With emphasis on meeting goals such as net-zero emissions, industry is increasingly managing its supply chain (SC) to reduce its life cycle emissions. With businesses preferring to purchase materials from suppliers that have a smaller life cycle impact, suppliers are also starting to reduce the life cycle impact of their activities.

Many efforts have focused on developing systematic methods for the design of supply chains of chemical products as summarized here. Eskandarpour et al.

T. Ghosh
Resources and Sustainability Group in the Strategic Energy Analysis Center at the National Renewable Energy Laboratory, Golden, CO, USA

B. R. Bakshi (✉)
School for Engineering of Matter, Transport and Energy, School of Sustainability, School of Complex Adaptive Systems, Arizona State University, Tempe, AZ, USA

Lowrie Department of Chemical and Biomolecular Engineering, The Ohio State University, Columbus, OH, USA
e-mail: bhavik.bakshi@asu.edu

B. R. Bakshi (ed.), *Engineering and Ecosystems*,
https://doi.org/10.1007/978-3-031-35692-6_17

[1] provide a detailed review of all articles in this domain and list them into the different groups to which they belong. Out of those, 84 articles are listed to have some sort of environmental dimension within their purview. A vast majority of these papers deal specifically with the supply chain (SC) of biomass to liquid fuel conversion processes. De Meyer et al. [2] provide a detailed review of supply chains focusing specifically on management of biomass supply chains. Nikolopoulou and Ierapetritou [3] review sustainable supply chain design (SSCD) articles focusing on energy efficiency, waste management, and water management. Previously most articles used to focus only on the traditional economic objective for designing supply chains [4–6]. However, as awareness for environmental impacts has increased, researchers have increasingly combined the environmental dimension to supply chain design, thus developing multiobjective optimization problems. Hugo and Pistikopoulos [7] were one of the first studies to perform a multiobjective optimization by combining LCA with SC optimization for fuels. GuillénGosálbez and Grossmann [8] also combined LCA with SC design but introduced a stochastic Mixed-Integer Nonlinear Programming (MINLP) model to account for uncertainty in the design solutions. Mota et al. [9] used the ReCiPe indicator with SC design while attaching a social indicator to capture social, economic, and environmental sustainability. They applied it to SC of a Portuguese battery producer. Santibañez-Aguilar et al. [10] used a multiperiod, multiobjective optimization model to maximize profit and minimize environmental impact using the Eco-indicator 99 LCA technique for a biorefinery supply chain in Mexico. You and Wang [11] designed supply chains of biofuels by combining annualized cost with greenhouse gas emissions. Later, You et al. [12] extended the study by including economic input–output (EIO) models. However, the IO models were used for quantifying social impacts rather than to expand the system boundary of the problem. They applied it for the design of cellulosic biorefinery supply chains focusing on economic, environmental, and social sustainability. Yue et al. [13] demonstrated a life cycle optimization framework where they introduce the concept of functional units under economic and environmental criteria. Corsano et al. [14] proposed a MINLP optimization problem where a detailed process superstructure of ethanol plants is integrated within the overall SC model to perform sustainability analysis. Zamboni et al. [15] developed a multiechelon optimization framework which accounted for both environmental and economic dimensions and applied to the design of bioethanol factories in Italy.

Such efforts reflect the common blind spot of engineering by ignoring the role of ecological systems. They attempt to progress toward sustainability by reducing environmental impact of the supply chain but ignore the role and status of the natural systems that sustain all activities. As a result, and as described in this book, such efforts may continue to cause unintended harm by degrading ecosystems. In addition, they may not benefit from the likely ability of ecological systems to meet industrial needs in a manner that is economically and environmentally superior to conventional engineering solutions.

This chapter describes a general framework for the design of sustainable supply chains while accounting for their life cycle impacts and the role of ecosystems. It combines the theoretical framework of LCA described in Chap. 3, the multiscale process-to-planet (P2P) framework for hybrid modeling used on Chap. 16, and the techno-ecological synergy (TES) framework described in Chap. 12. This allows design of the manufacturing process as well as its associated upstream or downstream value chain and associated ecosystems. The resulting P2P-TES framework is applied to the design of a corn ethanol supply network to mitigate phosphorus runoff and harmful algal blooms (HABs). The case study focuses on the region around western Lake Erie and develops designs by determining the location and size of wetlands to intercept and treat farm runoff, choosing between farming with and without tillage, and location of biorefineries in the selected watershed. Despite its simplifications, the case study demonstrates the potential benefits of supply chain design in synergy with relevant ecosystems.

The next section provides a brief introduction to the frameworks of P2P and TES. This is followed by the methodology for integrating these frameworks to result in the framework of P2P-TES. The case study applies this framework to supply chain design for mitigating algal blooms in the western Lake Erie basin. Results are compared for designs by the conventional approach with only technologies, by including wetlands but as end-of-pipe solutions, and by including wetlands as integrated solutions. They convey the benefits of the framework and the need for more detailed spatial design with the P2P-TES framework.

17.2 Background

17.2.1 Process-to-Planet Framework

Integration of supply chain design with process design required the use and modification of a multiscale framework that could incorporate both of these components separately. The process-to-planet (P2P) framework [16] proved to be the rational choice for the foundation of this new modeling system. P2P framework consists of three separate scales: *equipment* for modeling the engineering manufacturing technology or process, *value chain* for modeling life cycle network of the manufacturing processes, and *economy* scale for using economic models to capture the life cycle that is not covered by the smaller two scales. Together these three scales allow the incorporation of better environmental impact assessment through hybrid LCA (economy + value chain) in sustainable process design. The underlying mathematical framework is described in Chap. 3. It is an expansion of conventional process-based LCA (life cycle assessment) matrix structure which is defined as,

$$Xm = f \tag{17.1}$$

where X is the technology matrix (TM) showing the activity network of the system, and m is the scaling variable or multiplier for determining the size of the activities needed to satisfy the final demand f. The P2P framework can be represented mathematically as,

$$\underline{\overline{X}}\{z\}\overline{\underline{m}} = \overline{\underline{f}} \tag{17.2}$$

$$\underline{\overline{H}}\{z\} \geq 0 \tag{17.3}$$

$$\overline{\underline{g}} = \underline{\overline{D}}\overline{\underline{m}} \tag{17.4}$$

where $\underline{\overline{X}}$ represents a multiscale or hybrid technology matrix that captures interaction between the economy, value chain, and equipment scales. Variable $\overline{\underline{z}}$ represents decision variables at any scale. In conventional sustainable process design, z represents decision variables of various equipment in the process flowsheet. Similarly, $\overline{\underline{m}}$ and $\overline{\underline{f}}$ represent the multiscale scaling vector or multiplier and multiscale final demand that contain the sizes of three separate scales and their respective final demands or flows to consumers. Overbars in these equations represent the economy scale, underbar represents the value chain scale, while variables without any bars are at the equipment scale. Combination of overbars and underbars together represents multiscale P2P matrices. Equation 17.3 collectively represents all mass flow, energy balance, and reaction equations that make up the process model of the industrial process with variables such as temperature, flow rate, equipment sizes, heat input, etc. Environmental impact from this multiscale framework is given by Eq. 17.4, where $\underline{\overline{D}}$ is the multiscale environmental interventions matrix containing impact information for activities at all three scales. It is multiplied with the scaling variables to obtain the total life cycle emission $\overline{\underline{g}}$ for a specified final demand $\overline{\underline{f}}$. The optimization framework for P2P design uses Eq. 17.4 as the objective function to be minimized subject to constraints of Eqs. 17.2 and 17.3. It has been applied to several case studies. Hanes and Bakshi [17] showed using the P2P framework how system boundary could be expanded to account for omitted emissions at the economy scale and obtain optimized solutions for process design. Ghosh and Bakshi [18] demonstrated how P2P multiobjective framework can be used to discover win-win solutions over conventional process-based LCA when applied to sustainable process design.

17.2.2 Techno-ecological Synergy

As introduced in Chap. 12, the techno-ecological synergy (TES) framework integrates technological process design and flows of materials(resources and wastes) to and from the environment [19]. Such a framework incorporates ecosystems as unit operations and their services as flows within the model so that necessary bounds and constraints imposed by ecosystems can be modeled and the technological process designed around these bounds. Bakshi et. al. [19] explained that the TES framework addresses environmental challenges in two ways, encouraging “less bad”

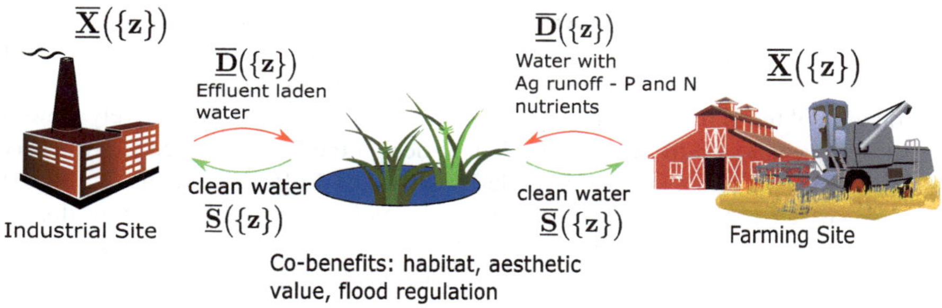

Fig. 17.1 Techno-ecological synergy with wetland ecosystems. (Reproduced with permission from [24])

through impact minimization (reducing raw material input and emission levels) while doing "more good" by means of ecosystem restoration and technological innovation as shown in Fig. 17.1. The TES framework has been applied to designing a residential system [20] and industrial manufacturing site [21] and has been proven to have both economic and environmental benefits—"win win" solutions. This is also described in various chapters in this book. Liu and Bakshi [22, 23] have developed a framework that combines TES and LCA to include ecosystem services in conventional life cycle assessment. The computational structure of the TES-LCA framework is expressed as,

$$\begin{bmatrix} A & C \\ D & S \end{bmatrix} \begin{bmatrix} m \\ m_e \end{bmatrix} = \begin{bmatrix} f \\ f_e \end{bmatrix} \tag{17.5}$$

where A is the technology matrix of process-based LCA containing economic product flows between technological activities, D is the environmental intervention matrix which indicates resource use or emissions associated with the technological activities, and S is the ecosystem matrix, representing flows between ecological modules in a similar manner to the technology matrix. While technological activity production values are provided in the A matrix, the uptake rate of emissions by ecosystems is provided in the S matrix. C is the management matrix representing the economic product flows from technologies to ecosystems for their maintenance. The objective for this single-scale framework is to determine f_e or total environmental impact while accounting for ecosystem services. Its focus is on the analysis of a given fixed system and does not involve any optimization or design.

17.3 Methodology

Motivation behind creating a modeling framework that can perform both process and supply chain design while accounting for ecosystem services and life cycle comes from visualizing the process-based LCA matrix structure as an upstream network of any process. Process-based LCA represents different activities that are connected with one another and supplies the necessary input flows for the production of the main product. If this upstream network can be alternatively viewed as a supply chain, modifying the problem to design the process-based LCA matrix structure evolves into a supply chain design problem. In fact, Weidema et al. [25] have discussed and compared value chain and supply chain and shown how closely they are linked to one another. An initial version of this framework was briefly described in Ghosh et al. [26]. The steps for modifying the P2P framework to the P2P-TES framework are sequentially listed below.

17.3.1 Modification 1: Including Ecosystems in P2P Framework

To include ecosystems at different scales, the P2P framework has to be expanded to account for flows to and from the ecosystems. Using the TES-LCA framework explained in Sect. 17.2 and combining it with the P2P framework to bring in design variables of process and supply chain, we develop the multiscale P2P-TES framework. While the single-scale TES framework is used only for analysis of environmental impacts at the value chain scale, the multiscale P2P-TES framework optimizes a given objective to solve for designs of engineering activities as well as upstream input raw material pathways. Along with three different scales of P2P framework, ecosystems at these scales are also integrated with technological activities. Flows between ecosystems and technological systems are explicitly included in this framework. The framework is mathematically expressed as shown in Fig. 17.2.

Ecological systems considered in TES can be forests, wetlands, pollinators, or other nature-based solutions. While sequestration of carbon dioxide can be considered a global ecological service analogous to the economy scale of P2P, deposition and capture of sulfur dioxide particulate matter can be regional services provided at the value chain scale. Wetlands help in treating water locally, which contains effluents, heavy metals, organic nutrients and provides clean water. This service can be obtained at the local site just beside an industry at the equipment scale or may be combined together for watersheds at the value chain scale. These ecosystems are components of the $\underline{S}$ matrix. Ecosystems at the equipment scale are represented as unit operations using

$$S(\{z, b\}) \geq 0 \tag{17.6}$$

At the value chain scale, ecosystems are represented like technological activities are modeled, as linear models with inputs and outputs proportional to each other in some fixed ratio. The difference is that emission flows from technological systems are inputs to these activities. Such database for ecosystems is not yet available

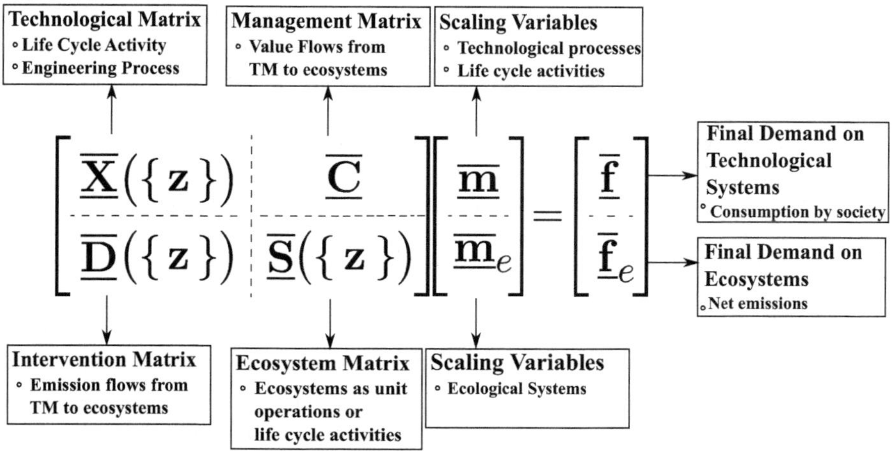

Fig. 17.2 Multiscale matrix structure, equations, and representation of the P2P-TES framework. (Reproduced with permission from [24])

but can be obtained by averaging several ecosystems over a large regional area, such as a watershed. To build models of ecosystems at the economy scale, the first step is to obtain monetary values of various ecosystem services provided by nature. There are some databases that provide such information [27]. Next, these ecosystem economic valuations need to be integrated with input–output model economic flows that reduce the net economic throughput which in turn results in lower environmental impact through Environmentally Extended Input–Output analysis (EEIO) calculations. Every component of the large P2P-TES matrix is explained as follows:

- $\overline{\underline{X}}(\{z\})$ represents the network of technological activities in the system and is known as the technology matrix (TM). This is similar to the technology matrix used in the conventional process-based LCA analysis with product flows making up the rows of the matrix and activities or processes making up the columns. This matrix represents the complete flow of materials within the system.
- $\overline{\underline{D}}(\{z\})$ represents the emissions matrix, included separately to denote the flows from technological systems to ecosystems shown in Fig. 17.1. The interventions matrix in conventional process-based LCA is diagonalized and included at the bottom of the technology matrix as shown in Fig. 17.2 to connect these flows directly to the ecosystems.
- $\overline{\underline{S}}(\{z\})$ is analogous to the $\overline{\underline{X}}(\{z\})$ matrix except that it contains ecosystems at different scales. This matrix represents the ecosystem services obtained from natural systems, such as air quality regulation and water provisioning by forests, water treatment through wetlands, etc., which is given back to the technological systems as shown in Fig. 17.1. Just as in the technology matrix,

the rows represent flows whereas the columns represent processes. However, in this matrix, the flows are to and from the biosphere rather than between industries. The processes or activities in the columns are ecological systems rather than industrial processes.

- $\overline{\underline{C}}(\{z\})$, known as the management matrix, represents flows of materials that are necessary for maintenance of ecosystems from the technological systems. In addition to satisfying the final demand of society, the TES framework models flows that would be required by industrial processes to maintain the ecosystems to support human activities. Modeling this flow allows to keep track of any expenditure on maintaining ecosystems if necessary.
- $\overline{\underline{f}}$ is the final demand of goods and products by society from technological systems and $\overline{\underline{f}}_e$ denotes final demand on ecosystems or environmental impact at different scales. $\overline{\underline{f}}$ contains demand of useful products which are consumed by society, whereas emissions or environmental impacts are expressed as demand of ecosystem services in $\overline{\underline{f}}_e$.

Mathematically, the major difference between this equation and Eq. 17.5 is that the individual terms in the matrices are all multiscale rather than single scale. Also, these terms include variables of engineering process design z since the objective of this framework is to perform optimization for design solutions. The matrix structure enabling integration of these different components is shown in Fig. 17.2.

17.3.2 *Modification 2: Introduction of Spatial Variables*

The P2P framework is set up as an optimization problem for achieving different objectives as per the concerned stakeholder. Design variables are only restricted to the equipment scale where they are used to determine process parameters. In this framework, spatial decision variables are introduced into the three scales such that the problem can be solved for choices of suppliers, economic sectors, as well as search for locations for the primary process. These variables denoted as b can be binary when choices are mutually exclusive or continuous. With this modification, P2P-TES framework can now be mathematically represented as

$$\begin{bmatrix} \overline{\underline{X}}(\{z,b\}) & \overline{\underline{C}}(\{z,b\}) \\ \overline{\underline{D}}(\{z,b\}) & \overline{\underline{S}}(\{z,b\}) \end{bmatrix} \begin{bmatrix} \overline{\underline{m}} \\ \overline{\underline{m}}_e \end{bmatrix} = \begin{bmatrix} \overline{\underline{f}} \\ \overline{\underline{f}}_e \end{bmatrix} \tag{17.7}$$

In previous design studies, the $\overline{\underline{X}}$ matrix was fixed. It represented the network of technological activities in the system. The number of activities and their products are previously determined, and they always existed within the matrix. With this modification the $\overline{\underline{X}}$ matrix becomes variable. Activities within this matrix may or may not be included based on the values of spatial b variables. If for a certain activity, its corresponding binary spatial variable is 0, that activity is removed

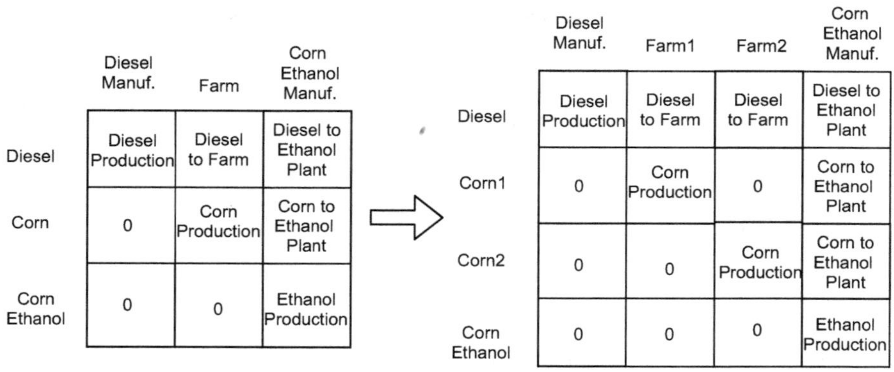

Fig. 17.3 Conversion of simple technology matrix to multiprocess spatialized technology matrix

from the technology matrix. The activity gets included when its corresponding b has a value of 1. An example for this spatialization modification is provided in Fig. 17.3. On the left, we have a simple technology matrix connecting three processes: production of diesel, corn, and corn ethanol. Corn is produced from one farm only and so are the other products. We modify this matrix to introduce two farms, farms 1 and 2, both of which can supply corn to the ethanol refinery. The choice of the farm actually supplying the corn can be controlled with the help of binary variables. Based on the optimization objective, the relevant activities are chosen by the algorithm. This is the basic logic of converting a simple technological matrix to a framework with multiple supply chain options. Due to the presence of spatial variables in the technology matrix, associated $\overline{\underline{D}}(\{z, b\})$, $\overline{\underline{S}}(\{z, b\})$ and $\overline{\underline{C}}(\{z, b\})$ also get modified with their corresponding b variables. Solving Eq. 17.7, the unknown scaling variables $\overline{\underline{m}}$ and b are determined.

17.3.3 Environmental Impact Assessment

The P2P-TES framework, built on the matrix structure form of LCA, is used for environmental impact calculation. Final demand matrix in Eq. 17.7 is demand of valuable products by society manufactured by the system under study. The first step is to solve for technological scaling variables $\overline{\underline{m}}$ using

$$\overline{\underline{X}}(\{z, b\})\overline{\underline{m}} = \overline{\underline{f}} \tag{17.8}$$

Emission flows from the system are given by

$$\overline{\underline{f}}_e = \overline{\underline{D}}(\{z, b\})\overline{\underline{m}} \tag{17.9}$$

If ecosystems are considered to take up emission flows from technological activities, then Eq. 17.9 gets modified into,

$$\overline{\underline{f}}_e = \overline{\underline{D}}(\{z, b\})\overline{\underline{m}} - \overline{\underline{S}}(\{z, b\})\overline{\underline{m}}_e \tag{17.10}$$

In an environmentally sustainable system $\overline{\underline{f}}_e$ should be 0, which denotes that there are no net emissions or demand of ecosystem services. This condition is mathematically represented as,

$$\overline{\underline{f}}_e = 0 \tag{17.11}$$

$\overline{\underline{m}}_e$ denotes the scaling variables to determine the size of ecosystems while $\overline{\underline{m}}$ represents the size of technological systems. These are analogous to scaling variables in conventional process-based LCA. The net environmental impact is obtained as,

$$\overline{\underline{g}} = \sum_n Q^n \overline{\underline{f}}_e^n \tag{17.12}$$

which represents the sum of final demand to ecosystems. Q is the characterization factor matrix. It contains conversion factors for impact categories such as acidification potential, global warming potential, etc., for every environmental flow. Using Eq. 17.12, the different environmental flows are lumped together by converting them into midpoint indicators for various impact categories. If required, the midpoint indicators can be summed up to calculate endpoint indicators. The P2P-TES framework not only includes direct environmental impacts in the life cycle but also considers the supply of ecosystem services to mitigate such impacts. Incorporation of ecosystem services allows design of technological systems such that supply and demand of these services are balanced and do not exceed limits, making the system unsustainable. Solving Eq. 17.11, as mentioned earlier, technological scaling variables $\overline{\underline{m}}$ are determined. Along with that, size of ecosystems required for mitigating impacts can be determined by solving for ecological scaling variables $\overline{\underline{m_e}}$. If these variables are known, then using Eq. 17.12, the net environmental impact can be calculated rather than minimized.

17.3.4 Supply Chain Transportation

Transportation is one of the main components of supply chain design. Without modeling of transportation distances, the P2P-TES framework cannot be utilized for supply chain design. Thus, a framework is needed to model transportation distances within the structure of the P2P-TES framework. The transportation matrix is introduced, which is similar to the technology matrix as shown in Fig. 17.4. The rows and columns represent the activities or processes in a system. The diagonal elements are zero because they represent the distance between same activities.

Fig. 17.4 Transportation matrix for P2P-TES framework

	Diesel Manuf.	Farm1	Farm2	Corn Ethanol Manuf.
Diesel Manuf.	0	D to F1	D to F2	D to C
Farm1	F1 to D	0	F1 to F2	F1 to C
Farm2	F2 to D	F2 to F1	0	F2 to C
Corn Ethanol Manuf.	C to D	C to F1	C to F2	0

The nondiagonal elements list the distances between the activities. For example, *D to F1* represents the distance between diesel manufacturing and Farm 1. *D to F2* represents the distance between diesel manufacturing and Farm 2. The transportation distances for the supply chain are arranged into a transportation matrix $\underline{\overline{T}}$ as shown in Fig. 17.4. Using this matrix, the transportation impacts for the supply chain can be calculated using Eq. 17.13. Distances can be converted to environmental impacts using greenhouse gas emission factors (m_e) or cost using relevant expense multiplication factors (m_c).

$$\text{Transportation cost} = \underline{\overline{T}}_{m_c} \tag{17.13}$$

$$\text{Transportation emissions} = \underline{\overline{T}}_{m_e} \tag{17.14}$$

17.3.5 Objective Functions

$$\textbf{Environmental objective} \quad Z_1 = \underline{\overline{g}} + \underline{\overline{T_{m_e}}} \tag{17.15}$$

$$\textbf{Economic objective} \quad Z_1 = \underline{\overline{X}} p_x + \underline{\overline{C}} p_c + \underline{\overline{T_{m_c}}} \tag{17.16}$$

The design solution is obtained by optimizing relevant objective functions depending upon the scope of the problem. Environmental objectives capture the life cycle impacts from all the different scales using Eq. 17.12. Economic objective functions depend primarily upon the stakeholders. The first term of Eq. 17.16 captures the costs of technological activities at all the different scales, from raw materials, transportation, and plant operation. The second term denotes the cost of maintenance of ecosystems paid by stakeholders. This can be cost of land or regular upkeep of forests or wetlands. The Z_3 term can be added depending on how transportation burden is included in the optimization formulation. It can represent transportation emissions or cost of transportation, depending upon the multiplication factors used to derive Z_3 as shown in Eq. 17.13. Including this term in the objective results in optimal supply chain design.

Optimization Formulation

$$\textbf{Minimize } Z_1, Z_2 \tag{17.17}$$

$$\textbf{subject to} \quad \begin{bmatrix} \underline{\underline{X}}(\{z,b\}) & \overline{\underline{C}}(\{z,b\}) \\ \overline{\underline{D}}(\{z,b\}) & \overline{\underline{S}}(\{z,b\}) \end{bmatrix} \begin{bmatrix} \overline{\underline{m}} \\ \overline{\underline{m}_e} \end{bmatrix} = \begin{bmatrix} \overline{\underline{f}} \\ \underline{f}_e \end{bmatrix} \tag{17.18}$$

$$H(\{z,b\}) \geq 0 \tag{17.19}$$

$$S(\{z,b\}) \geq 0 \tag{17.20}$$

$$\overline{\underline{m}} \geq 0 \tag{17.21}$$

$$\overline{\underline{m_e}} \geq 0 \tag{17.22}$$

The scaling variables are multiscale and relate to the technological and ecosystems at different scales. Presence of variables in the equipment scale allows the framework to be used for process design. In the value chain, it allows choosing between suppliers of raw materials and required inputs, which proves supply chain design capabilities. Variables at the economy scale can be used for policy design. In the current work, we explore only the supply chain and process design multiscale problems.

17.4 Case Study

As explained in Sect. 17.1, the northwestern region of Ohio and eastern Indiana is grappling with eutrophication of Lake Erie causing HABs and severely impeding drinking water supply. Several industries that depend on Lake Erie are also suffering due to this. One of the ways of addressing this environmental problem is to reduce the phosphorus load on the lake by using treatment wetlands to intercept and treat the runoff and by changing farming practices. To illustrate the P2P-TES framework and its potential for mitigating HABs, we design a biofuel supply chain network with the goals of minimizing economic and environmental objectives while choosing the "best" industrial, agricultural, and spatial decision variables.

17.4.1 Problem Description

Twenty-one counties in the northwestern part of Ohio are selected in the watersheds of the Maumee, Portage, and Sandusky rivers. The problem is solved at the county scale with information of corn production, corn usage by consumers as food, and phosphorus runoff from different counties. The location of farms, corn ethanol

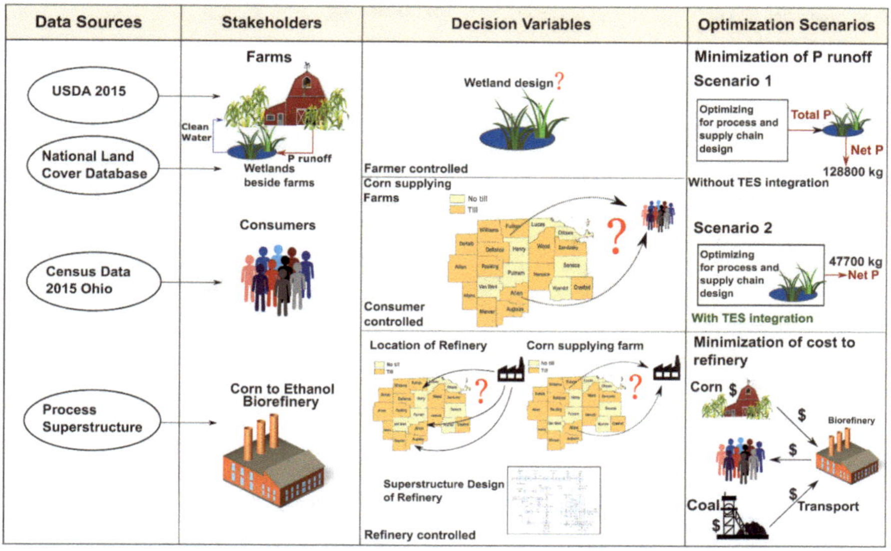

Fig. 17.5 Overview of case study describing data sources, stakeholders, decision variables of different stakeholders, and optimization scenarios explored in study. (Reproduced with permission from [24])

refineries, consumers, as well as ecosystems are assumed to be at the centroid of the counties. The framework described in Methodology is general, but for purposes of this case study, we are not considering the national economy scale within this problem due to its regional focus.

A complete overview of the system under study is shown in Fig. 17.5. There are three primary Stakeholders refineries, farms, and consumers. Farms produce corn resulting in agricultural runoff laden with phosphorus nutrients. Corn is consumed as food by consumers and as raw material by refineries for production of bioethanol. Bioethanol produced is used to satisfy the bioenergy demand of the consumers in that region. To reduce phosphorus in runoff water, ecosystem services provided by constructed wetlands are considered in this study. The wetlands are assumed to be present near the farms so that runoff water flow can be directed into them. The goal of the problem is to determine the location of biorefineries in this region, as well as farms from which both refineries and consumers source their corn supply. These solutions depend on the objective being considered, such as minimization of phosphorus runoff in water, minimization of cost for corn ethanol production, etc.

Corn production information is obtained from [28] for the different counties shown in Fig. 17.6. From USDA, it is found that all 21 counties considered had corn production. Thus, farms are present in all the counties. To explore the effect of different farming practices, counties are assumed to practice different forms of agricultural practices as shown using different colors for the counties in Fig. 17.5. The farming choices are expected to affect the supply chain choices from where

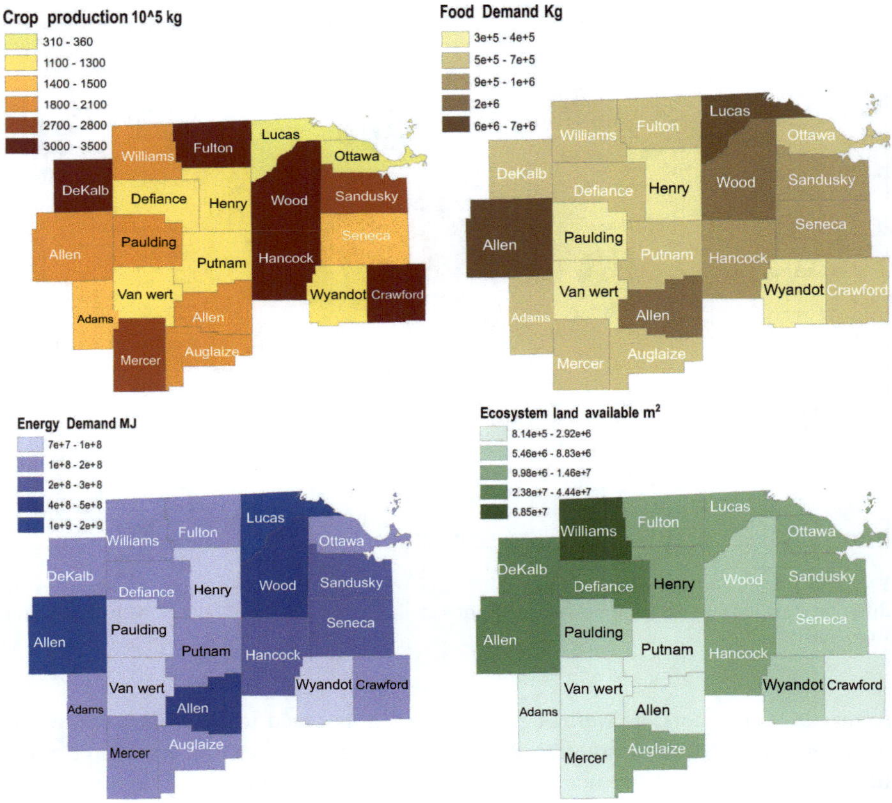

Fig. 17.6 County-scale information of crop production, biofuel energy demand, crop production, and ecosystem area available in 21 counties in Ohio. (Reproduced with permission from [24])

the biorefineries source their corn, as shown in Fig. 17.5 in the decision variables column. As no-till agriculture results in lower production of corn, it is considered that no-till corn is priced higher to recuperate losses for the farmers. However, no-till agriculture results in lower phosphorus levels in agricultural runoff.

Using the countywise population information [29] and statewise demand of corn as food and energy from biofuel [30], the countywise demand for ethanol-based energy and corn consumption by residents are derived. This information is mapped in Fig. 17.6. While different methods can be applied for reduction of phosphorus runoff, in this study we explore the use of wetlands for removal of phosphorus from agricultural runoff water. For this service, land needs to be available to be converted into wetland. For such purposes, in this study, we considered barren, swamp, and shrubland areas in these counties to be available for conversion. These converted wetland areas are also considered to be at the centroid of the counties. The available land for different counties is shown in Fig. 17.6. Amount of different types of available area is obtained from National Land Cover Database [31] through ArcGIS 10.6 software.

Bioethanol refinery location is determined through solving this problem as shown in Fig. 17.5. The fundamental engineering model [17] of the corn to ethanol plant is based on a 50-million-gallon-capacity plant. After obtaining the total bioethanol energy demand of consumers in this region from US Energy Information Administration (US-EIA) [30], it is observed that corn ethanol produced by two 50-million-gallon-capacity refineries would be enough to satisfy the corn ethanol demand of this entire region. Hence, the choice of binary variables is such that the refineries can be placed at any of the counties with a maximum of two biorefineries in the entire region for satisfying the demand of corn ethanol. The plant capacities have been assumed to be fixed at 50-million-gallon value because in real plant operation, a plant generally operates at its designed intended capacity. The biorefineries can choose their corn supply from any of the farms in the 21 counties. Along with design of the supply chain, the process design of the corn to ethanol refinery is also included in the system as shown in Fig. 17.5.

A major assumption in this study is that it does not consider trade with entities outside the system boundary of the selected region. In reality, a major part of the corn produced in this region would be exported outside these counties. Similarly, ethanol produced by the refineries would also be exported. However, as we are delimiting our system to this location, the consumption of corn and ethanol is limited to consumers in this region.

Due to the presence of multiple players (farms, consumers, and refineries) in this problem setup shown in Fig. 17.5, the optimization objectives need to be modified to cater to specific goals within the purview of the problem. The three major stakeholders in this problem are the consumers, mainly based in and around the region of Toledo in Ohio. These people are the most affected by the declining water quality in Lake Erie due to eutrophication. Through their elected representatives, consumers have the power to enact laws or policies to monitor the amount of phosphorus flowing into Lake Erie through the rivers. The second stakeholders are farmers. Their main goal is to produce and sell as much corn as possible to increase their revenue. However, the farmers have control over their farming methods that can help reduce phosphorus runoff into water streams—such as reducing phosphorus application, reducing corn production, less intensive agricultural practices, no-till agriculture, wetlands to treat water, catchment areas along riverbanks, etc. The third stakeholders are the corn ethanol refineries. All the refineries are considered together as a single entity. Major goal of refineries is to produce as much as bioethanol demanded by the users in this region. These refineries can improve their profits by buying cheaper corn from farmers, which is their main raw material. Supply chain decisions are also taken by the biorefinery agent. In this case study, our focus is on the design of corn ethanol refineries and their supply chain while minimizing environmental impact through phosphorus runoff reduction, which is directly affecting the consumers by degrading their water quality.

17.4.2 Wetland Ecosystem Models

Wetland ecosystems can break down and absorb pollutants like phosphorus and nitrogen nutrients, heavy metals, suspended solids. They can also reduce biological oxygen demand and destroy microorganisms like algae. Various industries such as paper and pulp mills, meat processing facilities, and petroleum refineries use constructed wetlands as treatment units to remove oil and grease, heavy metals, chemical oxygen demand, and other pollutants. These wetlands provide a cost-effective alternative to conventional treatment units since costs associated with their regulation, maintenance, and operation are relatively small. Viability of ecosystems for treating wastewater has been studied previously [32–34]. Wetland operations are modeled using first-order plug flow reactors. They have an exponential profile between inlet and outlet flows. These equations are used to find the surface area of wetlands needed to treat a certain concentration of wastewater based on reactor conditions such as temperature. However, for purposes of this study, it was difficult to use the Plug Flow Reactor (PFR) model, primarily due to lack of information about the phosphorus concentration in runoff water from farms at the scale of individual counties. Thus, to circumvent this problem, the first-order models are converted to zero-order models by plotting wetland treatment data from [35] and regressing a log-based model to develop a relation between wetland area and phosphorus uptake. The values obtained from the equation are found to be in concordance with experimental data obtained from wetlands already constructed in Ohio.

17.4.3 MINLP Problem Formulation

A multiobjective, nonlinear optimization problem is developed for solving this design problem. The 21 counties are defined by a set **C**. The set of 21 farms pertaining to every specific county is defined as **F**. Similarly, the sets of consumers and possible refineries in these counties are defined as **P** and **R**. Ecosystems are considered to be present alongside every farm and their set is denoted as $\mathbf{F_E}$. At each step, development of the P2P-TES models is illustrated in Figs. 17.7, 17.8, 17.9, 17.10, and 17.11.

Corn Ethanol Refinery

$$\underline{\overline{X}}_{k,k}(\{z,b\}) = fc_{eth}(H(\{z\}))\ s.t.\ k \in \mathbf{R} \tag{17.23}$$

$$H(\{z\}) \geq 0 \tag{17.24}$$

Equation 17.24 denotes the combination of fundamental models to describe the process superstructure of a technological activity such as a corn to ethanol conversion process. It comprises of mass, energy balances, and all other variables that determine

Fig. 17.7 Building the P2P-TES $\underline{\overline{X}}$ matrix: incorporating corn ethanol refinery operation

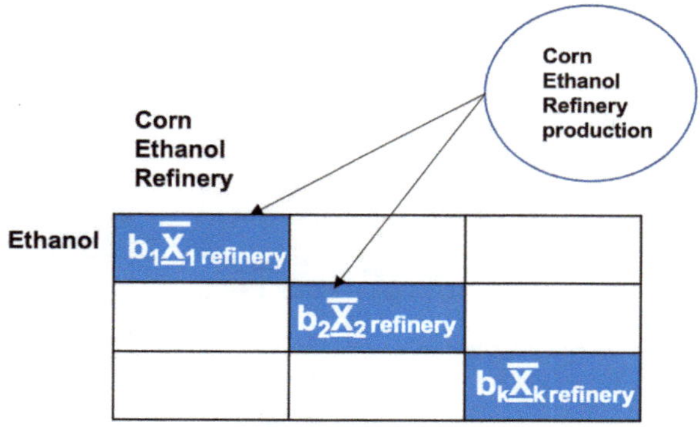

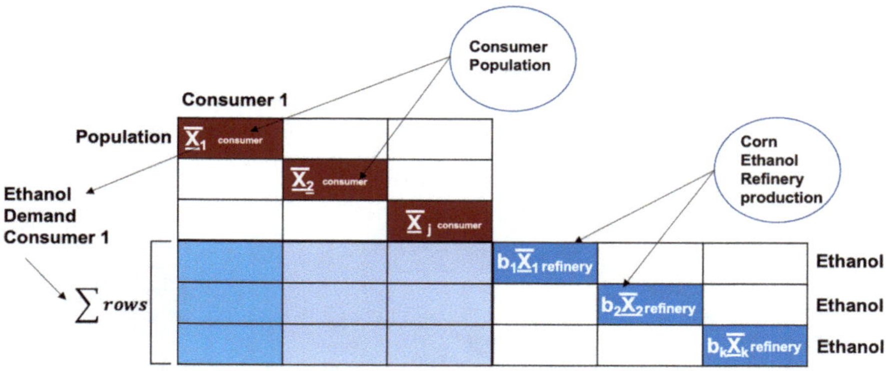

Fig. 17.8 Building the P2P-TES $\underline{\overline{X}}$ matrix: incorporating consumer population in the region

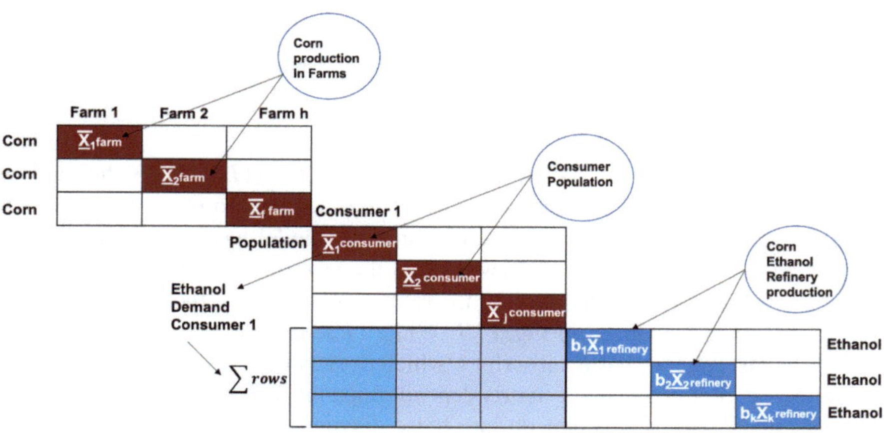

Fig. 17.9 Building the P2P-TES $\underline{\overline{X}}$ matrix: incorporating farm production in the region

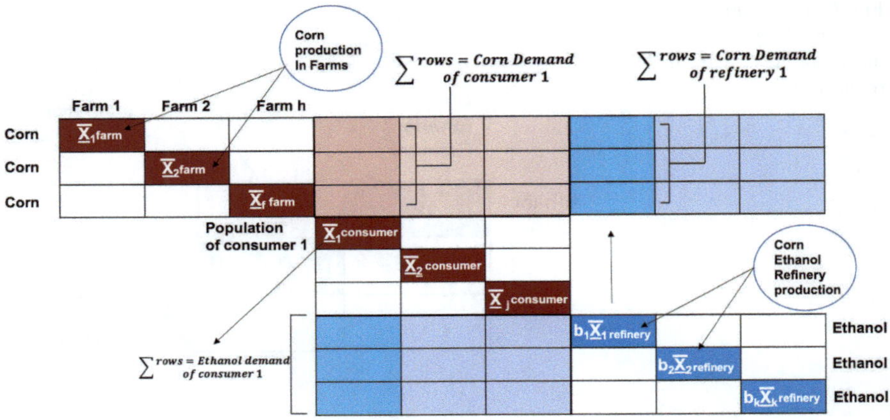

Fig. 17.10 Building the P2P-TES $\overline{X}$ matrix: incorporating corn demand flows of consumers and refineries

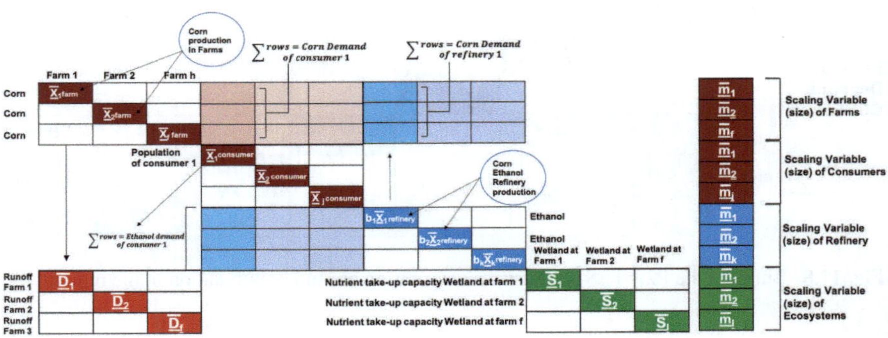

Fig. 17.11 Building the P2P-TES $\overline{X}$ matrix: incorporating phosphorus emissions and wetland ecosystems

the conversion of raw materials to a useful product. $fc_{eth}(H(\{z\}))$ is the outflow of ethanol from the process superstructure of the wet corn to ethanol technology represented as $H(\{z\})$. The conversion technology model has been adopted from Karuppiah and Grossmann [36]. Few parts of the process have been linearized for ease of optimization. Incorporation of the refinery operation in the multiscale P2P-TES $\overline{X}$ matrix is visualized in Fig. 17.7. While three refineries are filled up in the figure, the last refinery has a subscript k which depicts that the number of possible refineries incorporated in the $\overline{X}$ matrix depends upon the total number defined within the case study. The values that go into these diagonal cells of the matrix are the production quantities of each refinery.

Refinery Location

$$\exists!\underline{\overline{X}}_{k,k}(\{z,b\})\ :\ b_k = 1\ s.t.\ k \in \mathbf{R} \tag{17.25}$$

$$\sum_k b_k \leq 2 \tag{17.26}$$

Equation 17.25 represents the existence of a refinery if the binary variable b associated with that region is equal to 1. b_k and $\underline{\overline{X}}_{k,k}(\{z,b\})$ have a one-one mapping relation to each other. Equation 17.26 makes sure that a maximum of two refineries can be incorporated in this system. Thus, the variable k actually relates to two refineries only. In Fig. 17.11, this points to the ethanol production cell. Out of the 21 counties, only 2 will be chosen for refinery placement as explained in Sect. 17.4.1. The refineries are multiplied by the binary variable b in Fig. 17.7, which essentially provides the functionality of the optimization algorithm choosing the required number of refineries(2) from the k location choices.

Consumer The population of the 21 counties in the region are introduced into the $\underline{\overline{X}}$ matrix as shown in Fig. 17.8. The subscript j for the consumers belong to the set $\mathbf{P}$ of 21 consumers in the region. Only three are shown in Fig. 17.8 for brevity.

Consumer Demand of Bioethanol

$$BD_j = \sum_k \underline{\overline{X}}_{k,j}(\{z,b\}) \tag{17.27}$$

$$\sum_j \underline{\overline{X}}_{k,j}(\{z,b\}) \leq \underline{\overline{X}}_{k,k}\ s.t.\ j \in \mathbf{P}, k \in \mathbf{R} \tag{17.28}$$

where BD_j is the bioethanol demand of the jth consumer. $\sum_k \underline{\overline{X}}_{k,j}(\{z,b\})$ denotes the flow of ethanol as biofuel from sum of kth refineries to the jth consumer. The total demand of bioethanol is calculated from the population of a certain region as shown in Fig. 17.8. As shown in this figure, this equation refers to the row sum for every consumer column in the ethanol consumption part of $\underline{\overline{X}}$ matrix as shown in Fig. 17.8. This constraint makes sure that the amount of ethanol flowing from all refineries to consumers of a certain county equals the bioethanol demand of all consumers in that county. $\underline{\overline{X}}_{k,k}$ is the flow of ethanol output from a refinery obtained from the fundamental engineering model. Equation 17.28 ensures that total amount of bioethanol consumed from each biorefinery does not exceed its production.

Farm Figure 17.9 depicts the incorporation of farms in the 21 subregions considered in this study within the $\underline{\overline{X}}$ matrix. The diagonal cells of the matrix are filled with the total production quantity of each farm. The subscript f relates to the $\mathbf{F}$ set of 21 farms.

Consumer Demand of Food

$$FD_j = \sum_i \underline{\overline{X}}_{f,j}(\{z, b\})\ s.t.\ f \in \mathbf{F}, j \in \mathbf{P} \tag{17.29}$$

where FD_j is the food demand of the jth consumer. As shown in Fig. 17.10, Eq. 17.29 denotes the flow of corn as food material from fth farm to the jth consumer. This constraint makes sure that the amount of corn flowing from all farms to a specific consumer equals the corn demand of that consumer. The corn demand of a region is obtained from its population of consumers.

Corn Demand of Biorefinery

$$CD_k = \sum_i \underline{\overline{X}}_{f,k}(\{z, b\})\ \ s.t.\ f \in \mathbf{F}, k \in \mathbf{R} \tag{17.30}$$

$$\sum_k \underline{\overline{X}}_{f,k}(\{z, b\}) \leq \underline{\overline{X}}_{f,f} \tag{17.31}$$

where CD_k is the corn demand of the kth refinery. The value is calculated from the refinery process model and thus relates to the total production of corn as shown in Fig. 17.10. $\sum_i \underline{\overline{X}}_{f,k}(\{z, b\})$ denotes the flow of corn from sum of fth farms to the kth refinery. This constraint makes sure that total amount of corn from all selected farms to a specific refinery equals the corn demand of that refinery. Equation 17.31 relates the corn demand from the supply chain to the corn feed input flow in the process superstructure of the refinery. It makes sure that the total consumption of corn by refineries (columnwise k sum) from a particular farm f is less than or equal to the total production of that farm. This flow information relates to the corn consumption by refineries cell in Fig. 17.10. It must be noted that the flow information depends on whether a particular refinery in a county has been chosen to be located or not. If there is no refinery, then flows do not occur.

Energy Demand of Biorefinery

$$ED_k = \underline{\overline{X}}_{coal,k}(\{z, b\})\ s.t.\ k \in \mathbf{R}. \tag{17.32}$$

It is assumed in this problem that the refineries obtain its energy for operation from coal. However, coal supply is not considered within the system and is modeled to be supplied from outside the system boundary. To capture the transportation costs of coal to each refinery, an external coal supplying sector is introduced into the $\underline{\overline{X}}$ matrix. Since it is an external sector to the system boundary and incorporated separately, it has not been included in Fig. 17.10. The coal sector has been included in this problem for the sole purpose of modeling transportation of fuel and affects the design decision of biorefinery location. Environmental impacts from using coal are not captured in the case study due to the focus on nutrient runoff.

Environmental Impact

$$g_{phos} = \sum_{i} \underline{\overline{D_f}\overline{m_f}}\ s.t.\ f \in \mathbf{F}; \tag{17.33}$$

$$g_{phos} = \sum_{f} \underline{\overline{D_f}\overline{m_f}} - \sum_{l} \underline{\overline{S_l}\overline{m_{e_l}}}\ s.t.\ f \in \mathbf{F}\ , l \in \mathbf{F_E}\ ;\ f \rightarrow l. \tag{17.34}$$

where D_f denotes the phosphorus flow from the fth farm, and m_f is the scaling multiplier of that corresponding farm as shown in Fig. 17.11. Equation 17.33 is for the case where ecosystem services are not considered. Equation 17.34 integrates ecosystem services along with technological activities. The environmental impact flows (first term) are remediated by the water quality regulation services provided by the wetland (second term). As farms only considered to have wetlands beside them, ecosystem indices l are mapped one-to-one with farm indices f. Every value for the cells of the $\overline{D_f}$ is calculated based on the total production of the fth farm as shown in Fig. 17.11.

Wetland Ecosystems

$$PU_l = \underline{\overline{S}}_{l,l}(\{z, b\})\ s.t.\ l \in \mathbf{F_E}\ ;\ f \rightarrow l. \tag{17.35}$$

where PU_l is the phosphorus uptake by the lth wetland from agricultural water by the fth farm. f and l have a one-one mapping relation. The wetland information is inputted into the $\underline{\overline{S}}$ matrix shown in Fig. 17.11. PU_l value which is inputted in the diagonal cells of the $\underline{\overline{S}}$ matrix is estimated from wetland size and simple phosphorus take-up models. The b variable in the ecosystems can be used as a design parameter if the optimization routine is employed to determine location of ecosystems within the region for achieving certain level of phosphorus runoff.

Management of Ecosystems These flows are not modeled in this case study. If present, they would have indicated the flow of material from farms to the wetlands for their maintenance and efficient operation. Although empty, this is denoted as $\underline{\overline{C}}$ in Fig. 17.12.

Size of Processes

$$\underline{m}_f \leq 1\ s.t.\ f \in \mathbf{F} \tag{17.36}$$

$$\underline{m}_j = 1\ s.t.\ j \in \mathbf{P} \tag{17.37}$$

$$m_k = 1\ s.t.\ k \in \mathbf{R} \tag{17.38}$$

$$\underline{(m_e)}_l \leq 1\ s.t.\ l \in \mathbf{F_E} \tag{17.39}$$

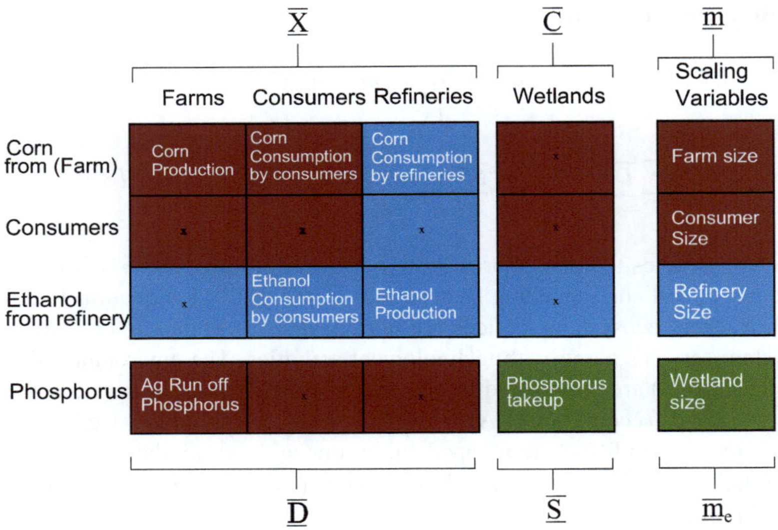

Fig. 17.12 Entire matrix for the corn ethanol problem. Flows in cells with x marks are not considered in this study. Consumers do not have any product that is sold to refineries or farms. (Reproduced with permission from [24])

As seen, from Fig. 17.11, the $\underline{\overline{X}}$ matrix contains information about the different technological activities. In this case study, information for farming activities are provided at their 2015 county-level production obtained from USDA [28]. Equation 17.36 limits farming scaling variables to 1, ensuring that maximum production of corn within the system model is less than or equal to the 2015 county-level corn production. The upper limit on the farming scaling variable denotes the constraint that total farm activity is lower than the maximum production of farm in a certain region. For refineries, the actual data for a 50-million-gallon ethanol plant is provided to the technology matrix. Thus, in Eq. 17.38 the scaling variable value of 1 makes sure that refineries in this study are fixed at the 50-million-gallon capacity. Consumer scaling variable of 1 in Eq. 17.37 fixes the size of consumers in a certain region to 2015 population levels. Population level and total farm production information are provided to the $\underline{\overline{X}}$ matrix. Similarly, Eq. 17.39 ensures that the size of wetlands in a certain region does not exceed the maximum available barren land area in that region.

Cost to Biorefinery

$$p_{cost} = \sum_{i,k} \underline{\overline{X}}_{i,k}(\{z,b\}) p_{corn} + \sum_{i,k} \underline{\overline{X}}_{i,k}(\{z,b\}) T_{i,k} p_{tc} + \sum_{k,j} \underline{\overline{X}}_{k,j}(\{z,b\}) T_{k,j} p_{te}$$

$$+\underline{\overline{X}}_{coal,k}(\{z,b\})T_{coal,k}p_{tcoal}$$
$$+\underline{\overline{X}}_{coal,k}(\{z,b\})p_{coal}\ s.t.\ i\in\mathbf{F},\ j\in\mathbf{P},\ k\in\mathbf{R} \tag{17.40}$$

Total cost to the biorefinery is obtained by adding the cost of corn (first term), the cost of transportation of corn from farms to refineries, the cost of transportation of ethanol from refinery to consumers (second term), the cost of coal transportation (third term), and the cost of coal (fourth term). Such a function forces transportation distances to be shorter and refineries to be more energy efficient. This was sufficient to solve for the locations of biorefineries and supplying farms. As land procurement prices and its spatial variation were not considered in this study, incorporation of capital costs, which is required to calculate net present value, even though important, is unnecessary for the goal of this study. p refers to prices. $\sum_{i,k}\underline{\overline{X}}_{i,k}(\{z,b\})$ represents the quantity of corn flow between farms and refineries. The transportation cost depends on the distances between the chosen supplier farms and the refinery and consumers whose information is contained in the transportation matrix T which contains distances between the centroids of different counties. Distance and quantities of flow are combined with necessary price parameters p_{tc} for corn, p_{te} for ethanol, and p_{tcoal} for coal to find total transportation cost.

17.4.4 Results

The optimization objectives of the problem are shown in Fig. 17.5 in the optimization scenarios section. It is a Mixed-Integer Nonlinear Programming (MINLP) problem. There are 8043 continuous variables, 21 discrete variables, and 7105 equations. The MINLP is solved using the BARON solver in the GAMS programming platform. Initially, the problem is solved for minimization of the two objectives separately. Implications of farming practices explained in Sect. 17.4.1 result in a trade-off between these two objectives. For every scenario, the process design and farm choices are observed and explored. Explanation of why changes occurred due to changing objective functions is provided. Next, the same problems are solved while including ecosystem services by considered conversion of free land available as explained in Problem Description into wetlands. The wetlands are expected to treat the excess phosphorus in the water. The objectives are then compared with each other to explore trade-offs and see if using wetlands and harnessing their ecosystem services result in "win-win" solutions for designing the biofuel network. Summary of different optimization results are provided in Table 17.1.

Minimization of Phosphorus Runoff The optimization problem is defined as

$$Z_1 = g_{phos} = \sum_i \underline{\overline{f}}(phos)_e^i\ s.t.\ i\in\mathbf{F} \tag{17.41}$$

Table 17.1 Summary of results for different optimization scenarios for the county-scale case study

Objectives	P2P-TES integrated design	Total phosphorus runoff (kg)	Net phosphorus runoff (kg)	Cost to refinery ($)	Farms chosen	Refinery locations
Minimization of phosphorus runoff	No. Wetlands considered to be used as end-of-pipe solution	175,900	128,800	9.71E+07	Lucas, Henry, Van Wert, Mercer, Putnam, Wood, Wyandot, Seneca, Ottawa	N/A
Minimization of phosphorus runoff	Yes	261,300	477,00	9.64E+07	Lucas, Williams, Henry, Van Wert, Putnam, Wood, Wyandot, Seneca, Ottawa, Dekalb	N/A
Minimization of cost to refinery	Yes	236,800	105,400	9.63E+07	Lucas, Fulton, Defiance, Paulding, Henry, Van Wert, Mercer, Auglaize, Putnam, Wood, Wyandot, Seneca, Ottawa, Sandusky, Allen, Dekalb	Sandusky, Van Wert

$$\text{subject to} \quad \begin{bmatrix} \underline{\overline{X}}(\{z,b\}) & \underline{\overline{C}}(\{z,b\}) \\ \underline{\overline{D}}(\{z,b\}) & \underline{\overline{S}}(\{z,b\}) \end{bmatrix} \begin{bmatrix} \underline{\overline{m}} \\ \underline{\overline{m}_e} \end{bmatrix} = \begin{bmatrix} \underline{\overline{f}} \\ \underline{\overline{f}_e} \end{bmatrix} \tag{17.42}$$

$$H(\{z,b\}) \geq 0 \tag{17.43}$$

$$S(\{z,b\}) \geq 0 \tag{17.44}$$

$$\underline{\overline{m}} \geq 0 \tag{17.45}$$

$$\underline{\overline{m_e}} \geq 0 \tag{17.46}$$

$$\sum_i \underline{\overline{D_i m_i}} - \sum_l \underline{\overline{S_l m_{e_l}}} \geq 0 \; s.t. \; l \in \mathbf{F_E} \; ; \; i \rightarrow l. \tag{17.47}$$

$$\underline{\overline{m_e}} \leq \underline{\overline{m}} \tag{17.48}$$

Two major constraints in this formulation are Eq. 17.47 and 17.48. Equation 17.47 models the assumption that phosphorus cannot flow between counties. This is a geographical constraint that makes sure that phosphorus runoff from one county cannot be treated by wetlands in another county. This is because phosphorus runoff can be only treated with nearby or adjacent wetlands within a small local region. The minimum possible phosphorus flow in a county can be 0. Equation 17.48 is an allocation determination constraint. Suppose a farm in a specific county has a scaling variable less than 1. That means in the context of the farm, the production of the farm is lower than its 2015 production value. Thus, the amount of ecosystems available for uptake of phosphorus flow is also limited by that number. This makes sure that even though a farm is not used or is used less than its maximum production, the amount of wetland for runoff treatment is attributed accordingly.

Without Ecosystems In this scenario, integration of ecosystems is not considered during optimization and design. Mathematically, this can be represented as modification of the above optimization formulation with

$$\underline{\overline{m_e}} = 0 \tag{17.49}$$

Minimization of g_{phos} results in optimization solutions that focus on improvement of the corn to ethanol transformation ratio through the process design of the plant. Along with that, no-till agriculture practicing farms are chosen by the biorefineries and consumers to source required corn. Without ecosystems, the optimization solution first selects all the counties that practice no-till agriculture. This is because for the same amount of corn production, no-till releases much less phosphorus to runoff compared to farming with tillage. However, the farms in these regions are not able to satisfy the demand of corn by biorefineries as well as consumers. Thus, along with the no-till farms, two tillage practicing farms from counties Wood and Mercer are chosen to satisfy the remaining corn demand as seen in Fig. 17.13. The reason these two counties are chosen is because the phosphorus outflow per kg of corn there is the least compared to other till agriculture practicing counties as determined by

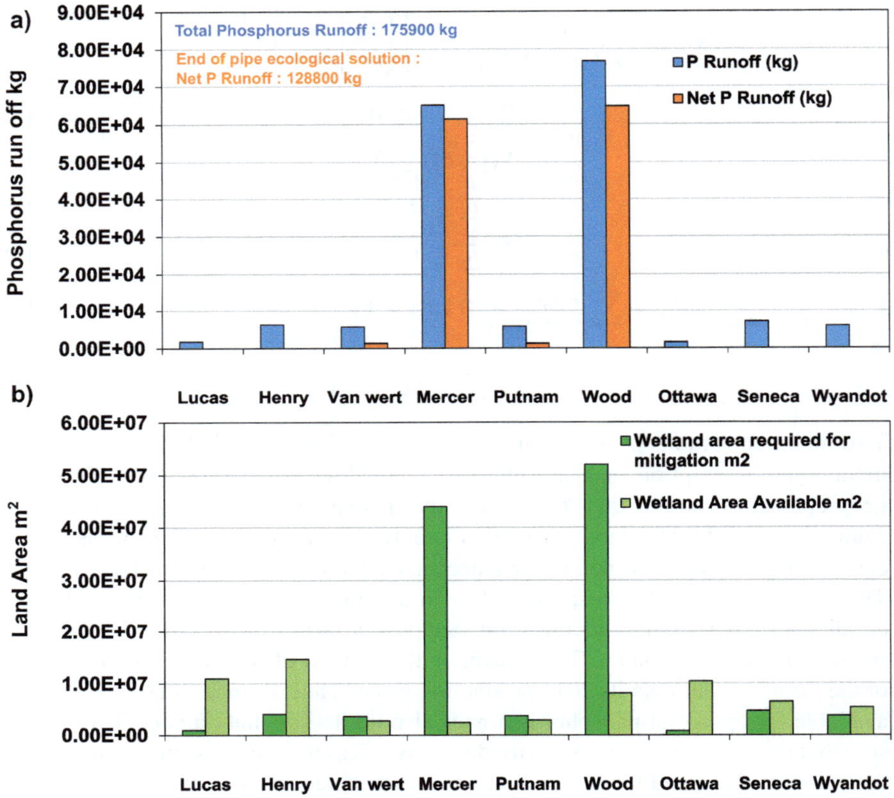

Fig. 17.13 (**a**) Total and net phosphorus runoff from different counties selected in supply chain design optimization. Net phosphorus runoff obtained from end-of-pipe solution. (**b**) Wetland area required for total phosphorus removal compared to barren land available for conversion to wetlands. (Reproduced with permission from [24])

the crop production and phosphorus runoff data. However, their individual runoff values are quite high compared to no-till farms as seen from Fig. 17.13a. The sum of all the blue bars in this figure gives the total phosphorus runoff of 175,900 kg. Figure 17.15a is a visual representation of the entire supply chain. The graph is read using the colored bands flowing between nodes of farms, consumers, and refineries. Source of flow is obtained by matching the color of the band with circumference segment of same color. Bands and segments of different colors denote destination of flows. In this problem, the choice of placement of refinery is not modeled. The objective function in this scenario is phosphorus runoff from the farms. The placement of biorefineries did not affect the objective function because biorefineries did not have any outflow of phosphorus in its wastewater. So only the choices of farms in this scenario are described in the results. This is the reason why refineries in Fig. 17.15a are generic and named 1 and 2. Corn ethanol production from two biorefineries are enough to satisfy the demand of biofuel energy from the consumers in that region. The width of the bands in Fig. 17.15a represent the fraction of corn being consumed from the farms by refineries and consumers. Farm Wood supplies the maximum of corn among all the farms. After all the no-till farms are used to their maximum capacity, Wood county is the next best farm choice with lowest phosphorus per unit mass corn according to available data. Mercer county is more phosphorus-intensive and occupies a much smaller percent of the network as it is not required to its full capacity and chosen only to satisfy the remaining demand of corn after all other selected farms have been completely used. The coal supplying sector is considered to be outside the system boundary.

Ecosystems as End-of-Pipe Solutions Rather than being a separate optimization scenario, this is just calculation of net phosphorus runoff given by the equation

$$g_{phos} = \sum_{i} \overline{\underline{f}}(phos)^{i}_{e} \; s.t. \; i \in \mathbf{F} \tag{17.50}$$

$$\overline{\underline{m_e}} = \overline{\underline{m}} \tag{17.51}$$

outside the optimization problem. In this analysis scenario, rather than optimizing, using the decision variables obtained in the without ecosystems scenario, Eq. 17.50 is calculated while using values of m_e obtained from Eq. 17.51. The results described in the without ecosystems scenario are for the situation where the problem is solved using a conventional supply chain framework like conventional P2P framework. It does not explicitly account for ecosystem services provided by wetlands. However, if after solving this problem, wetlands are harnessed for reduction of phosphorus runoff, the following results are seen. Using Eq. 17.51, the farms that are chosen from optimization are allowed to use their available land area for ecosystems to treat runoff. As seen from Fig. 17.13b, Van Wert, Mercer, Putnam, Wood, and all other counties do not have enough land area available for conversion to wetlands that can take up phosphorus. This results in a net phosphorus runoff of 128,800 kg from the entire region. Actual and net phosphorus runoffs are compared in Fig. 17.13a. The sum of orange bars in this figure gives the net phosphorus runoff.

Ecosystem Services Integrated into Design In contrast to the above scenario, the same problem is now solved with the P2P-TES framework that integrates ecosystem services explicitly into the design. This is basically solving the optimization problem with Eq. 17.41 to Eq. 17.48. Rather than choosing the no-till farms which have lower phosphorus runoff, farms with higher land area available for conversion to wetlands are chosen. This results in a much higher total phosphorus runoff of 261,300 kg as compared to 175,900 kg in the previous scenario. However, due to the availability of required area for wetlands in most of the counties, as seen from the bottom graph in Fig. 17.14a, the net phosphorus runoff is reduced to 47,700 kg by the wetlands. It is also shown in the top bar chart in Fig. 17.14a, which shows a huge difference between total and net phosphorus runoff. Williams, Dekalb, and Wood are three tillage practicing farm counties that are chosen in this supply chain design. All these

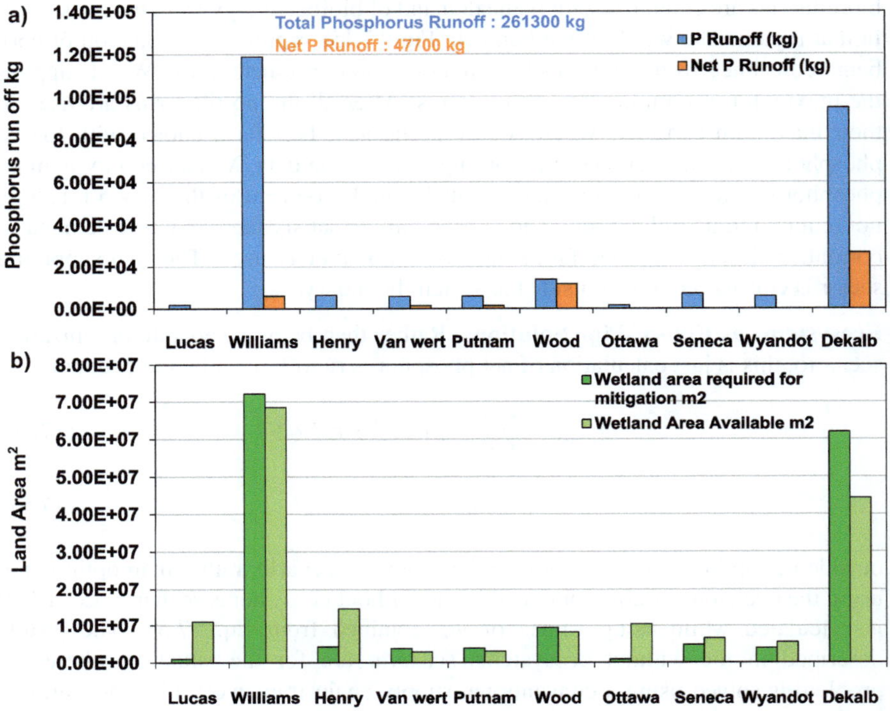

Fig. 17.14 (**a**) Total and net phosphorus runoff from different counties selected in supply chain design optimization. Net phosphorus runoff obtained using P2P-TES integrated framework. (**b**) Wetland area required for total phosphorus removal compared to barren land available for conversion to wetlands. (Reproduced with permission from [24])

counties have large amount of land available for wetland construction. Williams and Dekalb were not chosen in the previous solution because they have a high total phosphorus runoff as seen from Fig. 17.14a. They are chosen in this solution because the presence of large wetlands reduces the net phosphorus runoff significantly. This is a superior solution in terms of phosphorus runoff when compared to the previous one and could only be obtained by using P2P-TES integrated framework for design. Figure 17.15b shows the supply chain design for this optimization scenario. Dekalb and Williams farms supply large amount of corn as seen from the width of colored bands starting from them and going to refineries. The reason for this can be seen from Fig. 17.14b. Large amount of wetlands enable to take up lots of phosphorus from runoff, resulting in choosing these farms. Compared to the previous scenario, Wood is not chosen as much because of its relatively low availability of unused land for ecosystems. As mentioned in the without ecosystem scenario, in this optimization scenario, the location of refineries did not matter since phosphorus runoff is not affected due to that decision.

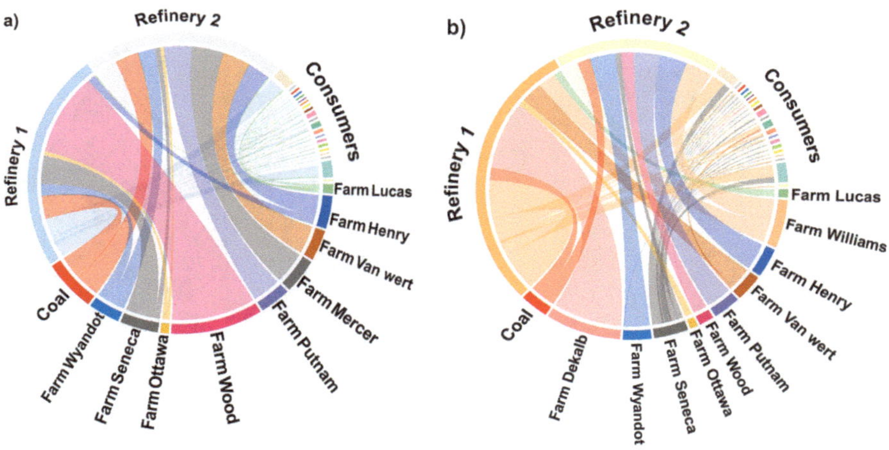

Fig. 17.15 Network graphs showing connection of flows between chosen farms, refineries, and consumers, and quantity of flows for (**a**) conventional design without ecosystems, (**b**) P2P-TES integrated design with wetlands. Source of flow is obtained by matching the color of the band with circumference segment of same color. Bands and segments of different colors denote destination of flows. (Reproduced with permission from [24])

Minimization of Cost to Biorefinery In this scenario, the objective function is changed to

$$Z_2 = p_{cost} \tag{17.52}$$

The transportation components whose cost is considered in this objective function are (1) transportation of corn from farms to biorefineries, (2) transportation of corn from farms to consumers for consumption as food, (3) transportation of produced bioethanol from refineries to consumption by consumers, and (4) transportation of coal to refineries. Along with these, coal and corn cost are also added. As mentioned in Problem Description, the price of corn from till agriculture is higher than that of non-till farms. Along with that, farther the transportation distances, more is the cost for transporting materials to and from the refineries. The prices from all these different sources are added up to give the total cost to the refineries for corn ethanol production. Equation 17.42–17.48 are part of the optimization formulation for this scenario. The results are presented in Fig. 17.16a. The biorefineries are placed at strategic counties that minimize the total cost from all sources mentioned above. Along with that, farms that have lower selling price for unit mass of corn are also chosen. Minimization of cost also results in changes in the process design of the refineries to become more energy as well as mass efficient. However, in this scenario, the presence of ecosystems does not affect the system since wetlands have only been considered to treat phosphorus and hence does not change the cost objective function. Figure 17.16b depicts the complete supply chain design. To lower transportation costs, refinery Van Wert sources majority of its corn

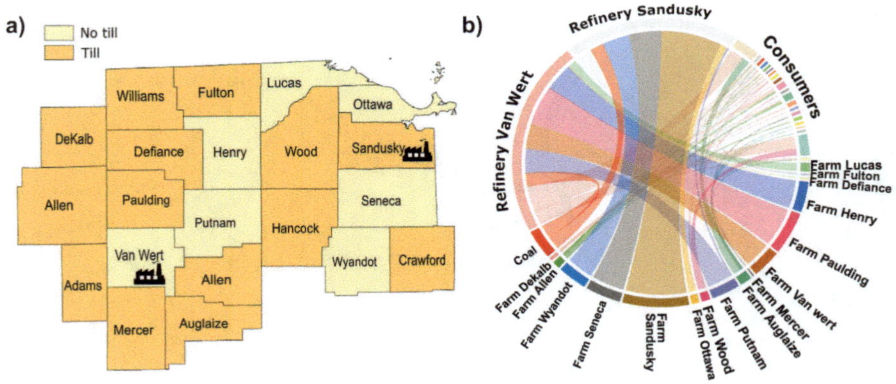

Fig. 17.16 Biorefinery placement and supply chain design for minimizing cost of corn consumption, energy consumption, and transportation of corn and coal to refinery and ethanol from refinery to consumers. The industry symbol shows the location of refineries. (Reproduced with permission from [24])

from nearby farms of Paulding, Henry, Putnam, and Van Wert itself. Similarly, refinery Sandusky sources its corn mostly from Ottawa, Seneca, Wyandot, and Sandusky itself. This information can be explored by tracing the colored bands starting from similar colored circumference segments and going to different colored segments. Due to lack of space, the names of consumers counties are not provided. However, results showed that refinery Van Wert supplied ethanol to consumers in surrounding counties of Lucas, Fulton, Williams, Defiance, Henry, Paulding, Van Wert, Mercer, Auglaize, Dekalb, and Allen. Similar connections are observed for refinery Sandusky, supplying surrounding counties with ethanol. All these connections are in agreement with the sole objective of reducing costs, of which transportation cost is a major contributor.

References

1. Majid Eskandarpour, Pierre Dejax, Joe Miemczyk, and Olivier Péton. Sustainable supply chain network design: an optimization-oriented review. *Omega*, 54:11–32, 2015.
2. Annelies De Meyer, Dirk Cattrysse, Jussi Rasinmäki, and Jos Van Orshoven. Methods to optimise the design and management of biomass-for-bioenergy supply chains: A review. *Renewable and sustainable energy reviews*, 31:657–670, 2014.
3. Amalia Nikolopoulou and Marianthi G Ierapetritou. Optimal design of sustainable chemical processes and supply chains: A review. *Computers & Chemical Engineering*, 44:94–103, 2012.
4. Ozlem Akgul, Andrea Zamboni, Fabrizio Bezzo, Nilay Shah, and Lazaros G Papageorgiou. Optimization-based approaches for bioethanol supply chains. *Industrial & Engineering Chemistry Research*, 50(9):4927–4938, 2010.
5. Aslıi Aksoy, Ilker Küçÿkoglu, Seval Ene, and Nursel Öztürk. Integrated emission and fuel consumption calculation model for green supply chain management. *Procedia-Social and Behavioral Sciences*, 109:1106–1109, 2014.

6. Alex J Dunnett, Claire S Adjiman, and Nilay Shah. A spatially explicit whole-system model of the lignocellulosic bioethanol supply chain: an assessment of decentralised processing potential. *Biotechnology for biofuels*, 1(1):13, 2008.
7. André Hugo and Efstratios N Pistikopoulos. Environmentally conscious long-range planning and design of supply chain networks. *Journal of Cleaner Production*, 13(15):1471–1491, 2005.
8. Gonzalo Guillén-Gosálbez and Ignacio E Grossmann. Optimal design and planning of sustainable chemical supply chains under uncertainty. *AIChE Journal*, 55(1):99–121, 2009.
9. Bruna Mota, Maria Isabel Gomes, Ana Carvalho, and Ana Paula Barbosa-Povoa. Towards supply chain sustainability: economic, environmental and social design and planning. *Journal of Cleaner Production*, 105:14–27, 2015.
10. José Ezequiel Santibañez-Aguilar, J Betzabe González-Campos, José María Ponce-Ortega, Medardo Serna-González, and Mahmoud M El-Halwagi. Optimal planning and site selection for distributed multiproduct biorefineries involving economic, environmental and social objectives. *Journal of cleaner production*, 65:270–294, 2014.
11. Fengqi You and Belinda Wang. Life cycle optimization of biomass-to-liquid supply chains with distributed–centralized processing networks. *Industrial & Engineering Chemistry Research*, 50(17):10102–10127, 2011.
12. Fengqi You, Ling Tao, Diane J Graziano, and Seth W Snyder. Optimal design of sustainable cellulosic biofuel supply chains: multiobjective optimization coupled with life cycle assessment and input–output analysis. *AIChE Journal*, 58(4):1157–1180, 2012.
13. Dajun Yue, Min Ah Kim, and Fengqi You. Design of sustainable product systems and supply chains with life cycle optimization based on functional unit: general modeling framework, mixed-integer nonlinear programming algorithms and case study on hydrocarbon biofuels. *ACS Sustainable Chemistry & Engineering*, 1(8):1003–1014, 2013.
14. Gabriela Corsano, Aldo R Vecchietti, and Jorge M Montagna. Optimal design for sustainable bioethanol supply chain considering detailed plant performance model. *Computers & Chemical Engineering*, 35(8):1384–1398, 2011.
15. Andrea Zamboni, Fabrizio Bezzo, and Nilay Shah. Spatially explicit static model for the strategic design of future bioethanol production systems. 2. Multi-objective environmental optimization. *Energy & Fuels*, 23(10):5134–5143, 2009.
16. Rebecca J Hanes and Bhavik R Bakshi. Process to planet: A multiscale modeling framework toward sustainable engineering. *AIChE Journal*, 61(10):3332–3352, 2015.
17. Rebecca J Hanes and Bhavik R Bakshi. Sustainable process design by the process to planet framework. *AIChE Journal*, 61(10):3320–3331, 2015.
18. Tapajyoti Ghosh and Bhavik R Bakshi. Process to planet approach to sustainable process design: Multiple objectives and byproducts. *Theoretical Foundations of Chemical Engineering*, 51(6):936–948, 2017.
19. Bhavik R Bakshi, Guy Ziv, and Michael D Lepech. Techno-ecological synergy: A framework for sustainable engineering. *Environmental science & technology*, 49(3):1752–1760, 2015.
20. Xinyu Liu and Bhavik R Bakshi. Extracting Heuristics for Designing Sustainable Built Environments by Coupling Multiobjective Evolutionary Optimization and Machine Learning. In *Computer Aided Chemical Engineering*, volume 44, pages 2539–2544. Elsevier, 2018.
21. Varsha Gopalakrishnan and Bhavik R Bakshi. Ecosystems as unit operations for local techno-ecological synergy: Integrated process design with treatment wetlands. *AIChE Journal*, 64(7):2390–2407, 2018.
22. Xinyu Liu, Guy Ziv, and Bhavik R Bakshi. Ecosystem services in life cycle assessment-Part 1: A computational framework. *Journal of Cleaner Production*, 2018.
23. Xinyu Liu, Guy Ziv, and Bhavik R Bakshi. Ecosystem services in life cycle assessment-Part 2: Adaptations to regional and serviceshed information. *Journal of Cleaner Production*, 2018.
24. Tapajyoti Ghosh and Bhavik R Bakshi. Designing biofuel supply chains while mitigating harmful algal blooms with treatment wetlands. *Computers & Chemical Engineering*, 126:113–127, 2019.

25. Bo P Weidema, Massimo Pizzol, Jannick Schmidt, and Greg Thoma. Attributional or consequential life cycle assessment: A matter of social responsibility. *Journal of Cleaner Production*, 174:305–314, 2018.
26. Tapajyoti Ghosh, Xinyu Liu, and Bhavik R Bakshi. Including ecosystem services in sustainable process design across multiple spatial scales. In *Computer Aided Chemical Engineering*, volume 44, pages 1837–1842. Elsevier, 2018.
27. TEEB. The Economics of Ecosystem and Biodiversity. http://www.teebweb.org/our-publications/teeb-study-reports/ecological-and-economic-foundations/, 2016.
28. Corn production 2015-2016 Ohio. https://www.nass.usda.gov/Statistics_by_State/Ohio/Publications/Annual_Statistical_Bulletin/Ohio%20bulletin%202015-2016.pdf.
29. Ohio countywise population 2015-2016 Ohio. https://www.ohio-demographics.com/countiesbypopulation/.
30. Ohio bioenergy demand 2015-2016. https://www.eia.gov/state/seds/seds-data-complete.php?sid=OH.
31. National Land Cover Database 2011. https://data.nal.usda.gov/dataset/national-land-cover-database-2011-nlcd-2011.
32. Robert H Kadlec. Overview: surface flow constructed wetlands. *Water Science and Technology*, 32(3):1–12, 1995.
33. Robert H Kadlec. An autobiotic wetland phosphorus model. *Ecological Engineering*, 8(2):145–172, 1997.
34. Robert H Kadlec and Scott Wallace. *Treatment wetlands*. CRC Press, 2008.
35. Robert H Kadlec. Large constructed wetlands for phosphorus control: A review. *Water*, 8(6):243, 2016.
36. Ramkumar Karuppiah, Andreas Peschel, Ignacio E Grossmann, Mariano Martín, Wade Martinson, and Luca Zullo. Energy optimization for the design of corn-based ethanol plants. *AIChE Journal*, 54(6):1499–1525, 2008.

Chapter 18
Demand and Supply of Air Quality Regulation Ecosystem Services

Michael Charles

18.1 Introduction

18.1.1 Criteria Air Pollutants

Among the needs of survival for any species is clean air to breathe. Unfortunately, more than 90% of the global population lives in a location which does not meet the World Health Organization's (WHO) standard for clean air [1]. More than half of the globe's population is exposed to air that does not meet WHO's lowest standard air quality target. In the United States alone, at least 166 million people reside in counties with levels of air pollution that can impact their health [2]. In response, the dangers of poor air quality have sparked action across the world in an effort to increase the access to clean air and decrease the health impacts and risks.

In the United States, the Clean Air Act was passed, enabling the Environmental Protection Agency (EPA) to regulate emissions which impacted human health and general welfare. It also required the adoption of the National Ambient Air Quality Standards (NAAQS), giving quantitative target concentrations of hazardous air pollution. Further, the US EPA identified six criteria air pollutants that determine air quality, are regulated by the NAAQS, and are reported by industry. These pollutants are ground-level ozone (O_3), particulate matter (PM), carbon monoxide (CO), lead (Pb), sulfur dioxide (SO_2), and nitrogen dioxide (NO_2).

Across the world, the dangers of air pollution vary along with the responding environmental regulations. In 1998, Beijing, China, declared war on air pollution. A rapid response to their poor air quality led to significant improvements in the

M. Charles (✉)
Department of Biological and Environmental Engineering within the College of Agriculture and Life Sciences, Cornell University, Ithaca, NY, USA
e-mail: mtc58@cornell.edu

B. R. Bakshi (ed.), *Engineering and Ecosystems*,
https://doi.org/10.1007/978-3-031-35692-6_18

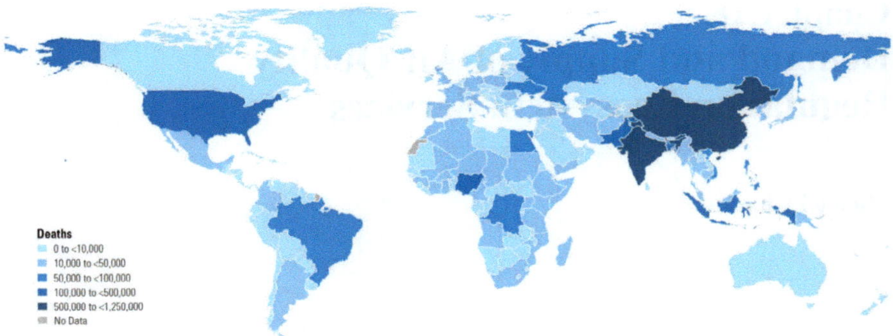

Fig. 18.1 Number of deaths per country attributable to air pollution in 2017 (Used with permission from Health Effects Institute [1])

concentrations of pollution in their air. According to a report by the United Nations Environment Programme, on-ground observation data shows that between 1998 and 2017, the annual average concentrations of SO_2, NO_2, and PM_{10} (particulate matter less than 10 μm in diameter) decreased by 93.3%, 37.8%, and 55.3%, respectively [3]. As many developed countries are seeing improvements in their air quality from associated regulations, many countries are still developing their technologies, economies, and capacity to implement such regulations.

Varying concentrations of air pollutants yield varying corresponding health impacts across the world. Air pollution exposure can have major health impacts on susceptible populations, including increased hospitalizations, disability, early death from respiratory diseases, heart disease, stroke, lung cancer, diabetes, and communicable diseases like pneumonia [1]. Further, the State of Global Air Report in 2019 states that air pollution accounts for 41% of global deaths from chronic obstructive pulmonary disease, 20% of deaths from type 2 diabetes, 19% of deaths from lung cancer, 16% of deaths from ischemic heart disease, and 11% of deaths from stroke. Figure 18.1 shows the differences across countries in total death count in 2017 attributable to air pollution exposure. This map shows air pollution hot spots in countries like China, India, the United States, Russia, and Brazil. These countries are also known to have some of the highest gross domestic products (GDPs) in the world along with some of the largest populations.

18.1.2 Carbon Dioxide and Greenhouse Gases

A discussion around air quality would be incomplete without including greenhouse gas (GHG) emissions and their role in global warming and climate change. The most common greenhouse gas discussed in the context of climate is carbon dioxide. The impacts and risks of greenhouse gas emissions vary from criteria air pollutants.

Rather than impacting human health through respiratory diseases, heart disease, and other impacts mentioned in the previous section, greenhouse gases yield risk to global warming which can unfold many other impacts that include human health, safety, and wellness. Often, the physical effects of global warming are referred to as climate change.

Climate change is an increasingly discussed issue and regularly researched by the Intergovernmental Panel on Climate Change (IPCC), which provides scientific assessments to the policy makers of the United Nations Framework on Climate Change (UNFCCC). Recent IPCC special reports have focused on *Climate Change and Land*, *The Ocean and Cryosphere in a Changing Climate*, and the impacts of *Global Warming of* 1.5 °C, all pointing toward the need for urgent action in addressing climate change.

There are many risks and potential impacts associated with climate change, all which vary in intensity depending on future carbon emissions and corresponding degrees of warming. Due to uncertainty in carbon activity and human response to the risks of climate change, potential impacts are often predicted by modeling scenarios of different degrees of global warming increase (i.e., *Global Warming of* 1.5 °C). In the IPCC Fifth Assessment Report (AR5), representative concentration pathways were also proposed, which describe different climate scenarios depending on the rate of increased concentrations of greenhouse gases over the twenty-first century [4]. These represent varied levels of urgency in global carbon emission responses, predicting mitigation impacts on atmospheric carbon and resulting warming effects. Some extreme scenarios also consider inaction or increased emissions. These different scenarios all predict increased global warming over time, resulting in many effects such as rising sea levels, heavy precipitation in some regions, drought in other regions, increased storm intensity, thawing of permafrost, increased forest fires, increased forced migration and invasive species, biodiversity-related risks due to habit change, and many others. Beyond physical climate risks, economic, political, sociological, and health risks are also predicted as resources lose security [5].

18.1.3 The Role of Nature

Although increasing industrialization and population contribute toward the increase in air pollution, the biosphere has processes which filter and clean the air on which we rely for survival. Vegetation across the Earth's surface adsorbs air pollutants through its leaf stomata and also sequesters carbon from the atmosphere and stores it in biomass and in the soil. These services contribute toward the regulation of air quality and are categorized as regulating ecosystem services per the Millennium Ecosystem Assessment [6]. Nature has been playing a vital role in regulating air quality longer than humans have existed, yet is often ignored in the solutions for mitigating air pollution impacts.

Most of the responses to poor air quality and climate change focus solely on technological and societal solutions for decreasing emissions, and therefore, decreasing the demand for air quality regulation imposed on nature. However, it is important to determine all opportunities toward achieving a sustainable planet: one in which ecological functions within the biosphere have the capacity to meet the increasing demands of human activity. This is one way to define absolute sustainability. With this mindset, it is not only important to consider solutions which decrease emissions but also consider how we can increase the capacity of ecosystems to provide their air quality regulating services. Without considering ecological solutions, the range of solutions becomes narrowed toward reliance of technology and can have unintentional negative environmental impacts. In fact, Lewis et al. argue in their recent *Nature* article that "restoring natural forests is the best way to remove atmospheric carbon" [7]. This article focuses on climate change in an international context and shows how reconsidering land use can provide real solutions to respond to global warming. Although ecological solutions for climate change can and should be considered at large scales (global, national, subnational, etc.), smaller scales may be needed to address hazardous air pollution (local, regional, etc.) and the local and regional health impacts. Similarly, studies have concluded that ecosystems provide economically competitive solutions for addressing air pollution mitigation as well [8, 9]. These results encourage sustainable design to not only include technological design but also ecological, or landscape, design as well.

One way to consider both technological and ecological systems in sustainable design is to implement a metric for absolute sustainability, one that considers both the **demand** and **supply** of the ecosystems service(s) in focus. One example of this from an analytical perspective is the concept of ecological footprints, which compares pollution emission rates to the amount of land that would be needed to mitigate pollution of an activity. On a global scale, many IPCC reports discuss the role of vegetation, specifically forests, and their capacity to sequester carbon dioxide through the process of photosynthesis. However, humanity imposes an increasing demand onto our biosphere and its ability to sequester greenhouse gases and regulate its surface temperature. Using 2018 data, it is estimated that humanity demands an equivalent of 1.7 Earths [10]. The Global Footprint Network provides an online tool to analyze and compare these ecological footprints by country, per capita or per total land area [11]. Figure 18.2 shows the role that human carbon emissions have played in the global ecological footprint, revealing that our global demand for carbon sequestration overshot the planet's capacity around the early 1970s. The two most significant impacts on the global ecological footprint in the last 50 years are attributed to croplands (orange) and carbon emissions related to human activity (blue). To calculate these values, quantifying both the demand (emissions rates) and supply (ecosystem service rates) of corresponding activities and land cover is necessary.

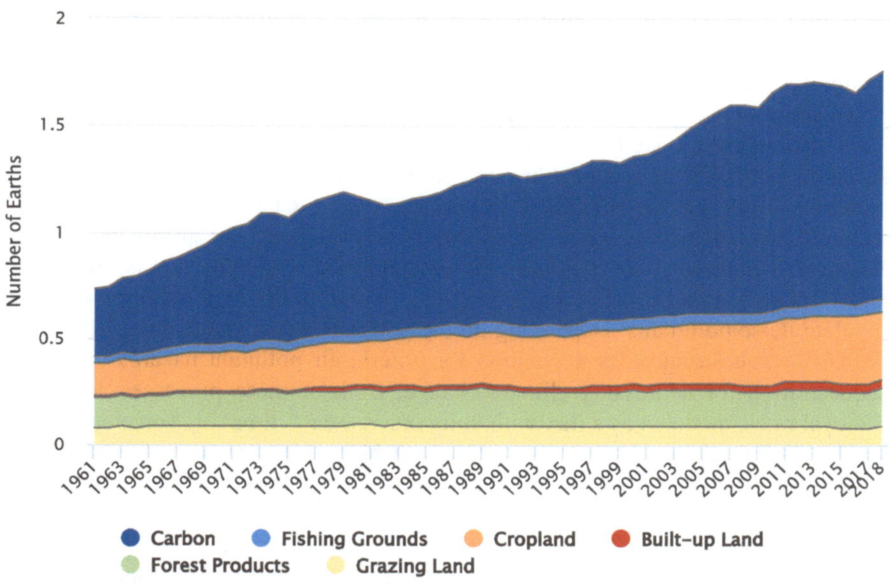

Fig. 18.2 World ecological footprint by land type from 1961 to 2018 using the Global Footprint Network's Open Data Platform (Used with permission from Global Footprint Network Open Data Platform [11])

18.1.4 Discovering Synergies Between Technology and Ecology

To begin comparing the flows between nature and pollution sources, the scale of interaction or the scale of interest (based on political or commercial boundaries) must be defined. Planting a tree across the world will not improve a local air quality hazard, and a business aiming to achieve sustainability goals may not receive credit for land-based projects off their property. Although ecological solutions for climate change should be considered at large scales (global, national, subnational, etc.), smaller scales may be needed to address hazardous air pollution and the local and regional health impacts.

To determine this scale of interaction, the term **serviceshed** or airshed has been used to discuss "the areas that provide specific ecosystem services to specific beneficiaries" [12]. From this definition and the concept of including nature in engineering, we can begin to think of industrial processes as beneficiaries. In the context of air quality regulation, the ecological serviceshed can be the areas of land which directly or indirectly remove pollutants from an industrial facility. Knowing this area is essential in developing techno-ecological synergistic designs in regard to air pollution. Charles et al. provided quantitative mathematical definitions of the air regulating serviceshed to determine this scale for a particular source of pollution [13]. These definitions rely on results from atmospheric dispersion models and provide spatially explicit boundaries of all the areas that satisfy each mathematical definition of serviceshed. Other applications have simplified the serviceshed as a particular distance from a point source based on Gaussian plume modeling, assuming a circular serviceshed based on a radius.

To continue exploring the role nature has in engineering, more detailed studies have focused on case studies and single-site designs to determine the role ecological solutions can play in sustainable engineering with respect to air pollution. The techno-ecological synergy (TES) concept described in Chap. 12 suggests that both technological and ecological solutions should be considered for sustainable design. With consideration of air pollution, the demand of the system (D_k) is the rate of air pollutant k emitted, while the supply (S_k) is the ability of local ecosystems to regulate those pollutants. Depending on the pollutant risks, this ecological process is likely to be through dry deposition for criteria air pollutant hazards or carbon sequestration for addressing climate change. Both of these processes are active removal rates, modeled as a flux with units [mass area^{-1} time^{-1}]. Some ecosystem services are static, such as carbon storage, which is not an active removal rate but rather an accumulation of previously removed pollution. The removal of both criteria air pollutants and carbon dioxide occurs from many different land use types whether the area be forest, shrublands, grasslands, or agriculture [14, 15, 16]. Further, carbon sequestration occurs both above ground and below ground, actively removing carbon from the atmosphere as plant structure grows above ground and in the soil as nutrients cycle underneath [17, 18, 19].

In understanding how nature can be incorporated into engineering methodology and how to better design sustainable processes, TES designs must be able to make decisions about both technology and ecology within the same framework, understanding trade-offs or "win-wins" between costs and emissions. Determining the role land and vegetation can play in response to poor air quality and climate change requires data which assesses both supply and demand of the air quality regulating ecosystem services. One such metric, described in Chap. 12, is the sustainability index which represents sustainability of ecosystem services at a selected scale, subtracting the demand of a service from the ecological supply. This index provides an absolute metric, rather than a relative one, because it is based on the physical carrying capacity of the biosphere's ecosystems and not on changes from past activity. Further, the sustainability index (V_k) incorporates ecosystems and their services within a function that can lead as a design objective or constraint. As an example, a common constraint would be for the sustainability index to be greater than zero, meaning that the supply of ecosystem services is greater than the demand.

Incorporating ecological models into a TES design framework for air quality regulation also requires technological models of process unit operations used for air pollution control or entire process models. The technological models often quantify the demand while ecological models can be used to valuate the supply. With synergistic systems that include both technology and ecology, sustainable design can achieve absolute sustainability rather than designing for incremental improvements. This chapter will introduce methods for collecting and producing this type of data along with some of the applications and early conclusions of understanding the role of ecological solutions in addressing air quality and climate change issues.

18.2 Assessment and Analysis

In order to explore the inclusion of ecosystems within sustainable design, we need to be able to quantify the interactions between technology and ecology. As introduced, the techno-ecological synergy concept compares the demand and supply of ecosystem services. In this context, the demand for ecosystem services is equivalent to the emissions from a given activity or source and the supply is the amount of removal, or uptake, of those emissions from the landscape. In the context of air quality regulation, the supply and demand values vary with each chemical species present in the flue gas.

Similar to any design, emissions sources and a time span must be identified; however, TES requires the additional identification of spatial boundaries of the landscape. Essentially, the following questions must be asked: What process or processes am I interested in? Which chemical species will I consider? How much land is available? What region am I interested in? How long of a time span will I consider? Are there dynamics at smaller time steps that I am interested in? All these questions are important for the assessment and must be determined before quantifying.

Case studies can range from a single manufacturing site to a city to the entire planet. Likely, assessments for engineering and design work will occur at smaller scales, but policy work, like international climate change negotiations, may require global assessments of emissions and nature's regulatory capacity. This section will discuss tools, databases, and other methods for determining these supply and demand values for air pollution of both criteria air pollutants and carbon dioxide.

18.2.1 Quantifying the Demand—Emissions

Quantifying the emissions, or the demand for air quality regulating services, is typically the easier valuation between demand and supply. The emissions rates of all the chemical species that are being evaluated are the desired demand values. These rates should have units of [mass time^{-1}]. These emissions can be quantified from a range of sources, data types, and process models.

Depending on the source, direct measured data may be available from the owner of the operation. If only the total stack flow rate is available for a flue gas stack or similar point source, then composition information will also be needed for the source to determine the chemical species-specific flow rates. Further, most countries have national inventories of emissions data for both criteria air pollutants and greenhouse gases. Other than directly measured firsthand data or national inventories, emissions rates can also be estimated or valuated using process flow sheets, engineering model equations, design and operation software, or literature of similar processes.

National Databases (the United States) For credible data regarding emissions within the United States, the Environmental Protection Agency (EPA) provides a few different databases. For air pollutants, the EPA's National Emissions Inventory [20] provides point sources, nonpoint sources, on-road sources, non-road sources, and event sources. Point sources include stationary sources such as industrial emissions and many other facility-based information. Nonpoint sources include county-level or tribal reservation data. On-road sources are used to estimate transportation impacts associated with driving on the road, while non-road sources include marine vehicles or other mobile sources not on the road. Event sources include sporadic occurrences such as wildfires or volcanic eruption. This collection of data consistently contains criteria air pollutants but also includes many other emitted species for different sources.

If climate change impacts are being analyzed, the Green House Gas Reporting Program (GHGRP) provides self-reported greenhouse gas (GHG) emissions data for over 8000 facilities and can be found using the EPA's FLIGHT (Facility Level Information on Greenhouse Gases Tool) [21]. Most of the data provided by the EPA uses annual averages and typically does not include higher resolution dynamics like seasonal or monthly variation of emissions. Global greenhouse emissions are reported by both the US EPA and National Aeronautics and Space Administration (NASA) along with many other agencies, organizations, and tools for visualization. Outside the United States, national inventories exist for both criteria air pollutants and greenhouse gases and are typically published by government-affiliated agencies like the European Environment Agency for Europe [22, 23, 24]. More details are provided in the chapters in Part III of this book, with air emissions being the focus of Chap. 10.

Control Technology In comparing the supply and demand of air quality regulation ecosystem services, it is often of interest to compare nature-based solutions with existing technological counterparts. In these cases, quantifying the amount of pollutant uptake that can or does occur from technological equipment and the related costs is necessary. Specific to air pollution, determining appropriate technological options for the pollutants present in the flue gas is essential. The typical criteria air pollutants studied in most air pollution literature are SO_2, NO_x, particulate matter, and ozone. Technologies like scrubbers, filters, and reactors are used to separate these pollutants from the flue gas for disposal. With a specific focus on air quality control equipment, the US Environmental Protection Agency has published the Air Pollution Control Cost Manual [25], which can be used to generate cost equations for equipment like selective catalytic reactors, baghouse filters, and flue gas desulfurization units. For assessment purposes, parameters for the design equations either need to be known or estimated based on existing literature values. For design, different levels of detail may be used at different stages. For example, approximated costs may be the first iteration before sizing and tuning the parameters of the unit. Therefore, it is important to obtain the design equations for the appropriate context of your application. Another tool that exists specifically for air pollution removal is the Control Strategy Tool (CoST) [26]. This tool can help produce cost estimates for removing SO_2, PM_{10}, NO_x, and volatile organic compounds (VOCs) along with determining best available technologies. Estimating costs using tools or equations is essential for sustainable design; however, for analysis studies, site-reported data may be available without any calculations needed.

18.2.2 Quantifying the Supply—Ecosystem Air Quality Regulation

Quantifying the supply of air quality regulating ecosystem services can be much more difficult than the demand. This is because ecological processes introduce complex dynamics as they provide services. The capacity of their services changes with intra-annual seasons along with inter-annual growth. Quantifying the supply of ecosystem services requires defining clear spatial and temporal boundaries. Yet, even as those boundaries are defined, it is inevitable that other services, or co-benefits, will be left undervalued in one's assessment. Although many ecosystem services exist, this section will focus solely on the uptake of criteria air pollutants and greenhouse gases through dry deposition and carbon sequestration, respectively.

Dry Deposition The supply of air quality regulating ecosystem services can be quantified with multiple tools and models depending on application. Chapter 10 introduces some of the fundamental models used to quantify vegetative ecosystem services. This chapter will build on previous material by providing review and introducing related tools and datasets. As a review, dry deposition is a process of pollutant uptake that takes place on the leaf surfaces of vegetation. The rate of the process is dependent on characteristics of the land, meteorology, and ambient pollution in the region. The flux of a chemical is directly related to the concentration of that species, written as

$$F = v_d \, C \tag{18.1}$$

where F is the flux, v_d is the deposition velocity [distance time^{-1}], and C [mass volume^{-1}] is the concentration at the surface [14]. Ambient concentration data of various regions can be determined using measured data from air quality monitors. For data in the United States, the EPA has a tool named the Interactive Map of Air Quality Monitors [27], which provides individual map layers for different chemicals monitored by these stations.

The deposition velocity is a value which is dependent on the inverse sum of three resistances: aerodynamic resistance (R_a), quasi-laminar boundary layer resistance (R_b), and canopy resistance (R_c) [14].

$$v_d = (R_a + R_b + R_c)^{-1} \tag{18.2}$$

The three resistances are based on a combination of meteorological, land surface, and ecological variables. Further reading on the UFORE-D (Urban Forest Effects Model, dry deposition component) can be found in the "UFORE Methods" document at the cited link [15], providing both the equations associated with the dry deposition calculations and many assumptions commonly used in modeling software.

Vegetation is incorporated into the calculation through the canopy resistance calculation, which calculates a mesophyll resistance dependent on leaf area index (LAI). LAI is a dimensionless number which compares leaf area per ground area. Most software assume "forest" land types to have an average LAI of 7. If LAI is not assumed based on land cover data, such as the National Land Cover Dataset [28], direct LAI data can be acquired from the NASA MODIS instrument, a moderate resolution imaging spectroradiometer. However, using user-defined LAI data typically requires calculating the deposition velocity outside of many of the tools which calculate dry deposition flux.

There are many software which can be used to calculate dry deposition ecosystem services, such as the i-Tree software suite [29]. The i-Tree software suite has different tools, using different levels of data input to valuate the dry deposition fluxes and monetary benefit of the ecosystem services. County-level flux averages are used along with land area and tree cover percentage to determine the deposition rates [mass time^{-1}] of a given area space in i-Tree Canopy. Using i-Tree Design can yield species-specific deposition rates along with tree growth simulations.

Other tools which valuate dry deposition are atmospheric dispersion modeling software. In the United States, the Clean Air Act requires the EPA to identify models which are approved to be used in the Prevention of Significant Deterioration (PSD) program. The most recent updated list of recommended model was published in 2017 and is found in Appendix A of the Guideline on Air Quality Models [30]. Two examples of these software are AERMOD [31] and CALPUFF [32], and both can be used to model the spatial distribution of pollution from a set of sources, including the dry deposition of neighboring land cover. These software are open-source and written in FORTRAN. This type of software can reveal the spatial heterogeneity of ecosystem service value and deposition flux across a landscape.

Air pollution can also be transported through wet deposition and impact the quality of rainwater. However, for these ecosystem service analyses, only dry deposition will be considered as a service because it can be allocated to specific land areas and also does not impact regional water quality. Further, social costs and valuations of the removal of these pollutants are discussed in environmental economics, where the value of removing a ton of pollution is determined by the exposure of that pollution to people and potential health risks associated with the exposure. National accounts of these social values of removing air pollution from various land cover are discussed in Chap. 10. BENMAP [33] is a common tool for calculating these economic values as it overlays population density maps with spatial information of pollution removal. Other tools, such as i-Tree Canopy [29], can be used to quickly look up the county-scale average value for removing a ton of a number of pollutants, including carbon dioxide.

Carbon Sequestration Valuating the ecosystem service of carbon sequestration for climate regulation is important in assessing ecological solutions to respond to climate change and understand the gaps between our current state and one that characterizes absolute sustainability. Hazardous air quality impacts from gaseous and particulate chemicals impact areas locally and regionally, but climate impacts from greenhouse gas emissions impact the entire planet's ability to regulate its temperature. The location of ecological solutions must also be within these areas that are impacted, also known as the ecological serviceshed. This means that although solutions to poor air quality hazards can be responded to within a region, county, or country, the response to climate change will require a global solution.

To compare the supply and demand of the climate regulation ecosystem service for a predetermined analysis, many models and tools are available. The UFORE model previously mentioned for dry deposition calculations also has a carbon sequestration component, UFORE-C [14, 15]. These equations rely on the activity of the ecosystems which convert carbon dioxide to organic biomass. This vegetative growth sequesters carbon actively and stores it in its structure above ground and also in the roots and soil below.

The calculations of carbon sequestration rely on allometric growth functions per species or genus. This growth is then converted into stored carbon into the woody biomass and also the leaves, if the tree is not deciduous. In UFORE-C, a root-to-shoot ratio of 0.26 is used [15] to estimate the carbon stored below the surface. The calculations determine the marginal weight of annual tree growth and use an estimated carbon content of the biomass along with the stoichiometric relationship and molecular masses of C and CO_2 to determine the carbon dioxide sequestered from the atmosphere.

This model is available in the aforementioned i-Tree software suite. These tools can estimate both the rate of carbon sequestration in units of flow [mass time^{-1}] and also in estimated economic benefit. The UFORE methodology uses an associated economic value of $20.30 per ton of carbon sequestration based on estimated marginal social costs of carbon dioxide emissions [15, 34].

In order to use these tools, varying input data is required depending on availability. These inputs can include tree cover area or number of trees, species type(s), and diameter. Using a sample of your set of trees can also be one way to reduce the analysis time and still aim to characterize the trees of a particular area or region. With the right tool and input data, the rate of carbon sequestration can be estimated for the land within the scope of the analysis. This can be compared with the demand of the activity in the analysis for determination of sustainability and possible opportunities moving forward.

Other open-source modeling software, such as FORCARB2: An Updated Version of the U.S. Forest Carbon Budget Model, can be used for more advanced calculations [35]. FORCARB2 is a text-based software written in FORTRAN, similar to CALPUFF. A list of other tools and software for carbon-related estimating tools is available on the USDA Forest Service website as cited [36].

Including Land Use Change and Temporal Dynamics Quantifying and analyzing the supply and demand of ecosystems is an essential first step toward understanding the current state of ecological capacity compared to human activity. However, to explore solutions moving forward, land use change (LUC) scenarios must be generated. LUC scenarios provide hypothetical futures based on proposed initiatives for increasing ecosystem service supply, such as converting lawns to forests on one's property. In practice, corporations or institutions have specific projects they want to analyze as nature-based solutions and the impact they will have on increased ecosystem services production. However, large-scale studies (regional to global) may require multiple scenarios based on lower-resolution information, like land cover data. For either case, usually a "best-case" scenario is first conducted to determine if projects have potential.

A best-case scenario usually assumes that all land that is defined as available is restored to a forested state. A good assumption for the amount of canopy cover is to either use the county-level average or if you are using a spatial model, use the default leaf area index (LAI) value assigned for the forest land cover category. One way to model various scenarios is use a linear scale between the current land cover and the best-case. For example, one can determine the potential of restoring half of the land in the best-case and assume it would equal around half of the increased deposition. This assumption can be refined using spatial models, determining the areas of restoration that provide the most increased supply of air quality regulation. Another way to develop scenarios is to split up current land cover by land use and calculate the increase of deposition for the conversion of each existing land type such as grasslands, agricultural lands, or barren lands. Then, each project can be analyzed separately or as combinations of the proposed projects.

When considering land use changes, another important aspect to consider is the temporal dynamics. Ecosystems take time to mature and produce the ecosystem services that benefit the environment around them. This is important for both analysis and design approaches. Chapter 19 discusses the seasonal intermittency of nature and approaches to modeling these dynamics and designing TES systems accordingly. Some models, like the i-Tree suite, yield values at different time points (10 years, 20 years, 50 years) after the project is conducted. Other models, like the US Forest Service Forest Vegetation Simulator (FVS), simulate growth functions of forest stands over time. Temporal growth data can either remain tabular at each time point or be used to fit growth functions. For dry deposition of criteria air pollutants, the variable of interest is the increase in tree crown growth, leaf area index, or percentage of crown coverage. For greenhouse gases and carbon sequestration, the focus should be on the change of carbon in the forest stand, either just above ground or considering both above and below the soil. Often, these functions should be normalized between 0 and 1 to simulate the fraction of growth between the current land cover and future LUC scenarios. Figure 18.3 shows example functions for the dynamics of tree crown coverage (dry deposition) and change of carbon (carbon sequestration). Dry deposition rates often increase quickly over time as forest stands mature and then level out as growth stabilizes. On the other hand, carbon typically is emitted by young forests before any sequestration happens, then

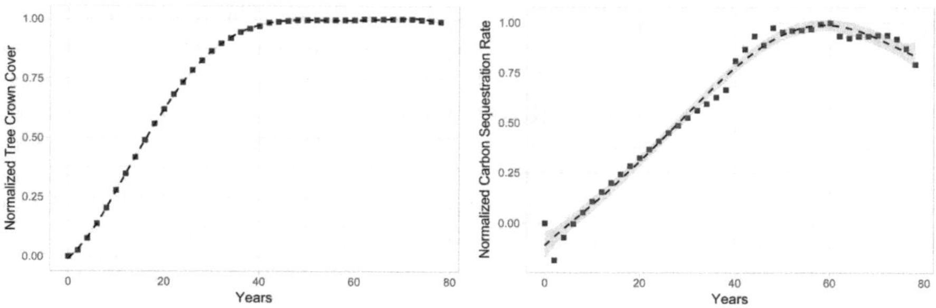

Fig. 18.3 Example growth dynamics of dry deposition rates (left) and carbon sequestration (right) (Adapted with permission from Charles and Bakshi [37])

active sequestration occurs as the trees grow in biomass and also deposit carbon in the soil. But as ecosystem growth begins to stabilize, the change of carbon begins to slow its rate as the carbon value of the forest transitions from active sequestration from the atmosphere toward storage in its biomass.

18.2.3 Applications

In application, several publications have conducted techno-ecological analyses, ranging in scale from a single manufacturing facility to a national assessment. The first that will be presented is the assessment of techno-ecological systems for a single site (biodiesel manufacturing) as compared to solely technological systems. Gopalakrishnan et al. [9] aimed to quantify multiple ecosystem services of a proposed ecological restoration project, focusing on air quality regulation, carbon sequestration, and water quality regulation [9]. The proposed restoration project included a 26-acre forest restoration and a 1.1-acre wetland. The article compared the demand and supply of ecosystem services required to regulate different chemical pollutants in both air and water. It concluded that once the forest restoration matured, the supply of air pollution regulation for NO_2, O_3, PM_{10}, and SO_2 was greater than the demand from the biodiesel process, including utility generation. However, for CO_2 and water quality ecosystem services, the demand was greater than the supply. The publication further explored the cost comparison between adding additional technological pollution control equipment versus the ecosystem restoration projects to achieve net-zero emissions. The techno-ecological approach yielded higher net present value (NPV) and return on investment (ROI) while also providing lower annualized costs at different years of analysis. One key notion made in this work is that technological systems depreciate in value over time, whereas ecological systems appreciate.

Following up on the single-site assessment, Gopalakrishnan et al. also produced two national assessments on the capacities and costs of ecosystems to uptake air pollutants of industrial facilities across the United States. The first compared the supply and demand of air quality regulation on a national scale, analyzing industrial sites across the United States [38]. The article explored the role of vegetation around nearly 20,000 sites from different industrial sectors and geographic regions. The criteria air pollutants NO_2, PM_{10}, and SO_2 were the species in consideration. Further, high-emitting facilities from sectors such as utilities and manufacturing were filtered from the study because the ecological restoration scenarios were deemed economically impractical. This study concluded that within a 500-meter radius, many low-emitting industrial sites either have enough current capacity or can invest in land restoration and forestation projects to mitigate their emissions—especially in the Southeast United States and particularly from industrial sectors of mining, quarrying, oil and gas extraction, transportation and warehousing, and management of companies and enterprises.

The second national assessment compares the economic competitiveness of nature-based solutions as compared to technological alternatives [8]. This assessment compared restoration scenarios for each county, comparing the current and potential deposition of SO_2, PM_{10}, $PM_{2.5}$, and NO_2. Restoration scenarios determine potential deposition, assuming that each county converts grasslands and shrublands to county-average canopy cover until county-wide ecological uptake is equal to industrial emissions or until all available land is reforested. The cost of restoration used county-specific site preparation and decadal management rates from a timber industry study conducted in the early 2000s [39]. For estimating technological costs, the Control Strategy Tool (CoST) database was used along with its associated Best Available Control Technology (BACT) tool. As demonstrated in Fig. 18.4, the results suggest that restoration costs are less than equipment costs for most counties in the United States (more than 75%); however, the

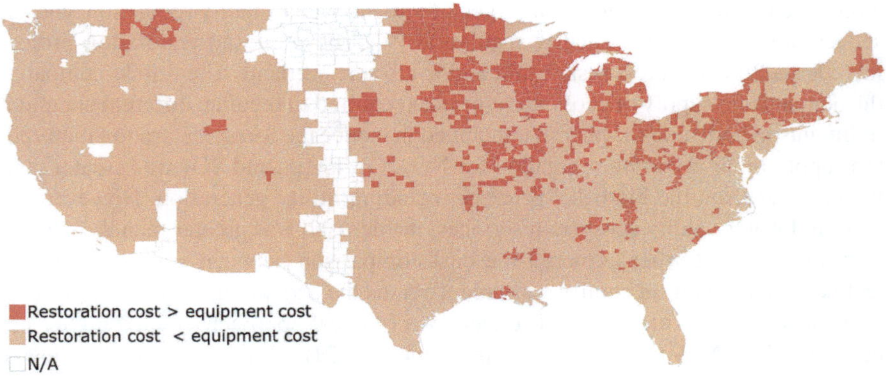

Fig. 18.4 U.S. map which shows the counties where the conversion of grasslands and shrublands to tree canopy cover is cheaper than the installation and operation of air pollution control technology to provide an equivalent amount of pollutant uptake. (Used with permission from Gopalakrishnan et al. [8])

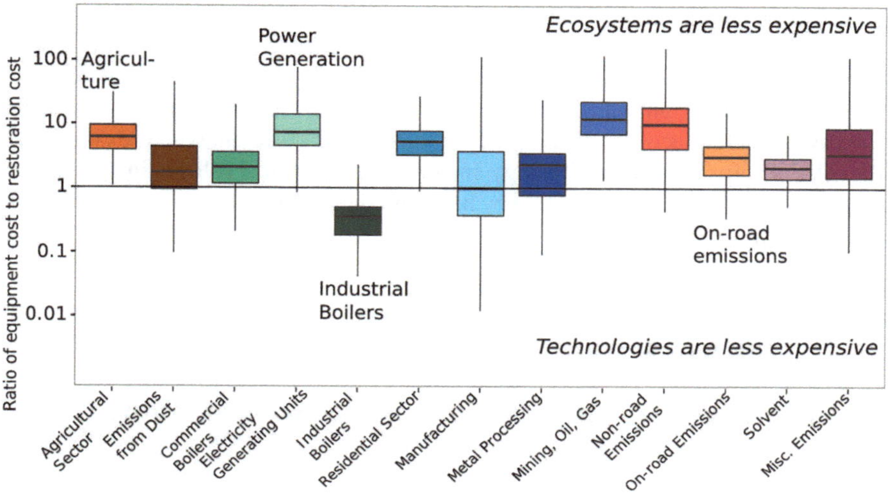

Fig. 18.5 The cost ratio between the annualized equipment cost for a given sector and the restoration cost of land in each county to uptake equal amounts of pollutant emissions. If the ratio is greater than one, restoration is less expensive than the best available pollution control technology for that sector. (Used with permission from Gopalakrishnan et al. [8])

opposite is true in many counties across the North where grasslands and shrublands are less available for restoration. Figure 18.5 shows the results of this study's analysis by industrial sector. Based on the median value of the ratio comparing equipment cost with restoration cost, nature-based solutions were concluded to cost less than technological approaches for all sectors except industrial boilers and manufacturing. This study provided support for techno-ecological design as it can provide environmental and economic improvements in many cases.

The previous assessments have all compared the supply and demand of air quality regulating services; however, all have assumed scales of restoration based on circular buffer zones or political boundaries (county lines). To explore spatially explicit interactions between pollution source and surrounding landscapes, Charles et al. presented mathematical definitions for air pollution servicesheds using atmospheric dispersion models [13]. This provided a framework for including dispersion models within techno-ecological design and generating spatially informed data. Utilizing atmospheric dispersion models enables one to connect the pollution of a particular point source with deposition values of areas surrounding a particular point. These models produce a set of points that contain surface-level concentration and dry deposition value, which can be used to create visual maps. It's these values that are required for applying the mathematical definitions presented in the work of Charles et al. [13]. These definitions refine the set of points from the dispersion results based on particular criteria: some based on concentration values, some on dispersion values, and others that utilize the supply and demand perspective of ecosystem services. The definitions based on supply and demand provide spatial analyses of the air pollution "footprints" of each chemical species. In this exploration of quantitative and reproducible serviceshed definitions, it was concluded that these servicesheds vary for different chemical species and are also dynamic in time, such as seasonal variations. The insights around dynamics, spatial and temporal, are essential for improving techno-ecological design approaches. All these analytical examples of the relationships between technology and design play critical roles in improving the methods of sustainable design, as they highlight the potential of ecological solutions for industry.

18.3 Design and Synthesis

18.3.1 Integrating Air Quality Ecosystem Services Within Sustainable Design

Quantifying ecosystem services is a valuable tool and can be essential for developing environmental policies and land management. However, these valuation methods can also provide valuable insight for sustainable design. Modeling and integrating ecosystem services within frameworks that include current process design methods can promote more holistic system design, as suggested by TES. As we discuss the integration of air quality regulating ecosystem services, we will discuss the importance of system boundaries for design, some basic building blocks of optimization frameworks, and present a few applications particular to air quality regulation. Although other design approaches exist, this chapter will focus on optimization programs as a strategy for sustainable design.

Defining the System Boundary In designing or constructing any system, it is important to define the boundary of the model, or in other words, which unit operations will be considered in designing the system for a particular goal. Ecologically, this could also mean defining a spatial boundary of the land area which is available for restoration, management, or land use change. It also requires determining available ecological solutions for the application and determining a method for valuation. On the technological side, either the full process model can be included or just the equipment units that parallel the ecological functions in focus. Using the full process model typically presents many more decision variables, design equations, and potential complications for the optimization program. Including more design parameters creates more challenges but also opens more opportunities and insights for the design process. On the other hand, including only the downstream air pollution control equipment can simplify the approach and can offer initial insight toward the opportunities of ecological solutions in the desired region. An example of this approach is shown in Fig. 18.6. In this case, three chemical species are considered (PM_{10}, SO_2, and NO_2) and the pollution control equipment and regional ecological uptake are considered one techno-ecological system. The control technology includes a baghouse filter (PM_{10}), a flue gas desulfurization unit (SO_2), and

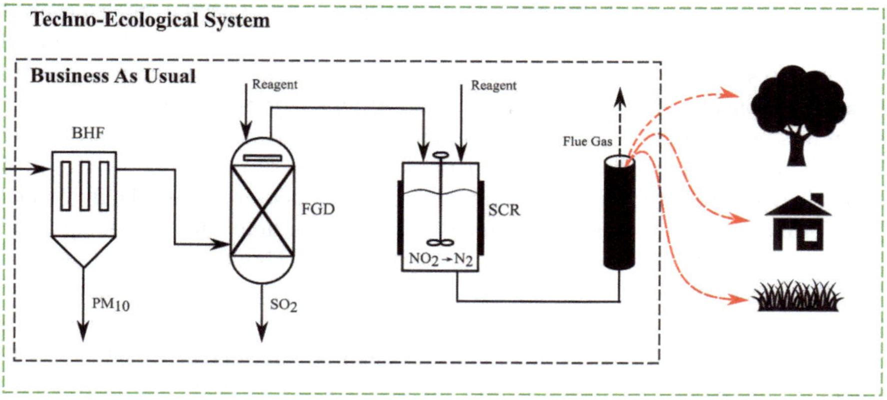

Fig. 18.6 Example system boundaries to compare conventional and techno-ecological design approaches (Used with permission from Charles and Bakshi [37])

a selective catalytic reactor (NO_2). This is a downstream case for a power generation station and omits the design variables involved in the combustion, transmission, or consumption processes.

Building Optimization Frameworks After the design problem has been defined, it is important to build out an approach to solve it. In the engineering disciplines, this typically involves an optimization program. In its simplest form, an optimization program must include an objective function, decision variables, and constraints. The objective function should qualitatively be understood by the time the problem has been defined. Programs can be single-objective or multi-objective depending on the interests of the stakeholders. In sustainable design, often two common objectives are to minimize costs (or maximize NPV) and to minimize emissions. This creates a multi-objective program; however, if an environmental target is known, then it can be written as single-objective. In this case, the cost becomes the objective function while the environmental target is defined as a constraint.

The decision variables define the specific parameters that are allowed to change as a result of the design process. Some example decision variables might be the size of the removal technology, the amount of land changed for increased uptake, the product output of the entire manufacturing site, or specific sites to be restored. These variables can be continuous or discrete and bounded or unbounded. Using discrete variables generally includes mixed-integer or disjunctive programming but is an important tool for including characteristics like toggling equipment use on or off and including spatial variables for ecological solutions. In the program, the decision variables are adjusted until the objective function reaches its maximum or minimum.

Constraints determine the feasible set of solutions to the program. This includes any bounds to the decision variables, physical limitations, economic budgets, and the design equations for the technological and ecological models. Decision variables bounds and economic budgets are fairly straight forward and need to be determined based on the case study being conducted. Physical limitations can include limited space for restoration projects, limited land types available for restoration, or with time steps considered, preventing the program from restoring the same space twice when it is not physically possible. Integrating design equations can be much more difficult and often is the bulk of the program code.

In chemical and process systems engineering, unit operations describe individual steps of the overall process such as reactors, separation units, heaters, pumps, and many other essential steps of an industrial-scale process. Unit design equations exist for all these technologies and are the basis of systems design and optimization. Lots of software exist for modeling processes of unit operations like ChemCAD, Aspen, SuperPro, and many others. Most integrate design and cost equations into the program and these equations can sometimes be found in technical guides or published papers supporting the software. However, these can sometimes be difficult to find due to incremental updates and adjustments that result in multiple documents to sort through. One approach is to build the process (or use a sample if available) and adjust the variables of interest to generate data and construct a surrogate model. If possible, using the source design equations is best practice as they generate less error and uncertainty. Depending on the detail of the design (determined by the system boundary), many of the same tools described in the *Quantifying the Demand* sections can provide the equations needed to implement these constraints.

As the objective functions, decision variables, and constraints are constructed together, an optimization program forms. The programs are typically labeled based on the types of equations in the constraints, linear versus nonlinear, and the type of variables present, continuous or mixed-integer. Most programs are labeled as linear programs (LPs), nonlinear programs (NLPs), mixed-integer linear programs (MILPs), mixed-integer nonlinear programs (MINLPs), or sometimes quadratic programs (QPs), and mixed-integer quadratic programs (MIQPs). The categories help determine best-fitting solvers. Solvers, modeling tools, and programming languages are all constantly evolving, so researching updates and new software is important for complex programs. Common programming languages for optimization include GAMS, Python, Matlab, Mathematica, and Julia. Each of these languages contains modeling tools (with their own syntax) for constructing optimization programs. The modeling tools then use solvers (such as IPOPT, Gurobi, CPLEX, CBC, BARON, etc.) to search for solutions. Compatibility and licensing are also important factors to consider before attempting to construct the optimization program. Although this chapter will not cover this, it is important to understand how to write optimization formulations for communicating concepts and frameworks.

18.3.2 Applications

The design of techno-ecological systems can be difficult due to the dynamics and complexities of the natural world. As more of these dynamics are captured, whether spatial or temporal, TES design becomes increasingly difficult. One particular challenge is capturing the intermittency of nature's regulatory capacity across seasons. This is discussed in detail in Chap. 19, but we will briefly discuss one of these design applications in this chapter as well. In application of TES design for air pollution, Shah et al.'s work [40] provided an approach for including ecosystems as unit operations in process design with inclusion of intermittency and growth dynamics [40]. This approach was applied to a chlor-alkali process, designing the TES system for 20 years of operation with variable production rates, technological utilization, and land area of ecological reforestation. To capture the growth and seasonal dynamics, FVS was utilized in this work and the optimization formulation was reported as a two-stage mixed-integer linear program (MILP). Different sustainability target scenarios were presented, concluding that all cases that included ecological solutions yielded cheaper than the "technology-only" case, which required selective catalytic reduction of all emissions. The concept of adjusting manufacturing production according to the seasonal capacity for air pollution uptake of local ecosystems is a unique design perspective made possible through the TES approach. This work noted an essential conclusion that without appropriate accounting for the benefits from ecosystems and novel policy planning, ecological solutions and TES design may struggle to flourish in industrial application.

Along with temporal dynamics, locality and spatial information is essential when it comes to air quality regulation. Air pollution inherently imposes varying risks from place to place based on the dispersion of pollutants from the source and throughout the atmosphere. One approach to solving spatially explicit TES design is to include atmospheric dispersion models within the design framework. The first framework to include spatial atmospheric dispersion models and air quality regulation of ecosystems within manufacturing site-level design was published by Charles et al. [37]. The proposed framework includes the generation of spatially explicit information and its inclusion within a mixed-integer optimization program. The program focuses on the same system boundary presented in Fig. 18.6, comparing air pollution control technology with the potential of spatially explicit forest restoration. The approach utilized the atmospheric dispersion information but did not integrate the model within the framework. Rather, it generated results using the dispersion program and used tables and mixed-integer variables to store information and toggle different land areas on and off as the solver searched for an optimum solution. The framework accounted for both the spatial heterogeneity of land cover in the region along with temporal dynamics, allowing pollution control technology to adjust its operation according to the long-term growth of the ecological restoration. Figure 18.7 shows results from this approach, applying budget constraints to the ecological initiatives to highlight areas of priority or those planted earliest. Spatial-temporal maps reveal solutions that suggest both *where* and *when* restoration projects should be implemented based on available resources. A given location's priority, in this context, is based on the increase in pollution uptake between the current state and restored state of land. This example shows the potential of integrating complex spatial and temporal dynamics for more informed landscape design and regional planning.

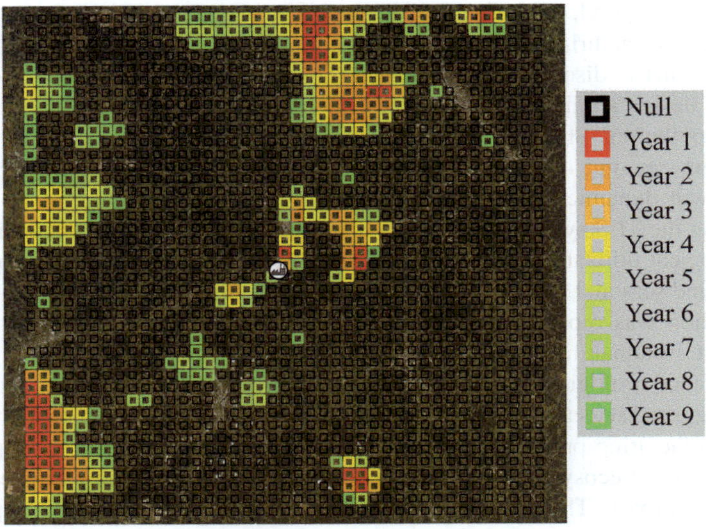

Fig. 18.7 Spatial design accounting for budget constraints and scheduling of ecological restoration projects (Used with permission from Charles and Bakshi [37])

The previous example is based on designing for maximum uptake from neighboring ecosystems; however, the threat of criteria air pollutants is directly related to human health. Including health risk assessments and population density data into spatially explicit sustainable design builds on a holistic approach that includes technology, ecology, and public health. Two common pollutants studied in epidemiological research are ozone and particulate matter (PM2.5 and PM10). These studies relate human exposure to these chemicals with incidences of disease, hospitalization, and fatalities in the form of health impact functions [41]. Population exposure is a key variable in these functions, a variable that is inherently spatially explicit. Tools like BENMAP [33] estimate the economic value of human health effects based on either positive or negative changes in concentration of pollutants like ozone and particulate matter. These economic benefits can be directly calculated for landscape changes and ecological restoration, which enables public health integration into cost objectives or constraints. Preliminary results of this extended design approach reveal that land-based restoration projects should target areas with high population and low vegetation cover [42].

In application to climate change and carbon emissions, Gopalakrishnan et al. [43] explored the TES concept to manage the biosolid pathways for wastewater treatment plans to achieve net-zero CO_2 emissions [43]. The model formulation included a multi-objective optimization problem which minimized costs and maximized net CO_2 supply (supply minus demand) across the entire system. This work quantified the total CO_2 emissions of various biosolid waste disposal methods associated with the anaerobic digestion at the water treatment plant. The disposal methods created different pathways for the biosolids including incineration, landfill, compost, and fertilizer. Including ecological restoration of forests and management variables, like extended timber cycles, provided the ecological supply of carbon sequestration. The work provided a framework for designing TES networks to utilize multiple pathways for biosolid waste management. The results showed that net-zero, and even net-positive, carbon emissions were possible but came at a higher cost than more relaxed emissions targets. The multi-objective approach presented curves that resembled the trade-offs between cost and carbon impacts. Utilizing a feedback system for timber harvests to provide biomass-based incineration for power production opened up the opportunity to improve the curve of trade-offs without the feedback system. The curve that represented the feedback system provided a "win-win" scenario compared to that without, meaning that each point yielded a lower cost and higher net carbon sequestration supply (both objectives considered). All the scenarios presented show different possibilities for applying the TES concept to design systems which are economically competitive and address air pollution. The number of possibilities will continue to grow as new design approaches, techniques, and case studies are explored.

18.4 Conclusions

Air pollution imposes risk upon human populations across the world, yielding disease and fatalities. These health impacts have led to many responses in the form of technology and policy. However, ecosystems are typically overlooked as solutions, even though they are the oldest means of air quality regulation. Throughout the chapter, potential synergies between technological and ecological systems were identified along with methodology for analyzing and designing such synergistic systems. Many tools, like i-Tree or FVS, were identified for valuating ecosystem services, particularly dry deposition of criteria air pollutants and carbon sequestration. If we wish to approach sustainable design from an engineering perspective, the ability to quantify these values creates a common language between ecological and technological systems, mathematics. At the same time, the inclusion of ecosystems and their services incorporates a more holistic perspective for approaching the concept of sustainability. Ecosystems provide complex spatial and temporal dynamics. However, the more we are able to simulate these dynamics in design work, the better we can include these systems as solutions and adjust industrial operations accordingly. Using TES to approach air pollution issues is an opportunity to decrease health risks in regional communities while increasing other co-benefits of nature, like carbon sequestration. Further, we've shown that many studies have concluded that TES approaches often yield "win-win" solutions, where costs also decrease along with environmental impacts.

Some highlights for using ecological solutions to address air quality issues are shown in the results of existing applications as follows:

- Although technology depreciates with time, ecological systems appreciate as they mature and increase their capacity for air quality regulation.
- Ecosystems provide multiple co-benefits simultaneously, whereas paralleling technological solutions are designed for a single purpose.
- Land restoration projects can be applied specifically to "hot-spots" of air pollution to protect local communities.
- In response to climate change, carbon sequestration of any ecological restoration project benefits communities on a global scale.

Although many advantages have been identified, many challenges are also present in the development of TES methodology and the inclusion of nature within engineering approaches. Engineering is based on control, stability, and precision, whereas nature is intermittent, evolutionary, and in many cases, mysterious. In developing TES design, the inability to control and confidently predict nature's inputs to the system provides a substantial challenge. Design and operation must retain flexibility and adapt to nature and the goods and services she produces. Further, with the many spatial and temporal dynamics and complexities, computational expense becomes a concern. The resolution of detail and simulation of complexities must be balanced with the time and ability to solve corresponding problems or programs. Integrating complex ecological or dispersion models within design frameworks that include technological design equations also poses a challenge. Many integrated design approaches will require communication between software, simplification of models, or teams of experts to optimize complex TES systems.

Challenges provide opportunities, especially within academia. The opportunities and potential of a TES approach to sustainable design overwhelmingly outweigh the challenges to be addressed. We must remember that air pollution and the associated health impacts are unintentional consequences of past solutions from the perspective of humankind. Therefore, it makes sense to attempt to extend this perspective to include ecosystems that have observed and adapted to this planet long before our existence. Climate change and air pollution present dangerous risks and we must constantly learn how to appropriately factor in the goods and services of nature into our industries, economies, and societies so that we may sustain all of our needs for survival in this biosphere.

References

1. Health Effects Institute. State of Global Air 2019. *Special Report. Boston, MA: Health Effects Institute* (2019).
2. American Lung Association. State of the Air 2016. *American Lung Association* (2016). arXiv:1011.1669v3.
3. UN Environment 2019. A Review of 20 Years' Air Pollution Control in Beijing. Tech. Rep., United Nations Environment Programme, Nairobi, Kenya (2019). https://wedocs.unep.org/bitstream/handle/20.500.11822/27645/airPolCh_EN.pdf?sequence=1&isAllowed=y.
4. Intergovernmental Panel on Climate Change. Ipcc fifth assessment report—synthesis report (2014).

5. Masson-Delmotte, V., Portner, H. O. & Roberts, D. *IPCC Global warming of 1.5 C* (2018). arXiv:1011.1669v3.
6. Ash, N. *et al. Ecosystems and human well-being: a manual for assessment practitioners* (2010). arXiv:1011.1669v3.
7. Lewis, S. L., Wheeler, C. E., Mitchard, E. T. & Koch, A. Restoring natural forests is the best way to remove atmospheric carbon. *Nature* (2019).
8. Gopalakrishnan, V., Ziv, G., Hirabayashi, S. & Bakshi, B. R. Nature-based solutions can compete with technology for mitigating air emissions across the united states. *Environmental Science & Technology* **53**, 13228–13237 (2019).
9. Gopalakrishnan, V., Bakshi, B. R. & Ziv, G. Assessing the capacity of local ecosystems to meet industrial demand for ecosystem services. *AIChE Journal* **62**, 3319–3333 (2016).
10. Global Footprint Network. World Ecological Footprint by Land Type. https://data.footprintnetwork.org/#/analyzeTrends?type=earth&cn=5001. [Online; accessed April 2022].
11. Global Footprint Network. Footprint Network Open Data Platform. http://data.footprintnetwork.org. [Online; accessed April 2022].
12. Tallis, H. *et al.* Working Paper: "Servicesheds" Enable Mitigation of Development Impacts on Ecosystem Services. Tech. Rep., New York: Mimeo (2012). http://www.naturalcapitalproject.org/decisions/servicesheds.html.
13. Charles, M., Ziv, G., Bohrer, G. & Bakshi, B. R. Connecting air quality regulating ecosystem services with beneficiaries through quantitative serviceshed analysis. *Ecosystem Services* **41**, 101057 (2020).
14. Nowak, D. & Crane, D. The urban forest effects (UFORE) model: quantifying urban forest structure and function. *Integrated Tools Proceedings* (1998).
15. United States Forest Service. UFORE Methods. https://www.itreetools.org/documents/53/UFORE%20Methods.pdf. [Online; accessed April 2022].
16. Gopalakrishnan, V., Hirabayashi, S., Ziv, G. & Bakshi, B. R. Air quality and human health impacts of grasslands and shrublands in the United States. *Atmospheric Environment* (2018).
17. Kimble, J. M. & Follett, R. F. *The potential of US grazing lands to sequester carbon and mitigate the greenhouse effect* (CRC Press, 2000).
18. Kimble, J. M., Lal, R., Birdsey, R. & Heath, L. S. *The potential of US forest soils to sequester carbon and mitigate the greenhouse effect* (CRC Press, 2002).
19. Lal, R., Kimble, J., Follett, R. & Cole, C. *The potential of US cropland to sequester carbon and mitigate the greenhouse effect* (Sleeping Bear Press, 1998).
20. United States Environmental Protection Agency. National Emissions Inventory. https://www.epa.gov/air-emissions-inventories/national-emissions-inventory-nei. [Online; accessed April 2022].
21. United States Environmental Protection Agency. EPA Facility Level Information on Greenhouse Gases Tool. https://ghgdata.epa.gov/ghgp/main.do#. [Online; accessed April 2022].
22. European Commission Joint Research Centre Data Catalogue. Emissions Database for Global Atmospheric Research (EDGAR). https://data.jrc.ec.europa.eu/collection/edgar. [Online; accessed April 2022].
23. European Environment Agency. EEA Greenhouse Gases - Data Viewer. https://www.eea.europa.eu/data-and-maps/data/data-viewers/greenhouse-gases-viewer. [Online; accessed April 2022].
24. European Environment Agency. National Air Pollutant Emissions Data Viewer 1990–2019. https://www.eea.europa.eu/data-and-maps/dashboards/necd-directive-data-viewer-5. [Online; accessed April 2022].
25. Daniel, M., Ravi, S., Paula, M. *et al.* EPA air pollution control cost manual. Tech. Rep., United States Environmental Protection Agency Office of Air Quality Planning … (2002).
26. United States Environmental Protection Agency. Cost Analysis Models/Tools for Air Pollution Regulations - Control Strategy Tool. https://www.epa.gov/economic-and-cost-analysis-air-pollution-regulations/cost-analysis-modelstools-air-pollution. [Online; accessed April 2022].

27. United States Environmental Protection Agency. Interactive Map of Air Quality Monitors. https://www.epa.gov/outdoor-air-quality-data/interactive-map-air-quality-monitors. [Online; accessed April 2022].
28. Multi-Resolution Land Characteristics Consortium. National Land Cover Database. https://www.usgs.gov/centers/eros/science/national-land-cover-database. [Online; accessed April 2022].
29. United States Forest Service. iTree Tools. https://www.itreetools.org/. [Online; accessed April 2022].
30. United States Environmental Protection Agency. Revision to the Guideline on Air Quality Models: Enhancements to the AERMOD Dispersion Modeling System and Incorporation of Approaches to Address Ozone and Fine Particulate Matter. *Federal Register* (2017).
31. American Meteorological Society/Environmental Protection Agency Regulatory Model Improvement Committee. AERMOD Modeling System. https://www.epa.gov/scram/air-quality-dispersion-modeling-preferred-and-recommended-models#aermod. [Online; accessed April 2022].
32. Exponent Engineering and Scientific Consulting. CALPUFF Modeling System. http://www.src.com/. [Online; accessed April 2022].
33. United States Environmental Protection Agency. Environmental Benefits Mapping and Analysis Program. https://www.epa.gov/benmap. [Online; accessed April 2022].
34. Fankhauser, S. & Pearce, D. W. The social costs of greenhouse gas emissions. *OECD, 2, RUE ANDRE PASCAL, 75016 PARIS(FRANCE)*. 2–23 (1993).
35. Heath, L., Nichols, M., Smith, J. & Mills, J. *FORCARB2: an updated version of the US Forest Carbon Budget Model* (2010).
36. United States Forest Service. Tools - Carbon - USDA Forest Service. https://www.nrs.fs.fed.us/carbon/tools/. [Online; accessed April 2022].
37. Charles, M. & Bakshi, B. R. Designing industrial landscapes for mitigating air pollution with spatially-explicit techno-ecological synergy. *AIChE Journal* **67**, e17347 (2021).
38. Gopalakrishnan, V., Ziv, G. & Bakshi, B. R. Role of Vegetation in Mitigating Air Emissions Across Industrial Sites in the US. *ACS Sustainable Chemistry and Engineering* (2019).
39. Bair, L. S., Alig, R. J. *et al. Regional Cost Information for Private Timberland: Conversion and Management* (DIANE Publishing, 2007).
40. Shah, U. & Bakshi, B. R. Accounting for nature's intermittency and growth while mitigating NO2 emissions by technoecological synergistic design—application to a chloralkali process. *Journal of Advanced Manufacturing and Processing* **1**, e10013 (2019).
41. Turner, M. C. *et al.* Long-term ozone exposure and mortality in a large prospective study. *American journal of respiratory and critical care medicine* **193**, 1134–1142 (2016).
42. Charles, M. *Applying Lessons from Nature to Advance Computational Sustainable Design: Designing Industrial Landscapes and Transitions towards Neutrality*. Ph.D. thesis, The Ohio State University (2021).
43. Gopalakrishnan, V., Grubb, G. F. & Bakshi, B. R. Biosolids management with net-zero co 2 emissions: a techno-ecological synergy design. *Clean Technologies and Environmental Policy* **19**, 2099–2111 (2017).

Chapter 19
Designing Dynamic Synergies Between Ecosystems and Manufacturing Processes

Utkarsh Shah and Bhavik R. Bakshi

19.1 Introduction

Since the dawn of the industrial revolution, advancement in technology has led to exponential improvement in quality of life and human well-being. This technological advancement is accompanied by rapid ecological depletion and resource utilization. Critical life-sustaining systems are at risk of disruption or collapse. Unprecedented growth in human well-being is no longer sustainable. Instead, we are facing the effects of climate catastrophe that has resulted in tremendous loss of life, livelihood, and property.

To reverse the climate breakdown, multiple governmental, academic, and industrial stakeholders comprising 70% of the global economy have committed to net-zero by 2050. The net-zero commitment has received significant criticism from environmental advocates and academia. Net-zero targets are criticized to follow a "burn now, pay later" approach. They rely on purchasing future carbon offsets and carbon capture technologies without addressing the lack of current carbon-regulating capacity of ecosystems. Such practices are widespread in other environmental disciplines as well. For example, cap-and-trade schemes for air quality regulation reduced air pollution in the USA on an annual aggregate basis. Despite these improvement, multiple cities in the USA still suffer from air quality

U. Shah
William G. Lowrie Department of Chemical and Biomolecular Engineering, The Ohio State University, Columbus, OH, USA

B. R. Bakshi (✉)
School for Engineering of Matter, Transport and Energy, School of Sustainability, School of Complex Adaptive Systems, Arizona State University, Tempe, AZ, USA

Lowrie Department of Chemical and Biomolecular Engineering, The Ohio State University, Columbus, OH, USA
e-mail: bhavik.bakshi@asu.edu

B. R. Bakshi (ed.), *Engineering and Ecosystems*,
https://doi.org/10.1007/978-3-031-35692-6_19

issues as cap-and-trade schemes fail to account for variable and intermittent nature of ecosystems' capacity.

Accounting for ecosystems' capacity to provide goods and services and respecting it are essential conditions for sustainability. They provide a cost-effective ability to regulate pollutants and can lead to socially, economically, and environmentally win-win solutions. A practical adaptation of ecosystems requires consideration of their inherently variable and intermittent nature. For example, wind- and solar-based renewable energy generators are deemed to be economically cheaper than conventional fossil fuels-based resources. But their intermittent and variable nature makes it difficult to integrate with the current power grid. Hourly variations in meteorological conditions significantly affect ambient air quality and ecosystems' ability to regulate it, yet air pollution policy fails to account for variability. Hence, accounting for ecosystems would require novel algorithms that can transform static technological operations into nature adaptive operations.

Homeostasis is the ability of a system to regulate its operating condition around a set point under external perturbance [1]. A familiar example of homeostasis is regulation of human body temperature, heart rate, and blood pressure. Regulating the room temperature or reactor temperature within a narrow range around "optimal" set points using feedback loops is a typical example of homeostasis in human-engineered systems. Quality, defined as the inverse of variability, is a core tenet of engineering design and operation. Typically, engineering systems would be designed to operate around an optimal set point with a regulatory feedback controller to minimize the effect of perturbances. An ensemble of homeostatic engineering system fails to emulate homeostasis of its component. Often, these complex systems are difficult to model around a set point or common objective as their nonlinear interactions lead to multiple stable operating points or lack of any stable operating points. Such complex systems are often characterized by operational regimes, that is, set of operational dynamics rather than a single operating point.

Conrad Waddington suggested the term homeorhesis to characterize the ability of a system to maintain specific dynamics as opposed to a certain operating point under external perturbance [2]. Ambient temperature and river water level are typical examples of homeorhetic systems that do not converge to any particular set point but rather stay within a "normal" range of operation. Typically, a complex system with multiple component homeostatic systems, if stable, is characterized by homeorhesis. Unlike homeostatic system, the homeorhetic systems do not have a predefined goal or target. Rather these systems are characterized by existence of wide operation range and regulating transition between multiple states [3]. The "best" operating point is a matter of conjecture not an inherent property of system.

Technological systems are designed and operated to maximize quality, defined as an inverse of variability [4]. Human aversion to variability increases the desirability of homeostatic operations. Humans prefer to operate nature in a desirable "optimum" mode of operation. Construction of dams on rivers to regulate the flow of water and facilitate shipping is a textbook example of human imposition of homeostasis on homeorhetic systems. Early national park management policy was to maintain "good" animal population by controlling "bad" animals, imposing homeostasis on ecological dynamics [5]. Often, implementation of homeostatic policy on inherently homeorhetic system leads to loss of functionality and resilience. Thankfully, the homeostatic policies like predator population control were abandoned in favor of policies like habitat protection and natural succession. The aggressive regulation of room temperature to a "comfortable" set point using Heating, Ventilation, and Air Conditioning (HVAC) is another example of homeostatic imposition on natural homeorhesis. The implementation of homeostasis on homeorhetic systems requires sophisticated technology and energy-intensive operations leading to unintended consequences.

In nature, synergies exist among homeostatic organisms and homeorhetic ecosystems. Avian migration is a commonplace example of adapting to seasonality and food supply. Nonmigratory species have developed a mechanism of torpor or hibernation. Several species of polar fishes enter a state of hypometabolism and reduced activity to survive the winters [6]. These species use antifreezing glycoproteins to avoid freezing of their bodily fluids. Alternatively, bears gain fat and body weight during summer and hibernate during cold winters, surviving on body fat accumulated previously [7]. Social species like ants forage for food and build a special shelter during summer, then wait out the winter using their stores while closing off the entrance to their mounds to conserve body heat. Biological homeostatic systems exist in synergy with their homeorhetic surrounding by nature (i) migrating out of their environment, (ii) manipulating their demand from ecosystems, (iii) creating reserves of energy, or (iv) exhibiting a mix of all the above behaviors. On the other hand, anthropogenic systems often manipulate their environment by means of intensive materials and energy use that is often unsustainable to maintain their homeostatic lifestyle.

Formalization of Techno-Ecological Synergy (TES) would encourage further investment in homeorhetic ecosystems and its integration with technological homeostatic systems. For achieving true synergy and avoiding unintended consequences, a conscious engineering effort is required to minimize the imposition of homeostasis on homeorhetic systems. Engineering systems need to emulate the intermittency and dormancy of natural homeostatic systems like trees and animal during winter months to synergize with homeorhetic ambient temperatures. For example, integration of solar and wind energy into the electric grid would require both supply-side and demand-side efforts to synergistically use excess energy during summer month and reduce the demand of energy during winter months.

This chapter describes the approach and application that establish dynamic synergies between technological and ecological systems. Instead of the conventional engineering approach of ignoring nature's variation or attempting to impose homeostasis, the TES approach adapts engineered systems to nature's homeorhesis. The resulting dynamic synergy is capable of meeting human needs in a manner that can be economically attractive to the corporation while minimizing damage to human health and the environment. It can also encourage nature-positive engineering decisions. The type of variability of ecosystem services addressed in this work is conveyed in Fig. 19.1. As shown in Fig. 19.1c, the ambient temperature varies

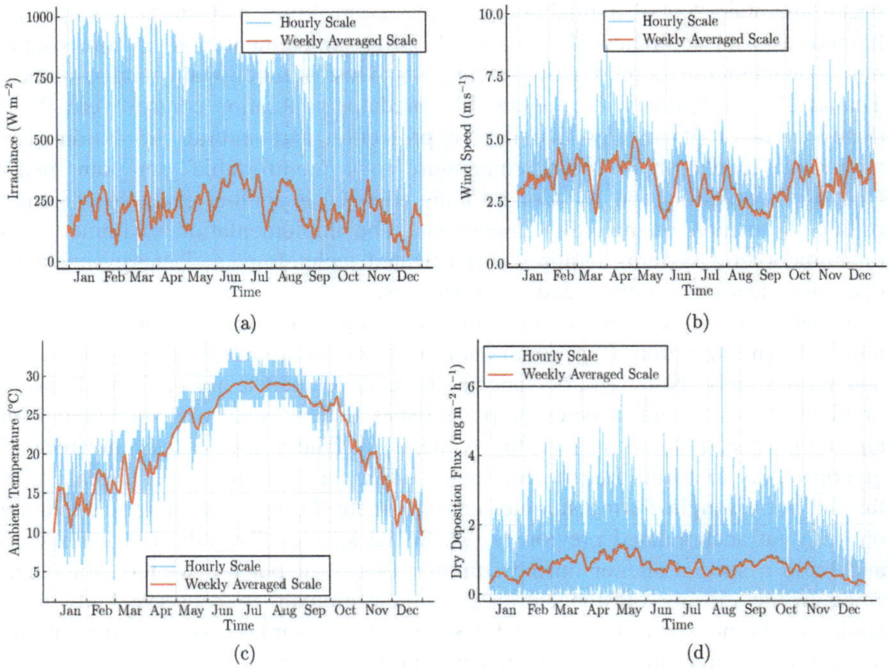

Fig. 19.1 Intermittency and variability in different ecological variables at Freeport, Texas (2009). Ecological variables do not follow a fixed set point but vary within a typical range as shown over the span of a year. This demonstrates homeorhesis. (**a**) Direct normal irradiance (DNI). (**b**) Wind speed. (**c**) Ambient temperature. (**d**) Ozone dry deposition flux for White Ash

seasonally and daily. Figures 19.1a and b depict the uncertain, variable, and intermittent nature of the availability of renewable resources. In the rest of this chapter, the innovations in grid management and effects of renewable energy penetration in energy grid are discussed in Sect. 19.2. Section 19.3 presents a case study on design and operation of techno-ecological synergistic system. The operation of chlorine manufacturing facility is varied in synergy with capacity of trees to aid ozone abatement depicted in Fig. 19.1d.

19.2 Dynamic Synergy Across Energy Grid

The high correlation between energy demand and ambient temperature [8] results in a homeorhetic energy consumption pattern. The hourly energy demand profile has a diurnal, weekly, and monthly seasonality. The conventional approach for meeting the homeorhetic loads has been to make enormous capital investment in energy generators with the objective of meeting ever-increasing peak energy demands by techniques such as supply-side management [9]. The fuel crisis of the 1970s and rising awareness about environmental impacts of fossil fuel-based energy generation forced the industry to consider alternatives to better manage the demand of electricity.

In the beginning, the emphasis was primarily on increasing energy efficiency. Under demand-side management programs, several utility companies offered their customers loans and rebates for reducing their demand by improving energy efficiency of their operation. Soon as energy efficiency of systems saturated, the marginal cost of reducing demand surpassed that of increasing supply. In order to improve synergy between homeorhetic supply and homeostatic demand, the utilities started deciding which load to shed. They introduced time-of-use rate where different time of day corresponds to different energy price. Time-of-use rates catalyzed energy storage service industry, as now they could store energy at cheaper rates and sell at a premium when rates were high. Real-time pricing, a sophisticated form of time-of-use, resulted in consumers shifting their energy use to off-demand periods and thus flattening the energy curve. In changing consumer demand behavior and utilization of storage, the electricity market achieved a synergy between homeorhetic consumer demand and homeostatic supply generation to reduce energy consumption and improve the profitability of the overall system. Employing real-time pricing also improved air quality around power plants as it effectively reduced peak nitrogen dioxide (NO_2) and sulfur dioxide (SO_2) emissions [10]. The global shift away from fossil fuels to reduce greenhouse gas emissions and gain energy security has incentivized further investments in renewable energy. Penetration of variable and uncertain renewable wind and solar generators in the supply mix disrupted the synergy.

The capacity of renewable electricity generators depends on homeorhetic natural systems like solar irradiance and wind velocities. Solar panels have an intermittent nature of power generation leading to reduced capacities during night. On the other hand, wind turbines have uncertain and variable production throughout the day with a fall in productivity during the night. In order to offset the uncertainty from renewable generation, fast flexible power generator using crude oil or natural gas and energy storage facilities are required to meet a constant homeostatic power load on a daily timescale. With an increase in penetration of renewables (beyond 60%), a need for seasonal storage arises. Renewables like solar and wind operate at peak capacities during spring and at lower than rated capacities during other time of year. Compressed Air Energy Storage, pumped hydro-storage, and compressed hydrogen storage are few of the options analyzed in the literature for their techno-economic feasibility [11]. Guerra et al. [12] showed that hydrogen storage systems can be cost-effective at weekly timescales at higher renewable penetration levels. A study by Demirhan et al. [13] demonstrated ammonia and methanol as storage species to be more economically attractive than hydrogen when compared at a high renewable penetration level and on a seasonal scale. To enable synergy between uncertain electric supply system and variable electric demand system, additional storage technologies need to be developed. The demand-side management literature needs to evolve beyond peak flattening and weekly real-time pricing to monthly and seasonal variations [14]. Chemical storage forms like methanol and ammonia will play a critical role in achieving this synergy as they can also be sold as value-added products. Expansion of energy storage technologies would lead to extension of energy demand synchronization problem to a mass integration problem. These are examples of integration between the needs of society with the homeorhetic behavior of renewable energy sources.

19.3 Dynamic Synergy for Air Pollution Abatement

To develop an approach for dynamic TES, we describe a mass balance problem, where dynamics of air pollution emission from a production facility are synchronized with the dynamic and intermittent air pollution absorption capacity of ecosystems [15]. Consider an existing chlorine production facility with an in-house coal-fired utility generator to satisfy its energy requirement. The energy requirement of chlor-alkali production is modeled as a piecewise linear function of chlorine production rate as carried out in Bree et al. [16]. The annual average production target of this chlor-alkali facility is 1 ton/h of liquified chlorine. If the annual average target is met, the market can buy any amount of chorine produced by the facility. The monthly production rate can vary from 0.5 to 1.3 ton/h. Burning of coal in the facility would lead to emission of nitrogen dioxide (NO_2), sulfur dioxide (SO_2), carbon dioxide (CO_2), and particulate matter (PM2.5) [17]. The facility is located in Freeport, Texas, which is part of the Houston-Galveston-Brazoria (HGB) ozone (O_3) nonattainment zone. In order to avoid the worsening of ozone levels in the region, this study explores techno-economic feasibility of TES design and operation for ozone precursor NO_2 abatement while considering "sustainability" of the ozone regulation ecosystem service.

Typically, NO_2 abatement technology can be partitioned into two categories: (i) precombustion methods like Low NOx Burner (LNB), Flue Gas Recirculation (FGR), and steam injection, and (ii) post-combustion methods like Selective Catalytic Reduction (SCR), Selective Non-Catalytic Reduction (SNCR), Electron Beam (EB), and Non-Thermal Plasma (NTP) [18]. To avoid design modification of the existing facility, in this example we consider only the post-combustion method of SCR for NO_2 abatement. The SCR unit is designed according to the procedure described in the United States Environmental Protection Agency (U.S. EPA) Air Pollution Control Cost Manual. The unit is designed with NO_2 efficiency of $\approx$ 100%. The concave capital cost and operating cost of an SCR unit are remodeled as a piecewise linear function of its size and monthly usage, respectively [19]. The SCR aims to reduce the emissions of NO_2, thus, reducing the demand for the air quality regulation service to be provided by ecosystems.

Alternatively, the supply capacity of ecosystems can be enhanced by investing in ecological restoration techniques like reforestation. In this study, we consider five native species for reforestation: (i) hardwood (fast growing, evergreen) species like American Elm and Southern Magnolia, and (ii) softwood (slow growing) species like Live Oak, Red Maple, and White Ash. These trees are considered to be planted as saplings and their natural growth dynamics through the project life span of 20 years is modeled using United States Department of Agriculture's (USDA) Forest

Vegetation Simulator (FVS) Southern (SN) variant [20]. This approximates the growth of a diverse forest with native tree species. At the start of the project, barren land with no capacity for air quality regulation is assumed to be available for reforestation. A reforestation density of 1000 saplings per acre is simulated in the Sam Houston National Forest with no human intervention and default parameter values for each species. Key structural properties like survival density, diameter (DBH), height, and crown ratio for trees over the project's lifespan are obtained from FVS simulation. The structural data are then used in iTree Eco v6 [21] to calculate the capacity of trees to absorb NO_2 and O_3. iTree provides hourly dry deposition flux for air pollution removal using hourly meteorological data from the National Climate Data Center (NCDC) and air pollutant concentration data from monitors located at Clover Field Airport. Hourly data are aggregated over a span of a month, and the iTree calculations is repeated for each set of structural data from FVS simulation, thus obtaining monthly dry deposition flux over a span of 240 months.

An overview of the mathematical model for TES design and operation optimization is presented in Model 19.1 below. The objective of minimization of total annualized production cost over the project life is described in Eq. 19.1a. The decision variables are the sizes of utility generator and SCR, reforestation area, hourly production rate of chlorine, and utilization rate of SCR. Equation 19.1b relates the electricity requirement to chlorine production rate using a piecewise linear function f. Equations 19.1c and 19.1d ensure fulfillment of power requirement and technological abatement need for a selected size of utility generator and SCR respectively. Equation 19.1f calculates the demand for NO_2 regulation ecosystem service, that is, NO_2 emission to the environment. Equation 19.1e determines the supply capacity of reforested area, that is, NO_2 absorption capacity of trees at each time period. Equation 19.1g calculates a sustainability metric $V_{NO_2,nm}$ as a difference in supply and demand of NO_2 regulation ecosystem service normalized by the demand. $V_{NO_2,nm}$ by definition is bounded between -1 and ∞. Negative V_{NO_2} indicates an excess of demand as compared to supply and ecological unsustainability. Since the case study is exploring ecologically sustainable design and operation, $V_{NO_2,nm}$ is constrained to be nonnegative for all production periods.

$$\min_{\substack{F_{nm}^{Cl_2}, F_{nm}^{SCR}, \\ Z^{SCR}, Z^{Util}, \\ A_t^{land}}} \quad C^{SCR} Z^{SCR} + C^{Util} Z^{Util} + \sum_{nm} \frac{O^{SCR} F_{nm}^{SCR}}{r^n} + \sum_t C_t^{plant} A_t^{land} \tag{19.1a}$$

$$s.t. P_{nm} = f(F_{nm}^{Cl_2}) \qquad \forall n, m \tag{19.1b}$$

$$Z^{Util} \geq P_{nm} \qquad \forall n, m \tag{19.1c}$$

$$Z^{SCR} \geq F_{nm}^{SCR} \qquad \forall n, m \tag{19.1d}$$

$$S_{NO_2^{eq},nm} = \sum_t J_{nmt} A_t^{land} \quad \forall\, n, m \tag{19.1e}$$

$$D_{NO_2^{eq},nm} = \eta P_{nm} - F_{nm}^{SCR} \quad \forall\, n, m \tag{19.1f}$$

$$V_{NO_2^{eq},nm} = \frac{S_{NO_2^{eq},nm} - D_{NO_2^{eq},nm}}{D_{NO_2^{eq},nm}} \geq 0 \quad \forall\, n, m \tag{19.1g}$$

$$X_{SCR,nm} = \frac{F_{nm}^{SCR}}{F_{nm}^{SCR} + D_{NO_2^{eq},nm}} \quad \forall\, n, m \tag{19.1h}$$

where,

$F_{nm}^{Cl_2}$	Production rate of chlorine for year n and month m
Z^{SCR} & Z^{Util}	Size of SCR unit and utility generator respectively
C^{SCR} & C^{Util}	Cost of SCR unit and utility generator respectively
O^{SCR}	Operating cost of SCR
C^{plant}	Cost of reforestation per unit area
$D_{NO_2^{eq},nm}$ & $S_{NO_2^{eq},nm}$	Demand and supply of NO2 regulation service.
P_{nm}	Power requirement for n year and m month
J_{nmt}	Dry deposition flux for n year, m month, and t tree species
η	NO_2 emission per unit power consumed
$X_{SCR,nm}$	Fraction of NO_2 treated by SCR

The dependence of dry deposition flux on local meteorology and pollutant concentration leads to inherent uncertainty in its calculation. In order to account for this uncertainty, we use meteorological and air pollutant concentration data from 2005 to 2015 to calculate monthly dry deposition flux over the span of 240 months. For each simulation, we hold the meteorological data year as constant. To simulate the uncertainty, we generate 100 equally likely random scenarios of dry deposition flux for 240 months by bootstrapping the available data for 11 years. The methods to calculate varying capacity of ecosystems are summarized in Fig. 19.2.

The objective of this study is to minimize total annualized cost of production of chlorine while ensuring that the demand for ecosystem service of NO_2 regulation is within the bounds of its supply and an annual average production target of chlorine at 1 ton/hr is met. A two-stage stochastic Mixed-Integer Linear Programming (MILP) model is solved where the first stage variables are size of the SCR and utility generator, reforestation area and species, while the second stage variables are chlorine production capacity and SCR utilization rate. First, we consider the baseline case of homeostatic operation using only the technological unit of SCR to treat all the NO_2 emissions. Using this approach, the total annualized cost of production over 20 years is USD 18.9 million and the SCR catalyst volume is

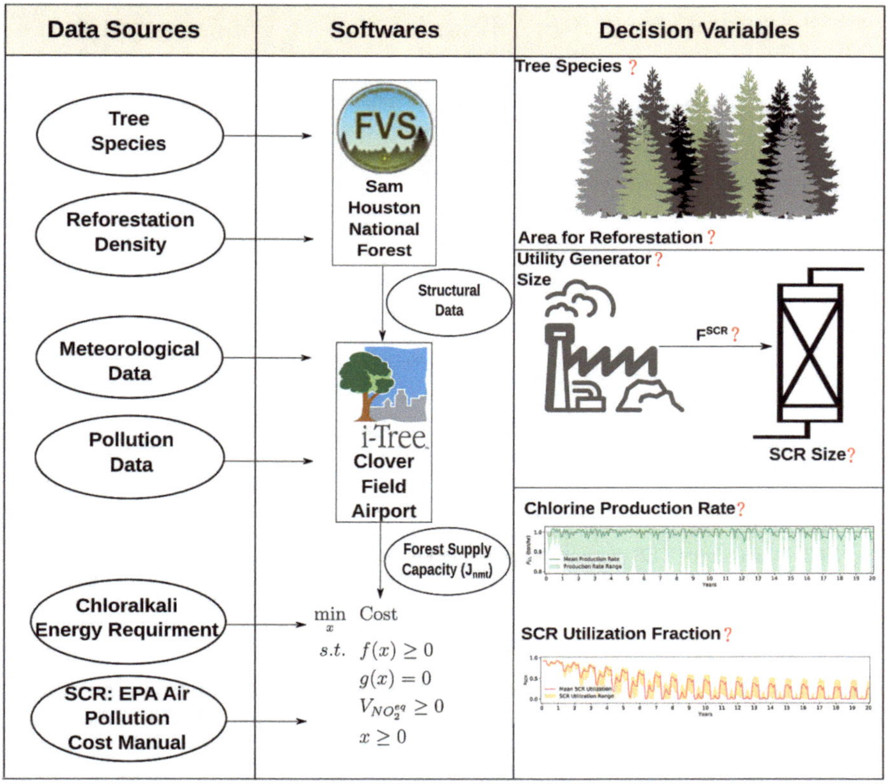

Fig. 19.2 Summary of methods used in the NO_2 abatement case study. (Reproduced with permission from [15])

5.2 m^3. Next, we study the TES scenario where investments in both technology and ecosystems are allowed. The homeostasis requirement on chlorine production rate is relaxed as long as the total annual production target is met. Results from the TES scenario are described in Fig. 19.3. The TES solution requires SCR catalyst volume of 4 m^3 and a land area of 32 km^2. The total annualized cost of this TES approach is USD 18 million, thus providing a cheaper solution as compared to the homeostatic technology-only solution. The optimal solution selects evergreen species of Southern Magnolia for reforestation and requires 32 km^2 area of reforestation.

Figure 19.3a shows the intermittency in chlorine production rate. During the leaf-on period of March to Oct, the supply capacity of trees is higher compared to the leaf-off period. This pattern of intermittent supply capacity is reflected in the low chlorine production rate of leaf-off and high production rate of leaf-on period. The utilization fraction of SCR is plotted in Fig. 19.3b. The growth of trees over time results in an increased ecosystem capacity and reduced reliance on expensive tech-

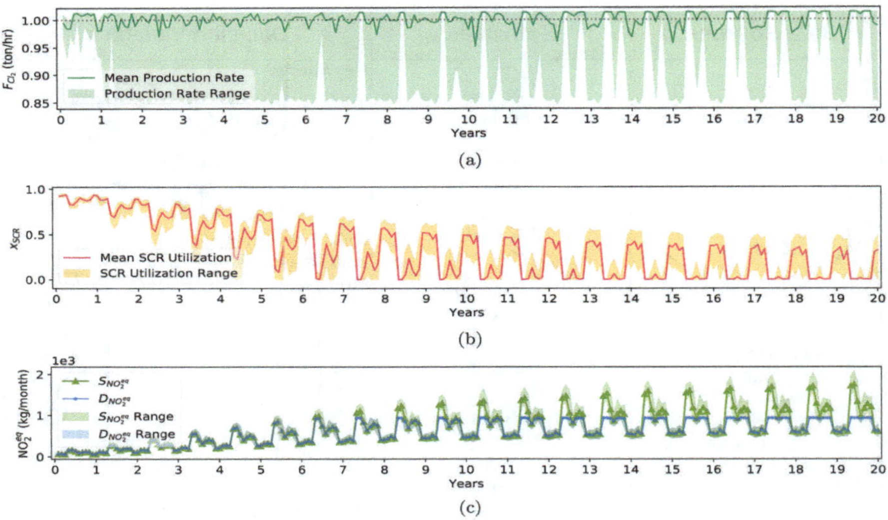

Fig. 19.3 Behavior of TES system. (**a**) Monthly production flow rate of Cl_2. (**b**) SCR utilization rate. (**c**) Supply and demand of air quality regulation service for NO_2^{eq}. (Reproduced with permission from [15])

nological system. For mature ecosystems, the SCR unit is used intermittently during winter to match low capacity of trees to absorb NO_2. Figure 19.3c demonstrates the gap between supply and demand of air quality regulation service. During the nascent years of plantation, the demand and supply are closely matched. As the trees mature over time, the supply capacity of trees surpasses its demand providing excess capacity and reduction in ambient concentration.

The total annualized cost of NO_2 abatement using SCR and reforestation strategy is $\$2400/t_{NO_2}$ and $\$500/t_{NO_2}$, respectively, over the course of project life. It should be noted that while the per ton operating cost of abatement for the technological option stays constant over the project life, the same doesn't hold true for reforestation strategy. Reforestation strategy cost depends on the age of trees in the reforested area and reduces over time as the forest matures. Vegetation also provides many cobenefits such as SO_2 and $PM_{2.5}$ abatement. Reforestation provides a habitat for wildlife and sequesters greenhouse gas emissions as well. Trees can help with storm water regulation, water quality regulation, and preventing water runoff. These cobenefits further increase the attractiveness of a TES solution.

Compared to conventional techno-centric solutions, TES solutions are economically superior. Techno-centric solutions often ignore their harmful effect on society. For example, when $V_k \leq 0$ the air quality in the vicinity of the production facility degrades and causes harmful health effect on society. Next, we present a study by Shah and Bakshi [22] that studies a trade-off between economic and societal cost of manufacturing and potential of TES to provide net-positive manufacturing solution.

19.4 Net-Positive Manufacturing Through TES

The ongoing ecological degradation has largely been caused by industrial revolution and relentless resource utilization. Increased scrutiny from stakeholders like customers, employees, and governments have highlighted net-negative societal impacts from current industrial practices. For sectors like power generation, refining, and agriculture, the harmful effects of their air pollution far exceed the benefits they provide to society, as described in Chap. 1. Such findings have increased pressure on industry to proactively manage and mitigate externalities from their activities. Multiple industries have concentrated their efforts to achieve net-zero or even net-positive impact process, where the benefits of an activity to society neutralize its harms or even exceed it. Net-zero greenhouse gas-emitting systems [23] or products that exceed impacts at micro- or macroscales [24, 25, 26, 27] are some examples of such effort. Using technology at low concentration of pollutants makes the goal of net-zero expensive, thus creating a trade-off between their economic cost and societal cost.

As demonstrated by the case study in Sect. 19.3, ecosystems can be an economically superior alternative to techno-centric solution for pollution abatement. Air quality regulation capacity of vegetation (Fig. 19.1d) depends on many intermittent and variable factors such as wind (Fig. 19.1b), temperature (Fig. 19.1c), solar irradiance (Fig. 19.1a), etc. The intermittency is present across timescale of minutes to years to decades. The case study in Sect. 19.3 accounted for months and decades scale. At shorter timescales, increased variability in meteorological characteristics can be observed in Fig. 19.1. Such short-term variability can lead to temporal hotspots of bad air quality when ecosystems capacity decreases at constant rate of pollutant emission. Ozone alerts in many metropolitan areas across the USA are examples of such temporal hotspots. Accounting for such short-term variability is necessary to prevent temporal hotspots when relying on nature-based solutions for pollution abatement.

We present a case study for design and operation of chlor-alkali manufacturing process that uses surrounding vegetation to mitigate ground-level ozone. With emphasis on the ability of techno-ecological synergy to improve air quality, cause less societal damage, and increase corporate profits. We demonstrate the ability of nature-based solutions to unlock net-positive manufacturing as compared to techno-centric approach whose best ability is to provide net-zero impact solutions. Exploiting the synergy between nature-based solution and technological systems leads to economically and socially win-win solution.

The case study presented in this section focuses on short-term hourly variability and intermittency in ecosystem capacity over a year as opposed to growth dynamics of nature-based solution over decades. This implies that chlorine production rate and SCR (if present) utilization rate change on hourly frequency instead of monthly frequency. On shorter timescales, the ambient concentration of air pollutants needs to be regulated. Here, we deviate from ecological constraint of Eq. 19.1g and use constraints based on hourly air quality.

The case study is modeled as functions of various decision variables to result in a multi-objective optimization problem that can be summarized as,

$$\min_{\substack{\text{Reforestation Area } (A^r) \\ \text{SCR Size } (S^{SCR}) \\ \text{Cl}_2 \text{ Production Rate } (F^{\text{Cl}_2})}} \quad \text{Cost of Production } (Z^{TAC})$$

$$\&$$

$$\min_{\substack{\text{Reforestation Area } (A^r) \\ \text{SCR Size } (S^{SCR}) \\ \text{Cl}_2 \text{ Production Rate } (F^{\text{Cl}_2})}} \quad \text{Health Impact Cost } (Z^{\text{Health}}) \tag{19.2}$$

$$\text{Chlorine production model and its power requirement} \quad P_t = f_1(F_t^{\text{Cl}_2}) \quad \forall\, t \in \mathbb{T} \tag{19.3}$$

$$\text{Supply-demand accounting for NO}_2 \text{ and O}_3$$
$$\begin{cases} D_{it}^E = f_2^i(P_t, S^{SCR}) & \forall\, t \in \mathbb{T},\ i = \{\text{NO}_2, \text{O}_3\} \\ S_{it}^E = f_3^i(A^r) & \forall\, t \in \mathbb{T},\ i = \{\text{NO}_2, \text{O}_3\} \end{cases} \tag{19.4}$$

$$\text{Air quality constraints} \quad C_{it}^f = f_4(C_{it}^0, D_{it}^E, S_{it}^E) \quad \forall\, t \in \mathbb{T},\ i = \{\text{NO}_2, \text{O}_3\} \tag{19.5}$$

$$\text{Economic and Health Impact Cost Calculations}$$
$$\begin{cases} Z^{TAC} = \sum_{t\in\mathbb{T}} f_5(F_t^{\text{Cl}_2}, A^r, S^{SCR}) \\ Z^{\text{Health}} = \sum_{t\in\mathbb{T}} f_6(C_{it}^f, C_{it}^0) \end{cases} \tag{19.6}$$

where A^r and S^{SCR} are first-stage decisions of reforestation and size of SCR, respectively. $t \in \mathbb{T} = 1, 2, \ldots, 8760$ represents each hour of a year for which operational decisions are to be optimized. P_t is hourly power requirement corresponding to hourly chlorine production rate $F_t^{\text{Cl}_2}$. The demand and supply of air quality regulation service for pollutants NO_2 ($i = 1$) and O_3 ($i = 2$) are represented by D_{it}^E and S_{it}^E respectively. C_{it}^0 and C_{it}^f are baseline and final (post-emission) ambient concentration of air pollutant i at hour t respectively. A complete mathematical formulation of the optimization problem is available in Shah and Bakshi [22].

The implementation assumes quasi-steady operation of chlor-alkali facility. For each hour t, the production rate is set constant, and electrochemical cell temperature, concentration, pressure, etc., are set in optimal configuration to minimize power requirement. Quasi-steady-state power requirements are derived using the cell model developed by Otashu and Baldea [28]. To improve tractability of large-scale multi-objective optimization, the nonlinear function mapping chlorine production rate to its power requirement is approximated as a piecewise linear function. Hourly coal combustion rate and pollutant emission rate are calculated from power requirement after accounting for generator's efficiency. Chlorine production rate changes are constrained using ramping rates to prevent impractical instantaneous change of production rate. A typical chlor-alkali facility takes 4 h to safely ramp up from minimum production rate to maximum production rate; the implementation allows for same. It also allows for production shutdown for a period of 24 h, which would result in zero emission rate and chlorine production. Although production shutdowns are permitted, annual production targets need to be met.

The study assumes that coal-fired generator is the sole source of NO_2 emissions in the area. SCR (if installed) can treat part of NO_2 emission and rest are emitted to the surrounding. In this study, the NO_2 emissions can alter the ambient concentration of NO_2 and O_3 in the vicinity of the plant. These emissions are attributed as demand for ecosystem services. iTree Eco v6 is employed to calculate the supply capacity of ecosystem for regulation of NO_2 and O_3 as,

$$S_{it}^{E} = v_{it}^{d} C_{it}^{f} A^{r} \quad \forall\, t \in \mathbb{T},\ i \in \mathbb{I} \tag{19.7}$$

where v_{it}^{d} is dry deposition velocity. C_{it}^{f} is ambient pollutant concentration. A^{r} is reforested area. This assumes an availability of 15 km^2 barren area for reforestation. The base case is assumed to have negligible (zero) ecological supply capacity. Only a single native species of 20-year-old White Ash is considered to calculate approximate capacity of restored forest ecosystem. Reforestation cost is set at \$75 km^{-2} [29, 30].

Real-time air quality is a function of ambient concentration of various air pollutants like SO_2, NO_2, O_3, etc. U.S. Environmental Protection Agency (EPA) translates these pollutant concentrations into a single index using Air Quality Index (AQI) [31]. AQI is used to communicate complex air quality information to lay public into a single index that ranges from 0 to 500 with zero being best. The case study uses retrospective ambient pollutant concentration data for year 2009 as baseline AQI. The post-emission AQI was constrained based on two scenarios: (i) If the baseline AQI is less than 50 (i.e., in "good" range), then post-emission deterioration of AQI by 10 is allowed or until the AQI stays in the "good" range. (ii) If the baseline AQI is beyond 50, then no deterioration in AQI is allowed. This prevents aggravation of bad (non-"good") air quality days.

The post-emission ambient concentration (C_{it}^{f}) of air pollutants is calculated as

$$C_{it}^{0} A^{s} H_{t} + D_{it}^{E} - S_{it}^{E} = C_{it}^{f} A^{s} H_{t} \quad \forall\, t \in \mathbb{T},\ i \in \mathbb{I} \tag{19.8}$$

Eq. 19.8 assumes an ideal continuously stirred tank reactor (CSTR) model for calculating post-emission ambient concentration. C_{it}^{0} is baseline ambient pollutant concentration. The control volume for CSTR is a product of mixing height, H_t and airshed area, A^s. Airshed refers to the area of impact due to chlor-alkali emissions. Typically, airshed is a spatiotemporally varying area. For the sake of simplicity, the study assumes a spatiotemporally invariant area of 15 km^2.

The change in air pollutant concentration due to emissions is used to calculate health impact and its monetary valuation using the U.S. EPA Environmental Benefits Mapping and Analysis Program–Community Edition (BenMAP-CE) model [32]. The study adapts methods described in Nowak et al. [33] and Chap. 10 to derive metrics from base case versus operational case. For each day, the change in concentration metric is calculated and translated into daily societal health cost. Finally, societal health cost is aggregated across a year to determine annualized health impact cost and minimize it.

To compare technological and ecological choices for minimal annualized cost of production, the study considers two scenarios:

- *Technocentric Scenario:* In this scenario, only investment in SCR is allowed along with variation in chlorine production rate and SCR utilization rate. The resulting optimal hourly operation is depicted in Fig. 19.4a. The optimal solution determines no investment in SCR and costs \$923 k with annualized societal cost of \$1.4 m. Due to lack of investment in ecological system, the optimal solution has no ecological supply capacity (S_{it}^E) and needs production shutdown. As ground-level ozone formation occurs in summer months of March–Sep, the production facility faces multiple shutdown to avoid aggravating non-"good" AQI during those months. To offset production loss due to shutdowns, the plant needs to operate at high production load through the rest of the year.
- *Techno-ecological Scenario:* In this scenario, investment is allowed in both reforestation and SCR along with variation in chlorine production rate and SCR utilization rate. The scenario allows for reduction in ecological demand by manipulating chlorine production rate as well as increasing ecological supply by investing in reforestation. The optimal solution determines 1 km^2 of reforestation area out of 15 km^2 available land to minimize annualized cost of production. The optimal cost of production is \$903 k with annualized societal cost of \$1.2 m. The optimal operation schedule is described in Fig. 19.4a. Compared to techno-centric scenario, the number of shutdowns has reduced from 51 days to 38 days. The reduction in production shutdown can be attributed to increase in ecological capacity, allowing the plant to operate on poor AQI days without aggravating them. Reduced shutdowns prevented operations at maximum production rate, improving energy efficiency and reducing production cost. Accounting for ecosystems leads to a reduction in annualized cost of production and annualized societal health impact cost, leading to a win-win solution.

The study further explores the trade-off between societal health impact cost and production cost for techno-centric and techno-ecological solutions. Often, societal and economical objectives are in conflicting nature. Improvement in the societal objective comes at a cost of deterioration in economical objective. Such trade-off is often quantified by a set of pareto optimal solutions that lie on a pareto front, where an infinite number of optimal solutions can exist. All solutions on a pareto front are optimal and considered equally good with regard to both objectives. Here, the study uses pareto front to quantify trade-off between techno-centric and techno-ecological solutions.

The pareto fronts for two scenarios are depicted in Fig. 19.5. Curve ABC shows pareto front for conventional or techno-centric scenario, where investment is only allowed in SCR along with change in operations. Solution A is presented previously in Fig. 19.4a. It minimizes annualized cost of product but has highest societal health

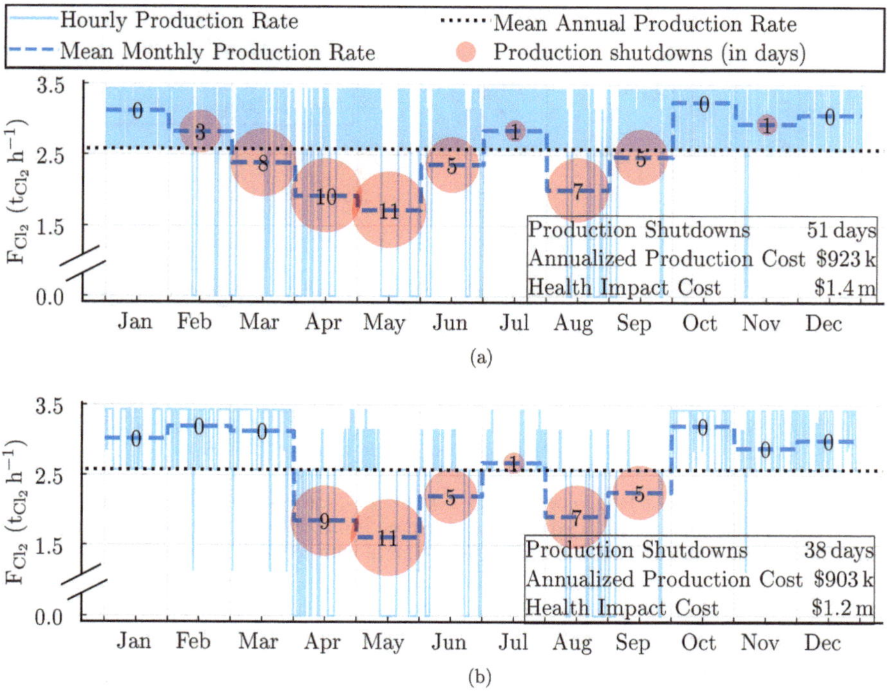

Fig. 19.4 Production schedule, shutdowns, and costs for various scenarios. (**a**) Production schedule for techno-centric solution. Production is higher in months that do not have poor AQI. Production shutdowns often occur on high AQI days. (**b**) Production schedule for techno-ecological solution. As compared to techno-centric scenario, production shutdowns are less frequent and product and social costs are smaller. (Reproduced with permission from [22])

impact cost among all the solutions on the pareto front. Moving from solution A to solution B, the cost of production increases with a decrease in societal health impact cost. The societal health impact cost is zero for solution C, lowest possible for techno-centric systems. At solution C, no pollutants are emitted and they are completed treated by an SCR with sufficient quantity. The zero societal health impact cost has highest annualized cost of production.

For techno-ecological scenario, investment in both reforestation and SCR is allowed along with operational changes. The trade-off between cost of production and societal health impact cost is depicted by curve DEH. Solution D has lowest annualized production cost and highest annualized societal cost among the solutions on pareto front DEH. The operational schedule for the solution D is described in Fig. 19.4b. To reduce annualized societal health impact cost and progressing from solution D to E, additional investment in ecosystems is required. Additional investment results in increased annualized cost of production and increased ecological supply capacity. The increased ecological capacity bridges the gap between supply

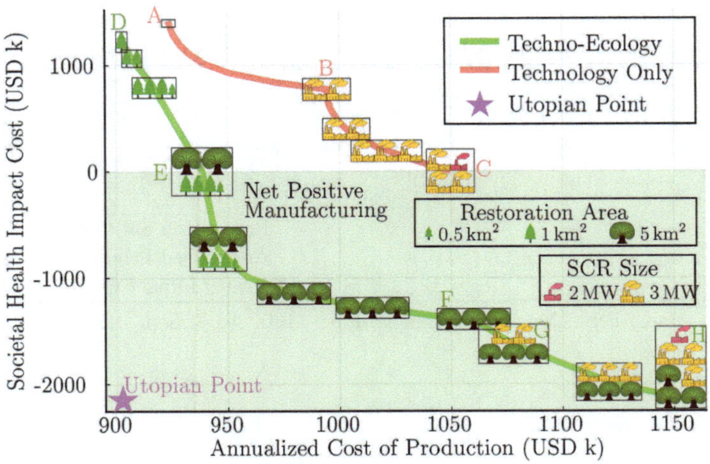

Fig. 19.5 Pareto fronts for techno-centric and techno-ecological systems. Adaptive manufacturing and techno-ecological synergy expand the design space from the techno-centric curve of ABC to the win-win design space of curve DEFGH. (Reproduced with permission from [22])

and demand of services, resulting in reduced deterioration of AQI and reduced societal health impact.

Solution E represents techno-ecological solution with zero societal health impact cost. Compared to techno-centric solution C, the annualized cost of production is lower for solution E at similar societal health impact cost. Additional investment in reforestation beyond solution E allows societal health impact cost to be negative. Negative annualized societal health impact cost is a benefit to society. In other words, solutions on pareto front beyond E (toward H) have net-positive societal benefit. In contrast, the best techno-centric solutions can do is provide net-zero societal cost solution C. The lowest annualized societal cost of \$$-2.2$ m is achieved by solution H. This solution invests in SCR to treat all its pollutant emissions and not contribute to ecosystem demand. It also reforests all 15 km^2 of available land area to provide additional supply capacity to improve AQI in the vicinity of the facility.

Ideally, one would like a solution to have $-\infty$ annualized cost of production with $-\infty$ annualized health impact cost. Such a solution point is referred as utopian solution. A pareto curve that is closer to utopian solutions is considered to be superior. Techno-ecological curve DEH is closer to utopian solution as compared to techno-centric curve ABC and is considered to be an economically and socially superior solution. TES can provide societally superior solution at the same annualized cost of production, resulting in a win-win solution.

Techno-ecological solution can be key to unlocking manufacturing solutions with net-positive impact. Their impact can be monetized using a payment for ecosystem services (PES) scheme [34], where the excess supplies and ecological cobenefits can be sold to meet emission abatement demands of neighboring entities. Integrating the PES benefits would further reduce the cost of reforestation and incentivize investments in ecosystems.

The potential benefits come at the cost of increasing variability in chlorine production operation and SCR utilization rate. The variability in chlorine production can be offset by using storage on the production site or consumer's end. A supply chain redesign may be required to deal with production variability and its impact downstream. Increased forest density can also lead to disservices such as pollen allergy and increased biogenic volatile organic compound (VOC) emissions. There is an urgent need for a multiscale life cycle assessment to ensure no unintended consequences of TES-based design and operation. The case study only considered a single species (White Ash) for reforestation. But in a practical application, the biodiversity of the reforestation attempt needs to be accounted as an additional objective along with cost of production.

The case study assumed instantaneous mixing of pollutant in air and ignored the spatial placement of trees problem. The spatial variation in pollutant concentration and its impact on the surrounding were also ignored. While Charles et al. [35] considered spatial placement of trees, they ignored the monthly temporal variation in forest supply capacity. Advanced algorithms and methods are required for spatiotemporal integration of TES design.

19.5 Conclusions and Future Work

Integration of nature-based solutions in energy production and manufacturing systems would lead to inherent variability and intermittency in production. Along with technological improvement in storage technology, a social engineering revolution is required for an operational and profitable synergy. Consumer demand patterns require modification to match the new production pattern arising from synergy with homeorhetic nature-based solution. Federal Energy Regulatory Commission and others [36] estimated an elimination 2000 power plant with typical peak capacity of 75 MW over a decade with full participation of consumers in a demand response program. McKenzie-Mohr [37] showed effectiveness of community-based social marketing for consumer's behavior change. In order to achieve a synergy between homeorhesis and homeostasis, engineering and social science need to transform toward greater convergence.

A synergy among steady-state tracking homeostatic technological systems and homeorhetic ecosystems have potential to provide economically and ecologically win-win solutions. Achieving this synergy in electricity grids and air quality regulation problem requires significant innovations in material and energy storage technology. Supply chains built around constant product flow-through would require a redesign to handle intermittence introduced by nature-based solutions. Finally, consumer consumption trends need to be social engineered to match the intermittency of nature-based solution.

The air quality regulation case study presented in this chapter was carried out at aggregated timescale of a month over a span of 20 years. The majority of electricity grid case studies are studied at minutes timescale over span of few weeks. In order to study these synergistic systems over longer time span, a timescale aggregation or decomposition is required. Algorithmic advances are required in simultaneous design and operation planning optimization.

Acknowledgments Partial financial support was provided by the National Science Foundation (CBET-1804943) and the Sustainable and Resilient Economy Program at OSU.

References

1. Cannon, W. B. The wisdom of the body. **1939**.
2. Waddington, C. H., *The strategy of the genes*; Routledge: 2014.
3. Waide, J.; Allen, T. A Hierarchical Concept of Ecosystems, 1986.
4. Montgomery, D. C., *Statistical quality control*; Wiley Global Education: 2012.
5. Leopold, A. S., *Wildlife management in the national parks*; US National Park Service: 1963.
6. Campbell, H. A.; Fraser, K. P.; Bishop, C. M.; Peck, L. S.; Egginton, S. Hibernation in an Antarctic fish: on ice for winter. *PLoS One* **2008**, *3*, e1743.
7. Tøien, Ø.; Blake, J.; Edgar, D. M.; Grahn, D. A.; Heller, H. C.; Barnes, B. M. Hibernation in black bears: independence of metabolic suppression from body temperature. *Science* **2011**, *331*, 906–909.
8. Yi-Ling, H.; Hai-Zhen, M.; Guang-Tao, D.; Jun, S. Influences of urban temperature on the electricity consumption of Shanghai. *Advances in climate change research* **2014**, *5*, 74–80.
9. Eto, J. The past, present, and future of US utility demand-side management programs. **1996**.
10. Holland, S. P.; Mansur, E. T. Is real-time pricing green? The environmental impacts of electricity demand variance. *The Review of Economics and Statistics* **2008**, *90*, 550–561.
11. Denholm, P.; Ela, E.; Kirby, B.; Milligan, M. *Role of energy storage with renewable electricity generation*; tech. rep.; National Renewable Energy Lab.(NREL), Golden, CO (United States), 2010.
12. Guerra, O. J.; Zhang, J.; Eichman, J.; Denholm, P.; Kurtz, J.; Hodge, B.-M. The value of seasonal energy storage technologies for the integration of wind and solar power. *Energy & Environmental Science* **2020**.
13. Demirhan, C. D.; Tso, W. W.; Powell, J. B.; Heuberger, C. F.; Pistikopoulos, E. N. A Multiscale Energy Systems Engineering Approach for Renewable Power Generation and Storage Optimization. *Industrial & Engineering Chemistry Research* **2020**, *59*, 7706–7721.
14. Blanco, H.; Faaij, A. A review at the role of storage in energy systems with a focus on Power to Gas and long-term storage. *Renewable and Sustainable Energy Reviews* **2018**, *81*, 1049–1086.
15. Shah, U.; Bakshi, B. R. Accounting for Nature's Intermittency and Growth While Mitigating NO_2 Emissions by Techno-Ecological Synergistic Design–Application to a Chloralkali Process. *Journal of Advanced Manufacturing and Processing* **2019**, *1*, e10013, https://doi.org/10.1002/amp2.10013.
16. Brée, L. C.; Perrey, K.; Bulan, A.; Mitsos, A. Demand side management and operational mode switching in chlorine production. *AIChE Journal* **2018**, https://doi.org/10.1002/aic.16352.
17. NETL *NETL Life Cycle Inventory Data Process Documentation*; tech. rep.; 2010, pp 1–8.
18. Skalska, K.; Miller, J. S.; Ledakowicz, S. Trends in NOx abatement: A review. *Science of the total environment* **2010**, *408*, 3976–3989.
19. Sorrels, J. L.; Randall, D. D.; Schaffner, K. S.; Richardson Fry, C.; Fry, C. R.; Richardson Fry, C. Chapter 2 - Selective Catalytic Reduction, 2016.

20. Dixon, G., *Essential FVS: A user's guide to the Forest Vegetation Simulator*; February, 2002, p 246, https://doi.org/10.1007/978-3-319-23883-8.
21. *iTree Eco User's Manual*; tech. rep.; 2018.
22. Shah, U.; Bakshi, B. R. Toward Nature-Positive Manufacturing by Adapting Industrial Processes to Pollution Uptake by Vegetation. *ACS Sustainable Chemistry & Engineering* **2021**, *9*, 16709–16718, https://doi.org/10.1021/acssuschemeng.1c05617.
23. Davis, S. J. et al. Net-Zero Emissions Energy Systems. *Science* **2018**, *360*, eaas9793, https://doi.org/10.1126/science.aas9793.
24. Renger, B. C.; Birkeland, J. L.; Midmore, D. J. Net-Positive Building Carbon Sequestration. *Building Research & Information* **2014**, *43*, 11–24, https://doi.org/10.1080/09613218.2015.961001.
25. Rahimifard, S.; Stone, J.; Lumsakul, P.; Trollman, H. Net Positive Manufacturing: A Restoring, Self-healing and Regenerative Approach to Future Industrial Development. *Procedia Manufacturing* **2018**, *21*, 15th Global Conference on Sustainable Manufacturing, 2–9, https://doi.org/10.1016/j.promfg.2018.02.088.
26. Norris, C. B.; Norris, G. A.; Azuero, L.; Pflueger, J., Structure of a Net Positive Analysis for Supply Chain Social Impacts In *Perspectives on Social LCA: Contributions from the 6th International Conference*, Traverso, M., Petti, L., Zamagni, A., Eds.; Springer International Publishing: Cham, 2020, pp 35–43, https://doi.org/10.1007/978-3-030-01508-4_4.
27. Zore, Z.; Cucek, L.; Kravanja, Z. Syntheses of Sustainable Supply Networks With a New Composite Criterion - Sustainability Profit. *Computers & Chemical Engineering* **2017**, *102*, Sustainability & Energy Systems, 139–155, https://doi.org/10.1016/j.compchemeng.2016.12.003.
28. Otashu, J. I.; Baldea, M. Demand Response-Oriented Dynamic Modeling and Operational Optimization of Membrane-Based Chlor-Alkali Plants. *Computers & Chemical Engineering* **2019**, *121*, 396–408, https://doi.org/10.1016/j.compchemeng.2018.08.030.
29. Nielsen, A. S. E.; Plantinga, A. J.; Alig, R. J. New Cost Estimates for Carbon Sequestration Through Afforestation in the United States. *Gen. Tech. Rep. PNW-GTR-888. Portland, OR: US Department of Agriculture, Forest Service, Pacific Northwest Research Station. 35 p.* **2014**, *888*.
30. Alexatos, A., Personal communication, TreeFolks, 2020.
31. EPA Technical Assistance Document for the Reporting of Daily Air Quality - the Air Quality Index (AQI), Technical Report EPA-454/B-18-007, 2018.
32. Sacks, J. D.; Lloyd, J. M.; Zhu, Y.; Anderton, J.; Jang, C. J.; Hubbell, B.; Fann, N. The Environmental Benefits Mapping and Analysis Program – Community Edition (BenMAP–CE): A Tool to Estimate the Health and Economic Benefits of Reducing Air Pollution. *Environmental Modelling & Software* **2018**, *104*, 118–129, https://doi.org/10.1016/j.envsoft.2018.02.009.
33. Nowak, D. J.; Hirabayashi, S.; Bodine, A.; Greenfield, E. Tree and Forest Effects on Air Quality and Human Health in the United States. *Environmental Pollution* **2014**, *193*, 119–129, https://doi.org/10.1016/j.envpol.2014.05.028.
34. Fripp, E., *Payments for Ecosystem Services (PES): A practical guide to assessing the feasibility of PES projects*; Center for International Forestry Research (CIFOR): 2014, https://doi.org/10.17528/cifor/005260.
35. Charles, M.; Bakshi, B. In *2018 AiChE Annual*, 2018.
36. Commission, F. E. R. et al. A national assessment of demand response potential. *prepared by The Brattle Group, Freeman Sullivan, & Co, and Global Energy Partners* **2009**.
37. McKenzie-Mohr, D. New ways to promote proenvironmental behavior: Promoting sustainable behavior: An introduction to community-based social marketing. *Journal of social issues* **2000**, *56*, 543–554.

Chapter 20
Assessing Techno-Ecological Synergies in the Life Cycle of Biofuels

Ying Xue, Ruonan Zhao, and Bhavik R. Bakshi

20.1 Introduction

In this chapter, we describe the framework for including the role of ecosystems in the technique of life cycle assessment (LCA), which is among the most popular approaches for sustainability assessment. LCA determines the contributions of life cycle stages to various environmental indicators and points out the opportunities for improving the largest contributors or hotspots. As described in Chap. 3, LCA aims to consider the direct and indirect interactions of products and processes with the environment, including use of resources and impact of emissions. This approach has been standardized and various software packages and datasets are available. Despite its popularity for sustainability assessment, until recently, LCA kept ecosystems outside its system boundary. It focused primarily on human-designed or technological systems, as illustrated in Fig. 20.1 in Chap. 3, while ignoring the role played by ecological systems in supporting these activities. The resulting ignorance about the role of ecosystem goods and services and the status of ecosystems to supply them means that traditional LCA is unable to help meet the basic requirement of sustainability which is to stay within nature's carrying capacity. Even though LCA encourages reduction of environmental impact, it is best for comparing alternatives and choosing the less bad product. Conventional

Y. Xue · R. Zhao
Chemical and Biomolecular Engineering, The Ohio State University, Columbus, OH, USA

B. R. Bakshi (✉)
School for Engineering of Matter, Transport and Energy, School of Sustainability, School of Complex Adaptive Systems, Arizona State University, Tempe, AZ, USA

Lowrie Department of Chemical and Biomolecular Engineering, The Ohio State University, Columbus, OH, USA
e-mail: bhavik.bakshi@asu.edu

B. R. Bakshi (ed.), *Engineering and Ecosystems*,
https://doi.org/10.1007/978-3-031-35692-6_20

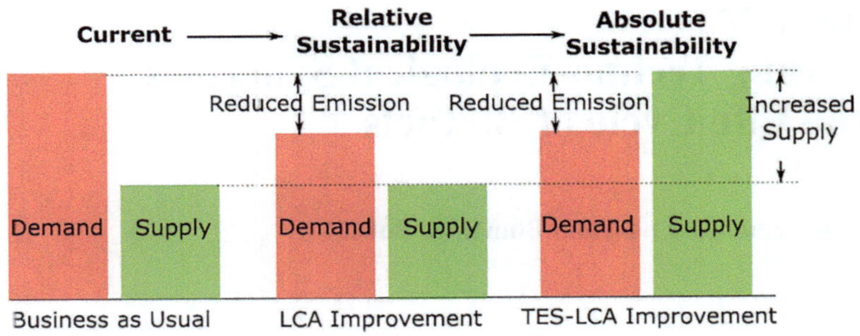

Fig. 20.1 Relative and absolute environmental sustainability (Liu and Bakshi 2018)

LCA does not provide insight into whether the system being analyzed is operating within nature's capacity and how close it is to the ability of relevant ecosystems to supply the goods and services used by the activity. The framework of TES-LCA (Liu and Bakshi 2018) is among the recent approaches for overcoming this shortcoming and is described in this chapter and illustrated by an application to biofuels.

The development of traditional petroleum fuels has brought many environmental problems due to its complex processing procedures and combustion performance. Replacing fossil fuels with biofuels—fuels produced from renewable organic material—has the potential to reduce some undesirable aspects of fossil fuel production and use, including emissions of conventional and greenhouse gas (GHG), depletion of exhaustible resources, and dependence on unstable foreign suppliers. In addition, biofuels are of considerable interest because they are made from renewable resources. Many studies have compared environmental impacts of biodiesel (BD) made from different raw materials or different processing technologies (Panichelli et al. 2009; Huo et al. 2009). LCA is a commonly used method for guiding decision-making which quantifies environmental impacts associated with a product's life cycle from raw material extraction through material processing, manufacture, distribution, and use phase to final disposal or recycling (Finnveden et al. 2010). LCA expands the system boundaries by including related upstream and downstream activities, aiming to reduce the environmental impacts associated with the life cycle of corresponding products (Schmidheiny and Stigson 2000). Applications of LCA on biofuels such as biodiesel and bioethanol have been widely studied. Panichelli and Gnansounou (2008) studied the LCA of soybean biodiesel in Argentina and found that the Argentinean pathway resulted in the highest GWP, nonrenewable energy consumption, aquatic ecotoxicity, and human toxicity compared with the reference biofuel pathways. Pradhan et al. (2011) studied the energy LCA of soybean biodiesel by fossil energy ratio (FER) and concluded that FER improved significantly from 3.2 in 1998 to 5.54 in 2006. To ensure the biodiesel is renewable, higher FER is expected. Huo et al. (2009) studied the energy use and GHG emissions of soybean biodiesel by LCA using the GREET model and concluded that biodiesel could have huge advantages in reduction of fossil energy use (>52%), petroleum use

(>88%), and GHG emissions (>57%) compared with petroleum fuels. It is worth mentioning that most of the LCA research discussed above is conducted in North America or Europe. Studies in the European or North American context can provide significant results, but with uncertainty associated with variable input parameter values, making it difficult to compare. Another feature of LCA studies is the wide range of values of net energy balances and net greenhouse gas emissions (GHG) for a given biofuel. For instance, Concawe et al. (2004) showed that compared to conventional fuels, rape methyl ester can reduce the GHG emissions in the range of 16–63%. This chapter describes the TES-LCA framework and applies it to assessing biofuels. This illustrates the methodology of TES-LCA for assessing absolute environmental sustainability. Results convey the benefits of including ecosystems and considering nature's carrying capacity in life cycle assessment.

20.2 TES-LCA Methodology

As described in Chap. 3, conventional LCA involves four steps: goal and scope definition, inventory analysis, impact assessment, and interpretation. Since it does not account for the role of ecosystems and their carrying capacity to sustain the life cycle of products/processes, conventional LCA only permits the quantification of relative sustainability which compares one system versus the other. The outcome of a relative assessment greatly depends on the choice of the reference (Bjorn et al. 2013). Figure 20.1 illustrates the concepts of relative and absolute environmental sustainability (AES). Relative sustainability metrics are useful for comparing multiple alternative systems, identifying the hotspots and opportunities of improving eco-efficiency. However, these metrics only evaluate whether the system is "comparatively good"; the role of nature is ignored. Thus, relative sustainability metrics focus on reducing environmental impact, as shown in the middle two bars of Fig. 20.1. Accounting for nature's carrying capacities, AES metrics inform stakeholders whether their activities stay within ecological thresholds. Besides, AES metrics help in quantifying or designing the life cycle not only by reducing impacts but also by protecting and restoring ecosystems, as shown by the right-most pair of bars in Fig. 20.1. The framework of techno-ecological synergy in LCA (TES-LCA) is based on conventional LCA but includes the role of ecosystem services (ES) and allows quantifying absolute environmental sustainability (AES). This framework calculates AES metrics which compare environmental impacts to an external list of environmental carrying capacities.

Planetary boundary (PB) is one of the most popular frameworks for absolute environmental sustainability assessment. As described in Chap. 2, it defines boundaries for nine earth system processes (Rockstrom et al. 2009) whose transgression could result in irreversible and large-scale environmental changes. The PB framework defines the "safe operating space" (SOS) for human development representing the region within which human perturbations respect earth systems. In parallel with methods based on the PB framework, the TES framework has also been used for

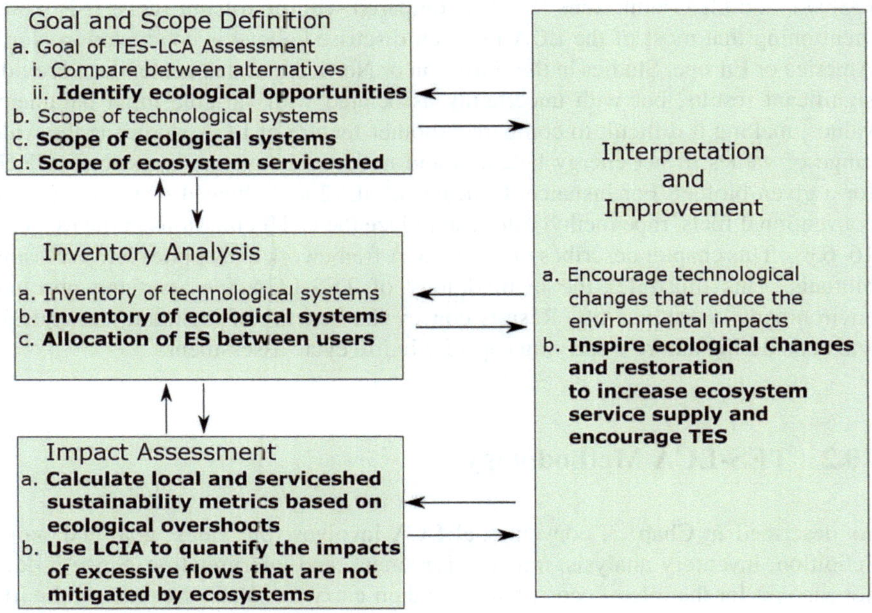

Fig. 20.2 Steps in the TES-LCA methodology. ((Liu and Bakshi 2018) Reprinted with permission)

AES assessment. For local context, methods based on the PB framework rely on direct downscaling of global or regional boundaries, while TES-based methods quantify the capacity of nature through local biophysical models which are more robust, can be done at different spatial scales, and have high geographical resolution.

The approach of TES-LCA extends conventional LCA by accounting for the contribution and carrying capacity of ecosystems (Liu and Bakshi 2019). Like conventional LCA, TES-LCA also has four steps; however, TES-LCA expands the system boundaries to include ecological systems, while conventional LCA only includes technological systems. This means goods and services provided by ecosystems would be considered across the life cycle of processes or products. The method of TES-LCA is shown in Fig. 20.2.

As shown in Fig. 20.2, the bold font in each step represents the steps in TES-LCA that are in addition to conventional LCA to account for the role of ecosystems. For step 1, the goal of TES-LCA is to firstly compare environmental impacts among alternatives, which is the same as conventional LCA. In addition to this, the ecological opportunities for their restoration or protection need to be identified. To enable fair comparison, a functional unit should be defined. The scope of technological systems can be obtained by selecting processes across the life cycle of corresponding product from raw material extraction through material processing, manufacture, distribution, and use phase to final disposal or recycling (Finnveden et al. 2010). In TES-LCA, multiple scales of ecosystem services are considered. The number of scales is decided by users. There is a trade-off between the number of scales and amount of data required. As more scales are included in the model, more regional data are needed, and the result is more precise. For instance, a two-level TES-LCA model considers local and serviceshed scales. The local scale includes services provided by the ecological systems in the vicinity of the technological systems. For instance, for a biodiesel manufacturing plant, the trees, lawn, and soil at the factory site can absorb carbon dioxide and air pollutants. Therefore, these are considered as local ecosystem services for this plant. The serviceshed for each ES is defined as the region providing ESs to specific users (Tallis et al. 2012). This is the largest ecosystem scale that is considered in TES-LCA and is relevant for determining absolute environmental sustainability. For the carbon sequestration service, its serviceshed is the entire planet's atmosphere because carbon dioxide is a global flow. However, for air pollutants such as sulfur dioxide which can only be transported within a limited geographical area, the serviceshed for sulfur dioxide regulation is at a sub-global or regional scale. Similarly, for the water-provisioning service, the serviceshed is the watershed of the water body to which water flows or is withdrawn. The supply of ecosystem services at any scale is quantified using biophysical models at the appropriate spatial scale.

The second step of the TES-LCA methodology is inventory analysis, which includes three sub-steps. The first sub-step is analyzing the inventory of technological systems, which may be interpreted as quantifying demand for the k-th ecosystem service. Demand represents emission and resource use of technological systems (Bakshi et al. 2015). The second sub-step is about the inventory analysis of ecological systems. Ecological inventory represents the ecosystem's ability to provide the k-th ecosystem service. Supply, denoted by S, represents the ecological capacity to mediate/sequester/absorb impacts (Bakshi et al. 2015). Conventional LCA databases such as ecoinvent Version 3 (Wernet et al. 2016) do not include information about S. Thus, such supply data need to be compiled into these databases in the future. In decision-making, the supply information needs to be partitioned among stakeholders when an ecosystem service is used by multiple users within a selected serviceshed, and this is the third sub-step. Proportional allocation splits the supply based on chosen quantities such as money, population, or area. Allocating an ES among multiple users is analogous to determining their right of use. Private land ownership implies that the landowner owns all the ecosystem services, while for public land ownership, all users inside that area are considered to "own" the ecosystem services equally. Currently, PB-based methods apply public ownership in downscaling, while the TES-LCA framework combines the approaches based on public and private ownership. Figure 20.3 illustrates the allocation of supply in a two-level TES-LCA framework.

As shown in this figure, the total supply inside this region is $S_A + S_B + S_C$. With the assumption of private ownership, only supply from public land can be allocated. Region C is public land; thus, S_C can be allocated between processes A and B. Regions A and B are private; thus, S_A and S_B belong to processes A and B, respectively. Total supply that process A can get is $S_A + S_C \times P$, where

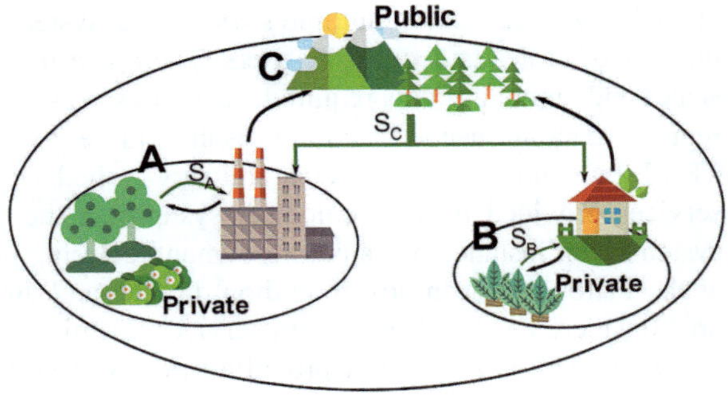

Fig. 20.3 Allocation considering private and public ownership (Xue and Bakshi 2022)

P represents the sharing or partitioning principle. Similarly, process B could get $S_B + S_C \times (1 - P)$. Commonly used sharing principles allocate ES in proportion to population, current environmental impact (demand for ES), gross value addition, etc. For the case study in this chapter, we apply the demand-based sharing principle, which in other words means the more the process demands (emission, resource use), the more supply gets allocated to this process.

The third step of the TES-LCA methodology is impact assessment. The absolute environmental sustainability metric of TES-LCA compares the supply and demand at each spatial scale; for the k-th ES, the absolute sustainability metric can be defined as:

$$V_{i,j,k} = \frac{S_{i,j,k} - D_{i,j,k}}{D_{i,j,k}} \tag{20.1}$$

Here, $S_{i,j,k}$ and $D_{i,j,k}$ represent supply and demand of the k-th ecosystem service at scale j for techno-ecological system i. We define $j = 1, 2, 3\ldots, J$ with $j = 1$ representing the smallest scale. When $V_{i,j,k} < 0$, ecosystem services cannot meet human demand indicating unsustainability and $V_{i,j,k} > 0$ indicates sustainability. Equation 20.2 represents the requirement of absolute sustainability at the serviceshed scale:

$$V_{i,J,k} \geq 0, \forall i, \forall \mathrm{k} \tag{20.2}$$

Here, J represents the largest scale, which in TES-LCA is the serviceshed scale. Equation 20.2 represents the situation where humans stay within nature's carrying capacity at the serviceshed scale. Ideally, environmental sustainability requires:

$$V_{i,j,k} \geq 0, \forall i, j, k \tag{20.3}$$

Here, demand is less than supply at all scales. This is a strong sustainability requirement. The final step in the TES-LCA method is the interpretation and improvement based on the results obtained in step 3. LCA can only identify tech-

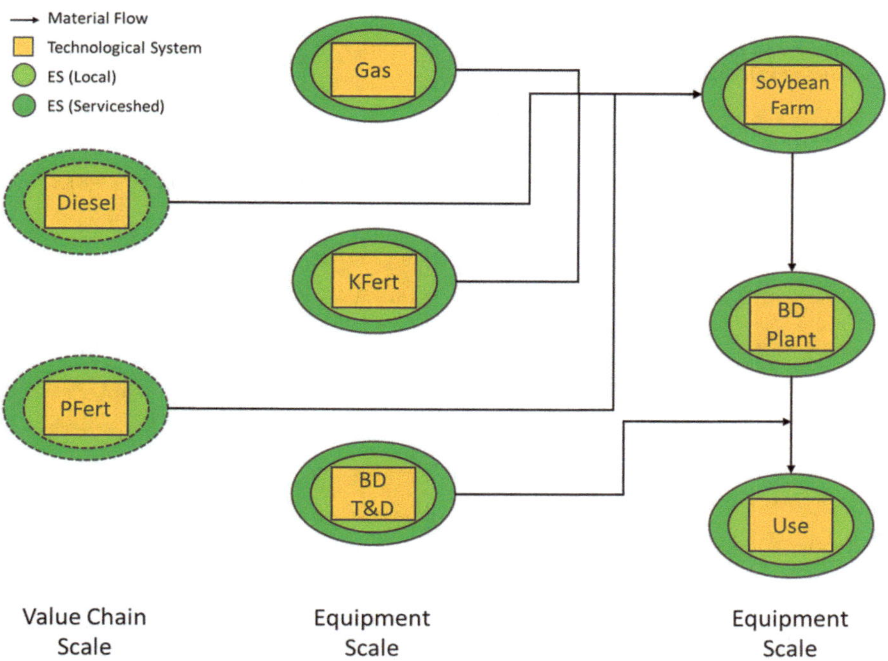

Fig. 20.4 System boundary in soybean biodiesel life cycle. Note: Kfert potassium fertilizer, Pfert phosphorus fertilizer, Biodiesel T&D biodiesel transportation and distribution

nological improvements. However, TES-LCA can also inspire ecological changes along with encouraging harmony between natural and human systems.

20.3 TES-LCA of Soybean Biodiesel

20.3.1 Goal and Scope Definition

The goal of this study is to quantify life cycle environmental impacts of soybean biodiesel and capacities of ecosystem services to mitigate these impacts. The functional unit is 100 kilometers (km) traveled by the vehicle. The fuel efficiency is 4.77 liter/100 km (GREET 2021). For ecological systems, we consider the ESs of carbon sequestration and water provisioning.

As shown in Fig. 20.4, the system boundary covers eight activities, and they are considered at different spatial scales.

Within the system boundary, six activities, namely, farming, biodiesel plant, biodiesel transportation and distribution, use phase, gasoline blendstock, and potassium fertilizer, are considered at equipment scale because of their site specificity.

Table 20.1 Location information of selected facilities at value chain scale

Facility type	Label	Latitude	Longitude
P fertilizer	Pfert 1	41.12	−89.34
	Pfert 2	43.44	−83.75
Diesel manufacturer	Diesel 1	37.21	−76.45
	Diesel 2	30.29	−93.14
	Diesel 3	32.11	−81.13

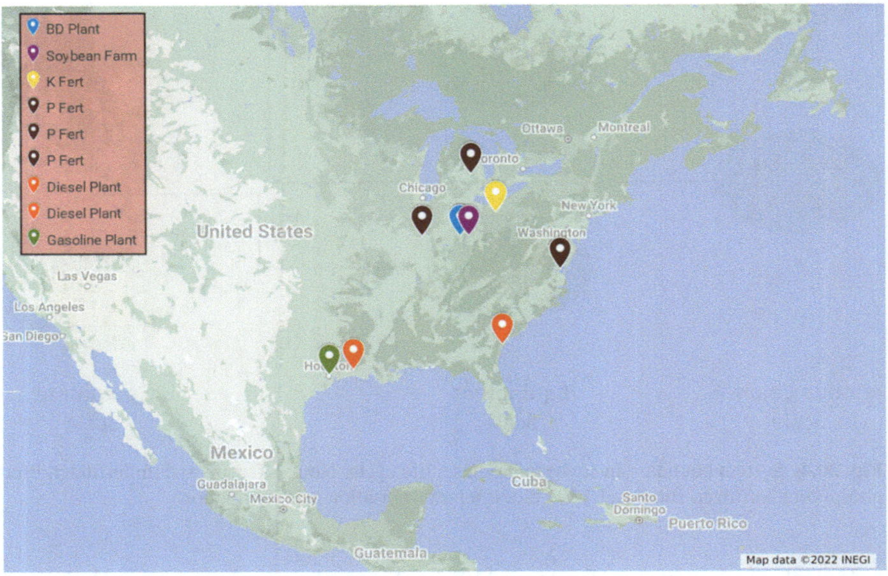

Fig. 20.5 Locations of each activity included in the system boundary

Diesel and phosphate fertilizer manufacturing are modeled at the value chain scale. For activities modeled at the value chain scale, multiple facilities across the USA are chosen and then averaged. This step is analogous to constructing life cycle inventory databases in conventional LCA, but in TES-LCA, both technological and ecological flows are considered. These facilities are chosen because their information (emission and water consumption) is available from both greenhouse gas emissions (USEPA 2011a) and National Emissions Inventory (NEI) databases (USEPA 2011b). Ideally, the information of demand from ecosystem services should be available for all facilities inside the system boundary considered. The location information of the chosen facilities is listed in Table 20.1 and plotted in Fig. 20.5.

For the farming phase, two farming practices, namely, conventional or continuous tillage (CT) and no tillage (NT), are compared. To maintain the growth of soybean, potassium and phosphorus fertilizer are two necessary nutrients. Trucks for transportation and other agricultural equipment for normal farming activities are powered by diesel and gasoline. The harvested soybeans are then transported to the biodiesel plant which mainly contains two parts: soybean oil extraction and soybean oil transesterification (Bulent Koc et al. 2011). Furthermore, the finished product (biodiesel) is transported, distributed, and finally used in vehicles.

For each activity modeled at the equipment scale, except farming, the spatial size of the relevant ecosystem is defined as the campus around the specific facility. Since information about land ownership is not readily available, a region with a radius of 500 meters is assumed to approximate the campus for each site. For farming phase, soybean can convert carbon dioxide into organic matter through photosynthesis. Thus, the farmland itself can supply some ecosystem services. Therefore, it is reasonable to assume that the farmland itself is the relevant local ecosystem for the farming phase. Trucks and other agriculture machines used to maintain the normal growth of soybean emit carbon dioxide inside the farm, and this part of emission is considered as the demand of the farm. Since carbon dioxide is a global flow (Steffen et al. 2015), the serviceshed for carbon sequestration service is assumed to be global. For the water provisioning service, the serviceshed is the watershed in which that activity is located. For transportation of soybean and biodiesel, and final use phase of biodiesel, the supply of ecosystem services at local and serviceshed scales is assumed to be zero. This is justified under the assumption that ecosystem supplies for these activities are extremely small after allocation.

20.3.2 Inventory Analysis

In this case study, region-specific demand data (emissions like carbon dioxide, carbon monoxide, etc. and resource use such as power, water, etc.) for each facility, except the farming phase, is provided by NEI and USEPA. Average emission and resource data from the GREET database (GREET 2021) are used when regional specific data are unavailable. The emission factors for fuel combustion are obtained from the GREET model.

For the farming activity, Environmental Policy Integrated Climate (EPIC 2017) model has been applied to simulate a specific farm located in Hamilton County, Ohio. The EPIC model takes soil and weather data as inputs and simulates the annual yield, fuel consumption, and carbon cycle data. The farm size is set to be 1 hectare, and continuous soybean operation with no irrigation is assumed. The data for farming management, such as types and amounts of fertilizers and pesticides, can be acquired from the Agricultural Resource Management Survey (ARMS) (USDA 2010a) and GREET (GREET 2021). Detailed operations are assumed to use default values in EPIC. Information about local soil properties of Hamilton County can be obtained from the Web Soil Survey (USDA 2010b). Data about weather conditions can be obtained from the National Centers for Environmental Prediction (NCEP 1979–2014). In the current study, the simulation is based on a 20-year period from 1995 to 2014. A 12-year pre-run is applied to adjust soil properties. Table 20.2 shows the farming inputs and yields for conventional tillage and no tillage planting modes.

Table 20.2 Inputs and yields for conventional and no-tillage methods

Inputs	Conv. tillage	No tillage	Unit	Data source
Diesel	120	70	L/ha	EPIC
P Fert	55	50	Kg/ha	ARMS
K Fert	100	106	Kg/ha	
Pesticide	1.4	1.8	Kg/ha	
Seed	89	82	Kg/ha	
Soybean yield	2523	3119	Kg/ha/yr	EPIC

Table 20.3 Data source and spatial boundary of local and serviceshed scale

	Local	Serviceshed	Data source
Carbon sequestration (farm)	1 hectare (farmland)	Global	EPICGlobal carbon budget
Carbon sequestration (other activities)	500 meter radius circle	Global	iTree CanopyGlobal carbon budget
Water provisioning (farm)	1 hectare (farmland)	Watershed	USGS geodata portal
Water provisioning (other activities)	500 meter radius circle	Watershed	USGS geodata portal

With properly defined local weather data, soil properties, and management practices, the EPIC model generates results about the cycling of carbon, nutrients, and water. These data provide inventory about the supply of carbon sequestration and water-provisioning services during the farming phase. Considering the fact that the amount of water from precipitation is sufficient for soybean growth, no-irrigation is assumed. Precipitation is used as the water supply. Water demand is approximated by its evapotranspiration rate, which is also available from the EPIC model. The CO_2 sequestration data and water-provisioning data for each ecosystem at different scales build up the ecological data inventory. Table 20.3 summarizes the data sources for all activities inside the system boundary.

For the water-provisioning service, the 8-digit hydrologic unit code (HUC) is used to identify the watersheds for selected processes. Table 20.4 lists HUC for processes inside the system boundary. The data can be obtained from Geo Data Portal (USGS 2010a). With regard to the ES of carbon sequestration, the planet is assumed to be the serviceshed. The supply of carbon sequestration at the global scale can be obtained from the Global Carbon Budget (Le Quere et al. 2012).

Since conventional tillage and no-tillage farming have different requirements for fertilizer and diesel, the input data are considered separately. FLIGHT (EPA 2020) provides the annual GHG emission data for each specific facility. To get the final result, the production capacity of that facility is also needed. For this case study, the production capacity information was found on the companies' websites. TES-LCA is a linear model, the supply and demand data for each process should be normalized by the functional unit. For instance, annual GHG emission for the diesel manufacturing plant will be first converted to kg CO2eq/L diesel then further

Table 20.4 8-digit hydrologic unit code (HUC)

Process	Location	Watershed (HUC8)
Soybean farm	Hamilton, OH	05090203
Potassium fertilizer	Clinton, OH	05090202
Gasoline blendstock	Harris, TX	12040104
Biodiesel plant	Hamilton, OH	05090203
BD transportation and distribution	–	
Use stage	–	
Phosphate fertilizer	Marshall, IL	07130001
	Tuscola, MI	04080205
Diesel manufacturing	York, VA	37215607
	Calcasieu parish, LA	08080206
	Chatham, GA	03060109

Note: For transportation-related activities such as soybean transportation, biodiesel transportation, and distribution and use stage, water-provisioning service is not considered

Table 20.5 Supply data at different scales for carbon sequestration

	Supply at Serviceshed
Carbon sequestration (kg CO2 eq/yr)	9.58E12
Water provision (m^3/yr)farm, BD manufacturing	1.36E9
Water provision (m^3/yr)P fertilizer	1.39E9
Water provision (m^3/yr)K fertilizer	1.10E9
Water provision (m^3/yr)Diesel	1.77E8
Water provision (m^3/yr)Gasoline blendstock	1.21E9
Water provision (m^3/yr)T&D, use	0.00

be normalized to kg CO2eq/100 km. The unit conversion factors are found from GREET (GREET 2021).

The supply (ES) data for carbon sequestration and water-provisioning services at their serviceshed scales are listed in Table 20.5. For processes at value chain scale, their serviceshed supply data are averaged over different watersheds in which the processes are located.

20.3.3 Impact Assessment

The overall result sums up the environmental impacts (demand) of all processes inside the system boundary and is compared with the total supply. Since this is a two-scale TES-LCA study, local and serviceshed AES metrics $V_{i,1,k}$ and $V_{i,2,k}$ for the carbon and water ES are calculated through Eq. 20.1.

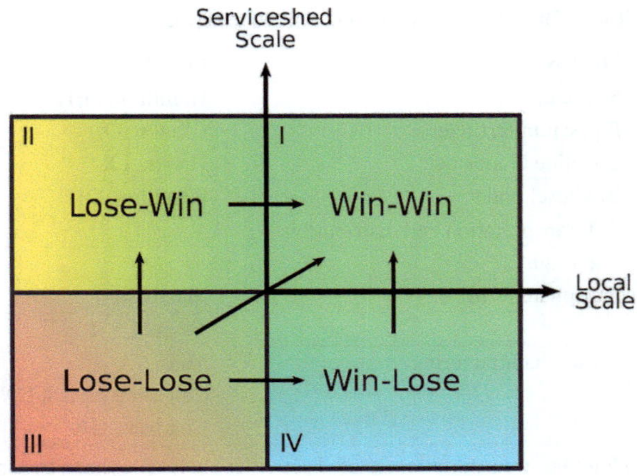

Fig. 20.6 Local and serviceshed scale metrics. Note: arrow indicates expected direction of improvements toward sustainability

20.3.4 Interpretation

In this case, the two-scale TES metrics could be visually understood by plotting the local and serviceshed scale metrics. As illustrated in Fig. 20.6, the local sustainability metric is plotted on the *x*-axis, with serviceshed sustainability metric plotted on the *y*-axis.

In Fig. 20.6, Quadrant I represents the best situation where corresponding activities are sustainable at both local and serviceshed scales. At the opposite extreme, Quadrant III represents the worst case: activities in this quadrant are unsustainable at local and serviceshed scales. Improvements for systems inside the third quadrant could be made through increasing local supply, reducing demand or both. Scenarios in Quadrants II and IV represent islands of unsustainability and sustainability, respectively (Wallner et al. 1996).

20.4 Results

20.4.1 Carbon Sequestration Service

Different farming practices, with and without tillage, are compared in this case study. Many factors such as crop type, crop rotation, local weather, fertilizer and pesticide usage, etc. will affect the supply of carbon sequestration. Figure 20.7 shows CO2 emission (demand) and sequestration (supply) at life cycle scale for biodiesel.

According to the IPCC standard, since bioenergy systems operate within the fast domain of the carbon cycle, biogenic CO2 from combustion is not considered as GHG emission. To understand carbon emission at each stage, Fig. 20.7 not only

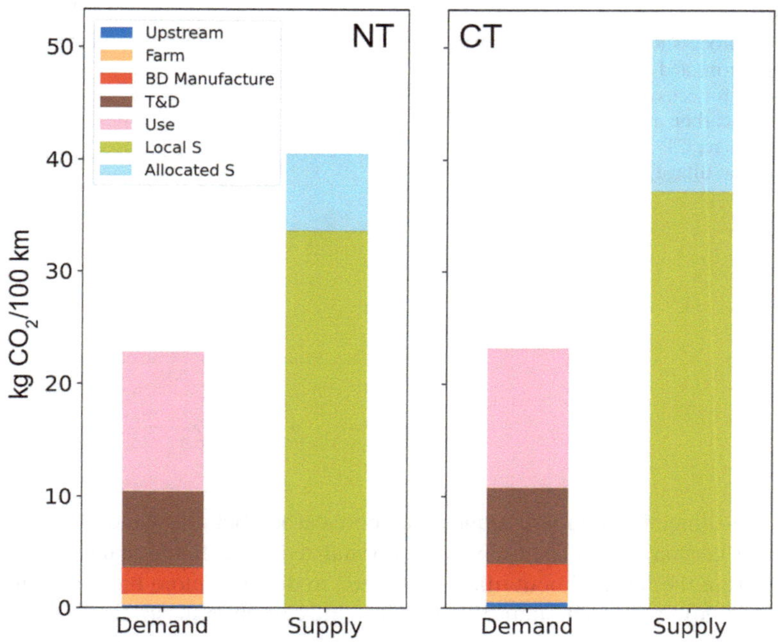

Fig. 20.7 Demand and supply of carbon sequestration ecosystem service for soybean biodiesel. Upstream emissions are from fertilizer, diesel, and gasoline (T&D transport and distribution, Local S local supply, Allocated S allocated supply from coarser scales, NT no-tillage, CT conventional tillage)

shows the GHG emission at each phase across the product's life cycle scale but also depicts the emission from the use phase (biogenic carbon). In the path of soybean biofuel production, soy oil and biodiesel are the main products, while soy meal and glycerin are coproducts. We allocate the demand and supply of carbon for related processes (farming, BD manufacturing) in proportion to energy content. For each process, the total supply includes local supply inside its campus (local supply) and allocated supply from the global scale. As discussed in Sect. 20.2, only public supply can be shared among users. At the global scale, we define that the amount of carbon sequestrated in oceans is publicly owned which belongs to everyone. The value of the public carbon budget at global scale is 9.58 billion ton/yr. This public carbon budget will then be allocated to each process based on its emission. The allocated part is illustrated by the light blue block in Fig. 20.7. The considerable amount of local supply is represented by the green block. From the bar on the left for both no-tillage and conventional tillage practices, we can see that the use phase of biodiesel is the biggest contributor to the overall global warming potential impact. As can be seen in Fig. 20.7, for two different farming practices, both of their emissions are lower than the total sequestration. Thus, $V_{i,j,k} > 0$ indicating absolutely sustainable for the carbon sequestration service. For two exact same

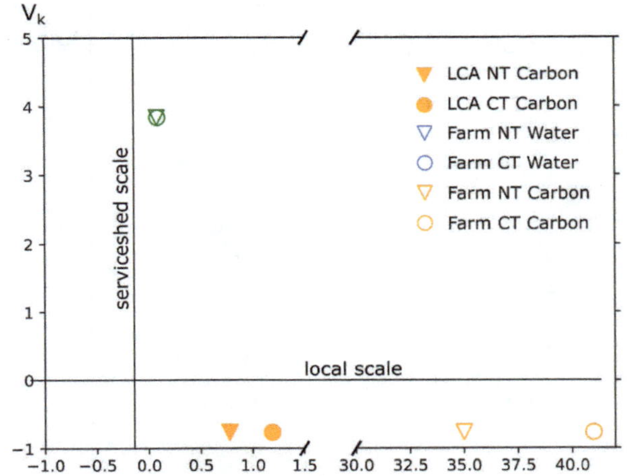

Fig. 20.8 Sustainability metrics at process level (soybean farm) and at life cycle scale for ecosystem services of carbon and water (NT no-tillage, CT conventional tillage)

farms, no-tillage farming will sequestrate more carbon but also yield more soybean. When the carbon demand and supply are normalized to per functional unit (100 km), supply from the farm without tillage will need to displace more biodiesel emission compared farm with continuous tillage. Thus, the carbon sequestration bar (right) of the no-till farm is smaller than farm with conventional tillage. One thing to notice, the absolute sustainability result shown in Fig. 20.7 only represents a specific case considered in this study. It does not account for spatial variability due to soil type, farming practices, etc. and does not represent the general average situation captured in other soybean biodiesel life cycle assessments.

The soybean farming process and biodiesel use phase are further studied in detail. The soybean farm has large local supply, while the use phase only has allocated supply. Local supply provides region-specific ecological information, and allocated supply reflects the holistic situation of larger spatial scales. For the farming and vehicle operation phases, neither local nor allocated supplies could be ignored. The total supply is larger than the demand for these two processes. The absolute sustainability metric at global scale is the same for any process: $V_{i,J,k} = -0.469$ where $J = 2$, k : *Carbon*. This absolute sustainability metric ($V_{i,J,k}$) is calculated by comparing the global CO2 emission versus global CO2 sequestration. Figure 20.8 shows the local sustainability metrics for the soybean farm and the whole system. For water-provisioning service, the soybean farm is at the most ideal status, meaning it is sustainable at both local and serviceshed scales. For carbon sequestration service, more actions toward sustainability such as reforestation and technological advances for reduction of GHGs should be done at local and global scale. The difference between different farming practices is more obvious in the carbon sequestration service.

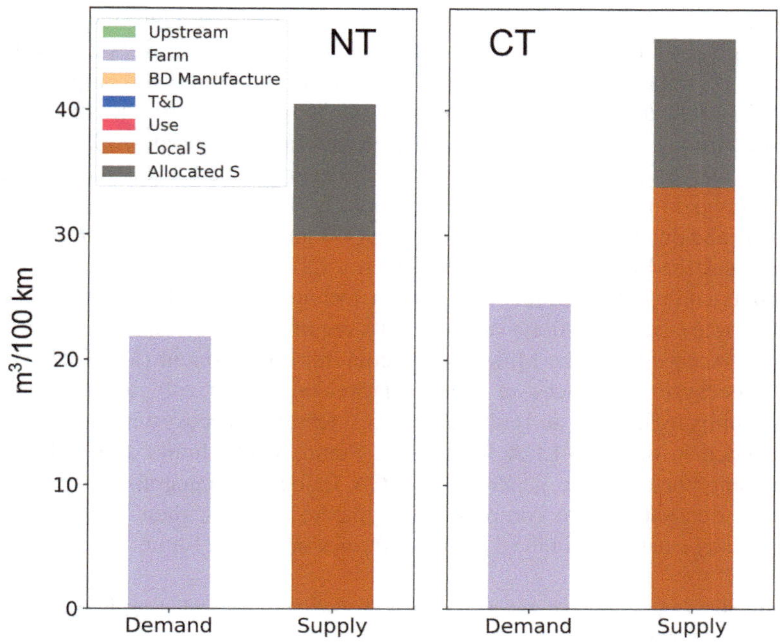

Fig. 20.9 Water-provisioning service for biofuel life cycle (T&D transport and distribution, Local S local supply, Allocated S allocated supply from coarser scales, NT no-tillage, CT conventional tillage)

20.4.2 Water-Provisioning Ecosystem Service

Similar analysis is done for the ES of water provisioning. As we can see from Fig. 20.9, most of the water is demanded by the farming phase, since most of the water leaves the crop through evapotranspiration. Thus, in this case study, precipitation is assumed as the water supply, and evapotranspiration is considered as the water demand of the farm. Water supply is estimated at each watershed scale and allocated to each process based on their water consumption. For the whole life cycle of biodiesel, water demand is smaller than the water supply indicating sustainability at local and serviceshed scales. This is also seen in Fig. 20.8. The difference between non-tillage and tillage farming is relatively small.

20.5 Conclusions and Future Work

This chapter applies a systematic methodology, TES-LCA, to evaluate the absolute environmental impacts of soybean-based biodiesel from the life cycle point of view.

Based on the framework of ecosystem services, TES-LCA bridges the gap between technological and ecological modules. The TES-LCA framework quantifies the capacity of nature through biophysical models. Environmental impacts of the selected system are compared with nature's carrying capacity for the ecosystem services of carbon sequestration and water provisioning. This framework is multiscale in nature and covers spatial scales from local to serviceshed scales. This multiscale nature brings in high geographical resolution. It identifies potential hotspots in the life cycle and quantifies spatial specific supply, demand, and characterization factors which break the limitation of geographical resolution.

Understanding the connection between technological and ecological systems, and their relevance to absolute sustainability, requires consideration of all ecosystem services. According to the Millennium Ecosystem Assessment (MEA 2005), there are 24 ecosystem services in total and divided into 4 categories: supporting, provisioning, regulating, and cultural services. Ideally, all ecosystem services should be included in the TES-LCA framework. Future work should aim at building a more comprehensive and general TES-LCA framework integrating all ecosystem services. Ecosystems are complex and dynamic systems; their temporal aspects should also be captured in the TES-LCA framework in the future.

Acknowledgments This work was supported by the National Science Foundation [grant numbers SBES-1739909 and CBET-1804943].

References

Bakshi, B. R., G. Ziv, and M. D. Lepech. 2015. Techno-ecological synergy: A framework for sustainable engineering. Environmental Science & Technology 49(3): 1752–1760.

Bjørn, Anders, and Michael Z. Hauschild. Absolute versus Relative Environmental Sustainability: What can the cradle-to-cradle and eco-efficiency concepts learn from each other?. Journal of Industrial Ecology 17.2 (2013): 321–332.

CONCAWE (Oil Companies' European Association for Environment, Health and Safety in Refining and Distribution, but the acronym is derived from "CONservation of Clean Air and Water in Europe"), Joint Research Centre of the EU Commission, and European Council for Automotive R&D, 2004. Well-to-Wheels Analysis of Future Automotive Fuels and Powertrains in the European Context, Version 1b, January, 60 pp., http://ies.jrc.cec.eu.int/Download/eh.

Göran Finnveden, Hauschild, M. Z., Ekvall, T., Jeroen Guinée, Heijungs, R., & Hellweg, S., et al. (2010). Recent developments in life cycle assessment. Journal of Environmental Management, 91(1), 1–21.

Huo, H. , Wang, M. , Bloyd, C. , & Putsche, V. . (2009). Life-cycle assessment of energy use and greenhouse gas emissions of soybean-derived biodiesel and renewable fuels. Environmental Science & Technology, 43(3), 750–756.

Koc, A. Bulent, Mudhafer Abdullah, and Mohammad Fereidouni. Soybeans processing for biodiesel production. Soybean-application and technology 19 (2011): 32.

Le Quere C, Andres RJ, Boden T, Conway T, Houghton RA, House JI, Marland G, Peters GP, Van der Werf G, Ahlström A, et al. 2012. The global carbon budget 1959–2011. Earth System Science Data Discussions 5(2): 1107–1157.

Liu, X., Bakshi, B.R., 2018. Ecosystem services in life cycle assessment while encouraging techno-ecological synergies. J. Ind. Ecol. https://doi.org/10.1111/jiec.12755.

Liu, Xinyu, and Bhavik R. Bakshi. Ecosystem services in life cycle assessment while encouraging techno-ecological synergies. Journal of Industrial Ecology 23.2 (2019): 347–360.

MEA. 2005. Ecosystems and human well-being: Wetlands and water. Washington, DC: World Resources Institute.

National Centers for Environmental Prediction. https://www.noaa.gov/jetstream/ncep. Accessed: 2022-05-10

Panichelli, Luis, and Edgard Gnansounou. GIS-based approach for defining bioenergy facilities location: A case study in Northern Spain based on marginal delivery costs and resources competition between facilities. Biomass and Bioenergy 32.4 (2008): 289–300.

Panichelli, L. , Dauriat, A. , & Gnansounou, E. (2009). Life cycle assessment of soybean-based biodiesel in Argentina for export. The International Journal of Life Cycle Assessment, 14(2), 144–159.

Pradhan, A. , Shrestha, D. S. , Mcaloon, A. , Yee, W. , Haas, M. , & Duffield, J. A. . (2011). Energy life-cycle assessment of soybean biodiesel revisited. Transactions of the ASABE, 54(3), 1031–1039.

Schmidheiny, S. and B. Stigson. 2000. Eco-efficiency: creating more value with less impact. World Business Council for Sustainable Development.

Steffen, W., K. Richardson, J. Rockstro öm, S. E. Cornell, I. Fetzer, E. M. Bennett, R. Biggs, S. R. Carpenter, W. de Vries, C. A. de Wit, C. Folke, D. Gerten, J. Heinke, et al. 2015. Planetary boundaries: Guiding human development on a changing planet. Science 347(6223): 1259855.

Tallis, Heather, et al. A global system for monitoring ecosystem service change. Bioscience 62.11 (2012): 977–986.

U.S. Environmental Protection Agency. https://www.epa.gov/. Accessed: 2022_05–10

USDA. 2010a. Economic research service: Arms farm financial and crop production practices. http://www.ers.usda.gov/ data-products/arms-farm-financial-and-crop-production-practices/ tailored-reports-crop-production-practices.aspx. Accessed March 2016.

USDA. 2010b. National agricultural statistics service. https://www.nass.usda.gov/. Accessed: 2022–03–20

USEPA. 2011a. Greenhouse gas emissions from large facilities. https://ghgdata.epa.gov/ghgp/main.do Accessed: 2022-03-20

USEPA. 2011b. National emissions inventory data. https://www.epa.gov/air-emissions-inventories/2011-national-emissions-inventory-nei-data. Accessed: 2022-03-20.

USGS. 2010a. Geo data portal. http://cida.usgs.gov/gdp/. Accessed: 2022-03-20.

USGS. 2010b. Water use in the United States. http://water.usgs.gov/watuse/. Accessed: 2022-03-20.

Johan Rockstrom, Will Steffen, Kevin Noone, Asa Persson, F Stuart Chapin III, Eric Lambin, Timothy M Lenton, Marten Scheffer, Carl Folke, Hans Joachim Schellnhuber, et al. Planetary boundaries: exploring the safe operating space for humanity. Ecology and society, 14(2), 2009.

Wallner, Heinz Peter, Michael Narodoslawsky, and F. Moser. Islands of sustainability: a bottom-up approach towards sustainable development. Environment and Planning A 28.10 (1996): 1763–1778.

Wang, Michael, et al. Summary of Expansions and Updates in GREET® 2021. No. ANL/ESD-21/16. Argonne National Lab.(ANL), Argonne, IL (United States), 2021.

Wernet, G., Bauer, C., Steubing, B., Reinhard, J., Moreno-Ruiz, E., and Weidema, B., 2016. The ecoinvent database version 3 (part I): overview and methodology. The International Journal of Life Cycle Assessment, [online] 21(9), pp.1218–1230. Available at: <http://link.springer.com/10.1007/s11367-016-1087-8> [Accessed 01/13/2022].

Chapter 21
Designing for Resilience and Sustainability: An Integrated Systems Approach

Joseph Fiksel and Bhavik R. Bakshi

21.1 Risk Management Challenges in an Age of Turbulence

For both human communities and business enterprises, the global risk landscape has become more volatile and uncertain due to increasing economic complexity and environmental turbulence. Catastrophic disruptions may emerge from unforeseen interactions among a variety of stresses and shocks and thus are difficult to forecast with any confidence. These threats have risen dramatically, for example, the frequency of natural disasters has increased tenfold since the 1960s (Institute for Economics and Peace 2020). Moreover, economic globalization has created long supply chains with critical interdependencies that may create vulnerabilities in times of crisis.

In the commercial world, most large enterprises have placed greater emphasis on improving their risk management and business continuity processes. However, these established practices are inadequate to cope with the mounting challenges that companies face including both natural disasters and anthropogenic trends, such as geopolitical unrest. Even before the COVID-19 pandemic, companies were experiencing greater frequency and severity in the occurrence of supply chain disruptions (McKinsey Global Institute 2020). Similarly, many urban areas

J. Fiksel (✉)
The Ohio State University, Columbus, OH, USA

The George Washington University, Washington, DC, USA
e-mail: fiksel.2@osu.edu

B. R. Bakshi
School for Engineering of Matter, Transport and Energy, School of Sustainability, School of Complex Adaptive Systems, Arizona State University, Tempe, AZ, USA

Lowrie Department of Chemical and Biomolecular Engineering, The Ohio State University, Columbus, OH, USA

B. R. Bakshi (ed.), *Engineering and Ecosystems*,
https://doi.org/10.1007/978-3-031-35692-6_21

around the world, especially in coastal regions, have been struggling to ensure environmental security and reliability of operations.

We can distinguish two major types of challenges (Fiksel 2015):

- *Gradual stresses*, such as climate change, sea-level rise, population growth, urbanization, infrastructure deterioration, pipeline corrosion, deforestation, natural resource depletion, and rising income gaps between the poor and the wealthy. Some types of gradual change, notably sea level rise, may not raise concerns until severe consequences become evident.
- *Sudden shocks*, such as hurricanes, tsunamis, industrial accidents, power failures, economic collapses, pandemics, terrorist attacks, and political upheavals. In some cases, small-scale disruptions, such as a facility structural failure or a regulatory policy change, can trigger a cascade of events that develop into a crisis.

When these stresses and shocks occur simultaneously and interact with one another, the results can be disastrous. For instance, in 2012, Hurricane Sandy pounded the northeastern US coastline, which had gradually become more vulnerable to flooding due to rising sea level. As a result, much of New York City and New Jersey lost power and water service for weeks, and economic losses totaled about $70 billion (Gibbens 2019).

In the case of the COVID-19 pandemic, the global scale and swift spread of the virus was unimaginable until it actually unfolded. Beyond the immediate toll in terms of both human lives and economic impacts, the long-term consequences will be felt for many years to come. Some business sectors, such as cinemas, may be permanently diminished, while others, such as remote conferencing, may thrive. Indeed, past catastrophic events, such as the 1984 methyl isocyanate release in Bhopal, India, the 2001 attack on the World Trade Center in New York, and the 2011 tsunami that destroyed the Fukushima nuclear plant in Japan, represent historical inflection points that altered market dynamics and corporate strategies around the world.

The World Economic Forum (WEF) publishes an annual report called the Global Risks Report (WEF 2016). An international group of experts representing the financial industries, risk management leaders, economists, and others, collaborate to develop a global assessment of the threats that confront the world today—including ecological, economic, political, and social forces. As mentioned above, these threats can take the form of sudden shocks or slow-moving stresses. WEF has increasingly recognized that it is challenging to anticipate these global risks because their consequences are interconnected. For instance, the combined effects of climate change and political instability have brought some countries closer to critical thresholds for other types of risks, such as political upheaval and economic collapse.

An emerging lesson for risk management is the futility of focusing on one crisis at a time, because the interplay of shocks and stresses can generate cascading consequences, giving rise to "systemic" risks. Existing management tools, such as risk analysis, are inadequate for understanding or predicting the collective impact of these complex forces upon a business enterprise. In particular, estimating the likelihood and magnitude of cascading risks in complex economic networks is a dubious exercise. WEF has observed that the volatility, complexity, and ambiguity of the global economy calls for a resilience imperative, which will require strong collaboration among diverse stakeholders, including business, government, and civil society (WEF 2016).

Beginning in the 1990s, an academic-industry consortium—the Treadway Commission—developed the concept of enterprise risk management (ERM), an approach commonly used in Fortune 1000 corporations (COSO 2004). Indeed, many companies appoint a Chief Risk Officer to oversee the ongoing implementation and monitoring of ERM. Recognizing that some level of risk is inescapable, the process involves the following steps:

- Identifying the portfolio of risks that may affect various business units
- Determining the corporation's "risk appetite" for each line of business
- Prioritizing potential threats or events that represent material risks
- Assessing the likelihood and magnitude of significant risks
- Responding to incidents that may occur
- Using risk control strategies, including insurance, to achieve the appropriate level of risk

While the conventional ERM approach is adequate for preserving business continuity in the face of routine operational risks, it has a number of key limitations in the context of the new challenges described above, specifically the following (Fiksel 2015):

- Risks cannot always be anticipated; in fact, rare events that can cause considerable damage are often unpredictable. One of the most common triggers is simple human error—caused by fatigue or distraction. Thus, it is impractical for companies to investigate all the potential risks that may be hidden in their global supply chains.
- Risks may be hard to quantify. Even if risks are identifiable, the lack of reliable statistical information makes it difficult to assess the most significant threats, namely, low-probability, high-consequence events. When focusing on business goals, managers may underestimate remote risks that they have never experienced.
- Adaptation may be needed to remain competitive. Conventional business continuity practices are aimed at returning to "normal" operations. Instead, companies should strive to learn from disruptions, understand the root causes, and adapt their assets and business models to overcome potential weaknesses.

Accordingly, we argue that both companies and communities need to enhance their risk management processes by building adaptive capacity, especially in the areas of supply chain continuity and resource management, including energy and water.

As shown in previous chapters, industrial enterprises and urban communities are heavily dependent on the availability of ecosystem services. However, most risk management practitioners do not account for potential loss of ecosystem integrity. Environmental forces such as increasing droughts, severe storms, and sea level rise, amplified by industrial resource extraction and urban development, are posing new threats to basic ecosystem services and amenities. These threats will only increase based on the current trajectory of economic development and population growth. For instance, the US Government Accountability Office (GAO) has estimated that over 30% of the thousands of US facilities that handle hazardous chemicals are increasingly exposed to disruption due to natural hazards that are exacerbated by climate change (GAO 2022). We argue that risk management should be extended to account more explicitly for ecosystem-related exposures and that additional work is needed on developing relevant tools and processes to inform risk management decisions.

21.2 Working Toward Resilience

To operate effectively in the face of an increasingly turbulent environment, companies and communities will need to anticipate and embrace change rather than resisting it (Fiksel et al. 2015). From a strategic perspective, industry and government leaders can consider three main strategies for coping with turbulent change:

- Resist change by hardening defenses and trying to maintain stability
- Anticipate change by preparing for disruptions based on experience and foresight
- Embrace change by designing a resilient organization that is capable of adapting to unexpected challenges

In the past, managers have pursued stability as the desired state of affairs. When a disaster strikes, the first instinct is to overcome the shock, assist the victims, and return to a stable equilibrium as soon as possible. But if the quest for stability is futile, then the best strategy may be to accept change as inevitable and improve the capacity for rapid response and adaptation. We argue that to achieve consistent success in the future, companies and communities must become more resilient. Accordingly, we have defined *resilience* as "the capacity to survive, adapt, and flourish in the face of turbulent change and uncertainty" (Fiksel 2006).

We can distinguish between two main types of resilience (Economist Intelligence Unit 2021):

- *Operational* resilience is the process of "bouncing back" from a disruption—similar to the concept of business continuity. This is a tactical approach that requires advance planning and real-time decision-making.
- *Strategic* resilience is the process of "bouncing forward" by adapting to a changing environment and improving responsiveness to sudden disruptions. This requires long-term thinking and learning from the collective experience of similar organizations.

In the business world, experience has shown that large enterprises tend to lose their resilience as they grow and mature—they become vulnerable to surprises and slow to recover from disruptions. Companies that emphasize stability may cling to outmoded practices and proven technologies, may fail to question their assumptions, and may have blind spots that hamper their recognition of external change. As a consequence, they are unable to react to external challenges until they reach a state of crisis and require a drastic intervention.

Conversely, companies that embrace change are better positioned to identify and seize emerging opportunities more nimbly than their competitors. Innovative companies in the USA and abroad have begun to view resilience as a source of competitive advantage (Fiksel 2015). Such companies have developed new business processes to supplement their established risk management protocols, including continuous monitoring of external situations and strategic capabilities for agility and adaptation. Thus, they are able to thrive in a constantly changing environment, consistently delivering shareholder value.

In spite of the turbulence around them, resilient companies find a way to survive and prosper by maintaining high performance while being alert and prepared for emerging challenges. They accept the inevitability of surprises and are able to adapt gracefully, sometimes transforming their very structure. In the words of Andrew Grove, former CEO of Intel: "Bad companies are destroyed by crises; good companies survive them; great companies are improved by them" (Yu 1998). Indeed, as described in the following sections, progressive companies such as Dow-DuPont, DHL, Toyota, and others have adapted to turbulent change by developing resilience strategies and often have benefited by discovering new business opportunities.

Likewise, governmental organizations at the local, state, and federal levels have begun to recognize the urgent need for improved resilience and are developing new initiatives to incorporate these into policy, planning, and service operations. For instance, following Hurricane Sandy, both the city and the state of New York developed plans for improving the resilience of New York and the surrounding region. These included multiple recommendations for rebuilding the communities that were victimized by Hurricane Sandy and for increasing the resilience of infrastructure and buildings (New York City 2016).

At the federal level, several agencies, including the Federal Emergency Management Agency (FEMA) and the National institute for Standards and Technology (NIST), have actively promoted the concept of resilience to floods and other natural disasters. More recently, the US Government Accountability Office (GAO) has published a framework for taking action to increase disaster resilience throughout the nation (GAO 2019).

Another bold effort was the launch of the 100 Resilient Cities initiative in 2013, focusing on the environmental and social factors that enable a city to remain healthy, vibrant, and diverse. Over the course of 6 years, the Rockefeller Foundation awarded $1 million apiece to 100 selected cities around the world, with the intent of providing four types of pragmatic support (Rockefeller Foundation 2015):

- Establishment of a new position in city government, a Chief Resilience Officer

- Leveraging available expertise for development of a robust resilience strategy
- Access to solutions, service providers, and partners to support the resilience strategy
- Membership in a global network of cities who can learn from and help each other

This initiative envisioned the development of resilient cities that provide disaster protection, especially for vulnerable populations, improved health and economic opportunities for residents, and a flourishing business environment with reduced risk. Many of the 100 cities found new insights through sharing of experiences and lessons learned. While the ultimate results will take years to emerge, the program demonstrated that true resilience is not just about responding to disasters but also about dealing with stresses such as unemployment, urban violence, and food or water shortages. One of the most important lessons was that resilient cities can turn tragedy into opportunity by adopting a "build back better" mentality.

21.3 Strategies for Fostering Resilience

The Center for Resilience at The Ohio State University, founded in 2005, partnered with a wide range of companies to explore practical applications of enterprise resilience concepts. The center's research found that leading organizations pursue a variety of resilience strategies. As illustrated in Fig. 21.1, these can be divided into four types, depending on the magnitude and abruptness of change (Fiksel 2015):

- First is *steer and adjust*: In dealing with routine fluctuations, which usually have discernable patterns, companies can learn from experience and respond effectively. An example is inventory management based on seasonal demand forecasting.
- In situations where disruptions are more abrupt, such as natural disasters or derailments, companies require a *sense and respond* strategy. The risk management process must include emergency preparedness, alertness to early warning signals, and flexible response capacity to ensure business continuity. An example is DHL's rapid response to a cloud of volcanic ash that grounded air traffic in most of Europe.
- Over time, fundamental changes in the business environment may gradually erode a company's viability or competitive advantage, calling for a strategy of *adapt and transform*. Trend forecasting and scenario planning can help to identify important paradigm shifts, such as the advent of autonomous vehicles or alternative energy sources.
- Finally, in the face of catastrophic disruptions, the appropriate strategy is to *survive and flourish*. "Business as usual" may no longer be viable, and disaster recovery is merely a survival strategy. To flourish in a turbulent world, companies must develop adaptive capacity and embed resilience into their business processes. For instance, in the wake of the Fukushima disaster, Toyota

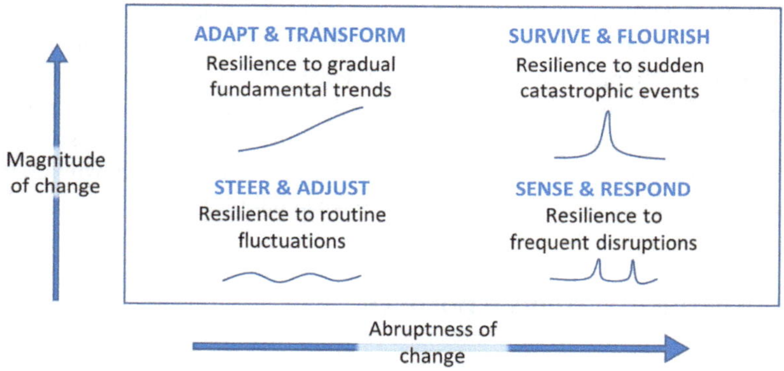

Fig. 21.1 Enterprise resilience strategies

took drastic steps to limit future business disruptions to no more than 2 weeks (Industry Week 2014).

An important consideration in development of resilience strategies is the concept of "inherent" resilience. This goes beyond simply adding layers of protection or redundancy, which often incur extra costs. Rather, inherence is an intrinsic property woven into the design of important assets and business processes, thus offering "no regrets" in the absence of disruptions. An example of inherent resilience in supply chain management is geographic dispersion of multiple suppliers and manufacturing facilities, placing them closer to key customer locations. This strategy can reduce logistical costs and lead times while providing the capacity for business continuity if a single facility is suddenly disabled. Another example, discussed at length in a later section, is the adoption of a "circular" business model.

A key insight from the pursuit of enterprise resilience is the need for "systems thinking" to understand the interdependencies between corporations and the broader environment in which they operate. For instance, a well-known set of interdependencies is the "water-energy-food" nexus, implying that disruption of one critical resource flow can generate shortages of other resources (Walker 2020). As discussed below, systems thinking can help to anticipate the hidden or unintended consequences of key decisions, such as introducing new policies, technologies, and business practices. For instance, a frequent consequence of unconstrained land development is the degradation of important ecosystems such as wetlands, thus threatening the long-term sustainability of global economic systems and human communities.

More generally, it is important to recognize the relationship between resilience and sustainability from a business perspective. Sustainability is a long-term concept—considering how decisions that governments or companies make today will influence the well-being of both present and future generations. Sustainability goals such as those advanced by the United Nations are based on an idealistic view of productive harmony between humans and the environment (United Nations

2022). Resilience is more of a real-time concept—considering how companies or communities can overcome unexpected disruptions while building an organization that embraces change. In practice, resilience and sustainability are closely intertwined. A sustainable system is generally more resilient to disruptions, and a resilient system is more sustainable in the long run. In other words, sustainability and resilience are distinct but mutually reinforcing. However, as shown below, there may also be conflicts between resilience and sustainability.

21.4 Taking a Systems Approach

To help guide decision-making in complex systems, we have developed a comprehensive approach called the Triple Value (3V) framework, which enables integrated modeling and analysis of economic, social, and environmental systems and reveals vulnerabilities that may cause disruptions (Fiksel et al. 2014). As illustrated in Fig. 21.2, the 3V framework explicitly represents dynamic flows among three nested domains—economy, society, and the environment—and helps to identify the potential impacts of alternative decisions or choices upon the resilience and sustainability of both human and ecological resources. Systems thinking teaches us to be conscious of our dependencies on environmental systems. Progressive organizations will strive to achieve sustainable operations—a dynamic equilibrium in which resource flows are balanced with economic and social well-being. However, if ecological systems are fragile or threatened, the resilience of human systems may be compromised.

The key interactions among the three domains in Fig. 21.2 are as follows:

- *Enterprises* need both human and environmental resources to fulfill economic demands. They extract resources from the environment, including energy,

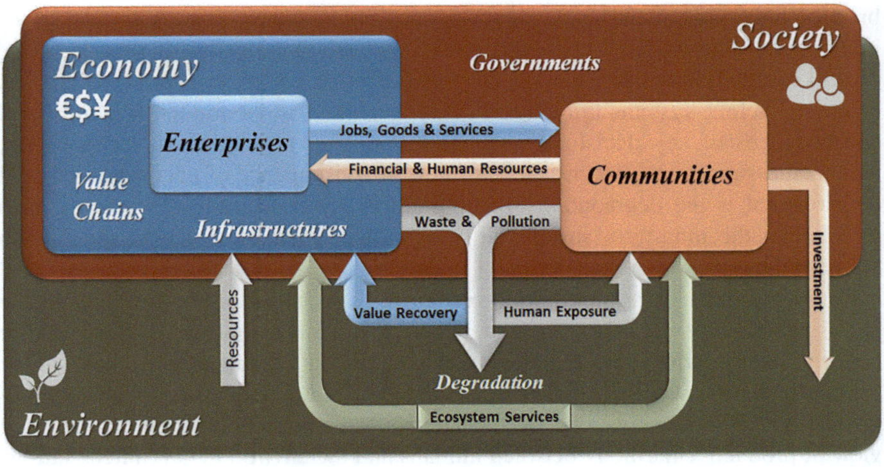

Fig. 21.2 The Triple Value framework (simplified overview)

materials, water, and food. Their productive assets are *built capital* including infrastructure and *intellectual capital* including technologies. They add economic value through manufacturing and supply chain operations, delivering products and services that provide value to society. However, they generate industrial wastes that may either be recycled for value recovery or deposited into the environment. Enterprises also generate shareholder value, create jobs, and generally improve community prosperity.

- *Communities* consume products and services and also generate wastes. They benefit directly from natural amenities, e.g., clean air, clean water, recreation, and psychic enjoyment. Human health and well-being may be affected by changes in economic or environmental flows, such as hazardous wastes and greenhouse gas emissions. *Human and social capital* deliver economic value to industry by providing essential workforce skills and market stability. In addition, governments can generate environmental value by protecting and restoring environmental resources.
- *Environmental* systems contain reservoirs of natural resources, including renewable resources (e.g., forests) that can be replenished over time, nonrenewable resources (e.g., petroleum), and finite environmental media (e.g., air, water, and land) that may become degraded. Ecological productive capacity is known as *natural capital*, and the flow of ecosystem goods and services delivers value to both industry and society.

The 3V framework has been used as a foundation for integrated assessment of strategic decisions in order to understand both direct and indirect impacts, including unintended consequences. For instance, the US Environmental Protection Agency developed a 3V modeling toolkit to evaluate strategies for mitigating nitrogen pollution in southern New England, accounting for both economic development and quality of life (Fiksel et al. 2014).

As shown in Fig. 21.2, business enterprises are directly dependent on natural resources. Most companies require land for siting their operating facilities. Most manufacturing companies require raw materials, including minerals, water, biomass, fuels, and other commodities, that are extracted from the environment. In addition, from a product life cycle perspective, companies are dependent on the availability of environmental resources, including ecosystem services, to support their extended supply networks. Of particular importance is the continuous availability of utilities and infrastructure, which are heavily dependent on environmental resources and are especially vulnerable to environmental disturbances such as earthquakes.

At the same time, there are a variety of environmental pressures that can impact business continuity and impose additional costs or delays. These include risks of noncompliance with regulatory restrictions on environmental releases or discharges of waste and emissions, as well as requirements for occupational and public health and safety. Technological failures or human errors can lead to incidents such as chemical spills, pipeline leaks, or accidental process failures that can result in significant costs and loss of goodwill. Moreover, enterprises may be liable for environmental problems caused by other parties due to negligence or noncompliance.

Practically speaking, it is impossible to eliminate all the risks that industrial systems encounter in a complex business environment. Indeed, experience has shown that mechanistic systems based on strict logical rules cannot cope with events that the designers failed to anticipate. Engineered systems, including electronic devices, buildings, and utility networks, are vulnerable to sudden failure or collapse. They are generally brittle—the opposite of resilient. Technological advances such as artificial intelligence can help to improve robustness, but engineering solutions tend to focus on known challenges rather than preparing for the unexpected.

In contrast, resilient organizations are able to avoid failure because they behave like living organisms, sensing, responding, and adapting to change. In the natural world, resilience is seen everywhere from individual cells to entire ecosystems. Similarly, human beings possess extraordinary resilience at many different scales, from individual people to metropolitan areas to entire cultures. Humans have the unique advantage of foresight. Rather than letting natural selection take its course, they can quickly adapt to a changing environment by redesigning their institutional and technological processes.

21.5 Sustainability, Resilience, and Circularity

As mentioned above, sustainability and resilience may not always be synergistic. There are situations where sustainability and resilience are opposing rather than reinforcing. Figure 21.3 illustrates several different situations in the context of supply chain management, with specific indicators representing resilience and sustainability.

- *Lower left*: In an age of turbulence, business as usual is neither sustainable nor resilient—innovative strategies are required.
- *Upper left*: Leaner production methods may reduce waste and ecological impacts, but achieving resilience often requires significant investment in reserve capacity.
- *Lower right*. Supply chain managers can expand their inventory buffers to help ensure business continuity, but this is more costly and tends to increase ecological resource consumption. It is challenging to become both lean and agile.
- *Upper right*. For supply procurement, local sourcing is an increasingly popular option that requires less energy, stimulates local economies, and avoids the risks of importing goods from distant sources.

The 3V framework is helpful for quantifying the benefits of an increasingly popular business approach called "circular economy," which strives to eliminate waste in industrial supply chains and promises to enhance both resilience and sustainability (Stahel 2016). As illustrated in Fig. 21.2, there are many opportunities to recover

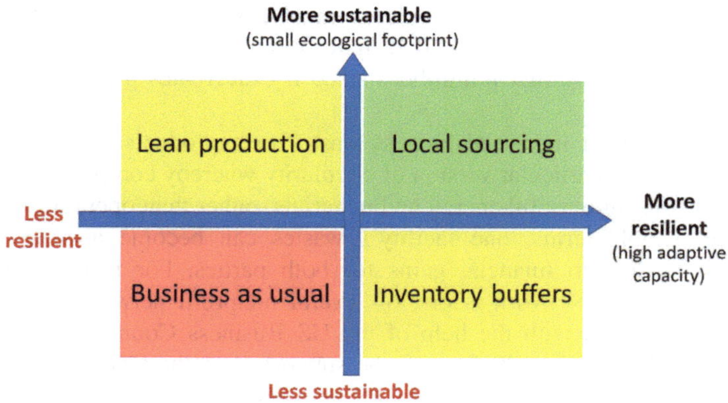

Fig. 21.3 Supply chain management synergies and trade-offs

value from both industrial and post-consumer waste streams, offering substitutes for nonrenewable materials, including fuels. Shifting from the traditional linear "cradle to grave" business model of industrial systems to a closed-loop model resembles the cyclical flows of natural ecosystems. In nature, there is no waste—one creature's wastes become another creature's nutrients. Thus, rethinking conventional product and process technologies can lead to the discovery of innovative pathways for transforming waste materials and energy into economically valuable resources. This approach reduces the need for virgin natural resources, mitigates environmental pollution and greenhouse gas emissions, and alleviates the burden on limited landfill space.

Due to population growth, urbanization, and poverty reduction, the demand for natural resources is projected to double within a few decades (OECD 2019). At the same time, global waste is expected to grow to 3.40 billion metric tons by 2050, more than double population growth over the same period (Kaza et al. 2018). Thus, reuse of industrial and consumer wastes represents an enormous market opportunity, estimated to be $4.5 trillion by 2050 (Lacy and Rutqvist 2015). The direct economic benefits of circularity, whereby waste materials replace virgin materials as inputs to industrial processes, include lower feedstock costs, lower waste disposal costs, lower environmental liabilities, and supply chain risk reduction (Fiksel et al. 2021).

In pursuit of sustainability, progressive companies have sought to achieve "zero waste" by finding alternative uses for discarded materials and closing the loop in their supply chains, thus increasing the recycled content of their products. For instance, automotive manufacturers have been aggressively increasing the recycled content of metals and plastics in vehicles and have discovered innovative uses for organic wastes such as plant fibers, coffee bean chaff, and potato peels (Taub 2021). Circularity not only offers economic benefits and reduces ecological impacts but also tends to increase business and community resilience by reducing dependence upon scarce natural resources and long-distance supply chains. The US Army has recognized this opportunity and pledged to work toward "net zero" at all of its

US installations in order to minimize its vulnerability to shortages of critical water and energy supplies (U.S. Army 2013). Indeed, the Department of Defense views climate change as a "threat multiplier" since it exacerbates global challenges to national security.

The conversion of industrial process wastes to byproducts, called *by-product synergy* (BPS), is a particular version of circularity whereby companies collaborate to convert wastes into useful energy and materials, rather than operating as isolated entities. In simple terms, one facility's wastes can become another facility's feedstocks, leading to financial gains for both parties. For instance, the Ohio Byproduct Synergy Network is one of several nonprofit networks developed in the USA and abroad with the help of the US Business Council for Sustainable Development. Founded in 2009, with the support from the Center for Resilience at Ohio State and the US EPA, this BPS network sponsors collaboration among regional companies to convert industrial wastes into feedstocks for other processes, effectively turning waste into profit while creating local jobs and strengthening supply chain resilience. This business-to-business approach is much more efficient and environmentally benign than collection and reuse of municipal solid waste.

Despite these benefits, there may be trade-offs involved in the practice of circularity. While waste elimination is arguably more sustainable, in certain situations it may compromise resilience. For instance, waste recovery and reuse may be economical only within a local region due to transportation costs. While local sourcing decreases environmental impacts, it also implies less dispersion of the supply network and tends to increase the potential for weather or infrastructure disruptions. Another downside is that the availability and quality of waste streams may be less consistent than virgin materials. Moreover, virgin materials often cost less than recycled materials; for example, when governments provide subsidies for resource extraction industries, such as mining, the costs of environmental impacts are typically not internalized into resource prices. Therefore, as described below, analytic methods are needed to support decision-making about balancing resilience with other goals.

21.6 Making the Business Case for Resilience

The main barrier to broader adoption of resilience strategies is concern about financial expenditures. Current risk management practices typically are based on loss avoidance, and there is a tendency for business leaders to underestimate the likelihood of theoretical losses while giving priority to profit-oriented expenditures. Therefore, it is important to emphasize the value proposition for resilience, including direct benefits in terms of shareholder value, as well as indirect benefits in terms of strategic positioning and stakeholder satisfaction. The direct benefits can be captured explicitly though an accounting tool that is favored by chief financial officers, which measures "economic value added" (EVA) as follows:

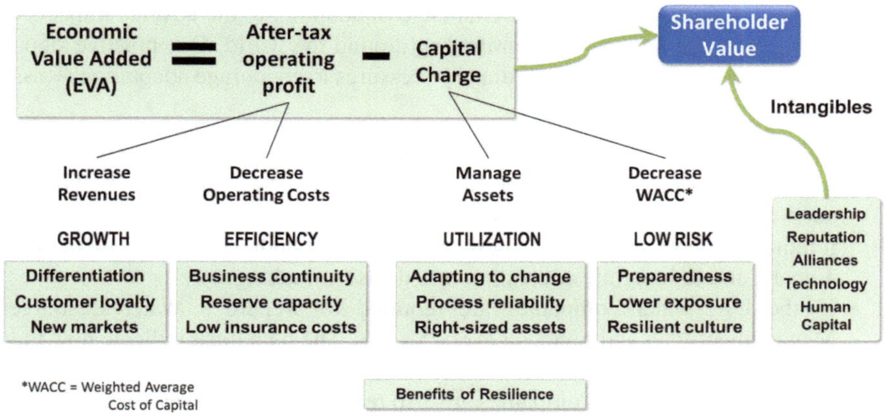

Fig. 21.4 How resilience contributes to shareholder value

$$\text{EVA} = \text{After-Tax Operating Profit} - \text{Capital Charge}$$

Thus, economic value can be created either by increasing cash flow or by reducing the capital costs required to generate cash flow. Note that the bottom line, corresponding to the profit and loss statement, is just one part of EVA. The other part is concerned with return on assets, as reflected in the balance sheet. Figure 21.4 presents the four main strategic *levers* by which a resilient company can increase EVA (Fiksel 2015):

- *Top-line growth*: increase revenues and expand market share by being more agile than competitors and building customer loyalty
- *Operating efficiency*: ensure business continuity and reduce costs of disruptions
- *Asset utilization*: reduce complexity through process simplification and operational flexibility
- *Risk reduction*: reduce financial exposure through incident prevention and preparedness for crises

Moreover, shareholder value is heavily influenced by *intangibles* such as leadership, reputation, alliances, technology, and human capital. These are at least as important as EVA, because they reflect future expectations rather than past performance (Low and Kalafut 2002).

Among the most important factors driving the need for enterprise resilience are global warming and climate turbulence, which are causing increased severity and frequency of storms, hurricanes, wildfires, droughts, and floods. According to a 2020 survey of over 300 companies, building supply chain resilience was one of the top three management priorities (ASCM 2021), and the top-ranked strategy for building resilience was making supply chains more socially and environmentally sustainable. Yet most businesses and nongovernmental organizations have focused on reducing carbon emissions. While emission reduction is essential, it is equally important to address the immediate challenge of adapting to the adverse impacts of climate change that are already manifesting around the world. One positive step is that governments have initiated regulatory pressures to encourage adaptation (Russo et al. 2021). For instance:

- The California Public Utilities Commission, which regulates energy, water, telecommunications, and public transportation services, has been working on adaptation since 2013 and has gradually tightened its requirements. Since 2020, it requires energy utilities to conduct vulnerability studies, to update them every 4 years, and to integrate adaptation into decision-making.
- In the UK, water companies are required to prepare a Water Resources Management Plan and review it every 5 years. The plan must indicate how they will ensure sufficient water resources to meet current and future demand over a minimum of 25 years. Companies are also required to submit a Climate Change Adaptation Action Plan to the UK Government every 4 years.
- European Union. A new strategy on climate change adaptation was adopted in February 2021, aiming for climate resilience by 2050. The principal objectives are smarter adaptation, informed by robust data and risk assessment tools; faster adaptation, focused on developing and rolling out solutions; more systemic adaptation, mainstreamed into all relevant policy fields; and driving international action through stronger global engagement.

To help raise awareness of climate risks and enterprise adaptation strategies, the French government has published a guide to business decision-making for climate adaptation that provides examples of three major types of decision methods (Russo et al. 2021):

- *Scenario planning* is a commonly used qualitative method for envisioning possible futures and testing company business models to develop a resilient strategy. For instance, Lendlease, a $15 billion Australian real estate group, developed four scenarios based on different levels of climate projections to 2050 and hypotheses of the underlying socioeconomic, technological, and environmental changes. These scenarios were translated into narratives and videos and were shared with 200 company managers during regional strategy workshops. The scenarios were a powerful driver of employee commitment and made it possible to integrate climate risks into both short- and long-term planning and decision-making processes.
- *Robust decision-making* is a computer-based method for testing alternative company strategies against a multitude of possible future scenarios generated by predictive simulation techniques. For instance, Anglian Water, a British water and wastewater services provider, implemented a robust investment strategy, with the objective of ensuring an adequate water supply-demand balance over the period 2020–2045. For the water supply component, a base-case investment plan was defined and then compared with approximately 60 scenarios generated by computer simulation to define alternative plans. By combining a multicriteria optimization model with stress-testing against extreme scenarios, the company

developed a strategic plan that minimizes the costs of climate adaptation while being sufficiently robust to cope with many possible futures.
- *Flexible adaptation pathways* is an approach for developing alternative adaptation trajectories and postponing investments by combining and sequencing actions over time. For instance, Los Angeles Metro is susceptible to various climate change risks, including heat waves, flooding, extreme rainfall, and fires. To make the system more resilient and to integrate such climate concerns into future facility development, LA Metro uses the flexible adaptation approach to progressively improve the outcomes of its climate strategy as knowledge improves and the impacts of climate change materialize.

21.7 Supply Chain Resilience Characterization and Analysis

The greatest threat to enterprise resilience is disruption of supply chains, which are only as strong as their weakest link and are generally outside the control of the purchaser. Globalization of trade has accentuated these vulnerabilities by creating longer supply chains with decreased visibility. As a result, modern supply chains are exposed to a broad variety of threats, including climate variability and natural disasters, interruption of energy and transport services, and political and economic fluctuations (ASCM 2021). In such a volatile business environment, "lean" strategies such as "just-in-time" manufacturing are no longer viable, and many firms are striving to increase their supply chain agility and buffer capacity. Some have reversed the trend toward offshoring by restoring domestic operations (Fiksel et al. 2015).

There is a broad range of options for improving supply chain resilience, and appropriate strategies vary depending on a company's specific characteristics. It is difficult and costly to gather quantitative data on global supplier characteristics and to monitor factors that may influence potential disruptions. In addition, making fundamental changes in supply chain facilities and infrastructure may require significant investments that require analytic justification. However other initiatives, such as improved supplier-customer collaboration, can be accomplished with existing resources while yielding ancillary benefits in terms of loyalty and trust.

A variety of methods and tools have been developed to support business decision-making regarding resilience improvement and risk reduction. These include:

- *Resilience indicators* for specific types of systems (products, processes, or assets) to enable system comparison, monitoring, and adaptive management. As illustrated below, qualitative indicators based on subjective assessment can support strategy development, although the results are necessarily approximate. In addition, key performance indicators (KPIs) can be used to quantify specific aspects of enterprise performance at a business or enterprise level and to measure progress in resilience improvement.

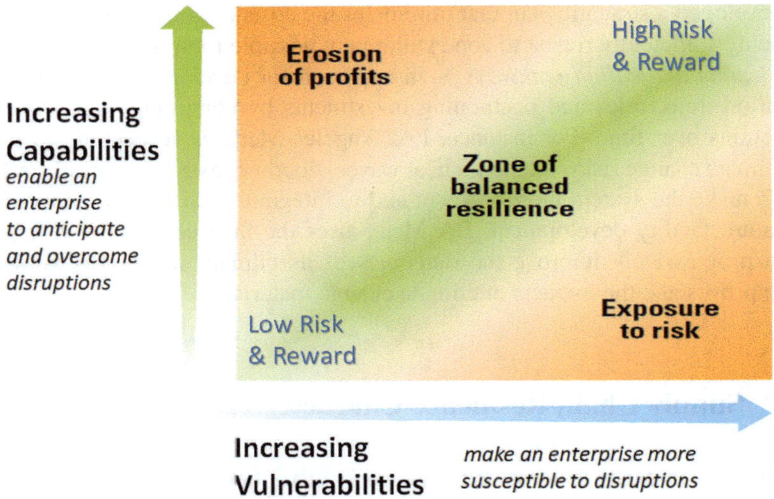

Fig. 21.5 The SCRAM™ approach: balancing resilience and capabilities

- *Predictive analytics* based on cause-effect logic enable companies to estimate the impacts of hypothetical disruptions, including unexpected feedback loops. An example discussed below is Dow Chemical's use of system dynamics combined with stochastic simulation. Even when probabilistic forecasting is not possible, scenario-based planning can help to identify enterprise vulnerabilities, anticipate potential disruptions, and develop resilience strategies. In addition, existing quantitative methods such as risk analysis and cost-benefit analysis are helpful for analyzing investment decisions, provided that adequate data are available.

Beginning in 2005, a multidisciplinary research team at Ohio State collaborated with a number of companies in diverse industries to develop a comprehensive process for supply chain resilience assessment and management (SCRAM™). This process engages a cross-functional team within the company to develop a resilience profile based on their qualitative judgments. The underlying SCRAM™ concept is illustrated in Fig. 21.5: the more a company's *vulnerabilities* increase, the more the company is *exposed to risk*. To counteract those exposures, companies can develop a variety of resilience *capabilities* that enable them to mitigate those risks. By identifying key business vulnerabilities and building appropriate capabilities, companies can prevent significant disruptions and achieve an acceptable level of risk (Pettit et al. 2013).

Through in-depth research, including numerous case studies, the Ohio State team identified six major categories of enterprise *vulnerabilities*:

- Turbulence—this can range from currency fluctuations to major natural disasters

Table 21.1 Supply chain capabilities and corresponding resilience indicators

Capabilities	Measurable factors	Qualitative factors
Flexibility: Sourcing	Supplier agility, alternate sources	Contractual options
Flexibility: Manufactg	Modularity, versatility, scalability	
Flexibility: Fulfillment	Distribution & service agility	
Capacity	Reserves, back-up resources	
Efficiency	Productivity, asset utilization	Quality, standards, maintenance
Visibility		Status monitoring, info exchange
Adaptability	Order re-routing ability	Gaming, innovation, learning
Anticipation	Forecasting effectiveness	Risk management, preparedness
Recovery	Equipment downtime	Crisis management, mitigation
Dispersion	Decentralized assets, markets	Distributed leadership, authority
Collaboration	Postponement of orders	Coordination, partnerships
Organization	Workforce flexibility	Adaptive, resourceful culture
Market position	Market share	Brand strength, customer loyalty
Security		Systems & procedures
Financial strength		Reserves, insurance, diversity
Product stewardship		Design, auditing, communciation

- Deliberate threats, including lawsuits, strikes, and industrial espionage
- External pressures, including regulatory changes, social movements, and competition
- Resource limits, including availability of raw materials, energy, water, or infrastructure
- Connectivity—this refers to the complexity of the supply and distribution networks
- Sensitivity of products or processes that require highly controlled environments

Similarly, the research team identified 16 major categories of capabilities. These are listed in Table 21.1, along with associated indicators that can be used to assess resilience in terms of either quantitative, measurable factors or qualitative, subjective factors.

Every line of business has a different pattern of vulnerabilities, so resilience strategies must be developed at the level of an individual business unit and in many cases at the level of a specific product family. Accordingly, the Ohio State team developed a detailed questionnaire that enables the business team to perform a self-assessment of its supply chain resilience—including both its vulnerabilities and the capabilities that it has to meet these challenges. Based on multiple applications of the SCRAM™ framework, a key insight emerged—in many cases companies have over-invested in resilience, so reducing their exposure to risk may actually result in diminishing returns and *erosion of profits*. Thus, as shown in Fig. 21.5, there is a zone of balanced resilience, where a company has deployed the right portfolio of capabilities to offset its specific vulnerabilities.

The typical result of the SCRAM™ process is a set of strategic recommendations for improving selected resilience capabilities. In some cases it may be possible to reduce key vulnerabilities, although these are often beyond the control of an individual company. These strategic recommendations can then be investigated through more detailed quantitative analysis in order to develop a business case for action. Examples of strategic objectives for resilience improvement include the following:

- Improving surge capacity and backup capacity
- Creating greater flexibility in sourcing and manufacturing
- Building agility and adaptability in responding to challenges
- Developing ability to anticipate and detect signals of change
- Dispersion of assets and resources geographically
- Improving supply chain visibility based on information technology
- Enhancing diversity of personnel backgrounds and perspectives
- Adopting security measures and broad intelligence gathering
- Strengthening market position and customer loyalty

The SCRAM™ approach was fully adopted at Dow Chemical, prior to the company's merger with Dupont. Under the leadership of top supply chain executives, Dow implemented the SCRAM™ process for more than 20 of its global business units, achieving significant business benefits. For instance, for the Glycol Ethers P-Series family of products, a cross-functional team identified several disruption scenarios for further analysis: production site shutdown, raw material supply outage, and internal raw material allocation shortage. They developed a simulation model to test the consequences of these scenarios and were able to confirm a 95% service level with their existing capabilities. This analysis led to right-sizing fixed assets and working capital, representing a $1.1 million savings for this business and a 500% return on modeling effort. The Dow team was recognized as a finalist in an innovation award competition by the Council for Supply Chain Management Professionals (McIntyre and Hemmelgarn 2011).

In another application involving significant capital investment, Dow utilized SCRAM™ together with simulation tools to implement a Design for Resilience process on behalf of Sadara Chemical Company, a joint venture of Dow and Saudi Aramco. The project involved designing the largest single chemical complex ever constructed in one phase, consisting of 26 manufacturing units occupying about 8 square kilometers. The company overcame a number of challenges, including a hostile desert environment, lack of transportation infrastructure and supplier network, and absence of a qualified workforce. The complex became fully operational in 2016, producing more than 3 million metric tons of product annually (Zavitz 2017).

21.8 Importance of Accounting for Ecosystem Services

Since the mid-twentieth century, most nations have benefited from economic and human development, but corresponding to these trends there have been significant negative impacts on ecosystems, which provide vital services to humanity. As described in Chap. 1, the results of this "win-lose" relationship between human activities and natural systems are clearly visible today, as degradation of ecosystems across the world has reduced their ability to provide goods and services that are essential for human well-being. According to the Millennium Ecosystem Assessment (2005), 15 out of 23 ecosystem goods and services are degraded or highly degraded. Biological diversity has dwindled due to species extinction as ecological habitats change over time. The World Economic Forum (2020) estimates that USD $44 trillion of economic value generation, which is more than half of the world GDP, is at risk due to ecological degradation.

In most cases, ecological degradation has gradual effects on human well-being. For instance, Chap. 6 describes the effect of pollinator loss on reduced crop yields and diminished nutritional content in nuts, fruits, and vegetables. If natural pollinators are absent, many flowers will produce fruit by self-pollination, but their nutritional value is lower than the products of pollination (Hoehn et al. 2008). Loss of biodiversity and species defaunation have ripple effects on dispersion of seeds and survival of many plant species. Moreover, global warming has not only disrupted existing ecosystems but also has increased climate volatility. These gradual ecosystem impacts may not become a concern for society-at-large until they reach a tipping point and cause a major disruption. Beyond this tipping point, the effects of ecological degradation often manifest as sudden shocks due to "natural" events such as hurricanes, heat waves, floods, and other extreme events.

Many engineering approaches to natural resource management aim to maximize the efficiency and predictability of the system. For instance, dams and canals have been utilized for water resource management since they reduce the effect of natural variability in the hydrological cycle and allow water to flow in a more efficient manner than the meandering flow of rivers. Similarly, the lumber industry has replaced old-growth forests with uniform plantations that expedite harvesting of wood. While such engineering approaches have been successful in reducing short-term fluctuations, eventually they erode the adaptive capacity of resource systems and make them more vulnerable to external perturbations. When subjected to extreme events, many engineered systems suffer from extensive disruption or collapse due to a lack of inherent resilience. One example mentioned earlier was the economic devastation caused by Hurricane Sandy, which was amplified by a gradual rise in sea level. In order to safeguard human communities and inform environmental policy, it is important to develop tools that account for the value of ecosystem services and the threats that they face.

Accounting for ecosystems and their services has at least two benefits for enhancing the resilience of human-designed systems. First, this approach provides opportunities to learn from nature about managing the trade-offs between efficiency and resilience. Over the millennia, due to evolutionary pressures, ecosystems have developed successful mechanisms for ensuring long-term resilience. Nature's emphasis is not on the efficiency of individual activities but rather on the effectiveness of the overall network. For instance, phenomena such as photosynthesis, food chains, and ecological succession have low efficiency. Photosynthesis converts only 4–8% of solar energy into biomass, many trees produce far more fruit than needed to germinate new trees and merely discard their leaves each autumn, and many predators do not fully consume their prey. However, the overall ecological system has high efficiency due to intense cycling of all materials and reliance of the system on renewable solar energy (Bakshi 2019). Mimicking such characteristics of ecosystems could help in developing sustainable and resilient supply networks, and this observation has inspired "industrial ecology" initiatives such as the circular economy movement mentioned above.

A second benefit of ecosystem accounting is enabling the explicit inclusion of ecosystem services in the design of technological systems. Since ecosystems are inherently resilient, integrated human-natural systems can also be more resilient (Zuniga-Teran et al. 2020). Applications of integrated design include the use of green infrastructure such as wetlands and bioswales for treating urban runoff and industrial waste (Vymazal 2011; Gopalakrishnan and Bakshi 2018), planting of trees for mitigating air pollution (Charles and Bakshi 2021), and cultivation of oyster reefs and mangroves for reducing the effect of storms (Scyphers et al. 2011). A systematic protocol for such integrated design is presented below.

In one example, a Dow Chemical plant in Texas replaced a conventional wastewater treatment system with a much less costly engineered wetland, saving an estimated $250 million over the life of the system (Fiksel 2015). This approach also benefited the local community, since the wetland provided habitat for deer, bobcats, and birds, as well as educational opportunities for schools. Moreover, the system proved to be resilient in several ways—it was decoupled from price changes or shortages in materials and energy, it did not degrade over time if properly managed, it could not fail suddenly, and it was not vulnerable to human error since it required virtually no supervision or maintenance. However, unlike conventional engineered systems, it required specialized scientific knowledge and could not be standardized and replicated easily. As described in Chap. 13, Dow followed up this pioneering effort by partnering with The Nature Conservancy (TNC) in 2011 to develop a methodology for assessing both the business value and environmental value of nature-based solutions.

Recent research has helped to quantify the value of ecosystem services based on economic input-output modeling and network analysis. For instance, an environmentally extended input-output model was used to study the direct and indirect effect of this pollinator loss on sectors of the US economy (Chopra et al. 2015). Animal pollinators such as insects, hummingbirds, and bats are essential for the reproduction of 75% of flowering plants that produce seeds (angiosperms), and studies over the last several years have shown a significant loss of pollinators in the USA and across the world (Goulson et al. 2015). Pollination by wild and managed

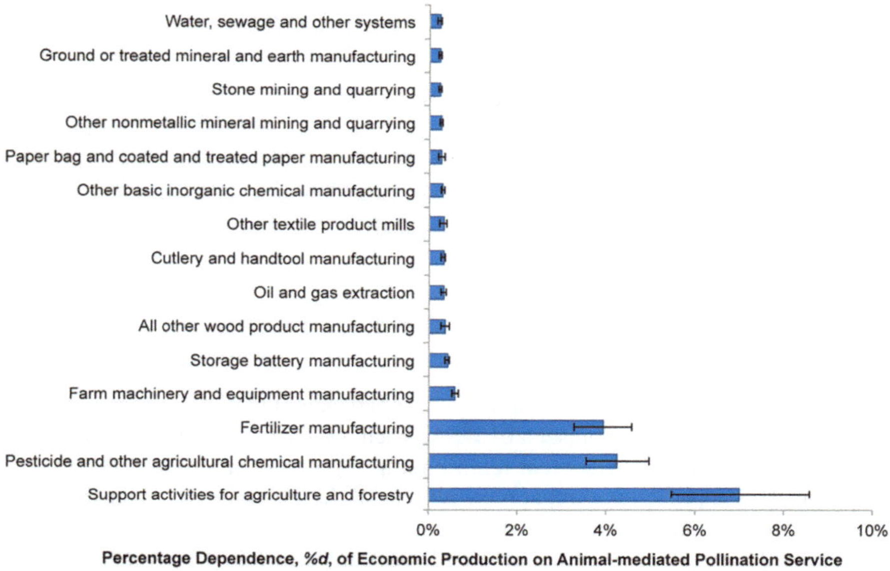

Fig. 21.6 Potential economic impacts by sector of loss of pollination services. (Reproduced with permission from Chopra et al. (2015))

populations in the USA is estimated to be worth USD $20 billion in 2010 (Chopra et al. 2015), and this value is estimated to be about ten times larger globally.

In addition to these direct contributions, pollinators indirectly affect a large number of economic sectors. The key findings of the Ohio State study are presented in Fig. 6.1 in Chap. 6, which shows the sectors that could have the largest monetary loss due to degradation of pollination services. Several sectors such as Support Activities for Agriculture and Fertilizer Manufacturing are not surprising; however, other sectors such as Wholesale Trade, Oil and Gas Extraction, and Management of Companies appear mainly due to a large indirect impact. Figure 21.6 shows the potential impact of such losses on various industry sectors in terms of percentage reduction in economic activity.

21.9 Design for Resilience in Industrial Systems

Whether a company is designing a new system or improving an existing system, it is useful to have a defined process that can guide cross-functional teams in considering whether the system is sufficiently resilient and designing appropriate interventions. We argue that designing for resilience should be a consideration in every innovation process, including product development, capital expansion, and supply chain configuration. As shown in Fig. 21.7, we define a protocol consisting

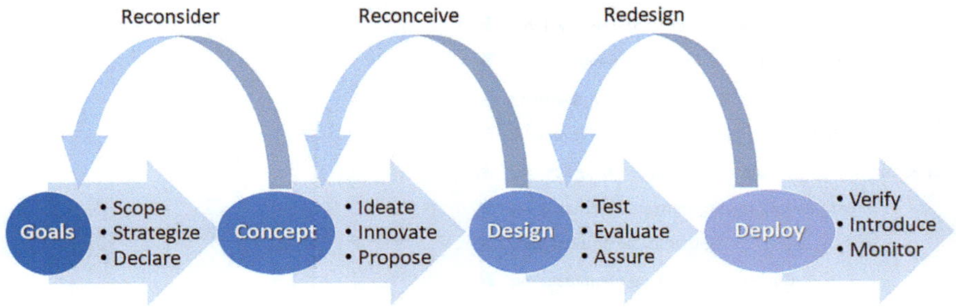

Fig. 21.7 Design for resilience (DFR) iterative protocol

of four principal stages—goal definition, concept development, design specification, and system deployment (Fiksel 2015). At each stage, the innovation team may discover problems or barriers that require returning to an earlier stage for another iteration. This approach is compatible with typical design processes that are used in practice, such as simultaneous engineering, whereby multiple design perspectives are explicitly integrated. Traditional "DFX" disciplines such as design for reliability, design for manufacturability, and design for safety can be augmented by design for resilience (DFR). (Note that the protocol is comparable to the "greenprinting" approach for design of urban systems, described in Chap. 14).

DFR will necessarily have a broad scope in order to assess how external phenomena (e.g., consumer behavior, resource scarcity) may affect the performance of the overall system. At one extreme, if there are no important interactions anticipated with external systems, the protocol reduces to a standard "stage-gate" process commonly used in industry. At the other extreme, the external system interactions may be so significant that the majority of the design effort is spent trying to understand these systems, including impacts and boundary constraints. Due to such system-level considerations, no technology can be deemed intrinsically resilient or sustainable. Recycled materials are not necessarily preferable to virgin materials if there is not an efficient recycling infrastructure. Renewable or bio-based materials are not necessarily preferable to inorganic materials if their physical or chemical properties are not uniform. Biodegradable materials are not necessarily preferable to durable materials if those materials can be used for new applications. It all depends on the system design, boundaries, and requirements.

This DFR protocol prompts a design team to pose following questions about resilience:

- What is our baseline position—key resource flows and interdependencies?
- What are the emerging opportunities for creating value within the broader system?
- What are the emerging threats to availability of the key resources that we will require?
- What innovations in technologies, processes, or business models might be helpful?

- What are the unintended economic, environmental, or social impacts of possible design innovations or interventions?
- How can we ensure that the design remains viable as conditions change in the future?

The *goal definition* stage of the DFR protocol involves defining the present and future needs of stakeholders, developing a strategy for addressing those needs, and declaring the overall goals for a design initiative or system intervention. This is a critical stage in the process because it defines the scope, context, linkages, and boundaries of the system to be considered and explicitly identifies the relevant stakeholder interests. For instance, introduction of an alternative energy technology should consider not only its affordability and the expected reduction in greenhouse gas emissions but also the potential for unintended consequences such as displacement of jobs or disruption of community lifestyles. Taking a systems view may also reveal unexpected benefits, such as increased resilience to power failures.

The *concept development* stage seeks innovative approaches that promise to meet the strategic goals. As described in Chap. 12, opportunities may exist to achieve techno-ecological synergies by integrating natural processes into the overall system design, for example, green infrastructure can be used to treat industrial wastewater. Recent advances in genomics, materials science, nanosystems, and information technology can contribute directly to resilience by increasing the efficiency and adaptability of existing products and processes. For instance, increased use of electronic communication and virtual meetings reduces the need for more costly physical transportation while enabling businesses to function even when physical infrastructure is disrupted. A critical step in concept development is the definition of testable system requirements related to the design goals. For instance, system characteristics such as diversity and adaptability may not have an obvious relationship to typical business performance measures but may contribute to the system's longevity and ultimate success.

The *detailed design* stage involves creation and testing of proposed designs to assure that they meet the stated requirements. This involves definition of performance indicators, evaluation of alternative solutions relative to the current baseline, and consideration of trade-offs and synergies. Here, the process shifts from conceptual analysis to quantitative assessment and evaluation of the expected costs, risks, and benefits for various stakeholder groups. Chapter 17 demonstrates a methodology for quantifying the benefits of techno-ecological synergies based on an advanced version of life cycle assessment. As mentioned earlier, the choice of resilience *indicators* is critical for assessing trade-offs among different alternatives and for analyzing uncertainties and sensitivity to key assumptions. For instance, Cisco is utilizing a resiliency index to evaluate new product introductions so that design teams can assess choices about supply chain partners and components. This allows Cisco to build supply chain resilience into the design of the product, rather than trying to de-risk the supply chain after the product launch.

Finally, the *deployment* stage involves verification, release, introduction, and monitoring of the system. To assure both resilience and sustainability, a company needs to consider the broad implications of an innovative system upon all of the enterprise stakeholders, including customers, employees, shareholders, regulators, public interest organizations, and the media. Monitoring the initial results and reactions is important to enable system refinement and adjust accordingly. For instance, biotechnology companies have developed genetically engineered pest-resistant crops under the banner of sustainable agriculture. Proponents claim that this technology will reduce pesticide use, increase agricultural productivity, and lower consumer costs, but opponents are concerned about unforeseen health and environmental impacts and long-term resilience of agricultural assets such as biodiversity and soil quality. Thus, history has shown that it is wise for designers to consider not only the direct benefits of a technological innovation but also the socioeconomic system into which it will be introduced.

Application of DFR is fundamental to the work of IBM's Smarter Cities program, established in 2011 to enhance the functioning of urban areas. Cities are perhaps the most complex and turbulent of all human systems, yet they remain extraordinarily resilient. Like living organisms, cities have survived, adapted, and flourished through the centuries, overlaying different cultures, lifestyles, and technologies in a rich and evolving mosaic. Thus, cities are a nexus of change, where social, economic, and environmental pressures are intensified, and the challenges of resilience and sustainability converge. Over the years, IBM has developed and commercialized a broad range of information technology applications that contribute to urban resilience, including weather forecasting, flood modeling, and structural monitoring for levees. A striking example is IBM's partnership with the City of Rotterdam, one of the world's largest and busiest ports. To defend against flooding from the North Sea, Rotterdam has erected a massive system of levees and flood barriers, but in recent years, it became apparent that these defenses were insufficient and that a more integrated approach was needed. IBM helped the city to introduce new strategies, such as expanding the use of floodplains and overflow reservoirs. Innovative sensor networks and citizen engagement, together with advanced forecasting and decision-support systems, enable a holistic view of emerging threats leading to a rapid, coordinated response (Fiksel 2015).

21.10 Climate-Resilient Process Design

Climate change is perhaps the greatest challenge to which human activities need to adapt. Human civilization thrived during the geological epoch of the Holocene, which lasted for 12,000 years and had relatively minor variation in global average temperature compared to previous epochs. Scientists now believe that we have entered a new geological epoch, the Anthropocene, whose global average temperature is close to exceeding the range of the Holocene. Thus, the future is likely to exhibit much greater extremes in global temperatures and associated climatic impacts such as droughts, floods, and heat waves. While actively working toward mitigation of greenhouse gas emissions, nations must also cope with rising impacts that have already manifested and are not easily reversed.

Over the last few centuries, engineering innovation has accelerated dramatically but has largely taken natural resources for granted, keeping environmental issues outside the boundary of system design. This includes the implicit assumption of stationarity, i.e., that future climate conditions will be similar to the past (Milly et al. 2008). Therefore, engineered systems tend to ignore the increasing variability due to climate change, and the resulting designs tend to be less resilient. Below, we describe two case studies demonstrating the design of chemical and manufacturing processes that are climate resilient. Both studies are located in the Muskingum River Watershed (MRW) in Ohio.

For this research effort, climate change scenarios are developed based on general circulation models that project future levels of precipitation and temperature (Blodgett et al. 2011) for various representative concentration pathway (RCP) scenarios (van Vuuren et al. 2011). These RCP scenarios reflect policy regimes that represent different degrees of climate forcing. RCP 8.5 assumes a worst-case scenario of high emissions due to lack of GHG mitigation policies, resulting in 4 °C of global warming above pre-industrial levels by 2100. A second scenario considered in this work is RCP 4.5, which corresponds to a 2 °C temperature achieved through reduction in atmospheric concentration by 2045 and stabilization by 2080. The hydrology of MRW is modeled by means of a SWAT model; further details about the modeling and design methods are described in Lee (2020).

Heat exchanger network design Heat exchanger networks (HENs) are used in chemical processes to recover waste heat and enhance energy efficiency. One likely effect of climate change on HENs is rising temperature of cooling water, which is commonly sourced from a local river or lake. If temperatures are too high, the process will be unable to cool some of the aqueous streams in the network sufficiently, which could affect the process energy efficiency and operational integrity. Such a situation is referred to as a "failure day." Future estimates of temperature in the MRW for RCP 4.5 and 8.5 are shown in Fig. 21.8. The ambient temperature of cooling water is proportional to this value.

Estimated failure days for conventional heat exchange designs are shown in Fig. 21.9, for both average and worst-case scenarios in the two RCPs. (Average scenarios correspond to the "projected average" lines in Fig. 21.8a, b, and worst-case scenarios correspond to the "projected max" lines in these plots.) For the RCP 4.5 scenario, a conventional design will fail to operate for an average of 10 days per year until 2099. For the RCP 8.5 scenario, the number of failure days will average 26 per year in the late twenty-first century. In the worst-case RCP 4.5 and 8.5 scenarios, the HEN will fail to operate for 54 and 102 days on average every year. These results demonstrate the importance of designing HENs to be capable of handling a wider range of cooling water temperatures so that they are less vulnerable to climate change. Of course, designing for resilience may require significant capital investments that will need to be justified based on risk/cost/benefit analysis.

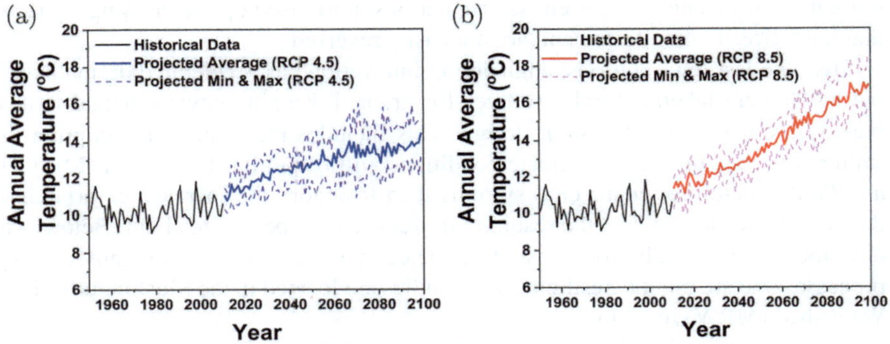

Fig. 21.8 Projected temperature in the watershed under two climate scenarios (Lee 2020)

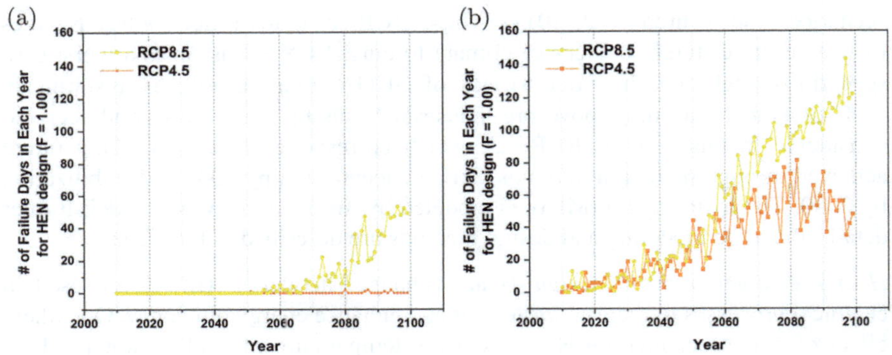

Fig. 21.9 Projected number of process failure days under two climate scenarios based on average (**a**) vs. worst-case (**b**) temperatures (Lee 2020)

Urea Manufacturing Systems This case study considers systems for the manufacture of urea fertilizer by reacting ammonia with carbon dioxide. Ammonia manufacture requires hydrogen, which may be obtained from natural gas by steam methane reforming or from water by electrolysis. Electricity may be generated from natural gas, wind, or sunlight. A climate-resilient urea manufacturing system needs to be economically feasible while adapting to the effects of climate change. For the RCP 4.5 scenario, Fig. 21.10 shows how the urea manufacturing process can adapt to the effects of climate change, such as reduction in water availability, as well as policy changes, such as limits on carbon dioxide emissions. The evolving strategy is determined by calculating the lowest-cost configuration for each year, following the flexible adaptation pathway approach mentioned above. In the base year of 2014, the system is entirely fossil fuel-powered and uses steam methane reforming to produce hydrogen. By 2040, solar power dominates, in order to reduce CO_2 emissions. Fossil fuels are entirely phased out by 2070, and electrolysis of water becomes the dominant route for producing hydrogen. By 2100, solar power is

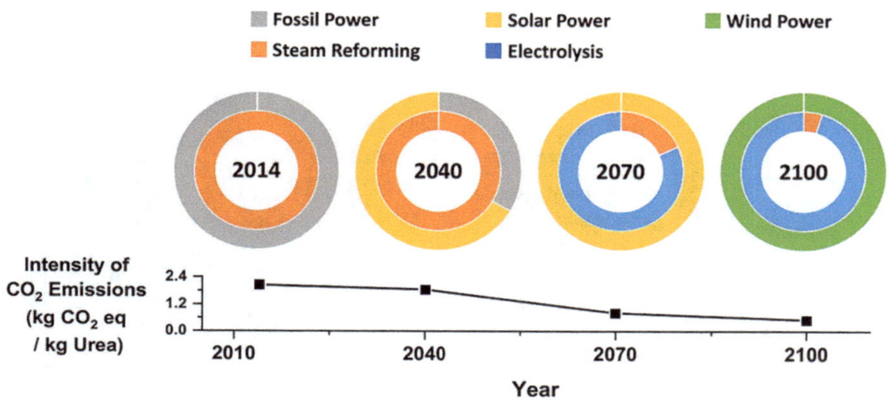

Fig. 21.10 Adaptation of the urea manufacturing process to climate change

replaced by wind, since increased electrolysis use results in dominant use of wind energy to conserve water that was previously used for cleaning solar panels.

21.11 Conclusions and Future Challenges

In the light of extreme climate volatility, recurrent military conflicts, the complex dynamics of global markets, as well as the threat of pandemics, it has become clear that we are living in a new normal. Mounting stresses combined with natural or human-caused shocks will continue to increase in frequency and intensity and may lead to unpredictable crises. Companies will need to adapt to these changes, rather than trying to maintain business as usual. The quest for enterprise resilience presents new challenges to traditional management patterns and beliefs.

Past innovations in industrial technologies and business processes have largely been aimed at improvements in operating efficiency—essentially achieving higher levels of output per unit of resource input. For instance, quality control and process improvement methods such as Six Sigma have sought to eliminate defects, minimize waste, and accelerate cycle time. However, such approaches toward process optimization fail to account for variability in the business environment, including both human and natural factors. Maximizing efficiency tends to reduce buffers and safety margins and thus pushes the overall system closer to critical thresholds. In a turbulent world, it is essential to balance efficiency with resilience. Moreover, to the extent that sustainability goals are linked to resource efficiency, there will be fundamental trade-offs between resilience and sustainability.

This paper suggests that future applied research should strive to extend the current practices of risk management by building enterprise resilience, especially with regard to supply chain operations. We have presented a variety of strategies,

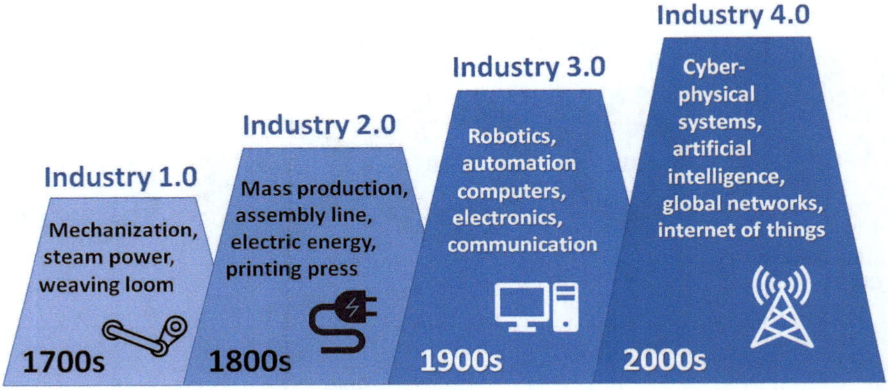

Fig. 21.11 Stages of innovation in industrial technologies

methods, and tools for managing enterprise resilience from a whole systems perspective, including qualitative and quantitative methods for analyzing existing vulnerabilities and investing in appropriate resilience capabilities. Building on the Triple Value framework, we describe a step-by-step protocol whereby companies can design for resilience, accounting for both economic and ecological consequences that will impact a firm and its stakeholders.

Given the momentum of ongoing global change, the importance of achieving resilience will undoubtedly grow. Not only are environmental and social pressures increasing, but the fundamental characteristics of industrial operations are rapidly evolving. In particular, as shown in Fig. 21.11, technological innovation is transforming the business landscape in what has been characterized as the "fourth industrial revolution" or Industry 4.0 (Mohamed 2018). Previous revolutionary changes are characterized as (1) mechanization of human labor, (2) introduction of mass production and electric power, and (3) emergence of digital technologies and automation. Industry 4.0 represents a new wave of innovation based on artificial intelligence and "smart" systems that are becoming pervasive in both industry and society.

Industry 4.0 offers many advantages, including facilitation of repetitive tasks, increased productivity, and ease of global collaboration and commerce. Modern technologies can arguably improve enterprise resilience by enabling companies to access vital information and adapt in real time to unforeseen disruptions. On the other hand, there may be a number of adverse hidden consequences, including cybersecurity threats, disruption of traditional jobs and communities, and exacerbation of income gaps between rich and poor. Moreover, the explosion of social networking and digital media has created the potential for abuse and misinformation, since viral dissemination of real or fictitious company failures can damage brand image and license to operate. Thus, social media pose extreme challenges to companies, including lack of visibility regarding information dissemination and lack of remedies to control the damage. Table 21.2 summarizes the potential impacts

Table 21.2 Potential impacts of Industry 4.0 on key attributes of enterprise resilience

Attribute	Benefits of Industry 4.0	Potential challenges
Cohesion	Continuous visibility and communication with business partners	Decreased loyalty and trust among employees and affiliates
Vulnerability	Smart technologies help to monitor changes and detect vulnerabilities	Complexity of cybercrime exposure may be overwhelming
Adaptability	Software-based capabilities can be upgraded and modified rapidly	Systemic vulnerabilities may exceed adaptive capacity
Efficiency	Automation enables quantum leaps in productivity of labor and capital	Autonomous equipment may introduce unforeseen vulnerabilities
Diversity	Global access to needed technical capabilities and human skills	Creativity of employee ideas and contributions may decline
Stability	Continuous monitoring can rapidly identify deviations from normal	Turbulence may increase due to rapid social and technological change
Recoverability	Response to disruptions can be faster and better coordinated	Profound social and environmental disruptions may be irreversible

of new information and communication technologies on some of the key resilience capabilities described earlier.

Another set of future challenges is posed by the historical conflict between economic growth and ecological integrity. Slow deterioration of ecosystem goods and services is resulting in erosion of system resilience, thus increasing the vulnerability of human systems to perturbations. At the same time, global turbulence due to environmental and geopolitical forces is increasing both the severity and frequency of extreme events. In response, businesses and academic disciplines need to recognize that taking nature for granted is a root cause of ecological degradation. Corporate profit calculations have typically not accounted for the loss of economic value due to environmental and societal damage from pollution and land use change. Academic disciplines such as economics have treated environmental damage as an externality that is outside the market.

Reversing this trend requires a shift in the prevailing paradigm, which ignores nature and assumes it to be an infinite source and sink. Such a shift will enhance enterprise resilience by creating awareness of the effects of ecological degradation, and the resulting protection and restoration of ecosystems will enhance overall system resilience. It will also enable businesses to tap into the estimated $10 trillion of business opportunities to develop a nature-positive world (World Economic Forum 2020). In such a world, business initiatives would begin with reversing the degradation of natural capital, followed by active protection and restoration of natural and biodiverse native ecosystems.

Finally, responding to the threats of climate change may be the most daunting challenge faced by both the public and private sectors. In pursuit of sustainability, many corporations and nations have pledged to become carbon-neutral in terms of their greenhouse gas emissions or "net-zero" in terms of their waste generation within a few decades. Many efforts in this direction focus on decarbonization by replacing fossil fuel-based energy with renewable energy. Capture, use, and sequestration of carbon dioxide is also an active area of research. As described above, developing a circular economy of products is an essential strategy for meeting net-zero goals. To realize these aspirations, decision-makers must address the formidable challenge of maintaining system resilience along with economic feasibility, ecological viability, and social desirability. Integrated conceptual frameworks such as the 3V approach will be needed to support the discovery and implementation of novel solutions that enable a resilient, sustainable, and equitable economy while avoiding hidden adverse consequences. In this age of turbulence, as new crises emerge and climate change continues inexorably, we shall need wise governance to remain resilient in pursuit of our economic and social goals. The cases presented here illustrate the importance of transdisciplinary, science-based systems thinking as an essential tool for strategy, policy, and decision-making in both the public and private sectors.

References

Association for Supply Chain Management (ASCM). *The Resilient Supply Chain Benchmark: Ready for Anything? Turbulence and the Resilience Imperative*. 2021. https://www.ascm.org/eiu-benchmark/ (accessed 19 Feb 2022).

Bakshi, B. R. *Sustainable Engineering: Principles and Practice, Cambridge University Press,* 2019.

Blodgett, David L., Nathaniel L. Booth, Thomas C. Kunicki, Jordan I. Walker, and Roland J. Viger. Description and testing of the Geo Data Portal: Data integration framework and Web processing services for environmental science collaboration. No. 2011-1157. *US Geological Survey*, 2011.

Charles, M. & Bakshi, B. R. Designing Industrial Landscapes for Mitigating Air Pollution with Spatially-Explicit Techno-Ecological Synergy. *AIChE Journal,* 2021, *67.*

Chopra, S. S., Bakshi, B. R. and Khanna, V. Economic Dependence of U.S. Industrial Sectors on Animal-mediated Pollination Service. *Environmental Science and Technology,* 2015, *49,* 14441–14451.

Committee of Sponsoring Organizations (COSO) of the Treadway Commission, 2004. *Enterprise risk management: Integrated framework.*https://www.coso.org/Documents/COSO-ERM-Executive-Summary.pdf (accessed 19 Feb 2022).

Economist Intelligence Unit. *Ready for Anything? Turbulence and the Resilience Imperative.* 2021. Report commissioned by the Association for Supply Chain Management (ASCM). https://www.eiu.com/n/ready-for-anything-turbulence-and-the-resilience-imperative (accessed 19 Feb 2022).

Fiksel, J. Sustainability and Resilience: Toward a Systems Approach. *Sustainability: Science, Practice, & Policy*, 2, no. 2 (Fall 2006): 14–21.

Fiksel, J, R Bruins, A Gatchett, A Gilliland, and M ten Brink. The Triple Value Model: A Systems Approach to Sustainable Solutions. *Clean Technologies and Environmental Policy*: Volume 16, Issue 4 (2014), pp. 691–702.

Fiksel, J. *Resilient by Design: Creating Businesses That Adapt and Flourish in a Changing World.* Island Press, Washington, DC, 2015.

Fiksel, J, M Polyviou, T Pettit and K Croxton. From Risk to Resilience: Learning to Deal with Disruption. *Sloan Management Review*, Winter 2015.

Fiksel J., P Sanjay and K Raman. Steps Toward a Resilient Circular Economy in India. *Clean Technology and Environmental Policy.* Vol 23, Issue 1, pp. 203–2018, 2021.

Government Accountability Office (GAO). *Disaster Resilience Framework: Principles for Analyzing Federal Efforts to Facilitate and Promote Resilience to Natural Disasters*. 2019. https://www.gao.gov/products/gao-20-100sp (accessed 13 April 2022).

Government Accountability Office (GAO). *Chemical Accident Prevention: EPA Should Ensure Regulated Facilities Consider Risks from Climate Change*. Report to Congress. Feb 2022. https://www.gao.gov/assets/gao-22-104494.pdf (accessed May 2, 2022).

Gibbens, S. Hurricane Sandy, Explained. *National Geographic*. Feb 2019. https://www.nationalgeographic.com/environment/article/hurricane-sandy (accessed 19 Feb 2022).

Gopalakrishnan, V. & Bakshi, B. R. Ecosystems as Unit Operations for Local Techno-Ecological Synergy: Integrated Process Design with Treatment Wetlands. *AIChE Journal,* 2018, *64*, 2390–2407.

Goulson, D.; Nicholls, E.; Botías, C.; Rotheray, E. L. Bee declines driven by combined stress from parasites, pesticides, and lack of flowers. *Science* 2015, 347 (6229), 1255957.

Hoehn, P.; Tscharntke, T.; Tylianakis, J. M.; Steffan-Dewenter, I. Functional group diversity of bee pollinators increases crop yield. *Proc. R. Soc. London, Ser. B* 2008, 275 (1648), 2283−2291.

Industry Week. *How to Build an Anticipatory Supply Chain*. June 2014.

Institute for Economics & Peace. *Ecological Threat Register 2020: Understanding Ecological Threats, Resilience and Peace*, Sydney, September 2020. https://www.visionofhumanity.org/wp-content/uploads/2020/10/ETR_2020_web-1.pdf (accessed 19 Feb 2022).

Kaza, S, LC Yao, P. Bhada-Tata, and F Van Woerden. 2018. *What a Waste 2.0: A Global Snapshot of Solid Waste Management to 2050*. Washington, DC: World Bank. https://openknowledge.worldbank.org/handle/10986/30317 (accessed 19 Feb 2022).

Lacy, P and J Rutqvist. *Waste to Wealth: The Circular Economy*. Advantage, New York/London: Palgrave Macmillan. 2015.

Lee, Kyuha, Sustainable Process and Supply Chain Design with Consideration of Economic Constraints, Climate Change, and Food-Energy-Water Nexus. Ph.D. dissertation, The Ohio State University, 2020.

Low, J and PC Kalafut. *Invisible Advantage: How Intangibles Are Driving Business Performance*. New York: Cap Gemini Ernst & Young, 2002.

McIntyre, J and S Hemmelgarn. How One Business Made its Supply Chain More Resilient. Presentation at the Annual Global Conference of the Council of Supply Chain Management Professionals for the 2011 Supply Chain Innovation Award. Philadelphia, PA, October 4, 2011.

McKinsey Global Institute. *Risk, resilience, and rebalancing in global value chains*. August 2020. https://www.mckinsey.com/business-functions/operations/our-insights/risk-resilience-and-rebalancing-in-global-value-chains (accessed 19 Feb 2022).

Millennium Ecosystem Assessment. *Living Beyond our Means: Natural Assets and Human Well-Being*. Island Press, 2005.

Milly, P. C. D., Betancourt, J., Falkenmark, M., Hirsch, R. M., Kundzewicz, Z. W., Lettenmaier, D. P. & Stouffer, R. J. Stationarity Is Dead: Whither Water Management? *Science,* 2008, *319*, 573–574

Mohamed, M. Challenges and Benefits of Industry 4.0: An Overview. International Journal of Supply and Operations Management, Vol.5, No.3, 2018.

New York City Housing Authority (NYCHA). *Recovery to Resiliency: NYCHA's Hurricane Sandy Recovery Program*. 2016. https://issuu.com/nycha/docs/recovery_to_resiliency_04.11.16 (accessed 13 April 2022).

Organization for Economic Cooperation and Development (OECD). *Global Material Resources Outlook to 2060: Economic Drivers and Environmental Consequences*. Paris, France. 12 Feb 2019.

Pettit, TJ, KL Croxton and J Fiksel. Ensuring Supply Chain Resilience: Development and Implementation of an Assessment Tool. *Journal of Business Logistics*, Volume 34, Issue 1, pages 46–76, March 2013.

Rockefeller Foundation. *City Resilience Framework*. 2015. https://www.rockefellerfoundation.org/report/city-resilience-framework/ (accessed 13 April 2022).

Russo, L, R.J. Daniel, and J Fiksel. How to make business decisions to adapt to climate change? Methods and case studies in France and internationally. ADEME, France, 2021. https://librairie.ademe.fr/changement-climatique-et-energie/4758-how-to-make-business-decisions-to-adapt-to-climate-change%2D%2D9791029717994.html (accessed 19 Feb 2022).

Scyphers, S. B., Powers, S. P., Heck Jr, K. L., & Byron, D. (2011). Oyster reefs as natural breakwaters mitigate shoreline loss and facilitate fisheries. *PloS one*, *6*(8), e22396.

Stahel, W. The circular economy. *Nature* 531, 435–438 (2016). https://doi.org/10.1038/531435a (accessed 19 Feb 2022).

Taub, E. *Building a More Sustainable Car, From Headlamp to Tailpipe*. New York Times, 9 Sept 2021. https://www.nytimes.com/2021/09/09/business/building-a-more-sustainable-car-from-headlamp-to-tailpipe.html (accessed 19 Feb 2022).

United Nations. *Sustainable Development Goals*. https://sdgs.un.org/goals (accessed 19 Feb 2022).

U.S. Army. Net Zero Energy Roadmap and Program Summary. 2013. https://www.asaie.army.mil/public/es/netzero/docs/FY13%20Army%20Net%20Zero%20and%20Energy%20Program%20Summary.pdf (accessed 19 Feb 2022).

van Vuuren, D.P., Edmonds, J., Kainuma, M. et al. The representative concentration pathways: an overview. *Climatic Change* 109, 5 (2011).

Vymazal, J. Constructed Wetlands for Wastewater Treatment: Five Decades of Experience. *Environmental Science & Technology,* 2011, *45*, 61–69.

Walker, S. Triple Threat: Water, Energy and Food Insecurity. *World Resources Institute*. Oct 2020. https://www.wri.org/insights/triple-threat-water-energy-and-food-insecurity (accessed 19 Feb 2022).

World Economic Forum (WEF). Global Risks Report 2016. https://www.weforum.org/global-risks/reports (accessed 19 Feb 2022).

World Economic Forum. New Nature Economy Report Series 2020. https://www.weforum.org/reports/new-nature-economy-report-series (accessed March 16, 2022).

Yu, A. *Creating the Digital Future: The Secrets of Consistent Innovation at Intel*. Free Press. 1998. p. 93.

Zavitz, D. Vice President, Supply Chain. Dow Chemical Company. Personal Communication. 2017.

Zuniga-Teran, Adriana A, Gerlak, Andrea K, Mayer, Brian, Evans, Tom P, Lansey, Kevin E. Urban resilience and green infrastructure systems: towards a multidimensional evaluation. *Current Opinion in Environmental Sustainability*, 44, 42–47, 2020.

Part V
Socio-economic Aspects of Accounting for Nature

Chapter 22
Environmental Markets

Alan Randall

22.1 Introduction

This chapter defines ecosystems services, shows how they acquire value and prices, and explains how the natural systems that produce them acquire asset value. Ecosystem service values and the asset values of natural systems relate in rigorous ways to the economic concept of welfare, and it is important to understand that welfare as defined in economics has been shaped by market logic. This has its advantages, especially in defining efficiency consistently in the public and private sectors. But conformity with market logic leaves welfare economics with some conventions – for example, the preferences of the well-off count for more in aggregate welfare calculations – that do not always accord with moral intuition.

Welfare change measurement has several applications in economics, including cost of living and standard of living indicators, and cost benefit analysis (CBA), which is used to evaluate whether a prospective policy or project is likely to improve welfare. Many of these policies and projects are connected in one way or another to both the private and public sectors, and the benefits and costs have market and nonmarket dimensions. CBA in many applications requires nonmarket valuation of active and passive use values for completeness, and the methods of nonmarket valuation are introduced and discussed.

The fundamental elements of welfare change measurement, price, and value can also be used in payment programs and markets that incentivize enhanced provision of ecosystem services. Here, several such programs are highlighted, and two, the US sulfur oxides cap-and-trade program and the Australian experiments with conservation auctions, are discussed in some detail. The exposition focuses

A. Randall (✉)
Sustainability Institute at The Ohio State University, Columbus, OH, USA
e-mail: randall.3@osu.edu

B. R. Bakshi (ed.), *Engineering and Ecosystems*,
https://doi.org/10.1007/978-3-031-35692-6_22

on ensuring that these payments and markets serve the public interest in providing ecosystem services efficiently. But we conclude by pointing out that the programs discussed here and an even larger set of prospective future programs provide opportunities for cost-savings and/or business expansion for private operators large and small.

22.2 Some Basic Concepts: Environmental Assets, ES, Value, and Price

The concept of ecosystem services Healthy and well-functioning ecosystems provide a variety of services that benefit humans who use them directly or by combining them with other kinds of inputs (Daily 1997). The Millennium Ecosystem Assessment (2005) identified three categories of ecosystem services (ES) that benefit humans directly:

- Provisioning services that include air and water, food, raw materials (not just long-familiar resources like minerals and timber but also genetic and medicinal resources that enable advances in science and medicine), energy, and ornamental plants.
- Regulating services as epitomized by the water and carbon cycles, which recycle water and carbon and decompose and detoxify wastes, cleansing and purifying them to support new life. Vegetation and wetlands help protect against flooding, and predator-prey relationships dampen fluctuations in populations of both kinds.
- Cultural services, such as land and water aesthetics, landscape amenities, atmospheric visibility, and biodiversity. These services improve the quality of human life by supporting nature study, sightseeing, and recreation and enhancing historical and cultural connections.

A fourth kind of ES – supporting services – help maintain ecosystem functions crucial to producing the first three.

While humans have found ways to privatize some ES and provide them via market institutions, many ES are public in some sense – e.g., while not everyone in a given region has equal access to clean air and green space, these services (once provided) are available in some degree to everyone – and many of them are non-marketed.

The ecosystems that produce these services and the environments that support these ecosystems can be viewed as environmental assets, that is, a kind of natural capital. Ensuring the adequacy and quality of ecosystem services provides strong motivation for maintaining and enhancing ecosystem assets. So, we begin by developing a simplified model of ES benefits and costs, values, and prices, by defining natural capital and ES. Imagine a natural asset N – note that we are now operating at the project scale of evaluation – producing a vector of ecosystem goods and services $\boldsymbol{E} = (E_1, \ldots, E_s)$ valued by people. The supply of each of these services in any time period, t, is a function of the attributes, $\boldsymbol{A} = (a_1, \ldots, a_s)$ of the natural capital which are uniquely determined by geological, hydrological, atmospheric, and ecological relationships:

$$\begin{aligned} E(t)_1 &= f_1\,(\boldsymbol{A}(t)) \\ &\ldots \\ E(t)_s &= f_s\,(\boldsymbol{A}(t)) \end{aligned} \tag{22.1}$$

The ecosystems that produce ES are coupled human and natural systems. People interact with ecosystems in many ways, most of which involve using ES and/or modifying natural capital purposefully or inadvertently. We start with humans as modifiers of natural assets: people may do this directly, for example, by reassigning land to other uses, diverting water, harvesting plants or animals, removing vegetation, disturbing soil for cultivation or mining, or improving soil with compost or fertilizers. They may also modify the natural asset as a side effect (expected or unexpected) of some other decisions, for example, disturbing land elsewhere for cultivation or mining or deposition of wastes in water upstream. For each kind of natural capital attribute:

$$\begin{aligned} a(t)_1 &= g_1\left(\boldsymbol{F}_{(t)}, \boldsymbol{X}_{(t)}\right) \\ &\ldots \\ a(t)_k &= g_k\left(\boldsymbol{F}_{(t)}, \boldsymbol{X}_{(t)}\right) \end{aligned} \tag{22.2}$$

where $\boldsymbol{F}$ is a vector of ecosystem structure and functions, for example, chemicals and chemical cycles and geological, hydrological, atmospheric, and ecological processes, and $\boldsymbol{X}$ is a vector of human-controlled inputs and activities, including harvesting effort.

Both $\boldsymbol{F}$ and $\boldsymbol{X}$ are subject to scarcity, and the attribute production functions are determined by the laws that govern natural systems and by technology. In this simplified model, the production system for ecosystem goods and services is now complete. Pause a moment to consider just how simplified this model is. We might reasonably expect that some of the individual equations in the sets 22.1 and 22.2 are nonlinear and perhaps not readily tractable; there are important interactions among some of the equations in these sets, e.g., high levels of harvest of E_{it} may affect production of E_i in later periods, and also production of other ES, and high levels of waste assimilation may have adverse impacts throughout the system; and these equations, which look simple and transparent, may be placeholders for much more complex and dynamic representations (Atkinson et al. 2012), perhaps of systems susceptible to regime switches that are difficult and costly to reverse, thus imposing big risks on incautious managers. Nevertheless, in order to communicate the basic concepts as simply as possible, we persist with these simplifications.

How do ES contribute to human well-being? Now consider the value of ecosystem goods and services to their human users. Each person, *j*, enjoys utility in each time period, *t*:

$$U(t)_j = U_{jt}\left[\boldsymbol{E}^a(t)_j, \boldsymbol{Z}^b(t)_j, \left(\boldsymbol{E}^b(t)_j\right), \boldsymbol{Z}^a(t)_j\right] \tag{22.3}$$

where Z is a vector of market goods and services. The ecosystem goods and services vector $\boldsymbol{E}$ is composed of $\boldsymbol{E}^a$, which are enjoyed directly, and $\boldsymbol{E}^b$ which are inputs into the production of $\boldsymbol{Z}^b$. Finally, $\boldsymbol{Z}^a$ are those $\boldsymbol{Z}$ produced independently of $\boldsymbol{E}$, e.g., goods manufactured in distant places. This formulation permits both direct demand for ecosystem goods and services $\boldsymbol{E}^a$ and demand for $\boldsymbol{E}^b$ derived from demand for $\boldsymbol{Z}^b$.

How are ES valued by individuals and society? Economic valuation is all about measuring changes in utility directly, or via some serviceable proxy, implying that value is derived entirely from satisfaction of human preferences, an assumption that has drawn a lot of critical attention. There are two conceptual approaches to deriving value from preferences, and, to expedite the exposition, I shall describe them in words rather than derive them mathematically.

Market demands and values Suppose the individual maximizes utility (Eq. 22.3) given a set of prices, and subject to budget constraint, *m*, measured in money. This generates a consumption bundle including $\boldsymbol{E}^a$, $\boldsymbol{Z}^b$, $\boldsymbol{E}^b$, and $\boldsymbol{Z}^a$, all in utility-maximizing quantities given income and the prevailing prices. Holding the budget constant, we can generate demand curves for particular goods and ES reflecting how quantities demanded vary with prices, $\boldsymbol{p}$, not only of the particular good or ES but also of other goods and ES that may be substitutes or complements in generating utility. So, demand for E_i is the function $\mathrm{D_i}$ $(\boldsymbol{E}, \boldsymbol{p}, m)$. Allowing the budget to vary, we can generate a demand map showing how demand for the Z_i or E_i varies with the size of budget (often called, loosely, income). Aggregate demands for a population can be calculated by summing individual demands, or estimated econometrically using aggregate data on quantities, prices, and income. Demand functions typically slope downward, indicating that consumers tend to purchase more if prices are lower.

To understand how consumption bundles vary with economic and environmental conditions, we need also to know something about supply. A production function for a good or service relates output quantity to quantity of various inputs and describes a given technology. Substitution among inputs is an option in some technologies, but in other technologies, inputs are used in fixed proportions. Given the technology, the cost, $\boldsymbol{q}$, of inputs is a key consideration in determining the supply that will be forthcoming at a given price for the product. When a producer can choose the kind and combination of outputs, as might a farmer deciding what crops to grow, the prices, $\boldsymbol{p}$, of this and competing outputs influence supply. The capacity of fixed plant (e.g., the size of the farm or factory) and the cost of expanding fixed capacity also

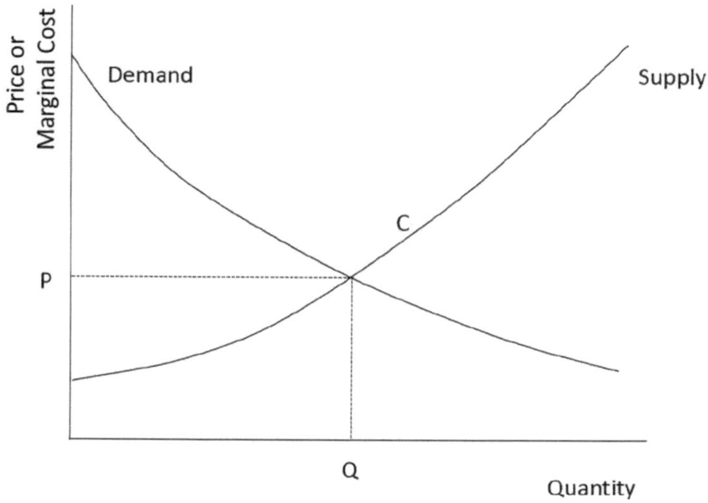

Fig. 22.1 Given demand D and supply S, the market clears at price P and quantity Q

matter. In the case of some but not all ES, production capacity is fixed by nature, and no purchased inputs are used in the production process. A supply function relates output to all of the considerations above that are relevant in a particular case. The simplest supply function is S($\boldsymbol{p}, \boldsymbol{q}| \ldots$), where the unspecified items taken as fixed include technology and capacity. Supply functions typically slope upward because marginal costs tend to increase as production is expanded. At equilibrium, point C (Fig. 22.1), supply and demand prices are equal at P, and the market clears where output Q is equal to quantity demanded. For quantity changes that are very small relative to the size of the market, the price P serves as a measure of unit value.

Measuring changes in welfare. Good information on demand and supply response to changes in market conditions and technology is sufficient for many private and public decision-making purposes (Fig. 22.1). But policy analyses often require information on whether a change would make people better off or worse off. To make such welfare judgments about a potential change in the supply of E_i, it is common practice to take the ordinary demand function $E_i(\boldsymbol{E}, \boldsymbol{p}, m)$ and two levels of supply $E_i(\boldsymbol{p}, \boldsymbol{q}| \ldots)$ where ... refers to various fixed factors such as technology and capacity, the baseline level, and the proposed level, and calculate the change in net economic surplus, i.e., the sum of the changes in consumers' and producers' surpluses. For a decrease in costs that shifts supply from S^0 to S^1, the approximate welfare change is the area $P^OP^1DC + AP^1D - BP^0C = ABCD$ (Fig. 22.2). Because the welfare change measure for consumers, P^OP^1DC, is derived from an ordinary or Marshallian demand function, it is called Marshallian consumers' surplus, CS^M.

However, this does not produce a precise measure of the welfare change, because welfare is changing as we move along an ordinary demand curve. Instead, minimize

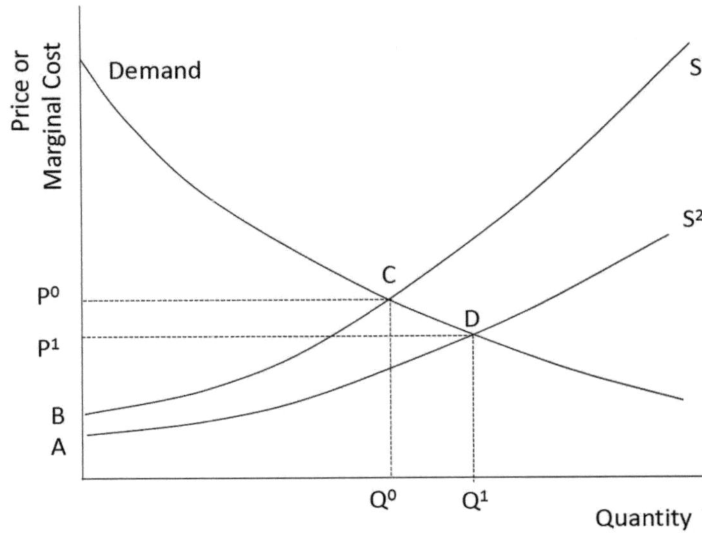

Fig. 22.2 The net welfare change due to an increase in supply is approximated by the area ABCD

expenditure, i.e., the sum of quantities taken of ***E*** and ***Z*** multiplied by their prices, to maintain a given level of utility u^*. The reference utility level is often the baseline level, which generates answers to questions about whether a given policy action to provide quantity Q^1 would make us better or worse off than we are at Q^0 in the absence of the policy (Fig. 22.3). But it can be a different level, perhaps aspirational as in the case of policies to combat hunger, where some analysts set the utility reference at a level that would induce adequate nutrition. Because the compensated demand functions, $E_i(\boldsymbol{E}, \boldsymbol{p}, u^*)$ are steeper than ordinary demand functions, the consumer welfare measures, called Hicksian compensated consumers' surplus (CS^H) generated by the expenditure minimization approach, differ systematically from those derived from utility maximization subject to a budget. Steeper compensated demands generate $CS^H \geq CS^M$ for welfare losses and $CS^H \leq CS^M$ for gains. In Fig. 22.3 a welfare gain is illustrated – for a public project where quantities are given and price may be implicit – and $CS^H = Q^0BEQ^1 < CS^M = Q^0BCQ^1$. For fairly modest quantity changes, the difference between CS^M and CS^H is quite small, and adjustments can fairly be called a "fiddling business." But for really consequential changes, e.g., elimination of a unique and highly valued environmental asset, the loss in welfare can be much greater when calculated using CS^H than using CS^M, and the use of the correct measure has substantial welfare implications (Hanemann 1991).

How are individual welfare gains and losses aggregated for a population? It is inevitable that some members of a population will gain more than others, or lose more than others, as a result of a proposed project or policy. So, there is continuing

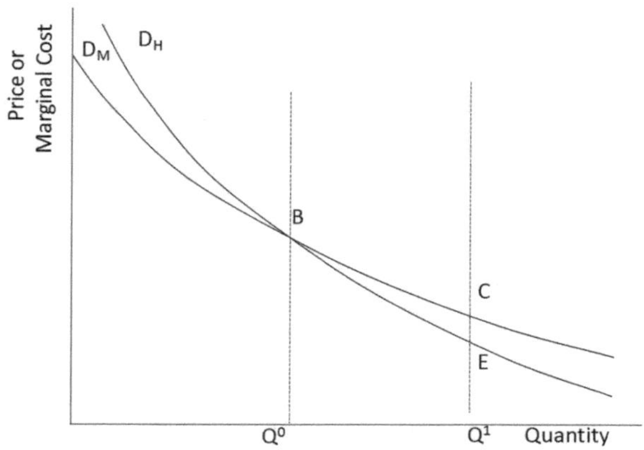

Fig. 22.3 Marshallian and Hicksian demands and consumers' surpluses (Q^0BCQ^1 and Q^0BEQ^1) for a public project that would increase quantity from Q^0 to Q^1

controversy about the right way to aggregate gains and losses across a population. Are there certain entitlements that should be respected? Should we pay greater attention to those who are less well-off, placing more weight on gains or losses that might come their way?

The market-based welfare measures, CS^M, usually are aggregated already because they are derived from market demand and supply. However, the correct welfare change measures are based on expenditure minimization, a process that spotlights individual welfare changes, thereby focusing attention on the distribution of gains and losses. The default method of aggregating gains and losses is simply to sum them across the relevant population without regard to individual differences in fortune. This greatly simplifies policy analysis – a proposal passes the test if the sum of benefits exceeds the sum of costs and losses – but it obviously avoids grappling with questions about what, if any, special consideration should be accorded to distributional concerns.

How well are equilibrium prices and welfare-relevant values reflected in markets? To begin, while economics students are well-versed in the virtues of prices as signals and incentives for efficient allocation of resources and budgets, we must concede that markets are complex systems and hence susceptible to instability, business cycles that are never explained fully by theories based on rationality assumptions, and bubbles in which asset prices overshoot rational levels and eventually collapse.

There are persistent price distortions and market failures of various kinds. Governments may distort prices, directly or via tax policy, to favor certain activities and discourage others. Market failures may persist, where prices are missing or exist but are systematically misleading. Examples include unregulated monopolies where beneficiaries are sheltered from competition, non-exclusiveness (where benefits cannot be restricted to those who contributed) and nonrivalry (where goods and services such as clean air which, once produced, can serve populations small or large). Much of the daily work of resource and environmental economists involves analyzing policies and estimating implicit prices that would improve efficiency and the welfare contribution of the nonexclusive and nonrival sectors.

How is production and demand for ES reflected in the value of natural capital? We have described ecosystems with the capacity to deliver benefits and/or incur costs over an extended period as natural assets. Now we formalize the notion of natural assets as a kind of capital (Costanza et al. 1997). Long-lived projects and policies that would modify and/or exploit ecosystems for human benefit can also be considered assets, in this case assets that complement the natural assets of the ecosystem. The value of an asset is the sum of net benefits delivered over time, reduced to *present value* by discounting at a standard reference rate of interest representing the reward earned by capital in alternative uses.

The capital value of natural capital N is obtained by integrating the net values of goods and service flows, discounted at the rate r, through time and summing the result across individuals, j:

$$V(N) = \sum_{\mathrm{j}} \int_{\mathrm{t}} V(t)_j \left(\boldsymbol{E}(t)_j\right) . e^{-rt} \mathrm{d}t \tag{22.4}$$

where $V(t)_j(\boldsymbol{E}(t)_j)$ is the value to j of the ES accessible to j at time t. Note the purposeful ambiguity in the term value: above, three concepts of value have been suggested: (i) unit price, and net economic surplus where consumers' welfare gain or loss is represented as (ii) ordinary Marshallian consumers' surplus and as (iii) Hicksian compensated consumers' surplus (Fig. 22.3 and accompanying discussion). Concept (iii) is the correct value measure, but (i) is equivalent when the quantity change is very small relative to the size of the market, and (ii) can be calculated from ordinary demand functions and provides a useful approximation of (iii) for modest quantity changes in ordinary (as opposed to unique and highly valued) goods and services.

Natural capital acquires economic value to the extent that the goods and services it provides are valued by people. Those goods and services are determined by natural capital attributes (Eq. 22.1), which are themselves determined ecosystem structure and functions and by the activities of people (Eq. 22.2). If ecosystem structure and functions were to be disturbed, say by changes in the $\boldsymbol{X}$ vector of human-controlled inputs, natural capital attributes could change, changing the $\boldsymbol{E}$ vector of the goods and services it provides and its capital value.

Discounting private and public benefits and costs The traditional motivation for discounting is to keep public and private investments in a rough balance, ensuring that public agencies do not gain privileged access to capital, thereby crowding out more productive private investments. The concept of discounting is relatively noncontroversial in the case of capital projects with expected service lives of no more than about 50 years and no further impacts after decommissioning.

But controversies may arise in application: for example, the exact reference rate of interest, r, is not always obvious, and real-world observable interest rates may be distorted by re-payment risk, expected inflation, and tax codes.

Discounting benefits and costs for future generations A more fundamental complaint arises in the context of discounting applied to projects and policies with very long-lived impacts: discounting in some sense devalues the future. This seems obvious: the present value of a large loss in 100 years' time is quite small, implying that such a prospect has little influence on today's decisions. But a correct interpretation distinguishes among motives for discounting. Suppose our objective is to provide present and future generations with equal well-being. Discounting future *utility* (well-being) – as an impatient society might – would fail future generations, thus offending the moral intuition for inter-generational equality.

But there is virtue in discounting to reflect the productivity of capital and the expectation of growth in productivity. In our example where intergenerational equality is the goal, future utility should not be discounted, but the proper inter-generational discount rate will be positive if we expect continued growth in per capita well-being. Discounting at any rate less than the expected rate of growth in per capita productivity would favor future generations at the expense of the present (Asheim 2010).

For projects and policies with very long-lived impacts, the appropriate discount rate is positive, but only to the extent that we expect higher per capita productivity in future generations. Even in that case, future utility should not be discounted, so the correct discount rate will be systematically lower than the rates we see in capital markets, e.g., ordinary construction projects.

To what extent can social and market values be reconciled? People are willing to sacrifice, within limits, to maintain and enhance ES; and doing so is costly. That much is noncontroversial and provides a motive to evaluate ES in monetary units. But we must pause to consider arguments that market prices, even if we succeed in getting all the underlying economics right, are unacceptable measures of value.

Economic valuation is a matter of estimating willingness to pay (WTP) for things we want individually or collectively and willingness to sell (or WTA, willingness to accept, i.e., the amount of compensation that would induce acceptance of a reduction) in the case of things we already have. This respects individual agency, as can be seen in the case of threatened reductions in $\boldsymbol{E}$: the correct measure of that loss is the compensation we would be willing to accept, as opposed (for example) to market valuations or compensation set by an arbitrator. Critics object that individual WTP and WTA are influenced strongly by individual budgets, with the consequence that the preferences of the well-off count for more in the aggregate assessment. Valuing ES this way ensures consistency with market values, one of the virtues claimed for cost benefit analysis. But it offends the moral intuition that everyone's environmental preferences should count equally.

More fundamentally, it can be argued that quantifying and valuing ecosystem services does not provide a complete account of the value of ecosystem assets and/or that human preferences cannot provide a sound guide to what really is good for people (Randall 2008, 2013):

- Preferences may be ill-considered and based more on appetites than moral reasoning, which would distort estimates of benefits and costs. The sense that preferences are impermanent may lead to a quest for some more enduring foundation for value.
- Immanuel Kant and the contractarians (e.g., John Locke) sought foundations for moral theory that seemed more secure and less malleable than preferences, Kant settling on obedience to universal principles and Locke on respect for rights. In both cases, this approach kicks the question upstairs: the principles and the rights themselves, and their foundational status, require justification.
- Kant also defended a stance that aesthetic judgments involve more than preference, appealing instead to reasoned discourse and shared agreement about aesthetic values.
- It has been argued from a variety of philosophical foundations that the economists' focus on human welfare is too limiting. Utilitarians have argued that animal welfare matters; some rights-based theorists have contended that rights should be extended to natural entities; and deep ecologists have worked to expand the set of entities that matter independently of human concern or patronage.

Environmentalists often object to the economists' persistence in viewing everything in terms of its monetary value. But I suspect that money is not the real issue: Money, after all, is just a convenient token of value in exchange. The real issue is that this monetary outlook assumes the economic fungibility of environmental entities: that they can be exchanged for, and substituted for, equivalents in ordinary goods and services and that tradeoffs across these two categories are meaningful.

In summary, critics argue that monetary valuation of ES is incomplete, not always consistent with other ways of valuing, and distorted by assumptions better fitted to the marketplace than to public policy.

22.3 Cost-Benefit Analysis (CBA)

CBA to guide social decisions Societies face a fundamental challenge: how to allocate resources in order to satisfy demands for goods and services. Market mechanisms work well for goods consumed by individuals and households acting independently, but some goods are inherently public in some sense. For instance, potable water for households is more affordable when a large cluster of households can be served by a single water system; live music is more affordable when many listeners can be assembled in a single concert hall; and while some locations emit vastly more greenhouses gases than others, the consequences are shared worldwide. The policy process is usually invoked at some point in the production and distribution of these kinds of goods and bads: for example, the provider may be a public agency, or private providers may be subject to regulation intended to ensure that they serve the public interest.

To help answer the resource allocation question, CBA has been developed to evaluate public projects and policies in terms comparable to the ways that markets evaluate private goods and services (Arrow et al. 1996).

The first law of thermodynamics applies to programs and projects: they cannot create something from nothing. Rather, at some expense, they transform an existing environment into a modified environment. It is helpful to think through this process. Consider a project Δ, which would change $\boldsymbol{X}$ to $\boldsymbol{X}^{\Delta}$, thus converting natural capital, N, to some "with project" state, N^{Δ}, at some conversion cost, $C^{\Delta} = \int_t C(t)^{\Delta}.e^{-rt}dt$ The proposed project would replace the "without project" stream of services, $\boldsymbol{E}$, with some "with project" stream, $\boldsymbol{E}^{\Delta}$. The net present value of such a project is:

$$\begin{aligned} PV\,(\Delta) &= \textstyle\sum_{j}\int_{t}\left\{\left[V(t)_j\left(\boldsymbol{E}(t)_j\right)\right] - C(t)^{\Delta} - \left[V(t)_j\left(\boldsymbol{E}^{\Delta}(t)_j\right)\right]\right\}.e^{-rt}dt \\ &= PV\left[\left(N^{\Delta} - C^{\Delta}\right) - N\right] \end{aligned} \tag{22.5}$$

where $PV(N)$ *and* $PV(N^{\Delta})$ are calculated as in Eq. (22.4). If $PV(\Delta)$ is greater than zero, the project will pass the benefit-cost filter.

Text Box 22.1 Principles of CBA: A Summary in Six Succinct Statements*

1. An existing environment can be viewed as natural capital, an asset-producing service that people value.
2. A proposed project or program is a proposal to modify that environment at some cost, changing the flow and mix of services it produces.
3. CBA compares (a) the value of the "with project" environment, minus the costs of transforming the environment into that state, with (b) the value of the "without project" environment. If (a) exceeds (b), the project will pass the BC filter.
4. For CBA, value information about the environments, N and N^{Δ}, and the goods and services, $\boldsymbol{E}$ and $\boldsymbol{E}^{\Delta}$, is equally useful, because the asset-pricing model (Eq. 22.4) specifies the relationship between asset values and the value of service flows.
5. Valuation using CS^H measures referenced to the baseline level of welfare will result in the proposal's passing the filter if, and only if, those who prefer $\boldsymbol{E}^{\Delta}$ are willing to pay enough to buy it (or, equivalently, its time stream of services) from those who prefer $\boldsymbol{E}$. But this quality is hypothetical: the actual purchase and the implied full compensation for losers are not required, a point often raised by critics of CBA.

(continued)

Text Box 22.1 (continued)

6. Accurate and reliable CBA is not just a matter of economics. The complex relationships among ecosystem structure and functions, human-controlled inputs, natural capital attributes, and the ecosystem goods and services provided (Eqs. 22.1 and 22.2) must be understood and quantified if the results of CBA are to be complete and reliable.

*Adapted from Randall (2007).

22.4 Nonmarket Valuation

When markets for a particular environmental good or service of interest E_i are absent or inadequate to generate accurate information about value, $V(t)(E_i(t)) = \Sigma_j V_j(t)(E_{ij}(t))$, various methods in nonmarket valuation may be used to fill that informational gap. For a proposed quantity change in ES, the goal is to estimate WTP aggregated across the relevant population for the given increment (or decrement) in ES and – for many purposes related to markets and/or payments for ES, M/PES – marginal WTP (i.e., the implicit price per unit) for changes in ES provision. A compensated demand function, as in Fig. 22.3, expresses the relationship between marginal WTP and quantity of ES.

Total value and the categories of value The nonmarket realm includes many natural and cultural goods and services that may be valued for their existence as well as for any direct use we may make of them. Thus, total economic value includes both active use value and passive use value. *Active use value* may include the value obtained from active outdoor recreation, or enjoying a scenic view from your home, and you may have paid something for these active use opportunities: the cost of a visit to a recreation site, or a premium price for a home with such nice views. Active use value may also include an option value for future use: you might be willing to pay something extra to ensure that the amenity will be there when you are ready to use it, or you may be willing to pay less than your active use value if you are uncertain of your future ability and enthusiasm for active use when the time comes.

Passive use value is derived from the pleasure of knowing that a unique and valuable amenity exists in good condition, independent of any potential for active use. Several motivations are plausible, including a sense that some part of nature is worth preserving even (or especially) in the absence of active use and a desire to pass some of nature's uniqueness and variety onto your descendants and/or future generations more generally.

22.4.1 Nonmarket Valuation Methods[1]

Two fundamental categories of nonmarket valuation methods are widely recognized: stated preference methods and revealed preference methods (Freeman 2003). Introducing some vector of nonmarket ES, $\boldsymbol{E}$, there are two ways to express the utility function. First, $U = U(\boldsymbol{E}, y)$ where y is the "numeraire" or the value of all goods and services other than $\boldsymbol{E}$. *Stated preference research* explores trade-offs between $\boldsymbol{E}$ and y in survey and/or experimental contexts. An alternative form for the utility function is $U(\boldsymbol{E}, \boldsymbol{Z})$ where $\boldsymbol{Z}$ is a vector of ordinary market goods. Where $\boldsymbol{p}$ is the vector of prices of $\boldsymbol{Z}$, the formulation $U(\boldsymbol{p}, \boldsymbol{E}, m)$ is equivalent to $U(\boldsymbol{E}, \boldsymbol{Z})$. *Revealed preference analysis* of $\boldsymbol{Z}$ and $\boldsymbol{p}$ generates evidence of the value of $\boldsymbol{E}$ in cases where $\boldsymbol{E}$ is enjoyed as a complement of some $\boldsymbol{Z}$ (e.g., travel to a recreation site or use of residential property). An additional approach to nonmarket valuation is *benefits transfer*, which seeks to avoid the expense of original nonmarket valuation work by adapting the findings of previous studies to estimate values relevant to a new potential project or program.

Here, the discussion of nonmarket valuation and the most frequently used methods in each category is introductory – there is a vast literature that develops, refines, and tests the various methods, and readers who intend to undertake nonmarket valuation work will need to dig much deeper into this literature.

22.4.2 Stated Preference Methods: Asking People About Their Valuations

For nonmarket estimation of active use values, stated preference methods compete with, or perhaps complement, revealed preference methods. But stated preference methods are applicable also for estimating passive use values and therefore total economic value.

Taking utility as $U(\boldsymbol{E}, y)$, we seek to explore the trade-offs people would make between $\boldsymbol{E}$ and y. The basic concept is the total value curve (TVC), which maps the $(\boldsymbol{E}, y)$ combinations that yield a given level of utility, in the case illustrated $U^0(\boldsymbol{E}^0, y^0)$. There are in principle a family of total value curves, each delivering a different level of utility, but welfare change measures based on the illustrated TVC are compensated at baseline utility U^0 and thus ideal for CBA.

Contingent valuation (CV) seeks to map the relevant total value curve (TVC) (Fig. 22.4). In a survey or experimental setting, changes in provision of $\boldsymbol{E}$ are described in substantial detail, the provision rule, i.e., the (hypothetical) arrange-

[1] In this section, the time dimension is suppressed for notational convenience, and the raw data may range from events (a recreational trip) to asset values (the price of house). However, practitioners need to be aware that the goal of CBA is usually to evaluate the welfare contribution of a time flow of ES and make the necessary adjustments.

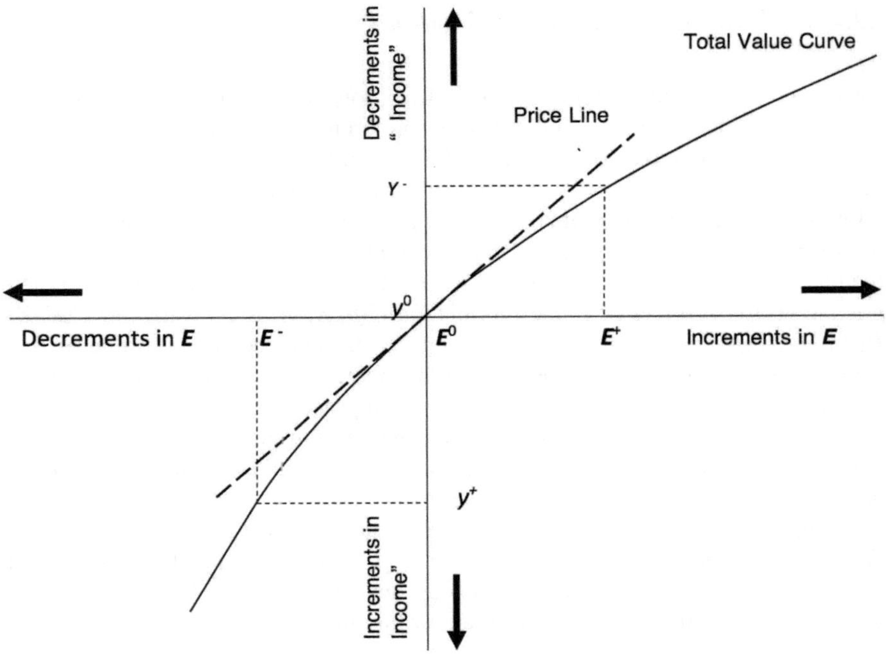

Fig. 22.4 The total value curve for baseline welfare, $U^0(\boldsymbol{E}^0, y^0)$

ments under which the change in E will be provided and the payment delivered, and the respondent is asked a series of questions designed to elicit her WTP or WTA (Mitchell and Carson 1989). Clearly, there are pitfalls to be avoided in this process: the whole exercise must be credible, there is quite a lot of information for the respondent to absorb, the respondent needs to identify her own WTP (or WTA) which likely is latent until the question is asked, and the provision rule should convey incentives for truthful response. Rules that allow a respondent to express a high WTP without any concern that the payment might actually be collected provide one example of inadequate incentives.

For changes in the quantity of public goods, one provision rule with good incentive properties offers a chance to vote in a referendum: if a specified change in ***E*** is offered and households like yours would pay $XX (i.e., a specified amount that is varied randomly across respondents) in higher prices and/or taxes if the referendum passes, would you vote yes or no? The favorable incentives for a truthful response arise from the plausibility of the "higher prices and/or taxes" consequence of the referendum passing (Hoehn and Randall 1987).

Early experience with CV revealed a set of puzzling phenomena in the responses:

- Anchoring. For most people, announcing one's WTP for some newly offered prospect is not a familiar task. Because a respondent's WTP is latent until she is challenged in this way, respondents may search for clues, and the $ amount offered in a contingent valuation exercise may be seen by the respondent as some kind of clue. Even after a series of questions seeking to identify the respondent's maximum WTP, the final announced WTP often shows sensitivity to the initial offer price. This problem is alleviated when a referendum provision rule is used.

- Scope, embedding, etc. CV responses often seem insufficiently sensitive to the scope of the change in $\boldsymbol{E}$ that is offered. But how sensitive should they be? For instance, respondents may quite reasonably offer only a modest WTP to increase the geographic scope of a change in $\boldsymbol{E}$ if they assume that the greater scope entails a larger population who would be asked to pay. Embedding is the term applied to cases where WTP for a package of changes is less than the sum of WTP for its parts evaluated one by one. This seems odd, but there is evidence that budget constraints are enough to induce some degree of embedding in datasets of actual purchases. CV practitioners are encouraged to avoid too little sensitivity to scope and too much to embedding, but realistic benchmarks are not always obvious.
- The different magnitudes of WTP and WTA. The TVC in Fig. 22.4 shows that WTA $(y^+ - y^0)$ is a little larger than WTP $(y^0 - y^-)$ for a similar change in $\boldsymbol{E}$ – as one would expect given Fig. 22.3 and accompanying discussion – but CV results routinely show much larger differences. Rational economic theory supports the TVC as illustrated in Fig. 22.4, and CV researchers were inclined to attribute the larger differences observed to measurement error – something in the CV process was not working and needed to be fixed. But Kahneman and Tversky (1979) argued that the data revealed something fundamental about human behavior – we are averse to losses, real or perceived, and will pay much more to avoid a loss than to realize a gain – and thereby achieved pioneer status in a newly emerging discipline of behavioral economics.

Literally thousands of CV studies have been conducted, the applied econometrics of CV has been explored and refined exhaustively (Habb and McConnell 2002), and CV remains among the techniques of nonmarket valuation commonly used today and acceptable in CBAs conducted for many government agencies and nongovernmental organizations. The opportunity to describe a potential project or policy and the provision rule in detail remains one of its most appealing characteristics.

Choice experiments (CE) offer respondents a series of pairwise choices among prospects differing in several attributes including price and analyze the responses (I would choose prospect 1, or prospect 2, or neither of them) to infer the underlying WTP (or WTA) for one or more particular attributes of interest (Hensher et al. 2005). Compared with CV, CE as a method of data collection has one major advantage: anchoring effects are much reduced because all of the prospects offered are priced by the researcher, and respondents are evaluating a limited number of prospects (e.g., park configurations) given the prices set.

Text Box 22.2 Choice Experiments: A Sample Question

An organization, perhaps a public agency or a for-profit firm, considering developing a multi-attribute recreational park seeks information about public WTP for various attributes and combinations thereof. In the example below, five attributes are considered along with a price, in this case an entry fee. A series of pairwise choices is offered sequentially to respondents. One of the many pairs of configurations is shown here. Upon completion of the exercise with an adequate sample of respondents, econometric methods are used to estimate the contribution of each attribute to WTP for entry to a prospective park.

	Attribute levels	
Attributes	Park Plan A	Park Plan B
1. Total acreage	450	320
2. Water acreage	80	115
3. Miles of hiking trails	65	40
4. Developed campsites	40	40
5. Picnic pavilions	3	6
6. Entry fee	$6	$10

Please indicate your preference between Park Plan A and Park Plan B (check one)

I prefer Park Plan A
I prefer Park Plan B
I prefer neither plan (that is, given this menu I prefer no new park)

Respondents are likely to regard choosing among different prospects offered at different prices as a familiar task. However, CE also has a serious disadvantage: the number of choice pairs offered tends to be quite large, and the competing prospects are described in much less detail than is common in CV. CE data is analyzed with random utility models (RUM) designed to study choices among different packages of multi-attribute goods. These models are rather complicated, and calibration requires researcher decisions that may influence the final results.

Since the mid-1990s, CE has grown in use relative to the older CV format (Hanley et al. 1998), and there is now a large literature on CE methods – conceptual models and empirical estimation techniques – and applications.

22.4.3 Revealed Preference: Inferring Value from Actions in Related Markets

Where utility can be expressed as $U(\boldsymbol{p}, \boldsymbol{E}, m)$ and $\boldsymbol{E}$ is enjoyed as a complement of some $\boldsymbol{Z}$, analysis of $\boldsymbol{Z}$ and $\boldsymbol{p}$ generates evidence of the value of $\boldsymbol{E}$. Techniques based on this premise are used to estimate the value of recreation trips via the travel cost method and the value of residential amenities via hedonic price analysis of residential property values. The values estimated are in the form of CS^M rather than compensated CS^H, but in many cases simple adjustments allow good approximations of CS^H.

The travel cost method (TCM) assumes weak complementarity between travel services and $\boldsymbol{E}$ at the recreation site, i.e., if the trip is not taken, the individual is indifferent to the level of $\boldsymbol{E}$. It is immediately obvious that TCM is addressed to active use values. Passive use values, if relevant, must be addressed in another way. To estimate WTP for visits to a particular recreation site, we estimate the relationship

$$\text{Trips}_j = \text{f}\left(\text{Cost}_j, \text{Substitutes}_j, \text{Demographics}_j, m_j\right) \qquad (22.6)$$

where Trips_j is the number of trips/year by j to this site, Substitutes_j is a vector of potential substitutes for the site, Demographics_j is a vector of demographic descriptors of j's household, and m_j is j's household income. Estimating Eq. (22.6) yields a demand curve for trips, which allows calculation of CS^M for discrete changes in recreation capacity at the site (e.g., expansion or elimination of the site and associated facilities).

To gather data for estimation of (Eq. 22.6), a survey of visitors is typically undertaken. Trips, demographics, and income are self-reported by the respondent; substitutes may be self-reported or specified by the researcher based on secondary data, and the cost of a trip is typically calculated by the researcher given respondent-reported data such as type of vehicle used and secondary data such as distance between j's residence and the site. The cost of a trip, especially, is subject to continuing controversy – e.g., about whether travel time has a cost and, if so, how does the hourly cost relate to the wage rate; how are costs of multiple destination trips to be attributed to specific sites; and whether apparently fixed items such as type of vehicle and location of residence might have actually been selected with some consideration of suitability for recreation activities and convenience of travel to preferred sites. The point is that that the cost of a trip is more complicated and less transparent than it sounds, and the cost numbers used influence the resulting estimates of WTP (Randall 1994).

For many kinds of recreational activities – hiking, hunting, fishing, swimming, boating, beach use, etc. – j enjoys a menu of potential sites varying in $\boldsymbol{E}$ but also in distance and cost of a trip. Researchers have responded with RUM that incorporate a variety of sites in a region, each with its package of attributes including $\boldsymbol{E}$ (Kaoru et al. 1995). RUM-TCM conceptualizes recreation site choice in a decision-tree framework: first decide to pursue, say, trout fishing; then decide to fish a river rather than a lake; and eventually choose a particular site on particular river. The researcher, for example, collects data on the choices individuals make as to where to go trout fishing and the characteristics of those sites such as distance from a visitor's home and site quality (which includes measures of $\boldsymbol{E}$). An estimated RUM allows calculation of the marginal implicit prices or marginal values for site characteristics $\boldsymbol{E}$ and CS^{M} values for nonmarginal changes in site characteristics (e.g., WTP for a nonmarginal change in catch rate).

Hedonic price analysis provides a way of inferring WTP for particular attributes from the prices of complex goods. A residence has many characteristics: its configuration (e.g., single or multiple family building), size, amenities, quality and condition, location (which implies something about distance to workplaces and amenities, neighborhood demographics and amenities, quality of schools, crime rates, etc.). Suppose

$$p_j = \mathrm{f}\left(\boldsymbol{H}_j, \boldsymbol{L}_i, \boldsymbol{D}_j\right) \tag{22.7}$$

where p_j is the price of j's residence, perhaps its assessed market value or perhaps the rental price; $\boldsymbol{H}_j$ is a vector of attributes of the residence; $\boldsymbol{L}_i$ is a vector of local neighborhood characteristics including $\boldsymbol{E}_j$; and $\boldsymbol{D}_j$ is a vector of demographics of j's household. Data for many of these variables can be drawn from public records, but it may be necessary to gather some primary data on $\boldsymbol{D}$.

If Eq. (22.7) can be estimated, and the estimated coefficient of some E_i of interest (say, green space in the neighborhood) is statistically significant, it will be possible to calculate the marginal implicit price of E_i and derive the hedonic demand function for E_i conditional on mean values of the other RHS variables.

Hedonic price analysis is now a well-accepted method (Smith and Huang 1995; Gibbons et al. 2014) – in an example of life imitating art, property assessors now routinely employ consultants with hedonic property price models to help update their valuations for properties that have not been sold in recent years. A large literature has accumulated, some of it suggesting and testing new conceptual and empirical refinements.

Hedonic price analysis can conceivably be used wherever complex multi-attribute goods are traded and data is adequate. For instance, hedonic wage analysis has been used to estimate the wage premium demanded by workers in hazardous jobs, and hedonic price analysis has been employed to tease out automobile buyers' WTP for marginal improvements in safety and fuel economy.

It is commonly argued that *revealed preference methods* have an in-built advantage, in that the underlying data is provided by actual purchaser decisions rather than stated intentions. But they also have a downside – they don't really reveal preferences. The purchaser decisions are real enough, but it requires some nontrivial assumptions and some substantial manipulation of the data in order to tease out the underlying attribute values.

While SP and RP are often framed as rival research programs in nonmarket valuation, they can serve complementary roles providing mutual benchmarks for tests of convergent validity. Furthermore, there is a modest literature on hybrid techniques that formally combine elements of both kinds of methods (Adamowicz et al. 1994).

Benefits transfer (BT) techniques are motivated by concerns in public agencies that new and original nonmarket valuation studies are too costly for routine use in public decision-making (Richardson et al. 2015). Assuming an adequate inventory of original studies for enough different kinds and bundles of $\boldsymbol{E}$, the working hypothesis is that benefits transfer – the application of existing knowledge about nonmarket values to new prospective projects at new sites – should be feasible, less expensive than original studies, and reasonably accurate and precise. One can imagine cases where a proposed project is so similar to a previous project with good original nonmarket value estimates that the values from the previous project can be used with, at most, minor adjustments. But these fortuitous cases are rare. Serious efforts to implement benefits transfer routinely are based on meta-analysis of previous studies. Suppose

$$V_i = \mathrm{f}(\boldsymbol{W}_i, S_i, \boldsymbol{E}_i, \boldsymbol{M}_i, R_i, \boldsymbol{D}_i, m_i) \tag{22.8}$$

where V_i is the value of wetland i; $\boldsymbol{W}_i$ is a vector of wetlands types (freshwater, saltwater, pothole, marsh, etc.); S_i is wetland size; $\boldsymbol{E}_i$ is a vector of $\boldsymbol{E}$ provided by wetland i; $\boldsymbol{M}_i$ is a vector of nonmarket valuation methods used to value wetland i; R_i is the geographic region where wetland i is situated; $\boldsymbol{D}_i$ is a vector of demographics in the "market area" around i; and m_i is the average household income in the market area. As implied by this formulation, valuation information on a wide variety of wetlands types in several regions, offering a wide variety of $\boldsymbol{E}$, evaluated by a wide range of methods, is assembled. The methods, $\boldsymbol{M}_i$, serve as control variables, accounting for the possibility that some kinds of methods tend to produce values systematically higher or lower than others. The remaining variables capture important attributes of the "transfer site" and the prospective wetland project for which benefit estimates are needed.

Wetland benefit studies are assembled and vetted for conceptual validity, competent execution, and adequate reporting of information essential to meta-analysis (Boyle et al. 2010). Inevitably, the studies selected for inclusion will have been conducted at different times, so income and value estimates are adjusted for inflation. Given a best estimate of (Eq. 22.8), BT proceeds by inserting the transfer site values for the relevant RHS variables and calculating its site value in total, or inserting values of particular E and calculating the value of their contribution. To my mind, the objective of meta-analysis for BT is not so much to select the estimated equation that achieves the best fit to the data (one can always improve fit by proliferating RHS variables), but rather the equation that provides the best estimates of parameters for key decision variables: if we improve and preserve a wetland of this type and size providing these $\boldsymbol{E}$ in this location in this region, what is our best estimate of the benefits that will accrue?

The literature on BT has grown substantially in recent years, and methods have been refined and improved. Not surprisingly, BT seems to work better for E_i that are in some sense more generic, e.g., reductions in pollution loads in streams and rivers, than for one-of-a-kind preservation projects for unique natural areas.

22.5 Markets and Payments for Environmental Services

The preceding discussion has been framed in terms of CBA and ways to estimate the prices and values of beneficial and adverse impacts of prospective projects and policies. As economists never tire of pointing out, prices serve dual functions, providing incentives for efficient production and prudent consumption and delivery systems for income and well-being. In this section, the focus is on the role of prices in incentivizing provision of $\boldsymbol{E}$ and maintenance and enhancement of the natural assets that provide them.

As early as the mid-1960s, economists had conceptualized three distinct approaches to securing environmental improvements in the public sphere: standards, taxes, and trading in pollution permits. The government could (i) regulate private and public activities, enforcing clear limits on allowable emissions, effluents, land-clearing, drainage, etc.; (ii) invoke price incentives to limit environmental damage by taxing emissions, etc., or subsidizing abatement; and/or (iii) set a cap on total emissions, etc., require permits for emitting, and encourage trading among emitters to reallocate effort toward the most efficient abaters. Approaches (ii) and (iii) explicitly invoke price incentives for contributions to environmental improvement. Cap-and-trade has proved more feasible politically than pollution taxes. As well as incentives to provide $\boldsymbol{E}$, taxes generate revenue for government. This may be considered a good thing by ordinary citizens, but perhaps not by polluters who would pay the tax. Furthermore, the business community has, in several well-publicized instances, convinced ordinary citizens that pollution taxes would be passed on to the consumers. In contrast, cap-and-trade policies are usually initiated by "grandfathering," i.e., giving pollution (emissions or effluent) permits to established polluters as opposed, say, to auctioning them off. This makes cap-and-trade budget-neutral to firms at the outset and explains the business community's preference for cap-and-trade rather than taxes.

22.5.1 Principles of M/PES

Payments for ecosystem services (PES) usually involves government offering payments to industrial firms, landholders, and perhaps public agencies at lower levels of government in exchange for additional production of $\boldsymbol{E}$. Markets in ecosystems services (MES) attempt to incentivize provision of $\boldsymbol{E}$ in ways that are cost-effective from government's perspective and encourage efficient allocation of effort among providers. The US sulfur trading program attempts to enhance those ***E*** that are damaged by sulfur oxides in the air, by capping total emissions and facilitating trading to price emissions and incentivize greater effort by the more efficient abaters. Conservation auctions contract with landholders who agree to provide, e.g., enhanced green space and wildlife habitat in return for specified payments.

For many ***E***, there is a strong local or regional dimension that limits the potential market for trading. *Banks, bubbles,* etc. are designed to facilitate trading by expanding the scope of markets. Real estate developers may gain permission to encroach on wetlands by making an offsetting contribution to a wetlands mitigation *bank* in the same hydrological region. To ensure a public benefit, offset ratios are set case by case, but always greater than 1 (the developer will pay to mitigate, say, three acres in the bank for each acre developed), and the wetland mitigation bank project will produce a larger and presumably more sustainable wetland that more than compensates for a series of small isolated losses due to development. Polluters within a designated *bubble* must meet abatement standards collectively and thus can trade among themselves to achieve least-cost abatement.

In design of M/PES, information on price, value, benefits, and costs serve roles that are familiar to those who have read this far. The government tries to formulate programs that provide positive net benefits and to avoid paying any provider beyond the point where marginal cost approaches the marginal benefit. In the case of trading mechanisms, government does not really need to know the providers' costs – their trading actions will be driven by their knowledge of their own costs. But M/PES also includes many *conservation auction* programs that contract with, e.g., landholders to incentivize enhancements to habitat, biodiversity, etc. Some sort of bidding-in procedure is utilized to assemble a set of willing providers at least cost to the public treasury. Government understands that providers have asymmetric information about the cost of providing ***E***, i.e., they know more about their own cost than does government. So, it tries to design bidding procedures that incentivize providers to reveal their true costs. A well-designed M/PES program needs a lot of information about benefits and costs, in total and at the margin, and some of this information is more accessible to interested parties (e.g., providers) than to government. So, design of MES programs has at least two objectives: to achieve an increment in ***E*** that results in positive net benefits to society and to incentivize its efficient provision and the revelation of providers' true costs.

22.5.2 Markets and Payments for Environmental Services: What Have We Learned?

All of the above M/PES mechanisms – pollution taxes, cap-and-trade, banks and bubbles, and conservation auctions – have been implemented in many places, at larger and smaller scales, with greater and lesser degrees of political acceptance, and with larger or smaller gains in efficiency and well-being. A comprehensive assessment is overdue but is far beyond the scope of this chapter (Salzman et al. 2018). Here I report briefly on two well-known but quite different programs: the cap-and-trade program for sulfur oxides SO_X in the USA and the Australian experiments with conservation auctions.

The US sulfur oxides, SO_X, cap-and-trade market Acid rain in the northeastern USA, attributed to coal-burning electricity generators further west, became a major environmental issue in the 1980s. Acid rain damages vegetation and aquatic life in lakes and streams, and the SO_X that causes it is a major air pollutant that damages human health. The 1990 Clean Air Act Amendments included a cap-and-trade program that set a limit on aggregate SO_X emissions and allocated SO_X emission allowances to power plants, which were then allowed to trade allowances (or permits) in an organized market. More efficient power plants were able to sell excess allowances to less efficient power plants. Less efficient power plants were thus incentivized, but not required, to invest in pollution control technology that would reduce the need to purchase allowances and perhaps even convert some buyers of permits into sellers. Each power plant had the flexibility to make its contribution toward meeting the aggregate pollution standard in the least-cost manner. Depending on one's perspective, the SO_X cap-and-trade program can be categorized as pollution control market or a market in ecosystem services, i.e., the ***E*** – such as water quality in streams and lakes, air quality, and human health that are enhanced when emissions are reduced.

The SO_X emission allowance trading program had its growing pains in the early 1990s, primarily because of suspicion and uncertainty over whether the program would work and regulators would stay the course. However, by 1995, the program had established its credibility and trades started to steadily increase throughout the rest of the decade as shown in Fig. 22.5.

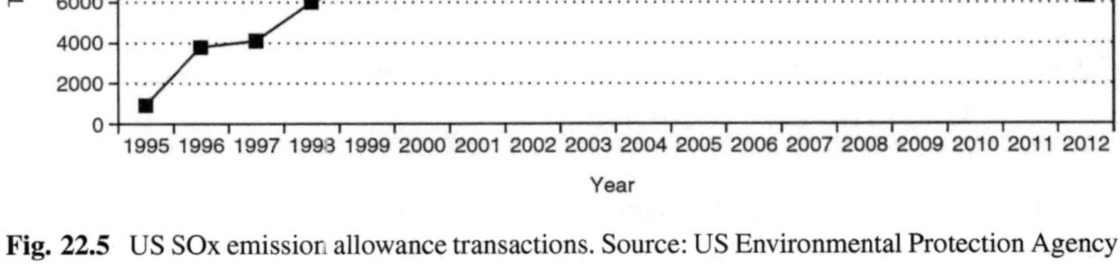

Fig. 22.5 US SOx emission allowance transactions. Source: US Environmental Protection Agency

Beginning in 2008, a series of court decisions undermined the SO_x market, culminating in a 2012 decision that set the program aside because, while the market successfully reduced national concentrations of sulfur oxides, it failed to prevent local and regional pockets of excess concentrations. A revised SO_x cap-and-trade program, restructured to resolve the issue of local and regional exceedances, was introduced in 2015. In consequence of this rather choppy recent history, most analyses of the performance of SO_x focus on the years up until 2008 when the status of the program was more stable.

By most criteria including the establishment of a well-functioning emission allowance trading market, pollution control cost savings, pollution reduction, and public acceptance, the SO_x trading program was considered a success. By the end of the 1990s, national emissions were reduced by almost 30 percent, a substantial accomplishment. By 2009, the dirtiest coal-fired plants had reduced emissions by 57 percent since the inception of the program, while cleaner plants had cut emissions by 14 percent. Compared to a uniform SO_x standard, the cap-and-trade program reduced abatement costs by around 50 percent and more than that according to some estimates (Burtaw and Szambelan 2009). Perhaps some of these gains from cap-and-trade were facilitated by other factors, e.g., deregulation of railroads, which reduced shipping costs and encouraged eastern power plants to burn low-sulfur western coal. A more rigorous recent analysis paying closer attention to specifying a realistic counterfactual (rather than just looking at changes relative to baseline) concludes that the cost savings attributable to cap-and-trade were large, even if not as great as had been suggested (Chan et al. 2015, revised 2017). These authors also suggest that the program may have induced some increases in health-related costs, because some allowances in low-population western states were sold for use in more populous eastern states. Nevertheless, there is broad agreement among scholars that the SO_x trading program has been a substantial success (Schmalensee and Stavins 2017). Among the pillars of its success are the relatively well-specified relationship between effort and results in SO_x reduction and the relative ease of monitoring emissions at the source: verification is reliable, simple, and inexpensive.

The Australian experience with conservation auctions The idea of conservation auctions has its roots in US agricultural conservation policy. The Conservation Reserve Program, introduced in 1985, contracts with participating landholders who promise specific practices, commonly to refrain from cropping, haying, or grazing particular acreage for a specified number of years. Program designers were well aware that contracting for effort is less desirable from their perspective than contracting for results. But they settled for effort, which is verified more readily than results and is preferred by landholders because government then bears the risk that effort underperforms in terms of results. The ultimate success of these programs depends on strong correlations between the practices contracted and results – reduced nutrient run-off and enhanced greenspace and wildlife habitat – but it is the practices that are monitored.

The agency aims to assemble the best portfolio of land to meet conservation priorities within a total budget constraint. There is obvious asymmetry of information.

Agency staff work with fairly transparent published criteria regarding conservation priorities to target regions for conservation, and they use estimates of nonmarket benefits to set a cap on the payments/acre they are willing to make to contract land in a given region. But they know less than the farmers who work the land about the opportunity cost of the land subject to potential conservation. The agency managers seek to meet their conservation objectives by contracting qualified land (from a conservation perspective that attends closely to ecosystems services) with the lowest opportunity costs at prices reflecting those opportunity costs (Hellerstein et al. 2015).

On a smaller scale and with quite specific targets, Australian authorities sought to contract land for biodiversity conservation in designated priority areas, e.g., high-elevation grasslands in the south-eastern highlands, where cattle grazing can have devastating impact on a unique and threatened native biota. The goal was to identify high-priority conservation areas and incentivize land owners to commit to a specific conservation program at prices reflecting the true opportunity cost of their land (typically, the producers' surplus from cattle grazing).

Over the years, excellent economists have been attracted to work on design of auction and related bidding mechanisms, to eliminate or reduce opportunities for bidders to game the system (Stoneham et al. 2003). For instance, the rules governing auctions for broadcast and broadband spectrum, valued in the billions of dollars, were designed by renowned economists. A collaborating group of federal and state agencies and research organizations in Australia assembled an excellent team of local and international economists with a mandate to solve the asymmetric information problem as it pertains to sign-ups in conservation markets. The best conservation auction procedures in operation these days are near-optimal in that respect. We know how to identify the prime land for conservation and invite landowners to commit it at prices that accord fairly closely with the true opportunity costs of the land.

The problem of actually getting the anticipated public benefits from the lands conserved has proven more intractable, for at least three reasons. *First*, pay-for-performance contracts would be ideal, but policy-makers often settle for "pay for prescribed practices" because practices typically are more observable than performance. In the case of high-elevation grasslands, the relevant practice is to exclude cattle and horses for a specified period and the desired result – improved habitat for native biota – is fairly well-correlated with exclusion of these large domestic animals. In other cases, e.g., practices intended to reduce nutrient loads in streams, the relationship between practices and results is more nebulous. *Second*, many kinds of conservation require long-term commitments in order to produce the maximum public benefit, yet land owners have good reasons to prefer shorter commitments. Conservation NGOs in many countries have been active in signing willing land owners to time extensions beyond their commitments to government programs, but there are limits to reliance on such voluntary provision of public goods. *Third*, it has proven difficult to sign up enough landholders to achieve program targets. Australian authorities have conducted a number of field experiments with conservation auctions, designating eligible areas and signing up landowners willing to commit to the required practices (Reeson et al. 2011; Whitten et al. 2013). Sign-ups have been disappointing, and some commentators have wondered whether program designers have been too effective at capturing the producers' surplus from committing to conservation – perhaps a more generous sharing of benefits with landholders might induce more sign-ups.

More generally, there is vast scope for expansion of conservation auctions, for example, in carbon capture and storage (CCS) in forests and farm fields. But the problem of additionality is pervasive: how can government be sure that it is paying for extra conservation, not just practices that farmers might implement anyway for their own good reasons (conservation tillage is a good example)? For CCS, given that forests and farm fields are carbon reservoirs rather than sinks, ensuring additionality requires monitoring and accountability for GHGs sequestered *and* released.

22.6 Concluding Comments

This chapter has defined ecosystems services, shown how they acquire value and prices, and explained how the natural systems that produce them acquire asset value. I have shown how value relates to welfare as economists define it while recognizing that the economic conception of welfare accords perhaps too much deference to markets in that, for example, the preferences of the well-off count for more. Cost-benefit analysis is welfare change measurement in action and, for completeness, requires nonmarket valuation of active and passive use values; and this chapter has provided introductions to the relevant theory and methods.

We have seen how market and nonmarket values can be invoked in payments and markets that incentivize provision of ecosystem services. The exposition has focused on ensuring that these payments and markets serve the public interest in providing ecosystem services efficiently. Nevertheless, there is much here that addresses, implicitly if not always explicitly, the concerns of those whose focus is mostly on economic and business opportunities. The cap-and-trade program for SO_X provided cost-saving opportunities for some of the nation's largest investor-owned electric utilities, builders and real estate developers have used wetlands mitigation banks to enable profitable development while complying with wetlands regulations, and landholders large and small have benefited from the Conservation Reserve Program and conservation auctions of various kinds. The scope for expanded private sector participation – e.g., in cap-and-trade programs for carbon emissions and land-based carbon capture and storage programs – has barely been glimpsed, but if these programs are designed to balance private incentives and the public interest, the possibilities are enormous.

References

Adamowicz W, Louviere J, Williams M. 1994. "Combining Revealed and Stated Preference Methods for Valuing Environmental Amenities," *Journal of Environmental Economics and Management* 26: 271–92.

Arrow KJ, Cropper ML, Eads GC, Hahn RW, Lave LB, Noll RG, Portney PR, Russell M. 1996. "Is There a Role for Benefit–Cost Analysis in Environmental, Health, and Safety Regulation?" *Science* 272: 221–2.

Asheim GB. 2010. "Intergenerational equity." *Annual Review of Economics* 2:197–222.

Atkinson G, Bateman I, Mourato S. 2012. "Recent advances in the valuation of ecosystem services and biodiversity." *Oxford Review of Economic Policy* 28: 22–47.

Boyle KJ, Kuminoff NV, Parmeter CF, Pope JC. 2010. "The Benefit-Transfer Challenges." *Annual Review of Resource Economics* 2:161-182.

Burtaw D, Szambelan SJ. 2009. "US Emissions Trading Markets for SO_2 and NO_x," RFF DP 09-40, Washington, DC: Resources for the Future.

Chan, HR, Chupp, A, Cropper, ML and Muller NZ. 2015 (revised 2017). *The Impact of Trading on the Costs and Benefits of the Acid Rain Program.* National Bureau of Economic Research Working Paper 21383.

Costanza R, d'Arge R, de Groot R, Farberk S, Grasso M, Hannon B, Limburg K, Naeem S, O'Neill RV, Paruelo J, Raskin RG, Sutton P, van den Belt M. 1997. "The value of the world's ecosystem services and natural capital". *Nature.* 387 (6630): 253–260.

Daily GC. 1997. *Nature's Services: Societal Dependence on Natural Ecosystems.* Island Press, Washington. 392pp.

Freeman AMIII. 2003. *The Measurement of Environmental and Resource Values.* 2nd ed. Washington, DC: Resources for the Future.

Gibbons S. Mourato S, Resende GM. 2014. "The Amenity Value of English Nature: A Hedonic Price Approach". Environmental and Resource Economics 57: 175–196.

Hellerstein D, Higgins N, Roberts N. 2015. "Options for Improving Conservation Programs: Insights from Auction Theory and Economic Experiments". *Amber Waves* February 2. http://www.ers.usda.gov/amber-waves/2015-januaryfebruary/options-for-improving-conservation-programs-insights-from-auction-theory-and-economic-experiments.aspx#.VgFlJ3tCWf4

Habb TC, McConnell KE. 2002. *Valuing Environmental and Natural Resources: The Econometrics of Non-market Valuation*, Cheltenham, UK and Northampton, MA, USA.

Hanemann WM. 1991. Willingness to Pay and Willingness to Accept: How Much Can They Differ? *American Economic Review* 81: 635–647.

Hanley N, Wright RTE, Adamowicz V. 1998. "Using Choice Experiments to Value the Environment: Design Issues, Current Experience and Future Prospects." *Environmental and Resource Economics* 11: 413–428.

Hensher DA, Rose JM, Greene WH. 2005. *Applied choice analysis: A primer.* New York: Cambridge University Press.

Hoehn JP, Randall A. 1987. "A satisfactory benefit cost indicator from contingent valuation." *Journal of Environmental Economics and Management* 14:226–247.

Kahneman D, Tversky A. 1979. "Prospect Theory: An Analysis of Decision under Risk" 47: 263–291.

Kaoru YV, Smith VK, Liu J. 1995. "Using Random Utility Models to estimate the recreational value of estuarine resources". *American Journal of Agricultural Economics* 77, 141–151.

Millennium Ecosystem Assessment. 2005. *Ecosystems and Human Well-Being: Synthesis.* Washington: Island Press. 155pp.

Mitchell RC, Carson RT. 1989. *Using Surveys to Value Public Goods: The Contingent Valuation Method.* Washington: Resources for the Future Press.

Randall A. 1994), "A Difficulty with the Travel Cost Method," *Land Economics* 70: 88–96.

Randall A. 2013. "Environmental Ethics for Environmental Economists", in Shogren J, Shortle J (eds.). *Encyclopedia of Energy, Natural Resource, and Environmental Economics,* 3: 25–32, Amsterdam: Elsevier.

Randall A. 2008. "Cost Benefit Analysis", *Encyclopedia of Environmental Ethics and Philosophy* 1, 189–193. Gale Cengage Macmillan.

Randall A. 2007. "Benefit Cost Analysis and a Safe Minimum Standard of Conservation", in Atkinson G, Dietz S, Neumayer E (eds.) *Handbook of Sustainable Development.* Cheltenham, UK: Edward Elgar. 91–105.

Reeson AF, Whitten SM, Williams K, Nolles K, Windle J, Rolfe J. 2011. "Adapting auctions for the provision of ecosystem services at the landscape scale." *Ecological Economics* 70: 1621–1627.

Richardson L, Loomis J, Kroeger T, Casey F. 2015. "The role of benefit transfer in ecosystem service valuation." *Ecological Economics* 115: 51–58

Salzman J, Bennett G, Carroll N, Goldstein A, Jenkins, M. 2018. "The global status and trends of Payments for Ecosystem Services."*Nature Sustainability* 1: 136–144.

Schmalensee R, Stavins RN. 2017. Lessons Learned from Three Decades of Experience with Cap and Trade, *Review of Environmental Economics and Policy* 11: 59–79.

Smith VK, Huang JC. 1995. "Can markets value air quality? A meta-analysis of hedonic property value models." *Journal of Political Economy* 103:209–227.

Stoneham G, Chaudri V, Ha A, Stappazzon L. 2003. "Auctions for conservation contracts: an empirical examination of Victoria's BushTender trial." *Australian Journal of Agricultural and Resource Economics* 47:477–500.

Whitten SM, Reeson A, Windle J, Rolfe J. 2013. "Designing conservation tenders to support landholder participation: A framework and case study assessment." *Ecosystem Services* 6: 82–92.

Chapter 23
Preventing Unintended Harm from Socioecological Interactions

Richard M. Gunton

23.1 Introduction

23.1.1 Ecosystems Are Always Included in Engineering

Engineering, like all human activities, takes place within a context that is not only physical and biotic but filled with human culture. The modern concept of "nature" all too often implies a separation of humans and our culture from the rest of reality, when in fact we are inextricably linked. From breathing to eating and from building to traveling, humans have no option but to interact synergistically with the rest of nature. Ecosystems are bound up in the life of a society: as humans we not only relate to our environment ecologically but experience and appreciate it culturally. Enhancing socioecological interactions is therefore an important goal for sustainable development, and this vision features prominently, for example, in the UN Sustainable Development Goals (UN Sustainable Development Platform 2015). Accounting for the rich complexities of the relationships on which these interactions depend calls for frameworks relating abiotic, biotic, and diverse cultural considerations (United Nations 1992). In our global village, projects must respect a diverse range of stakeholders' perspectives on all manner of ecological interactions if they are to be deemed successful in the broadest possible perspective.

The term "ecosystem" might evoke images of pristine forests, lakes, or traditional meadows, but in the broad sense of the term, ecosystems are everywhere. Within each person's gut, a microbiome of bacteria thrives as part of a healthy digestive system, constituting a human–bacterial ecosystem that influences our behavior (Enders 2015). All kinds of buildings support diverse kinds of ecosystems in the bacteria, fungi, lichens, and mosses that colonize exposed surfaces, not to mention

R. M. Gunton (✉)
University of Winchester, Winchester, UK

B. R. Bakshi (ed.), *Engineering and Ecosystems*,
https://doi.org/10.1007/978-3-031-35692-6_23

unwanted invertebrate inhabitants or the flora and fauna of water conduits. If we also consider that the composition of the atmosphere is shaped by the combined effects of all the living organisms on Earth, providing its oxidizing potential and its ongoing uptake of carbon dioxide, then it becomes clear that ecology is never really alien to engineering. This inevitable inclusion means that the question for engineering is not whether to consider ecosystem processes or not but how best to account for – or indeed integrate – them within the design of projects. Direct negative effects are necessarily considered as a matter of course; this book is primarily about adapting projects to exploit positive synergies. But this must be set in the context of an evaluation of the overall benefit of any adaptations, not just in financial terms but with regard to society at large. This chapter looks at how to reach a balanced integrative assessment of the benefits and harms that may accrue from a proposed engineering solution.

23.1.2 Outline of the Chapter

We begin by outlining some dangers in using the ecosystem service framework as a reference point in engineering, in Sect. 23.2. Section 23.3 then outlines a pluralistic evaluation framework for incorporating natural ecological processes into project design, management, and evaluation. Section 23.4 then looks at how such a framework can make a difference in the specific areas of hydrology and atmospheric interactions and agriculture. In Sect. 23.5, we sketch a protocol for implementing a pluralistic evaluation and look more closely at the challenge of overcoming the nature/engineering dualism. This is an ambitious vision, but our present global challenges and crises call for nothing less. The final section argues that the social and cultural challenges of our time may best be addressed when development projects are designed and evaluated in the broadest possible socioecological framework, with transparency and explicit recognition of ethical considerations.

23.2 From Services to Harmonious Synergy

Previous chapters have made clear the advantages of seeking ecosystem services to enhance engineering projects. However, when a project is modified so as to derive additional services from ecosystem functions, a balanced assessment must consider negative as well as positive effects of the proposed modification. This section indicates some potential dangers of focusing narrowly on measurable "services" and ends by pointing to the need for a broader framework.

23.2.1 Ecosystem Services: Real Benefits?

The concept of ecosystem services arises from an analogy with the human services economy. Just as customers may avail themselves of commercial services upon payment of an appropriate price, humans may be said to benefit from a range of services provided by ecosystems (Daily 1997) and perhaps ought to pay something for them, or at least to account for the potential cost of losing them. The term "services" is essentially a metaphor. Authors disagree as to whether ecosystem "functions," "outputs," or "benefits" are the best concept to define ecosystem services more precisely (Danley and Widmark 2016), and the microeconomics background of the term suggests that it is bound up with the concept of opportunity costs. For the purposes of engineering project design, it is reasonable to focus on the net benefits that a project's builders, owners, lessors, or users may derive from ecological processes that would otherwise have been more costly to obtain by other means. Such benefits, as earlier chapters have shown, may be of many kinds and will generally be measured in terms of financial net benefits to one or more of the abovementioned parties. But responsible engineering calls for a wider societal view on ecosystems, landscapes, wild places, and haunts. Real benefits must be assessed not only in terms of services but through a balanced audit that involves a wide range of stakeholders.

23.2.2 Service Users, Stakeholders, and Lovers

Whereas a commercial service is generally delivered by a business directly to a paying client – perhaps an engineering firm – with limited impact on other parties, ecosystems have broad effects that may be appreciated by some parties and suffered by others. This raises important issues of justice: suppose a service to one person or project is the blight of another? It is not straightforward to identify and measure an ecosystem's "services" in the abstract: a tree whose roots stabilize the slope on which a building sits may also deprive its occupants of natural light, or the beauty of a flowering meadow may be accompanied by hay fever for local residents. Multiple sets of people may be "served" by any particular ecosystem patch, over a range of timeframes: developers may reduce construction costs by clearing or "improving" vegetation that subsequent users of a facility might have appreciated if it had been left intact, for example. Another important contrast with commercial services arises in relation to **fungibility**:[1] whereas commercial services are largely treated as commodities, people may engage most strongly with the uniqueness of natural places. Indeed, the generic term "ecosystem" leaves little space for the love that people may have ***de re*** for specific places, trees, meadows,

[1] Terms in bold are explained in the Glossary at the end of the chapter.

ponds, etc. (O'Neill 2017). If a valley is flooded for a hydroelectric power scheme, there is no way to substitute for that particular valley with its contours, history, ecological communities, and inhabitants – even if some of the valley's hydrological, atmospheric, agricultural, and aesthetic functions may be substituted by services from elsewhere by **offsetting**. Thus, ecosystem service accounting can only be a partial accounting for the full range of ways in which humans and societies may appreciate natural places, and even within the sphere of fungible goods, it cannot yield a single analysis valid for all stakeholders (Gunton et al. 2017).

23.2.3 An Objective Assessment?

In a free market, the price of a service can be modeled as the intersection between a demand curve (how much consumers overall would buy at a range of given prices) and a service providers' supply curve (how much providers would collectively provide at different prices). In the case of ecosystems, there are no supply curves, and demand curves can only be modeled for certain market-oriented "seminatural" human enterprises such as farming (which is considered below). Therefore, other ways must be sought for evaluating ecosystem services in order to maintain the plausibility of the metaphor. For engineering, the most relevant consideration is the alternative cost of nonecological ways of achieving a function that ecosystems can provide: the opportunity cost of losing the ecosystem function. The relative costs of alternative ways of interacting with ecosystem functions should be an important part of designing cost-effective projects, and innovative interfaces with natural processes can certainly enhance an economy-wide financial evaluation of projects, at least where there is appropriate regulatory protection of common-pool resources. Previous chapters have shown how innovative approaches can improve the long-term profitability of projects while reducing negative externalities.

There is a risk, however, that monetary valuation of ecosystem functioning will obscure the diverse kinds of ethics that people hold and confuse the divergent interests of diverse stakeholders. The commodification of ecosystems into units of services inevitably means discounting much of what people really value about natural landscapes. If we commodify ecosystem services in monetary terms, there is prima facie a risk that the wealthy will eventually gain at the expense of the poor, especially if and when trading and financial speculation are introduced (O'Neill 2017). The type of instrumental value that can be priced and traded is a rather tightly bounded subset of "value" in the general sense (Spangenberg and Settele 2016). There are many reasons for resisting marketization – especially since alternative frameworks are available that can help avoid the problems of a "service" mentality.

For engineering, then, it is important to have a realistic approach to understanding how decisions in project design are likely to affect ecosystems and other processes alongside the prospects of the project itself. This should then be combined with an integrative approach to evaluating the positive and negative impacts of a project in the perspective of a wide range of stakeholders. A general evaluative framework is needed for combining these, and we now turn to sketch what this might look like.

23.3 A Pluralistic Evaluation Framework

If projects must be designed and evaluated with respect to multiple criteria and stakeholders simultaneously, a pluralistic framework is called for. To achieve this in a way that connects with the specifications and objectives of a project, we must combine objective evaluation of system functioning with subjective evaluation by stakeholders, and on each side a plurality of evaluation criteria should be considered.

The pluralistic evaluation framework (PEF) proposed by Gunton et al. (2022) has three pillars, concerning recognition of stakeholders, systems, and values (Fig. 23.1). Each pillar provides a suite of categories that provide a pluralistic perspective, and combining all three results in a template that can be used to guide the design of a project or structure and its post hoc evaluation. The suite of categories, known as **aspects**, is derived from the reformational philosophy framework pioneered by Herman Dooyeweerd (Dooyeweerd 1953a) and Dirk Vollenhoven (Vollenhoven 2005). These aspects arguably reflect the structure of the world as recognized in academic discourse, although the list is of course open to refinement.

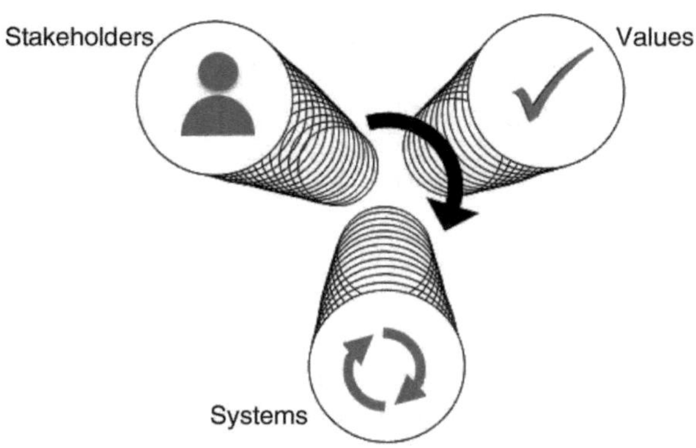

Fig. 23.1 Symbolic representation of the three pillars of the pluralistic evaluation framework. Functional groups of stakeholders are identified according to a suite of aspects (represented by layers in the "Stakeholders" column); then systems and processes created or modified by a project are identified according to a similar suite of aspects; and then a corresponding suite of modes of valuing is examined for each stakeholder–system relationship

Table 23.1 List of categories used in the pluralistic evaluation framework, with examples of their application to each of the three pillars

Aspect	Examples of stakeholder functional groups	Examples of system processes	Examples of positive (negative) values attributed
Ultimate	Religious/cultural groups	Ideology	Inspiring, sacred (unreliable)
Moral	Volunteer groups; NGOs	Public morality	Endearing, loved (despised)
Jural	Government; campaigners	Legislation	Just, equitable (inappropriate)
Aesthetic	Arts groups; Tourists	Fashion	Harmonious, enjoyable (ugly)
Economic	Businesses	Economy	Efficient, sustainable (wasted)
Social	Communities	Social dynamics	Sociable, welcoming (inhospitable)
Symbolic	Journalists	Discourses	Informative, significant (misleading)
Formative	Historians; Educators	Historical change	Developed, innovative (degraded)
Analytic	Scientists	Information systems	Distinctive, diverse (mixed-up)
Sensory	Mental healthcare providers	Emotional life	Stimulating, comfortable (unpleasant)
Biotic	Farmers	Ecosystems	Health-giving (toxic)
Physical	Resource managers	Hydrology; climate	n/a
Kinetic	Residents/commuters	n/a	n/a
Spatial	Local/dispersed	n/a	n/a
Numerical	Individuals/groups	n/a	n/a

23.3.1 Stakeholders

The first pillar of the PEF is a checklist for identifying a comprehensive set of stakeholders who may be affected by a project. Fifteen aspects of human functions or interests are outlined in Table 23.1, along with examples of stakeholder types whose interests may be characterized by each function. This classification is about the *roles* or functions of people, groups, and organizations; particular individuals will tend to fit into more than one category. The three basal aspects (numeric, spatial, and kinetic) evoke the basic dichotomies of individuals vs. groups, local vs. distant stakeholders, and resident vs. mobile stakeholders (e.g., tourists and commuters). The subsequent 12 aspects then point to typical interests and concerns that may identify particular groups of stakeholders, as illustrated in Table 23.1. We return to stakeholders when considering the third pillar below.

23.3.2 Systems

The second pillar calls for consideration of types of engineered system. Engineering challenges can readily be classified by the kinds of system they primarily interact with and the corresponding range of sciences they employ. Many conventional engineered artifacts are primarily physical systems, understood in terms of the physical sciences – from buildings and bridges to chemical plants and electrical devices. Some engineered systems are intrinsically biotic – from bioreactors to agricultural systems – and thus understood biologically as well as with physical sciences. Software and information systems, while dependent on physical hardware, are based on principles of logic and information science, along with elements of psychology. Within software engineering, language processing systems are informed by linguistic science as well as all the foregoing sciences. We may also recognize social engineering, economic engineering, and so on; or more commonly it is in multidisciplinary vocations such as architecture, business administration, and management consultancy that designers seek to manipulate higher-level systems. Each kind of engineering, in this broad sense, focuses on a range of systems, depending on the apparent laws of those systems – for example, gravity, electrical laws, and the properties of materials in many cases, but biotic, sensory, lingual, social, and economic processes in many cases too. A full suite of 12 aspects for considering kinds of systems that projects may create, modify, or interact with is outlined in the center of Table 23.1. Not all of these are prominent or necessarily important at present, but the philosophical framework behind the PEF suggests that each of them has potential and ought to be considered. In any case, multidisciplinary engagement across diverse sciences can go a long way to helping enhance the effectiveness of engineering projects and minimizing unintended side effects.

Within the systems perspective outlined here, there is a degree of natural ordering. Physical processes (e.g., erosion) determine the functioning of biotic ones (e.g., ecological succession), and biotic systems (e.g., vegetation) in turn feed back into the functioning of physical ones (e.g., watersheds). The biotic and psychological functioning of humans determine the possibilities for information systems, and information systems in turn shape the biotic and mental life of humans (increasingly so in the age of mobile devices), while interactions between information systems and physical systems will tend to be less direct. Information systems are part of historical development, the evolution of discourse, and especially social systems and dynamics (not least through social media), while all these domains of change feed back into the design of viable information systems. The ordering of these and the later aspects is more fluid, and again, less direct connections can be made between systems that are further apart in the sequence (see Brandon and Lombardi 2010 for more about this approach). In general, each category of system has its own intrinsic dynamics and also its interactions with neighboring kinds of system. But system analysis is only a foundation for a pluralistic evaluation framework.

23.3.3 Valuing

The third pillar of the PEF concerns the ways in which different stakeholders appreciate systems and indeed the world in general. The latter 11 of the aspects suggest modes of appreciation, or kinds of value that stakeholders may attribute to the functioning of any of the systems identified in the second pillar of the PEF. Stakeholders may make negative as well as positive judgements in each of these modes of appreciation, as indicated in the final column of Table 23.1. A given scenario may elicit positive judgements in some aspects and negative ones in others; sometimes stakeholders have a clear overall view of whether they are in favor of a scenario or against it, and in other cases thoughtful deliberation is necessary. There is an important social factor in how stakeholders appreciate situations, ranging from the general influence of social contexts during individuals' life histories to intentional processes of consultation and group decision-making. The modes of valuing outlined here may be useful as a checklist within such deliberation procedures, including outward-facing stakeholder consultations and the workings of internal committees. It should be pointed out that the social aspect, mid-way through the 11 aspects of valuing, should be considered independently of social deliberation processes. Individuals and social groups alike may consider the social benefits or detriments of a given scenario, not forgetting that even a degree of solitude can be a social good.

Within the context of human valuing, the natural ordering of the aspects is once again important. Modes of value at the lower (biotic) end of the spectrum tend to be compelling and non-negotiable, in that they concern health and safety, even life and death. Toward the higher (ultimate) end of the spectrum, modes of valuing tend to be more variable among individual people and cultures, reflecting religious and ideological traditions. Ultimate value commitments also tend to color the ways in which people value situations in earlier aspects, insofar as a vision of the ultimate meaning of life can shape what is perceived as good or bad.

These 3 pillars of stakeholders (with 15 categories), systems (12 categories), and stakeholders' values (11 categories) may be used, in the first place, as a checklist for assessing possible impacts of any plan, decision, or scenario. In Sect. 23.5, we look at how the PEF as a whole can be used as a decision-support tool.

23.4 Pluralistic Evaluation in Practice

This section looks at two cases of natural systems covered in earlier chapters of this book. Each subsection below briefly considers a group of natural and human systems that may be affected by a certain kind of engineering projects and then how a range of stakeholders may appreciate these with regard to particular benefits or losses that may be derived from them. Although space prohibits more detailed engagement with each topic, the aim here is to outline how a broad framework such as the pluralistic evaluation framework (PEF) may be useful when engineering projects are planned and evaluated. The focus, in line with that of this book as a whole, is on ecosystem processes – including both physical and biotic systems that function in the nonhuman natural world.

23.4.1 Hydrology and Atmospheric Dynamics

The cycles of water, carbon, nitrogen, and other substances are fundamentally physical processes that are profoundly modified by higher-level biotic and human processes. We will focus here especially on water, since this is an essential physical component or resource in many industrial processes and its local availability and quality have direct impacts on humans. The physics of water capture, transport, and loss are particularly important with regard to sustainability concerns, but the dynamics of water usage are not so much about physics: to understand these we must analyze biotic, social, economic, and political systems. The study of biology, and especially the concept of the ecosystem, lies at this interface (Fig. 23.2).

The ecosystem interface with hydrology is of concern for sustainability in all parts of the world and is the main focus of Chaps. 4, 9, 15, and 17 of this book. Projects that concern water treatment and provisioning for human uses can have either competitive or synergistic relationships with the functioning of ecosystems, as mentioned in Chap. 9. A set of questions must therefore be asked concerning system processes (the second pillar of the PEF): how do hydrological systems function with respect to other systems? Their interactions with other physical systems, such as the structural stability, thermodynamics, and chemical reactions of an installation, are routinely considered in engineering design, but the interactions with ecosystems are the focus of this book, and subsequent interactions with all kinds of human biological, psychological, and cultural dynamics are the particular concern of this chapter. Thus, we can consider a range of ways in which engineering projects concerned with hydrology might impact human life and also some ways in which human dynamics might feed back to affect hydrology.

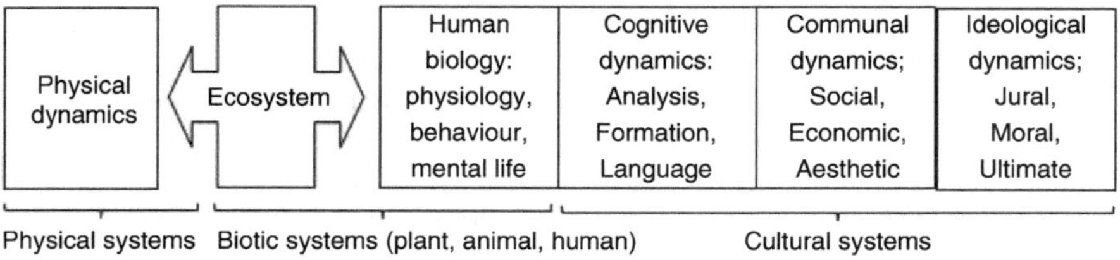

Fig. 23.2 How ecosystems sit at the interface between physical systems such as hydrology or climate and all kinds of human systems, conceived according to the second pillar of the pluralistic evaluation framework (Table 23.1)

For impacts on human life, it is routine to conduct risk assessments concerning human health and safety (largely concerning biotic processes), but a much wider kind of impact assessment is in view here. First, how will the project affect the sensations and behavior of culturally significant animal populations – including those of conservation concern, those where animal welfare is of concern, and pest species? Then, how will the project's hydrological impacts affect the sensory life of local human populations – considering all five senses and integrated effects on mental health? Moving to cognitive dynamics, what impacts will there be on education or on scientific research opportunities? Some engineering projects achieve a historic cultural significance that attracts media attention and influences discourse, such as the construction of dams, which provide an important case study (Nia et al. 2019). It may be expedient to design public engagement facilities, such as visitors' centers at reservoirs that may become important for wildlife (see https://www.essexwt.org.uk/visit/centres for an interesting set of such facilities).

We may also consider how engineering projects may produce unintended feedbacks from plant, animal, or human dynamics to hydrology. Ecological interactions are routinely considered – how algae, fish, and birds may affect water treatment processes, for example, but we must also look at changes in the complex behaviors of humans. These may arise from diverse kinds of human dynamics: social activities, recreational possibilities (especially important with reservoir construction), or economic opportunities, for example. Large-scale changes in human behavior may affect the quality of water or even the movement – especially if agricultural or industrial activities are involved. To predict such feedback effects, an analysis of potential stakeholders is important (the first pillar of the PEF). In a sense, humans must be brought into the ecosystem as responsive agents.

Once a set of stakeholders is identified, another set of questions to ask concerns valuing (the third pillar of the PEF): how do various stakeholders appreciate hydrological and atmospheric systems and how can this inform the design and evaluation of projects? This question cannot be avoided because to do so leads to the adoption of unstated ethical assumptions about how such physical-ecological systems should function. In other words, we must ask what the assumed goodness of "synergy" means in this case. Different stakeholders may consider engineered projects to work synergistically with ecosystems according to different criteria, some of which may sit in tension with each other. For instance, do we seek to treat waste water in ways that maximize affected habitats' biodiversity – and if so, do we care about all biodiversity, from microbial species richness through weed diversity to the presence of rare animals? Or are the relevant stakeholders more concerned about net carbon sequestration? How far do we prioritize the protection of particular habitats and ecosystems that may have historic, symbolic, and even religious value for certain stakeholders, and how far do we prioritize the sensory, economic, or aesthetic value of ecosystems such that novel or restored habitats may be deemed as good as, or better than, original sites that might be changed beyond recognition?

Humans' valuing and appreciation of carbon contrast with those of water in many ways. With growing awareness of human-induced climate change, carbon dioxide and other greenhouse gases such as methane have led to atmospherically available carbon being considered a public disutility in diverse ways. As greenhouse gases diffuse throughout the atmosphere, populations around the world suffer a range of aspects of climate change: altered seasonality of rains affecting food production and basic human biotic functions; flooding causing loss of life, property, and economic livelihoods; and higher temperatures affecting human health both directly and through changing patterns of diseases. These impacts concern human valuing in various aspects. But whereas hydrological impacts of a project must be evaluated in multiple context-specific ways, the carbon footprint of engineering projects is relatively simple to calculate and to feed into standard national and international protocols for assessing its contribution to multi-aspectual costs incurred by various populations and sectors of global society. In this way the carbon budgets associated with engineering projects should ideally be connected to a full analysis of impacts on each aspect of the lives of global citizens (e.g., not just impacts on health and livelihoods).

This PEF approach does not entail that all stakeholders' views are equally important for any given project. The provider of funding for the project (whether public or private) has a special authority over how it should be executed, and non-negotiable regulatory considerations will impinge at some points (issues to be considered in Sect. 23.5). But at the very least, an enlightened self-interest will prompt project designers to consider stakeholders' views in a nuanced way. One reason why this matters is because, as outlined above, the subsequent behavior of humans affected by a project can influence its success, either positively or negatively. More broadly, firms and authorities that commission and execute projects have an interest in their social reputation.

23.4.2 Agroecosystem Engineering

Farming is a form of ecological engineering that lends itself easily to the ecosystem services framework – if the farmed system itself is considered an ecosystem. The so-called agro-ecosystem is not so much the interface between humans and engineered systems (as in Fig. 23.2) but rather the engineered system itself as a provider of goods, in the form of agricultural foodstuffs. Farming systems may also provide other benefits and detriments to particular stakeholders, as outlined in Chap. 11, and the farming system itself may be designed to engage synergistically with a range of ecological processes, as outlined in Chap. 16.

If farming is the oldest form of engineering, it is also the archetypal form in which to seek synergies with natural processes. As mentioned in earlier chapters, agriculture has been developed in various ways to engage as closely as possible with natural ecological processes, from ancient forms of vegetable gardening to modern conceptions like **permaculture** and from the twentieth-century ideology of organic farming to contemporary concepts like **sustainable intensification**. The general term we will use in this chapter for adapting agricultural systems to make greater use of synergies with natural processes is "ecological intensification." Much of the material in this section is also applicable to forestry and some of it to fisheries also.

The impacts of farming systems on stakeholders need unpacking carefully. Merely cataloguing potential agro-ecosystem services, or indeed disservices, is likely to overlook the diverse and differential ways in which various stakeholders may appreciate or suffer from aspects of farming systems, whether conventional or more ecologically synergistic. Beginning with biotic effects, it is important to consider firstly the quality of agricultural produce itself. Pressures for productivity may adversely impact the nutritional quality of crops and livestock. There are also health and safety concerns for farm workers and local residents that arise from chemical inputs and the operation of various kinds of machinery, for example, systems using less chemical inputs for weed control sometimes make greater use of energy-intensive techniques and machinery such as in thermal weed control. Sensory effects are also important both via the quality of agricultural produce (color, taste, etc.) and by virtue of the farmed environment (e.g., considering either irritant chemicals or therapeutic experiences of workers). Moving on to human cultural considerations, most forms of ecological intensification result in greater spatial and temporal diversity in the farmed landscape, which tend to enhance human cultural processes. For instance, more diverse crop rotations, catch-crops, wildflower buffer strips, and larger hedges all tend to produce landscapes that are more cognitively, socially, and ideologically attractive for various stakeholders and only rarely have significant disbenefits for any stakeholders. This is not to say that obtaining real agronomic benefits by ecological synergies is straightforward in itself. The vagaries of weather and associated insect population dynamics and movements tend to produce a large variance in expected payoffs from any move toward ecological intensification.

In such a culturally significant and socially embedded arena as farming, changes to production systems can have far-reaching effects on a wide range of stakeholders at various spatial scales and temporal horizons. Figure 23.3 presents a framework for thinking about the objectives and possible impacts of moves toward ecological intensification, conceived here as a subset of interventions for sustainable intensification, and building on papers by Gunton et al. (2016) and Wigboldus et al. (2016).

The vision for sustainability illustrated in Fig. 23.3 subsumes ecological intensification mostly within its first layer: that of plant productivity. Strategies for improving soil fertility, ecological weed and pest control, and enhanced pollination, for example, can all fall within a farmer's orb of self-interest, although they may be tempered by business considerations as the system is opened up as far as a regular farm business (the seventh layer in Fig. 23.3). This is especially important with a long-term horizon in view (e.g., that of family landowners rather than tenant farmers or opportunistic profiteers). Moreover, actual farmers, like all humans, will have some conception of ultimate good – perhaps the common good in an ideological view – and this will shape their overall vision of good farming: their personal and communal ethics.

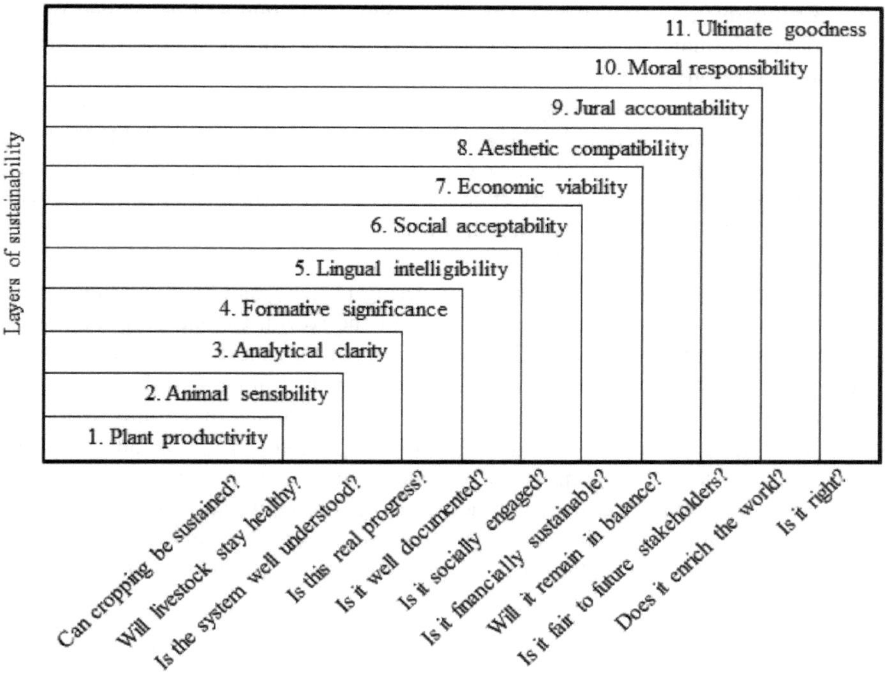

Fig. 23.3 The scope of sustainable intensification. Moving from the bottom left of the diagram toward the top right corresponds to opening up a land management system to additional layers of interest and thereby broadening the scope of its sustainability. Starting with a basic focus on plant productivity (1), as in gardening, additional considerations can qualify a project as, for example, innovative (4), socially embedded (6), and business-oriented (7). This far is sufficient for a farming system, but a more enlightened vision may entail aesthetic harmony (8), concern for others' rights (9, 10), and ultimately a commitment to the good (11)

23.5 Pluralistic Evaluation for Sustainable Engineering

23.5.1 Completing the Transition to Natural Engineering

We began this chapter by critiquing the traditional distinction between nature and humanity and between ecosystems and engineering. Nevertheless, we find the concepts of nonhuman ecological systems and processes extremely important. How can we maintain appropriate distinctions between human and nonhuman systems to conduct analyses like these and properly consider human impacts on so-called "natural" habitats like ecosystems where humans have minimal direct impact?

The philosophical concept that undergirds the structures recognized by the PEF, as outlined above, is that of enkaptic relationships. We can avoid creating a natural-plus-artificial dualism by saying that "lower" or "earlier" kinds of systems in the sequence of aspects outlined in Table 23.1 are caught up and transformed by higher ones. **Enkapsis**, a term derived from philosophy of biology and developed by the philosopher Herman Dooyeweerd (Dooyeweerd 1953b; Klapwijk 2008), means the wrapping up of one kind of system inside another one that transforms its meaning. Enkapsis therefore evokes the way in which the foundational functions of a system are given additional meaning by those of a later (higher) aspect (see Table 23.1 and Fig. 23.3) (Ouweneel 2014). Thus, we can look at the ways in which ecosystems are transformed for better or worse by changing the degree of human engagement with them. The value judgment of "better" or "worse" should, of course, be informed by a range of stakeholders' attitudes.

If we are to maintain that human engineering projects and technical innovation can be "goods" at all, we need such a view of the world. Considering nonhuman nature as an ultimate, unqualified good – as in the perspective of deep ecology (Curry 2011) – tends to entrench conflicts between the human and the nonhuman. In that view, the ideal is to minimize the impact of humans on certain regions of our planet, such as those where the illusion of pristine, virgin wildness can be maintained – while sacrificing other regions to human despoliation. If we avoid subscribing to such perspectives yet lack an ethical framework for seeing different aspects of the world as built upon each other and developed within each other for better or worse, we risk falling prey to unspoken ethics about how the ecological aspects of the world should function.

23.5.2 Decision-Making

At first glance, the list of values considered within the PEF is highly aspirational. Engineering projects are conducted primarily within financial and regulatory constraints, and when these dominate, there may appear to be little room for considering diverse stakeholders' ethical concerns that go beyond those required by relevant legislation. However, there are at least three reasons why project designers, managers, and evaluators may want to avoid such a narrow approach. First, as considered above, there are dynamic interactions among a wide range of cultural processes, and these affect the ways in which economic and regulatory criteria may be applied. Regulators, for example, in seeking the public interest, may pay special attention to firms that exploit loopholes in legislation and apply penalties or adjust the legislation in ways that benefit more public-spirited parties. This leads to a second consideration that the economic and regulatory criteria are themselves evolving in ways that cannot fully be foreseen. Economic realities may shift to favor firms and projects that have been designed with a more holistic or broader range of objectives in view, while regulatory criteria tend to be developed so as to seek greater realization of the public good. Third, humans invariably do have some notion of the good, and strict adherence to one or two overly narrow criteria may be personally unbearable or dehumanizing.

By this point it is clear that the nature of an evaluation, especially concerning potential societal and ideological impacts, will be colored by the worldviews and ideologies of the individuals and authorities that perform it. The normative practice approach (De Vries 2015; de Vries and Jochemsen 2019) is a framework that recognizes this, based on the same set of aspects as outlined in Table 23.1 and used in the PEF. The central insight of this approach is that in professional practices such as engineering, it is useful to separate the profession's own "constitutive norms," such as those of health and safety, teamwork, and economic efficiency, from the ethical "regulating context" within which engineers think and work. Being a good engineer primarily means working safely, productively, cooperatively, efficiently, etc., but, depending on the individual and the context, it may also mean being responsible, loyal, compassionate, etc. In short, good engineering is the practice of a range of virtues.

For planning and evaluating an engineering project in synergy with ecological systems, then, there is much to commend a multi-aspectual framework that is broad and transparent, not just ecologically but in terms of the whole of human life. The pluralistic evaluation framework sketched here offers at the very least a checklist for aspects of human life and culture that might be neglected, or thought to be of low priority, in the design of projects. It can also, however, be used as more than a checklist by recognizing the interdependencies among aspects of reality. The following procedure is suggested for an impact assessment of a project at the planning stage, with flexibility for adaptation according to the kinds of data available.

1. Identify relevant types of stakeholder along with system processes likely to be of concern to them that may be affected by the project. This step may need to be iterated with consultation of stakeholders to help identify additional system processes that the project might affect, which in turn might elicit additional stakeholders.
2. Consider the modes of valuing that might be relevant to each system process, for one stakeholder group at a time. Different kinds of stakeholder may be able to provide different levels of detail and possibly quantification in their evaluations within different aspects.
3. Identify scenarios to compare (e.g., under different kinds of proposed ecological synergy) and, where possible, describe these through modeling.
4. Elicit value assessments for each scenario or case from the stakeholders identified, for each relevant system process and aspect of value. The type of evaluation required, and availability of resources, will determine how much direct consultation of stakeholders is possible and how much must be imputed based on existing data and previous experience.
5. Complete the assessment, if appropriate, using multi-criterion optimization methods (Wątróbski et al. 2019) to explore the relative goodness of the scenarios. Each system process may be considered in turn to assess its overall improvement in the eyes of the relevant stakeholders, or each stakeholder may be considered in turn to assess their appreciation of the processes affected.

Such a procedure has similarities with participatory **systems mapping** (Lopes and Videira 2017), in which the causal relationships among systems and indicator variables are investigated through discussions among diverse stakeholders. The PEF adds structure to this and a value-explicit dimension. The use of this kind of structure with stakeholders has been described by Basden (2019), who offers further suggestions for operationalizing the use of Dooyeweerd's aspects. The pluralistic evaluation framework and its categories are described in more detail by Gunton et al. (2022).

23.6 Conclusion

We have sketched a view of ecosystems as not just omnipresent but enkaptically taken up into human affairs. This enkapsis is evident in the increasing concern for "ecological" ways of living at all levels of culture – a remarkable trend in view of the low awareness that most Western people have of natural ecosystems in daily life, the small amounts of time that modern city dwellers may spend in green spaces, and the low funding for ecology among the biological sciences. What we are seeing in contemporary culture is a growing awareness of the complex – we might say enkaptic – interrelationships between human lifestyles and ecosystems around the globe, epitomized by the transformation of environmental discourse to focus on the worldwide dispersion and impacts of CO_2 and other greenhouse gases. Engineering is rightly caught up in these ecological concerns, and it is appropriate that engineered projects of all kinds should be assessed on a broad range of criteria. These must include ecological criteria in the strict sense but also "environmental" criteria more broadly and indeed cultural criteria in the broadest possible sense. Our world of interconnected systems, from the physical and ecological through the societal and economic to the dynamics of ideology, demands nothing less than fully integrated design and evaluation of each innovation and project. We have pointed out here that the nature of evaluation, especially concerning potential societal and ideological impacts, will be colored by the worldviews and ideologies of the individuals and authorities that perform it. This makes it all the more important for evaluations to be transparent and clearly structured – which is the central benefit of the pluralistic evaluation framework outlined here.

Acknowledgments I would like to thank Yoseph Araya and Maarten Verkerk for helpful comments on drafts of this chapter. The development of the PEF benefitted from funding and support from the Centre for the Evaluation of Complexity Across the Nexus (www.cecan.ac.uk).

Glossary

Aspect (in reformational philosophy) An irreducible mode of functioning and meaning. Each aspect is an end-point in the process of abstraction, such that its meaning can be evoked but not defined. Fifteen aspects are classically posited (see Table 23.1), and any object or phenomenon functions in all of them, albeit only passively in some cases.

de re (in analytical philosophy) Literally "about the thing": used by O'Neill (2017) to evoke the way in which a thing (e.g., a person) may be valued as a unique individual that cannot be replaced. This is contrasted with *de dicto* ("about what is said"), in which something is valued according to its fitting a certain description, such that a substitute could be found (see **fungible**).

Enkapsis (in reformational philosophy) The involvement of one entity or system in another entity or system that transforms its functioning or meaning. In contrast to a part–whole relationship, an enkaptic relationship is one that links different **aspects**.

Fungible (in economics) Interchangeable with respect to value; substitutable

Offsetting A practice of compensating ecological loss at one site by creating substitute habitat of equivalent quality at another site. Biodiversity offsetting (where species richness and the presence of notable species are expected to be replicated or enhanced) is now enshrined for certain situations in the planning policies of some jurisdictions including the USA and Australia and appears to be an aspiration in others, including the UK.

Permaculture A paradigm of producing food and other natural products by designing a seminatural ecosystem in which humans participate with minimal disturbance

Shadow price A monetary value that is attributed in cases where no market price exists

Sustainable intensification Changes to a farming system that will maintain or enhance specified kinds of agricultural provisioning while enhancing the delivery of a specified range of other ecosystem services measured over a specified area and specified time frame (Gunton et al. 2016)

Systems mapping A process of analyzing a complex system to describe its components and boundary and to elucidate causal relationships among the components, often with respect to measurable variables

References

Basden, A. 2019. Foundations and Practice of Research: Adventures With Dooyeweerd's Philosophy. Routledge, Oxford.

Brandon, P. S., and P. L. Lombardi. 2010. Evaluating Sustainable Development in the Built Environment. Wiley-Blackwell, Chichester.

Curry, P. 2011. Dark Green or Deep (Ecocentric) Ethics. Pages 92–126 Ecological ethics; an introduction. Polity Press.

Daily, G. C. 1997. Nature's Services: Societal Dependence on Natural Ecosystems. Island Press.

Danley, B., and C. Widmark. 2016. Evaluating conceptual definitions of ecosystem services and their implications. Ecological Economics 126:132–138.

Dooyeweerd, H. 1953a. A New Critique of Theoretical Thought (4 Vols). Amsterdam.

Dooyeweerd, H. 1953b. A New Critique of Theoretical Thought (3): The Structures of Individuality of Temporal Reality. H.J. Paris.

Enders, G. 2015. Gut: The inside Story of Our Body's Most Underrated Organ. Greystone Books, Vancouver.

Gunton, R. M., E. N. van Asperen, A. Basden, D. Bookless, Y. Araya, D. R. Hanson, M. A. Goddard, G. Otieno, and G. O. Jones. 2017. Beyond ecosystem services: valuing the invaluable. Trends in Ecology and Evolution 32:249–257.

Gunton, R. M., L. G. Firbank, A. Inman, and D. M. Winter. 2016. How scalable is sustainable intensification? Nature Plants 2:16065.

Gunton, R. M., A. P. Hejnowicz, A. Basden, E. N. Van Asperen, I. Christie, D. R. Hanson, and S. E. Hartley. 2022. Valuing beyond economics: A pluralistic evaluation framework for participatory policymaking. Ecological Economics 196:107420.

Klapwijk, J. 2008. Purpose in the Living World? Creation and Emergent Evolution. Page (H. T. Cook, Ed.). Cambridge University Press, Cambridge.

Lopes, R., and N. Videira. 2017. Modelling feedback processes underpinning management of ecosystem services: The role of participatory systems mapping. Ecosystem Services 28:28–42.

Nia, M. G., M. F. Harandi, and M. J. de Vries. 2019. Technology Development as a Normative Practice: A Meaning-Based Approach to Learning About Values in Engineering—Damming as a Case Study. Science and Engineering Ethics 25:55–82.

O'Neill, J. 2017. Life beyond capital. Centre for the Understanding of Sustainable Prosperity, Guildford, U.K.

Ouweneel, W. 2014. Wisdom for Thinkers: An Introduction to Christian Philosophy. Paideia Press / Reformational Pub Project.

Spangenberg, J. H., and J. Settele. 2016. Value pluralism and economic valuation - defendable if well done. Ecosystem Services 18:100–109.

UN Sustainable Development Platform. 2015. Transforming Our World: The 2030 Agenda for Sustainable Development. United Nations.

United Nations. 1992. Convention on Biological Diversity. United Nations, Rio De Janeiro.

Vollenhoven, D. H. T. 2005. Introduction to Philosophy. Dordt College Press, Sioux Center, Iowa.

De Vries, M. J. 2015. Engineering as normative practice: Ethical reflections on tasks and responsibilities. Pages 12–21 Contemporary Ethical Issues in Engineering. Delft University of Technology, Netherlands.

de Vries, M. J., and H. Jochemsen, editors. 2019. The Normative Nature of Social Practices and Ethics in Professional Environments. IGI Global.

Wątróbski, J., J. Jankowski, P. Ziemba, A. Karczmarczyk, and M. Zioło. 2019. Generalised framework for multi-criteria method selection. Omega (United Kingdom) 86:107–124.

Wigboldus, S., L. Klerkx, C. Leeuwis, M. Schut, S. Muilerman, and H. Jochemsen. 2016. Systemic perspectives on scaling agricultural innovations. A review. Agronomy for Sustainable Development 36.

Part VI
Directions for the Future

Chapter 24
Outlook

Bhavik R. Bakshi

As highlighted throughout this book, ecosystems play a crucial and irreplaceable role in supporting all human activities, including engineering. We also identified some of the reasons why this critical link between human and natural systems has been ignored in most human decisions, including decisions made by engineers. Ignorance of the dependence of human activities on ecosystems makes it a root cause of the unsustainability of human activities. It is also a missed opportunity to develop innovative solutions for meeting human needs that can often be superior to solutions developed by conventional engineering.

This book contributes to efforts toward repairing the human-nature relationship by identifying the importance of nature and providing a variety of examples to convey that the human-nature relationship need not be antagonistic or win-lose, as it has been for many centuries and as illustrated in the top half of Fig. 24.1. As shown, improving human well-being has come at a cost of planetary well-being. Through several case studies, this book puts forth the argument that if done properly, the relationship between engineering and ecosystems can be transformed to become synergistic or win-win, as illustrated in the bottom half of Fig. 24.1. Here, improvement in human well-being results in simultaneous improvement in well-being of the biosphere. That is to say, it is possible to design and operate systems such that nature, industry, economy and society benefit together. Thus, by accounting for the role of ecosystems and respecting its limits, engineering can overcome its traditional antagonistic relationship with nature to develop a synergistic or mutually beneficial relationship. We argue that if engineers understand the role and

B. R. Bakshi (✉)
School for Engineering of Matter, Transport and Energy, School of Sustainability, School of Complex Adaptive Systems, Arizona State University, Tempe, AZ, USA

Lowrie Department of Chemical and Biomolecular Engineering, The Ohio State University, Columbus, OH, USA
e-mail: bhavik.bakshi@asu.edu

B. R. Bakshi (ed.), *Engineering and Ecosystems*,
https://doi.org/10.1007/978-3-031-35692-6_24

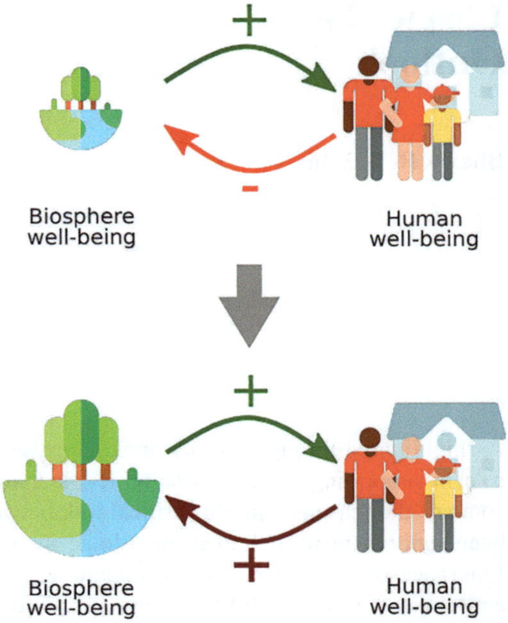

Fig. 24.1 The current human-nature relationship is antagonistic or win-lose, as shown in the top part of this figure. Benefits to human well-being degrade ecological well-being. This antagonistic win-lose relationship needs to transform to a synergistic or win-win relationship where humans and nature both simultaneously benefit from meeting their needs

function of ecosystems, gain knowledge about nature's complexity and behavior, and collaborate with ecologists, it will create many unique opportunities for win-win solutions like those described throughout this book.

Seeking synergies with nature presents huge opportunities for individuals, societies, economies, and the planet. The World Economic Forum in their New Nature Economy report estimates that by 2030 the "Great Reset" toward nature-positive economies could provide $10 trillion in business opportunities and 395 million jobs. As described in this book, nature-based solutions can mitigate air and water pollution, reduce the urban heat island effect, improve resilience to climate change, and enhance human well-being. They can help address issues of social inequities and environmental injustice by increasing vegetation cover in urban areas that tend to house lower income populations. This can help address past injustices such as the historic "red-lined" city blocks policy in the USA. Since many nature-based solutions have been known and practiced for centuries through traditional knowledge systems, adoption of such solutions can help rediscover these systems and contribute to addressing social issues in many marginalized and indigenous communities across the world. Examples of such efforts include the rediscovery of indigenous knowledge for managing fire-prone landscapes that are increasingly vulnerable to the effects of climate change. Another example is the use of traditional knowledge to manage water in arid landscapes by rainwater harvesting to recharge aquifers. Such knowledge about the benefits and mechanics of working with nature exists in traditional and indigenous societies and has been practiced for centuries.

However, over the last few centuries, this knowledge has been lost, ignored, and even actively discouraged. Seeking synergies with nature requires overcoming many barriers before its benefits can be realized in modern societies. Several such barriers and potential solutions are described in the rest of this outlook.

Currently, if landowners restore ecosystems on their property, they will bear the costs of restoration, but the benefits from the resulting ecosystem goods and services will be to them and to the rest of society. For instance, by creating a native forest or grassland on their property, a business would bear the cost of creating and maintaining the restored ecosystem. However, ecosystem services such as carbon sequestration, air quality regulation, and recreational opportunities will benefit society at large. From a conventional cost-benefit analysis point of view, such investment in ecosystem restoration is not likely to be economically attractive to the business or landowner because the direct monetary benefit of such an investment is relatively small. Thus, conventional economics may not incentivize nature-positive decisions. It is well-understood that environmental impact and ecological degradation are negative environmental externalities and are outside the market. In an analogous manner, nature-positive decisions result in positive environmental externalities, which are also outside the market, and are not reflected in market prices.

A large amount of work is directed toward overcoming this shortcoming of mainstream economics, as introduced in Chap. 22. This includes approaches from environmental economics that internalize the benefits of ecosystem services by monetizing them and including them in the market. This involves approaches such as environmental taxes, payment for ecosystem services, and emissions trading. Such approaches have already been successful in improving air quality, managing fisheries, and reducing the effect of excessive nutrients on water quality. However, for some ecosystem services such as climate regulation, approaches such as carbon taxes and emissions trading have had only limited success. Barriers to their success include politics associated with actual or perceived losses to corporations or regions and technical issues related to ensuring that approaches for offsetting emissions result in real change and not "greenwashing." Many practical challenges exist in verifying the benefits of claimed offsets. For instance, if trees are planted to sequester carbon dioxide, it is important to verify that they continue to grow and provide the carbon sequestration ecosystem service over many decades. Many schemes for offsetting emissions reduce potential future emissions but do not remove greenhouse gases that are already in the atmosphere. These include schemes for developing renewable energy, reducing methane emissions from landfills and cattle, and schemes to encourage climate-friendly farming practices. While such schemes help, reduction of existing atmospheric greenhouse gases is essential and should be prioritized. It is also important to ensure that ecosystem restoration efforts adopt a holistic view by focusing on the whole ecosystem and characteristics such as enhancing biodiversity. A reductionist view that focuses on a single objective like carbon sequestration may result in more ecological harm than benefits by shifting the burden to other impact categories or creating systems that are vulnerable to disturbances such as forest fires.

For engineers to design and operate synergies with nature, they need the appropriate knowledge and training. In addition, the attitude of engineers needs to shift from that of ignoring or dominating nature to accounting for it and respecting its limits. At present, such change seems difficult since most engineers are lacking in ecological literacy, and virtually no engineering curriculum includes education in ecosystem ecology. This needs to change. Courses on "Ecology for Engineers" need to be developed and become part of the curriculum. Such a course should cover the basic principles of ecology, ecosystem services, energy transformation (thermodynamics) in nature, and introduction to local species. Emerging principles of accounting for and mimicking ecosystems along with studies of the potential benefits and trade-offs from techno-ecological synergies like those in previous chapters of this book are also needed.

Such studies need advanced models of technological and ecological systems that can simulate spatial and temporal variation at multiple scales. Approaches are also needed that can use these models to find the best designs and strategies for real-time operation. This presents substantial modeling and computational challenges due to the multiscale character which contributes to the large size of the resulting optimization problem that needs to be solved to address spatiotemporal variations. Practical implementation of techno-ecologically synergistic designs will require advanced sensors for measuring relevant ecosystem characteristics such as air and water quality that can be used for addressing ecosystem variability and adjusting technological characteristics to adapt to nature. This requires advances in real time and automatic control strategies, methods for extracting information from data such as machine learning, and methods for distributed sensing and decision-making. Emerging technologies related to Industry 4.0 such as Internet of Things and digitalization are likely to help in meeting such goals.

Modern innovations are protected by the patent system, but such an approach may not work for innovations based on synergies between human and natural systems. This is because the underlying knowledge system in such synergies may not be new, therefore difficult to protect. This may not incentivize innovators driven by the profit motive to develop and implement systems based on seeking synergies with nature.

Despite the barriers and challenges, we take a guardedly optimistic view that the outlook for engineering to work with ecosystems to seek synergies toward a nature-positive world is excellent. We envision a future where including the role of ecosystems and accounting for its limits is a part of the engineers' toolkit and routine decision-making. We expect changes in this direction in other disciplines as well. For instance, in economics, we expect internalization into the market of environmental impacts and contributions from nature. This requires more than just accounting for the role of ecosystems in monetary terms since such accounting usually assumes substitutability between ecosystem services. New approaches are needed that also ensure that ecological boundaries are not violated. Eventually, we expect social and cultural changes that will make it socially unacceptable to implement nature-negative decisions for meeting human needs. Making progress toward realizing this vision of a nature-positive world will also help in meeting other global goals such as a socially equitable world with net-zero greenhouse gas emissions. Reinventing human enterprises in this direction presents many research, development, and implementation opportunities across all disciplines, including engineering. If done right, this is a rare opportunity to transform our future to become not only nature-positive, but also people-positive with net-zero emissions.

Subject Index

B. R. Bakshi (ed.), *Engineering and Ecosystems*,
https://doi.org/10.1007/978-3-031-35692-6

MIX
Papier aus verantwortungsvollen Quellen
Paper from responsible sources
FSC® C105338

If you have any concerns about our products,
you can contact us on
ProductSafety@springernature.com

In case Publisher is established outside the EU,
the EU authorized representative is:
Springer Nature Customer Service Center GmbH
Europaplatz 3, 69115 Heidelberg, Germany

Printed by Libri Plureos GmbH
in Hamburg, Germany